Advanced Applications of Fractional Differential Operators to Science and Technology

Ahmed Ezzat Matouk
Majmaah University, Saudi Arabia

IGI Global
PUBLISHER of TIMELY KNOWLEDGE

A volume in the Advances in Computer and Electrical Engineering (ACEE) Book Series

Published in the United States of America by
IGI Global
Engineering Science Reference (an imprint of IGI Global)
701 E. Chocolate Avenue
Hershey PA, USA 17033
Tel: 717-533-8845
Fax: 717-533-8661
E-mail: cust@igi-global.com
Web site: http://www.igi-global.com

Library of Congress Cataloging-in-Publication Data

Names: Matouk, Ahmed Ezzat, 1976- editor.
Title: Advanced applications of fractional differential operators to science and technology / Ahmed Ezzat Matouk, editor.
Description: Hershey, PA : Engineering Science Reference, an imprint of IGI Global, [2020] | Includes bibliographical references and index. | Summary: "This book examines fractional calculus and its application in science, engineering, and technology"-- Provided by publisher.
Identifiers: LCCN 2019048562 (print) | LCCN 2019048563 (ebook) | ISBN 9781799831228 (hardcover) | ISBN 9781799831235 (paperback) | ISBN 9781799831242 (ebook)
Subjects: LCSH: Engineering mathematics. | Fractional differential equations.
Classification: LCC TA347.D45 A37 2020 (print) | LCC TA347.D45 (ebook) | DDC 515/.35--dc23
LC record available at https://lccn.loc.gov/2019048562
LC ebook record available at https://lccn.loc.gov/2019048563

This book is published in the IGI Global book series Advances in Computer and Electrical Engineering (ACEE) (ISSN: 2327-039X; eISSN: 2327-0403)

British Cataloguing in Publication Data
A Cataloguing in Publication record for this book is available from the British Library.

Advances in Computer and Electrical Engineering (ACEE) Book Series

Srikanta Patnaik
SOA University, India

ISSN:2327-039X
EISSN:2327-0403

Mission

The fields of computer engineering and electrical engineering encompass a broad range of interdisciplinary topics allowing for expansive research developments across multiple fields. Research in these areas continues to develop and become increasingly important as computer and electrical systems have become an integral part of everyday life.

The **Advances in Computer and Electrical Engineering (ACEE) Book Series** aims to publish research on diverse topics pertaining to computer engineering and electrical engineering. **ACEE** encourages scholarly discourse on the latest applications, tools, and methodologies being implemented in the field for the design and development of computer and electrical systems.

Coverage

- VLSI Design
- Chip Design
- Power Electronics
- Applied Electromagnetics
- Sensor Technologies
- VLSI Fabrication
- Analog Electronics
- Microprocessor Design
- Algorithms
- Optical Electronics

IGI Global is currently accepting manuscripts for publication within this series. To submit a proposal for a volume in this series, please contact our Acquisition Editors at Acquisitions@igi-global.com or visit: http://www.igi-global.com/publish/.

The Advances in Computer and Electrical Engineering (ACEE) Book Series (ISSN 2327-039X) is published by IGI Global, 701 E. Chocolate Avenue, Hershey, PA 17033-1240, USA, www.igi-global.com. This series is composed of titles available for purchase individually; each title is edited to be contextually exclusive from any other title within the series. For pricing and ordering information please visit http://www.igi-global.com/book-series/advances-computer-electrical-engineering/73675. Postmaster: Send all address changes to above address.

Titles in this Series

For a list of additional titles in this series, please visit:
https://www.igi-global.com/book-series/advances-computer-electrical-engineering/73675

Open Source Software for Statistical Analysis of Big Data Emerging Research and Opportunities
Richard S. Segall (Arkansas State University, USA) and Gao Niu (Bryant University, USA)
Engineering Science Reference • © 2020 • 237pp • H/C (ISBN: 9781799827689) • US $225.00

Software Engineering for Agile Application Development
Chung-Yeung Pang (Seveco AG, Switzerland)
Engineering Science Reference • © 2020 • 330pp • H/C (ISBN: 9781799825319) • US $235.00

Applied Social Network Analysis With R Emerging Research and Opportunities
Mehmet Gençer (Izmir University of Economics, Turkey)
Engineering Science Reference • © 2020 • 120pp • H/C (ISBN: 9781799819127) • US $165.00

Novel Approaches to Information Systems Design
Naveen Prakash (Indraprastha Institute of Information Technology, Delhi, India) and Deepika Prakash (NIIT University, India)
Engineering Science Reference • © 2020 • 299pp • H/C (ISBN: 9781799829751) • US $215.00

IoT Architectures, Models, and Platforms for Smart City Applications
Bhawani Shankar Chowdhry (Mehran University of Engineering and Technology, Pakistan) Faisal Karim Shaikh (Mehran University of Engineering and Technology, Pakistan) and Naeem Ahmed Mahoto (Mehran University of Engineering and Technology, Pakistan)
Engineering Science Reference • © 2020 • 291pp • H/C (ISBN: 9781799812531) • US $245.00

Nature-Inspired Computing Applications in Advanced Communication Networks
Govind P. Gupta (National Institute of Technology, Raipur, India)
Engineering Science Reference • © 2020 • 319pp • H/C (ISBN: 9781799816263) • US $225.00

Pattern Recognition Applications in Engineering
Diego Alexander Tibaduiza Burgos (Universidad Nacional de Colombia, Colombia) Maribel Anaya Vejar (Universidad Sergio Arboleda, Colombia) and Francesc Pozo (Universitat Politècnica de Catalunya, Spain)
Engineering Science Reference • © 2020 • 357pp • H/C (ISBN: 9781799818397) • US $215.00

IGI Global
PUBLISHER of TIMELY KNOWLEDGE

701 East Chocolate Avenue, Hershey, PA 17033, USA
Tel: 717-533-8845 x100 • Fax: 717-533-8661
E-Mail: cust@igi-global.com • www.igi-global.com

To my family for their endless support.

Editorial Advisory Board

Table of Contents

Detailed Table of Contents

In this chapter, the basic concepts of fractional-order dynamical systems are presented, and the synchronization methodologies of fractional order chaotic dynamical systems are established using slide mode control techniques. Through observation of the different phase portraits and time-series graphs of fractional order finance systems through utilization of the fractional calculus and computer simulation, the authors have obtained that the lowest dimension of fractional order hyper chaotic finance system is 3.90, which is less than 4. Bifurcation diagrams and Lyapunov exponents of fractional order hyper chaotic finance system are calculated to justify the chaos in the systems. Synchronization of two identical fractional-order hyper chaotic finance systems are achieved using sliding mode control techniques.

This chapter is dedicated to studying the magnetic blood flow with uniformly distributed magnetite dusty particles (MDP) in a cylindrical tube. For this purpose, the two-phase fractional Brinkman type fluid model is considered. The fractional governing equations are modeled in the cylindrical coordinate system taking into consideration the magnetization of the fluid due to the applied magnetic field. The fractional governing equations are subjected to physical initial and boundary conditions. The joint Laplace and Hankel transform is employed to develop exact analytical solutions. The obtained solutions are computed numerically and plotted in different graphs. It is noticed that for a long time the blood and MDP velocities increase with increasing values of the fractional parameter. In contrast, this effect reverses for a shorter time. In the case of the magnetic parameter, both velocities are decreased with increasing values of the magnetic parameter.

This contribution sets out to extend the concept of helicity so as to include it in a fractional scenario with a low-level of fractionality. To accomplish this goal, the authors write down the left- and the right-handed Weyl equations from first principles in this extended framework. Next, by coupling the two different fractional Weyl sectors by means of a mass parameter, they arrive at the fractional version of Dirac's equation, which, whenever coupled to an external electromagnetic field and reduced to the non-relativistic regime, yields a fractional Pauli-type equation. From the latter, they are able to present an explicit expression for the gyromagnetic ratio of charged fermions in terms of the fractionality parameter. They then focus their efforts to relate the coarse-grained property of space-time to fractionality and to the (g-2) anomalies of the different leptonic species. To do this, they build up an axiomatic local metric derivative that exhibits the Mittag-Leffler function as eigenfunction and is valid for low-level fractionality, whenever the order parameter is close to 1.

The transformation properties of the fractional derivatives under spatial rotation in two-dimensional space and for both the Riemann-Liouville and Caputo definitions are investigated and derived in their general form. In particular, the transformation properties of the fractional derivatives acting on scalar fields are studied and discussed. The study of the transformation properties of fractional derivatives is an essential step for the formulation of fractional calculus in multi-dimensional space. The inclusion of fractional calculus in the Lagrangian and Hamiltonian dynamical formulation relies on such transformation. Specific examples on the transformation of the fractional derivatives of scalar fields are discussed.

In this work, a space-time fractional diffusion equation with spatial Riesz-Feller fractional derivative and Caputo fractional time derivative is introduced. The continuation of the solution of this fractional equation to the solution of the corresponding integer order equation is proved. Also, a very useful Riesz-Feller fractional derivative is proved; the property is essential in applying iterative methods specially for complex exponential and/or real trigonometric functions. The analytic series solution of the problem is obtained via the optimal homotopy analysis method (OHAM). Numerical simulations are presented to validate the method and to highlight the effect of changing the fractional derivative parameters on the behavior of the obtained solutions. The results in this work are originally extracted from the author's work.

Fractional calculus has applications in many practical problems such as electromagnetic waves, visco-elastic systems, quantum evolution of complex systems, diffusion waves, physics, engineering, finance, social sciences, economics, mathematical biology, and chaos theory. Theories of inequality are growing rapidly; the fractional version of Ostrowski's type inequality is one of them. In this chapter, the author firstly presents some generalized Montgomery identities for Riemann-Liouville fractional integrals, which are designed by using a new and special type of Peano kernels. Secondly, inequalities via convex function are also discussed. In addition, some general fractional representation formulae are studied for a function in terms of the fractional Riemann-Liouville integrals of different orders and its ordinary derivatives. Moreover, by utilizing Montgomery identities, some new fractional versions of Ostrowski's type integral inequalities are established. Some well-known results are deduced as special cases from the results are developed here.

In this chapter, the authors discuss the effectiveness of the invariant subspace method (ISM) for solving fractional partial differential equations. For this purpose, they have chosen a nonlinear time fractional partial differential equation (PDE) with variable coefficients to be investigated through this method. One-, two-, and three-dimensional invariant subspace classifications have been performed for this equation. Some new exact solutions have been obtained using the ISM. Also, the authors give a comparison between this method and the homogeneous balance principle (HBP).

In this chapter, the author introduces a concept of apparent measure in R^n and associates the concept of relative dimension (of real order) that depends on the geometry of the object to measure and on the distance that separates it from an observer. At the end, the author discusses the relative dimension of a Cantor set. This measure enables us to provide a geometric interpretation of the Riemann-Liouville's integral of order $\alpha\epsilon$ ├]0,1], and based on this interpretation, the author introduces a modification on the Riemann-Liouville's integral to make it symmetrical and then introduces a new fractional derivative that exploits at the same time the right and the left fractional derivatives.

Section 5
Applications of Fractional Calculus to Engineering

This chapter presents the application of fractional differential operator in modelling and control of a three-tank interacting level process. In cases where the usage of sensors for the measurement of primary variable, which is the level of third tank in present case, is physically or economically not feasible, the measurement of secondary variable (i.e., second tank level) is used to determine the level of third tank for control purpose, known as inferential control scheme. The process is modeled and linearized around

Foreword

Fractional calculus (FC) has been proposed more than 150 years ago. Its applications have been increased significantly in the 21st century. Recently a relation between FC and strong interactions has been discovered. Infinitesimally deformed oscillator algebra corresponds to a Heisenberg relation which may explain confinement. There is a correspondence between q-deformed oscillators and fractional calculus. This result has been confirmed experimentally by Herrmann who showed that $\alpha = 2/3$ corresponds to hadrons and $\alpha \sim= 3/4, 4/5$ correspond to mesons. Another relation between FC and earthquakes is highly expected. So far we have failed in earthquake prediction even for short range prediction.

Earthquakes are known to occur on fractures which are fractals. Fractals, by definition are continuous nowhere differentiable functions hence only fractional calculus can be used to study them. Earthquakes are expected to increase hence studying them is quite important.

There is an intrinsic relation between FC and epidemic modeling. Studying epidemics is an important problem. It is known that the present state of a patient depends on his/her past history. Therefore, mathematical models for epidemics need fractional calculus. Similarly, FC is related to economics, environmental science and several engineering applications.

On the other hand, there are still a large number of non-local phenomena unexplored and waiting to be discovered. Therefore, we expect new aspects of the fractional calculus and its applications.

The present collection of chapters takes a practical stand, by addressing the various issues in the form of self-contained introductory essays. The investigated matter mainly includes issues of mathematical biology such as the fractional-order prey-predator model involving Allee effect and models for epidemics such as the fractional-order model of antimicrobial resistance and waning vaccination. The issues of nonlinear dynamics and chaos which are ever more popular in all fields of research have also been discussed in the second part of this book. Applications of FC to physics have been discussed in third part such as the magnetic blood flow with uniformly distributed generalized magnetite dusty particles; extending the concept of helicity and the space-time fractional diffusion equation with spatial Riesz-Feller fractional derivative and Caputo fractional time derivative. Interdisplinary new topics in mathematics with elegant discussion have also been presented in the fourth part of the book. Applications of FC to engineering have also been discussed in the last part of this book which include interesting topics such as charge decay; dynamic route theory; flux decay; impasse point breaking; memristor; moments of a function; parasitic fractional element; three tank; interacting process; inferential control; fractional order control; model order reduction; genetic algorithm; FO-PI and FO-PID.

All these topics are of current interest. So, I believe that this edited book is as timely as it could possibly be, and I am sure that it will constitute a valuable tool to help both experts as well as graduate students, in developing their studies and interest in the rapidly evolving and important field of research in fractional calculus.

El-Sayed Mohamed El-Sayed Ahmed
Mansoura University, Egypt

Preface

Fractional-order calculus mainly couples with the problems of derivatives and integrals of non-integer order. As the extension of classical integer-order systems, fractional-order systems can date back to the 19th century. It has recently been found that fractional modeling provides more adequate and realistic description of the natural phenomenon. Therefore, in this edited book, novel fractional-order models which have useful applications to science and engineering, are presented. Dynamical analysis of such models are investigated such as local stability, bifurcations and existence of chaotic (hyperchaotic) attractors. Novel schemes for controlling and synchronizing chaos in some fractional-order models are also applied. Novel discretization methods are presented and applied to some fractional-order models to enhance the real description and adequacy of these models.

Indeed, non-local differential and integral operators are better candidate to handle complex dynamical behaviors of fractional derivatives which are also classified based on their kernels. It is also favourable in real world modeling since the non-locality property implies that the future dynamics of the fractional model also depends on its history states. Non-local property of fractional differential equations is useful in modeling population interactions possessing memories. Hence, non-local fractional differential operators are also better candidate to describe some biological, economic and social phenomena where memory effects are important. The Riemann-Liouville and Caputo types possess non-local operators with singular kernels, however the Caputo-Fabrizio type has non-local operator with non-singular kernel. Moreover, the Atangana and Baleanu fractional derivative has non-local operator with non-singular kernel. Furthermore, a modified type of Caputo-Fabrizio definition with a smooth kernel has recently been presented by Yépez-Martínez and Gómez-Aguilar. These fractional differential operators have been used for solving various interesting problems in science and technology. Although, there is no unique definition of the fractional derivative, the fractional differential operator can be classified as integro-differential operator that may provide an appropriate or realistic description of the natural phenomena. Throughout this book, two well-known non-local operators with singular kernels are used; The Caputo and the Riemann-Liouville fractional differential operators which are commonly used by scientists. Furthermore, these fractional differential operators are widely used in real applications.

This book is aimed to shed some light on the various applications of fractional calculus to science and engineering. The 15 chapters are based on recent and advanced developments of the subject. A few interesting highlights include:

- Nonlinear fractional differential equations and fractional partial differential equations
- Fractional models of biology such as antimicrobial resistance and waning vaccination models, and Prey-Predator models involving Allee effect

- Fractional models of physics such as the two-phase fractional Brinkman type fluid model
- Discretizing fractional order differential equations
- Fractional diffusion equation
- The transformation properties of the fractional derivatives under spatial rotation
- Fractional Coarse-Grained approach with axiomatic local metric derivatives
- Chaotic dynamics in fractional-order systems
- Chaos control in fractional-order systems
- Chaos synchronization in fractional-order systems
- The application of fractional derivative to the memristor and the memristor based circuits
- Integral of real order and relative dimension
- Fractional Riemann-Liouville integrals
- Fractional optimal control problems
- Fractional modelling of a three tank interacting level process

The book is a combination of new results analysis with a review of some previous works. The different topics are explained in details to understand the subject. The new results have been verified by analytical methods with various simulations and case studies. The advanced methods like the predictor-correctors scheme, Adomian Decomposition method, optimal homotopy analysis method (OHAM) and the invariant subspace method (ISM) are effectively used for solving fractional differential equations and fractional partial differential equations. On the other hand, some models of mathematical biology become more appropriate and realistic when it is modelled by difference equations and nonlinear fractional-order systems generally do not have analytic solutions expressible in terms of a finite representation of elementary functions. So, discretization schemes are provided to discretize some fractional-order biological models and investigate the resulting rich dynamics. In addition, methods of chaos control and chaos synchronization like linear feedback control and sliding mode control techniques are effectively applied to some fractional-order chaotic systems. Finally, a number of well-known results can be deduced as special cases from the results developed in this book.

The book is divided into five main sections as follows:

The first section is entitled 'Fractional-Order Calculus in Biological Models', and it covers the first three chapters. Chapter 1 introduces a simple fractional-order model for competition between drug resistant and drug sensitive bacteria, namely the fractional-order AMR model. Some results for local stability conditions in this model are provided. A discretization scheme has also been applied to the proposed fractional-order model to enhance the model's adequacy and accuracy. So, the discrete-time model is shown to exhibit rich dynamics such as existence of Neimark-Sacker and flip bifurcations, and chaotic attractors. Chapter 2 presents a discrete-time fractional-order prey-predator model involving Allee effect. The positive steady state of the proposed model is shown to be existed under certain condition. Local stability analysis in this model has also been performed in addition to investigation of local bifurcations such as flip and Neimark-Sacker bifurcations. Existence of chaos and its suppression via linear feedback control and hybrid control methods are also shown. Finally, in Chapter 3, the authors present a short review of a fractional-order control models and their applications to some mathematical bioscience. The effect of the control on the fractional models is also shown.

The second section is entitled 'Chaotic Dynamics of Fractional-Order Systems' and covers the chapters 4-6. In Chapter 4, discretization schemes for discretizing ordinary differential equations with piecewise constant arguments and fractional-order differential equations are introduced. The resulting discrete-time models show rich variety of dynamical behaviors including existence of chaotic attractors. Chapter 5 presents a new fractional-order hyperchaotic system which can be useful to chaos-based secure communications and text encryption. Moreover, the influence of the Caputo type fractional differential operator on controlling the proposed fractional-order hyperchaotic system to its equilibrium points, has been shown. In Chapter 6, the problem of chaos synchronization between two identical fractional-order hyperchaotic finance systems is investigated using sliding mode control techniques.

The third section in this book, which includes Chapters 7, 8, 9, and 10, has the title 'Applications of Fractional Derivatives to Physics'. In Chapter 7, authors study the magnetic blood flow with uniformly distributed magnetite dusty particles (MDP) in a cylindrical tube. The non-Newtonian Brinkman type fluid is considered for the blood flow, Maxwell's relations for the magnetic field effect, Newton's second law of motion for MDP and the Caputo fractional derivative for the generalization. Chapter 8 aims to extend the concept of helicity so as to include it in a fractional scenario with a low-level of fractionality. The authors show that by coupling the two different fractional Weyl sectors using means of a mass parameter, the fractional version of Dirac's equation can be reached which, whenever coupled to an external electromagnetic field and reduced to the non-relativistic regime, yields a fractional Pauli-type equation. So, an explicit expression for the gyromagnetic ratio of charged fermions in terms of the fractionality parameter can also be represented. In Chapter 9, the transformation properties of the fractional derivatives under spatial rotation in two-dimensional space and for both the Riemann-Liouville and Caputo fractional differential operators are investigated and derived in their general form. Finally, in Chapter 10, a space-time fractional diffusion equation with spatial Riesz-Feller fractional derivative and Caputo fractional time derivative is introduced. The iterative series solution for the fractional equation is obtained using the OHAM by the aid of the Riesz-Feller fractional operator property applied on trignometric functions. The advantage of using this technique enables us to estimate an approximation to the residual error.

Section 4, which contains Chapters 11, 12, and 13, is entitled 'New Trends in Applied Mathematics'. In Chapter 11, some generalized Montgomery identities for Riemann-Liouville fractional integrals are designed by using a new and special type of Peano kernels. Moreover, inequalities via convex function are also discussed. Furthermore, some general fractional representation formulae are studied for a function in terms of the fractional Riemann-Liouville integrals of different orders and its ordinary derivatives. Finally, by utilizing Montgomery identities, some new fractional version of Ostrowski's type integral inequalities are established. Chapter 12 investigates the effectiveness of the invariant subspace method (ISM) for solving fractional partial differential equations. Also, some new exact solutions have been obtained using the ISM. In Chapter 13, the author, Faycal Ben Adda, presents a concept of apparent measure in $\mathbb{R}^n$ and associates the concept of relative dimension (of real order) which depends on the geometry of the object to measure and on the distance which separates it from an observer. This measure enables us to provide a geometric interpretation of the Riemann-Liouville's integral of order $\alpha \in]0,1]$, and based on this interpretation the author introduces a modification on the Riemann-Liouville's integral to make it symmetrical, and then introduces a new fractional derivative which exploits at the same time the right and the left fractional derivatives.

The fifth section in this book has the title 'Applications of Fractional Calculus to Engineering', and it contains Chapters 14 and 15. In Chapter 14, four different schemes of inferential control strategy arising from the combinations of different nature of plant and controller in term of their order (integer or non-integer), are implemented for the level control of three interacting tank process. In Chapter 15, the application of fractional derivative to the study of memristor dynamic under the effects of parasitic fractional elements are reported, where it is shown that parasitic fractional elements cause flux and charge decaying along with impasse point breaking of the phase portraits of the memristor.

It is not necessary to go through the book in a linear way (from Chapter 1 to Chapter 15). The only relationship between the chapters is that all of them contain the recent developments and applications of fractional calculus from various fields of science.

Finally, the main purpose of this special edited book is to provide a platform for researchers to disseminate original research in the field of fractional calculus and its potential applications. This book is important for researchers in fields of applied mathematics, nonlinear sciences, physicists, engineers, decision-makers in health organizations and those who are interested in mathematical modelling.

Ahmed Ezzat Matouk
Majmaah University, Saudi Arabia

Section 1

Fractional-Order Calculus in Biological Models

Chapter 1
Fractional-Order Calculus in Antimicrobial Resistance and Waning Vaccination

E. Ahmed
Mansoura University, Egypt

Ahmed Ezzat Mohamed Matouk
https://orcid.org/0000-0001-5834-4234
Majmaah University, Saudi Arabia

ABSTRACT

In this chapter, a simple model for competition between drug resistant and drug sensitive bacteria is given. So, a model of antimicrobial resistance (AMR) and waning vaccination is presented. The model's steady states are obtained. The conditions of local stability of the equilibria are also derived via the fractional Routh-Hurwitz criterion. A discretization scheme has also been applied to the proposed fractional-order model to enhance the model's adequacy and accuracy of describing natural phenomena. So, dynamical behaviors of the resulting discrete-time model are studied such as local stability, existence of Neimark-Sacker, and flip bifurcations. Furthermore, existence of chaos in the discrete-time model is proved using the theorem given by Marotto. Chaotic attractors and routes to chaos are also depicted via various numerical tools.

INTRODUCTION

During the past few years, some diseases have reappeared although they were thought to have disappeared (Lewnard & Grad, 2018). A possible reason is antimicrobial resistance (AMR) and multidrug resistance (MDR) and waning vaccination (Lewnard & Grad, 2018; WHO Report, 2018). Mathematical models with fractional derivatives are developed for exploring several aspects of epidemics. Indeed, the fractional models are better candidate to describe biological phenomena since they are related with memory. Both AMR and MDR depend on the exposure time. Hence memory effects are crucial for both phenomena.

DOI: 10.4018/978-1-7998-3122-8.ch001

A commonly used definition of fractional derivatives is the Caputo one (Caputo, 1967; Podlubny, 1999) which is given as follows

$$D^{\alpha} f(t) = \frac{1}{\Gamma(n-\alpha)} \int_0^t (t-\tau)^{n-\alpha-1} f^{(n)}(\tau) d\tau, \tag{1}$$

where

$\alpha \in R^{+}, n-1 < \alpha < n \in Z^{+},$

$\Gamma(.)$ is the Euler's Gamma function and $f^{(n)}$ represents the nth-order derivative of $f(\tau)$. The Caputo fractional-order derivative has non-local operator since it is defined by integration. Indeed, non-local operators are better candidate to handle complex dynamics of the physical phenomena and they are sorted according to their kernels. Moreover, the Caputo fractional operator has singular kernel. For fractional-order differential equations, the non-local property implies that the next state of a system not only depends on its current state but also on its history states since the calculating time fractional derivative of the functions $f(t)$ at a specific time τ^*, requires all the previous history, i.e. all $f(t)$ from $t=0$ to $t=\tau^*$. So fractional-order derivative is more suitable to describe complex systems and biological systems since these systems are naturally related with long-range memory behavior. Therefore, fractional-order modeling is better than integer-order modeling of biological systems (Ahmed & Elgazzar, 2007; Elsadany & Matouk, 2015; Matouk & Elsadany, 2016; Al-Khedhairi et al, 2018), economic (Laskin, 2000), financial systems (Chen, 2008; Chen et al, 2011; Zhen et al, 2011; Hegazi et al, 2013b), psychological systems (Ahmad & El-Khazali, 2007), physical systems (Heaviside, 1971; El-Sayed, 1996; Kusnezov et al, 1999; Hilfer, 2000; Al-Khedhairi et al, 2019a; Al-Khedhairi et al, 2019b) and engineering systems (Sun et al, 1984; Petras, 2009; Matouk, 2009a; Matouk, 2009b; Matouk 2010; Matouk, 2011a; Matouk, 2011b; Hegazi & Matouk, 2011; Hegazi et al, 2011; Hegazi & Matouk, 2013; Hegazi et al, 2013a; Matouk & Elsadany, 2014; Matouk, 2016; El-Sayed et al, 2016a; El-Sayed et al, 2016b) where memory effects are very important. Furthermore, fractional derivatives have also useful applications in some interdisciplinary fields of science (Ichise et al, 1971; Bagley & Torvik, 1983; Bagley & Calico, 1991; Ben Adda F, 1997; Magin, 2006; Tavazoei et al, 2009; Tavazoei, 2010; Mainardi, 2010; Latif et al, 2013).

On the other hand, some mathematical models of epidemics become more realistic when they are represented by discrete-time equations. Moreover, the need for a discretization of fractional-order models arises from the fundamental realization that nonlinear fractional-order systems generally do not have analytic solutions expressible in terms of a finite representation of elementary functions (Matouk et al, 2015).

THE FRACTIONAL AMR MODEL

According to (Blanquart, 2019), competition between different species is an important factor in antimicrobial resistance. Hence, a simple model for competition between susceptible and resistant species is presented here as:

$$\frac{d^\alpha x}{dt^\alpha} = ax(1-x) - xy,$$
$$\frac{d^\alpha y}{dt^\alpha} = bxy - dy, 0 < \alpha < 1, \tag{2}$$

where susceptible species are denoted by x and resistant ones are denoted by y and a, b, d are all positive parameters. The AMR model (2) has the equilibrium points S_0=(0,0), S_1=(1,0) and $S_2 = (d / b, a(b-d) / b)$ as $d<b$.

Stability of the Fractional AMR Model

In this part, we will study local stability of the steady states S_0=(0,0), S_1=(1,0) and $S_2 = (d / b, a(b-d) / b)$ by means of the fractional Routh-Hurwitz criterion (FRH) (Ahmed & Elgazzar, 2007). An equilibrium point S_i, i=0,1,2 of (2) is said to be locally asymptotically stable (LAS) if

$$\left|\arg(\lambda_i)\right| > \alpha\pi / 2, \tag{3}$$

where λ_i is an eigenvalue corresponding to the equilibrium point satisfying the characteristic equation (Matignon, 1996):

$$\psi(\lambda) = \lambda^2 + c_1\lambda + c_2 = 0. \tag{4}$$

Then according to the FRH conditions, S_i is LAS if and only if

(i) $c_1 < 0,\ c_1^2 < 4c_2$ and $\left|\arctan\left(\sqrt{4c_2 - c_1^2}\right) / c_1\right| > \alpha\pi / 2,$ or

(ii) c_1>0 and c_2>0. (5)

For the equilibrium point S_0=(0,0), the characteristic polynomial $\psi(\lambda)$ is reduced to

$$\psi(\lambda) = \lambda^2 + (d-a)\lambda - ad = 0. \tag{6}$$

It is clear that the FRH conditions (5) are not satisfied which implies that S_0=(0,0) is not LAS. For the equilibrium point S_1=(1,0), the characteristic polynomial $\psi(\lambda)$ is reduced to

$$\psi(\lambda) = \lambda^2 + (a-b+d)\lambda + a(d-b) = 0. \tag{7}$$

Based on the part (ii) of the FRH conditions (5), S_1=(1,0) is LAS when $b<d$ and not LAS when $b>d$. For the equilibrium point $S_2 = (d / b, a(b-d) / b)$, the characteristic polynomial $\psi(\lambda)$ is reduced to

$$\psi(\lambda) = \lambda^2 + \frac{ad}{b}\lambda + ad\left(1 - \frac{d}{b}\right) = 0. \tag{8}$$

Since the point S_2 is only defined as $d<b$, it is clear that the part (ii) of the FRH conditions (5) is satisfied which implies that $S_2 = (d / b, a(b - d) / b)$ is LAS. This agrees with observations.

Numerical Results

An efficient method for solving fractional-order differential equations is the predictor-correctors scheme (Diethelm, 1997; Diethelm & Freed, 1999; Diethelm & Ford, 2002; Diethelm et al, 2002). This scheme is represented as follows:

Consider the following initial value problem (IVP)

$$D^{\alpha}\mathrm{X}(t) = f(t, \mathrm{X}(t)), 0 \le t \le T, \mathrm{X}^{(k)}(0) = \mathrm{X}_0^{(k)}, k = 0,1,\ldots,\lceil\alpha\rceil - 1, \mathrm{X} \in R^n, \tag{9}$$

and f is a continuous nonlinear vector function. Then according to (Diethelm & Ford, 2002), the IVP (9) is equivalent to Volterra integral equation of the second kind, described as

$$\mathrm{X}(t) = \sum_{k=0}^{\lceil\alpha\rceil-1} \frac{t^k}{k!}\mathrm{X}_0^{(k)} + \frac{1}{\Gamma(\alpha)}\int_0^t (t-\tau)^{\alpha-1} f(\tau, \mathrm{X}(\tau))d\tau. \tag{10}$$

Set

$$h = T / N, t_n = nh, n = 0,1,\ldots,N \in Z^+.$$

Hence, Eq. (10) can be represented as

$$\mathrm{X}_h(t_{n+1}) = \sum_{k=0}^{\lceil\alpha\rceil-1} \frac{t_{n+1}^k}{k!}\mathrm{X}_0^{(k)} + \frac{h^{\alpha}}{\Gamma(\alpha+2)} f(t_{n+1}, \mathrm{X}_h^p(t_{n+1})) + \frac{h^{\alpha}}{\Gamma(\alpha+2)}\sum \varphi_{j,n+1} f(t_j, \mathrm{X}_h(t_j)), \tag{11}$$

$$\varphi_{j,n+1} = \begin{cases} n^{\alpha+1} - (n-\alpha)(n+1)^{\alpha}, & j = 0, \\ (n-j+2)^{\alpha+1} + (n-j)^{\alpha+1} - 2(n-j+1)^{\alpha+1}, & 1 \le j \le n, \\ 1, & j = n+1, \end{cases}$$

$$\mathrm{X}_h^p(t_{n+1}) = \sum_{k=0}^{\lceil\alpha\rceil-1} \frac{t_{n+1}^k}{k!} x_0^{(k)} + \frac{1}{\Gamma(\alpha)}\sum_{j=0}^{n} \mu_{j,n+1} f(t_j, x_h(t_j)),$$

$$\mu_{j,n+1} = \frac{h^{\alpha}}{\alpha}((n+1-j)^{\alpha} - (n-j)^{\alpha}),$$

with error estimate

$$\max_{j=0,1,\ldots,N} \left|X(t_j) - X_h(t_j)\right| = O(h^p),$$

where $p=\min(2,1+\alpha)$.

The model (2) is numerically integrated using the predictor-correctors scheme with $a=1$, $b=10$ and $d=0.1$, and with fractional parameters $\alpha=0.99$, $\alpha=0.95$ and $\alpha=0.9$, respectively as shown in figures 1 to 3. It is evident that no chaotic attractors are expected in the integer-order model according to the Poincaré-Bendixon theorem.

Figure 1. Phase portrait of the fractional AMR model (2) using a=1, b=10 and d=0.1, and with fractional order α=0.99

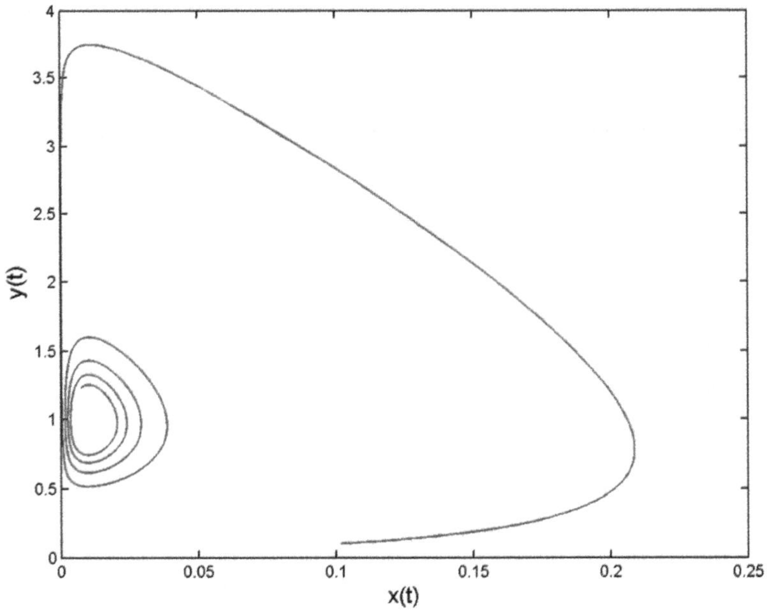

A DISCRETE-TIME FORM OF THE FRACTIONAL AMR MODEL

According to the discretization method (El-Sayed & Salman, 2013; El-Raheem & Salman, 2014), the fractional AMR model can be described as

Figure 2. Phase portrait of the fractional AMR model (2) using a=1, b=10 and d=0.1, and with fractional order α=0.95

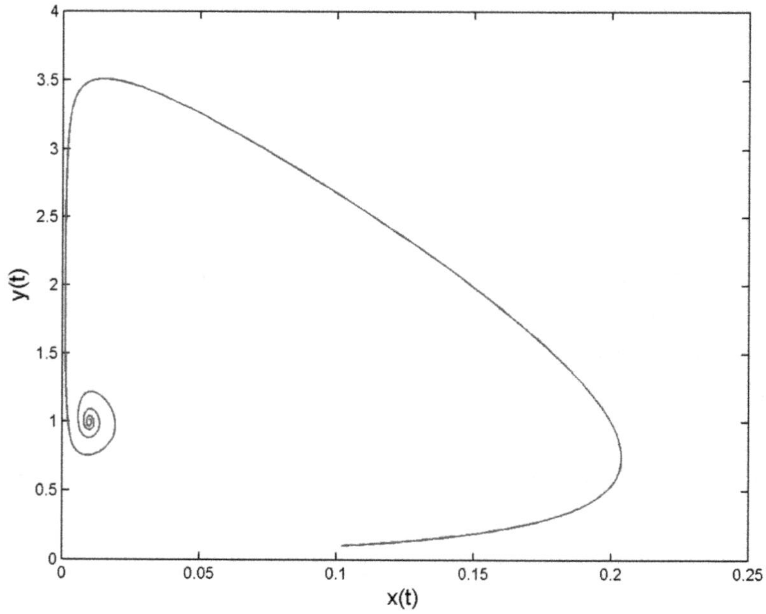

Figure 3. Phase portrait of the fractional AMR model (2) using a=1, b=10 and d=0.1, and with fractional order α=0.9

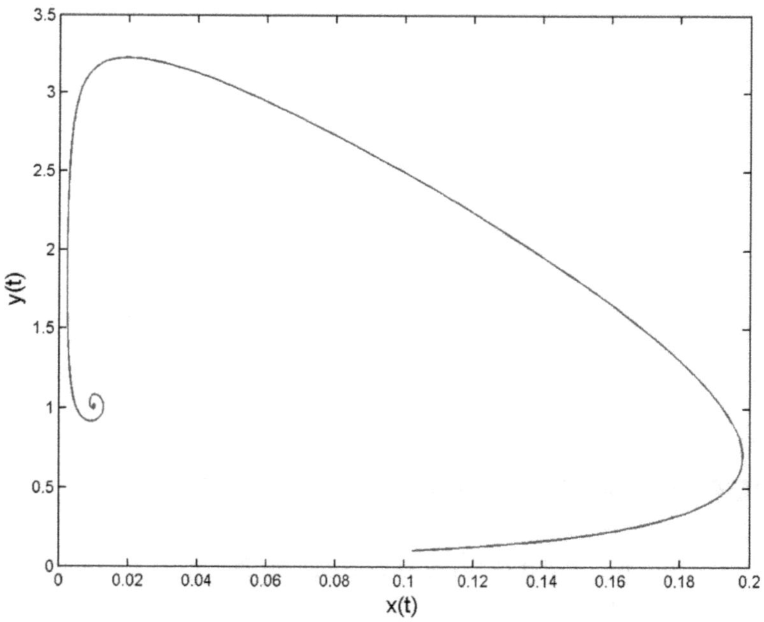

$$x_{n+1} = x_n + \frac{s^\alpha}{\alpha\Gamma(\alpha)}[ax_n(1-x_n) - x_n y_n],$$
$$y_{n+1} = y_n + \frac{s^\alpha}{\alpha\Gamma(\alpha)}[bx_n y_n - dy_n], \tag{12}$$

where $s>0$ is a discretization parameter. The system's steady states are also the points S_0, S_1 and S_2 defined above.

Local Stability of the Discrete-Time AMR Model

We next discuss the local stability of model (12). The Jacobian matrix of (12) is

$$J(x,y) = \begin{pmatrix} 1 + S[a(1-x) - ax - y] & -Sx \\ Sby & 1 + S(bx - d) \end{pmatrix}, \tag{13}$$

where $S = s^\alpha/(\alpha\Gamma(\alpha))$.

Since system (12) is represented by a two dimensional map, we have the following stability types:

1. A steady state S_i is LAS if each of corresponding eigenvalues has modulus less than one. In this case the point is also called a sink.
2. A steady state S_i is called source if each of corresponding eigenvalues has modulus greater than one. In this case the point is locally unstable.
3. A steady state S_i is called saddle if one eigenvalue has modulus greater than one and the other has an eigenvalue with modulus less than one.
4. A steady state S_i is non-hyperbolic if each of corresponding eigenvalues has modulus equals one.

Thus, we have the following discussion:

- The equilibrium state $S_0=(0,0)$ has the eigenvalues $\lambda_1=1+Sa$, $\lambda_2=1-Sd$ and is neither sink nor non-hyperbolic; It is source if $d>2/S$, however it will be saddle if $d<2/S$.
- The equilibrium state $S_1=(1,0)$ has the eigenvalues $\lambda_1=1-Sa$, $\lambda_2=1+S(b-d)$ and is not non-hyperbolic. It is sink if $2/S>a$ and $d>b$. S_1 is source if $2/S<\min\{a,d-b\}$ or $a>2/S$, $b>d$. Furthermore, S_1 is saddle if one of the following cases hold: $a,2/S<d-b$; $a<2/S$, $b>d$; $a>2/S$, $d>b$ or $d>b+2/S$.
- The equilibrium state $S_2=(d/b,\ a(b-d)/b)$ is sink if

$$\begin{cases} (i)\, 1 - TrJ + DetJ > 0, \\ (ii)\, 1 + TrJ + DetJ > 0, \\ (iii)\, DetJ < 1, \end{cases} \tag{14}$$

where

$$J = \begin{pmatrix} 1 - Sad / b & -Sd / b \\ Sa(b-d) & 1 \end{pmatrix}.$$

Bifurcations of the Discrete-Time AMR Model

Firstly, we will discuss the existence of Neimark–Sacker (NS) bifurcation in the discrete-time AMR model (12) which requires existence of a pair of complex eigenvalues λ_i, i=1,2 such that $|\lambda_i|=1$, i=1,2 at the critical value of the dynamical parameter. Thus, the equilibrium state $S_2=(d/b,\ a(b-d)/b)$ is only candidate to study existence of NS bifurcation in system (12). To find the indicated critical value of the dynamical parameter, we need to apply the following criterion (Wen, 2005):

Let $\bar{S}$ be a steady state of the nth-order discrete-time system whose eigenvalue equation is given by

$$\psi_\theta(\lambda) = \lambda^n + c_1\lambda^{n-1} + \ldots + c_{n-1}\lambda + c_n, \tag{15}$$

where $c_k=c_k(\theta)$, $k=1,\ldots,n$, and θ is the bifurcation parameter. Define the sequence of determinants $\Delta_0^\pm(\theta) = 1, \Delta_1^\pm(\theta), \ldots, \Delta_n^\pm(\theta)$ such that

$$\Delta_k^\pm(\theta) = \left\| \begin{pmatrix} 1 & c_1 & c_2 & \ldots & c_{k-1} \\ 0 & 1 & c_1 & \ldots & c_{k-2} \\ 0 & 0 & 1 & \ldots & c_{k-3} \\ \ldots & \ldots & \ldots & \ldots & \ldots \\ 0 & 0 & 0 & \ldots & 1 \end{pmatrix} \pm \begin{pmatrix} c_{n-k+1} & c_{n-k+2} & \ldots & c_{n-1} & c_n \\ c_{n-k+2} & c_{n-k+3} & \ldots & c_n & 0 \\ \ldots & \ldots & \ldots & \ldots & \ldots \\ c_{n-1} & c_n & \ldots & 0 & 0 \\ c_n & 0 & \ldots & 0 & 0 \end{pmatrix} \right\|, k = 1, \ldots, n. \tag{16}$$

So, NS bifurcation occurs at a critical parameter θ_0 provided that

1. the following eigenvalue assignment holds

$$\Delta_{n-1}^-(\theta_0) = 0, \psi_{\theta_0}(1) > 0, (-1)^n \psi_{\theta_0}(-1) > 0, \Delta_{n-1}^+(\theta_0) > 0, \Delta_k^\pm(\theta_0) > 0, k = n-3, n-5, \ldots, 1$$

(or 2) when n is even (or odd, resp.),

2. the transversality condition $d\Delta_{n-1}^-(\theta_0) / d\theta \neq 0$,
3. the nonresonance condition $\phi \neq \cos(2\pi/j)$ or resonance condition $\phi=\cos(2\pi/j)$ where j=3,4,5,… and

$$\phi = 1 - 0.5\psi_{\theta_0}(1)\Delta_{n-3}^-(\theta_0) / \Delta_{n-2}^+(\theta_0).$$

Based on the above criterion with the parameters a=11, d=3, s=0.1, α=0.9, the discrete-time AMR model (12) exhibits NS bifurcation at the critical parameter b_0=10.64. Figure 4 shows the creation of limit cycle around the point $S_2=(d/b,\ a(b-d)/b)$ using the above-mentioned parameter values and b=10.65.

Also, the first route to chaos in the discrete-time AMR model (12) via NS bifurcation is depicted in figure 5. To study the direction of this bifurcation, we put the discrete-time AMR model (12) with the above-mentioned parameters in the general form

$$\mathrm{X}_{n+1} = f_{\theta}(\mathrm{X}_{n}). \tag{17}$$

The point $S_2=(d/b,\ a(b-d)/b)$ is not origin, so the above system is transformed as

$$\mathrm{Z}_{n+1} = J((0,0),\gamma)\mathrm{Z}_{n} + \frac{\mathrm{B}(\mathrm{Z}_{n},\mathrm{Z}_{n})}{2!} + \frac{\mathrm{E}(\mathrm{Z}_{n},\mathrm{Z}_{n},\mathrm{Z}_{n})}{3!} + O\left(\left\|\mathrm{Z}_{n}\right\|^{4}\right), \tag{18}$$

where $(\mathrm{Z},\gamma) = (\mathrm{X} - S_2, \nu - \nu_0)$,

$$\mathrm{B}_i(u,v) = \sum_{j,k=1}^{n} \left.\frac{\partial^2 \mathrm{Z}_i(\xi)}{\partial\xi_j \partial\xi_k}\right|_{\xi=0} u_j v_k,\ \mathrm{E}_i(u,v,w) = \sum_{j,k,m=1}^{n} \left.\frac{\partial^3 \mathrm{Z}_i(\xi)}{\partial\xi_j \partial\xi_k \partial\xi_m}\right|_{\xi=0} u_j v_k w_m,\ i = 1,2,$$

and the matrix J is the Jacobian calculated at the origin O and is given as

$$J(O) = \begin{pmatrix} 0.59400443 & -0.03690859608 \\ 11.00000001 & 0.9999999998 \end{pmatrix}. \tag{19}$$

Hence, J has two complex conjugate eigenvalues $e^{\pm I\beta_0}, I = \sqrt{-1}$ on the unit circle. The direction of NS bifurcation is determined by (Kuznetsov,1998):

$$l_1(0) = \mathrm{Re}\left(\frac{\sigma_{21}e^{-I\beta_0}}{2}\right) - \mathrm{Re}\left(0.5\sigma_{20}\sigma_{11}\frac{e^{-2I\beta_0} - 2e^{-I\beta_0}}{1 - e^{I\beta_0}}\right) - \frac{\left|\sigma_{11}\right|^2}{2} - \frac{\left|\sigma_{02}\right|^2}{4}, \tag{20}$$

where

$$\begin{aligned}
&\sigma_{20} = \left\langle \delta, \mathrm{B}(\eta,\eta)\right\rangle, \quad \sigma_{11} = \left\langle \delta, \mathrm{B}(\eta,\overline{\eta})\right\rangle, \quad \sigma_{02} = \left\langle \delta, \mathrm{B}(\overline{\eta},\overline{\eta})\right\rangle, \\
&\sigma_{21} = \left\langle \delta, \mathrm{E}(\eta,\eta,\overline{\eta})\right\rangle + 2\left\langle \delta, \mathrm{B}(\eta,(I_2 - J)^{-1}\mathrm{B}(\eta,\overline{\eta}))\right\rangle + \left\langle \delta, \mathrm{B}(\overline{\eta},(e^{2I\beta_0}I_2 - J)^{-1}\mathrm{B}(\eta,\eta))\right\rangle \\
&+\frac{e^{-I\beta_0} - 2}{1 - e^{I\beta_0}}\sigma_{20}\sigma_{11} - \frac{2}{1 - e^{-I\beta_0}}\left|\sigma_{11}\right|^2 - \frac{e^{I\beta_0}}{e^{3I\beta_0} - 1}\left|\sigma_{02}\right|^2,
\end{aligned}$$

where I_2 is the identity matrix of size 2 and the complex vectors δ and η satisfy

$$J(\gamma_0)\eta = e^{I\beta_0}\eta, J(\gamma_0)\overline{\eta} = e^{-I\beta_0}\overline{\eta}, J^T(\gamma_0)\delta = e^{-I\beta_0}\delta, J^T(\gamma_0)\overline{\delta} = e^{I\beta_0}\overline{\delta}, \left\langle \delta, \eta\right\rangle = 1. \tag{21}$$

Thus, the vectors δ and η are computed as

$$\eta = (-0.05490688297 - 0.01845429802I, I)^T,$$
$$\delta = (-9.106326439, -0.1680508619 + 0.4999999999I)^T. \tag{22}$$

So, the functions B and E defined in Eq. (18) are also computed as

$$\mathrm{B}(\xi,\rho) = \begin{pmatrix} -S[2a\xi_1\rho_1 + \xi_1\rho_2 + \xi_2\rho_1] \\ bS[\xi_1\rho_2 + \xi_2\rho_1] \end{pmatrix}, \mathrm{E}(\xi,\rho,\zeta) = \begin{pmatrix} 0 \\ 0 \end{pmatrix}. \tag{23}$$

The quantities σ_{ij} appearing in (20) are computed as

$$\sigma_{20} = 0.02901619277 - 0.07775370295I,$$
$$\sigma_{11} = 0.05263300600 + 0.02570115083I,$$
$$\sigma_{02} = 0.1819529318 + 0.02635140127I,$$
$$\sigma_{21} = (-0.33 - 0.5I)\times 10^{-10}. \tag{24}$$

According to (Kuznetsov,1998), the NS bifurcation is supercritical since $l_1(0)$= -0.005266421<0, i.e. a stable closed invariant curve near $S_2=(d/b,\ a(b-d)/b)$ exists after the NS bifurcation.

Secondly, we will discuss the existence of flip (period-doubling) bifurcation in the discrete-time AMR model (12). For this reason we seek one eigenvalue equals -1 and the other lies inside the unit circle. This eigenvalues assignment is satisfied around S_1=(1,0) using the parameters b=0.3, d=1, s=0.01, α=0.1 and the critical parameter a_0=3.015578717. The second route to chaos in the discrete-time AMR model (12) via flip bifurcation is depicted in figure 6.

Finally, we need to determine the stability of period 2 orbits result from flip bifurcation as follows: The multilinear functions B and E are also given as

$$\mathrm{B}(\xi,\rho) = \begin{pmatrix} -S[2a\xi_1\rho_1 + \xi_1\rho_2 + \xi_2\rho_1] \\ bS[\xi_1\rho_2 + \xi_2\rho_1] \end{pmatrix}, \mathrm{E}(\xi,\rho,\zeta) = \begin{pmatrix} 0 \\ 0 \end{pmatrix}. \tag{25}$$

Also based on the above selection of parameters and the critical value a_0, the vectors δ and η are computed as

$$\eta = (1,0)^T,$$
$$\delta = (1, 0.4318574845)^T. \tag{26}$$

Therefore, the type of flip bifurcation is obtained by determining the sign of the quantity $C(0)$ (Kuznetsov,1998) which is defined as

$$C(0) = \frac{\langle \delta, \mathrm{E}(\eta,\eta,\eta) \rangle}{3!} - \frac{\langle \delta, \mathrm{B}(\eta,(J - I_2)^{-1}\mathrm{B}(\eta,\eta)) \rangle}{2!}. \tag{27}$$

Here, the quantity $\mathrm{B}(\eta,(J - I_2)^{-1}\mathrm{B}(\eta,\eta))$ is also calculated as

$$\mathrm{B}(\eta,(J - I_2)^{-1}\mathrm{B}(\eta,\eta)) = (-2a^2S^2, 0)^T. \tag{28}$$

So if $C(0)>0$, there exist stable period 2 orbits result from flip bifurcation; however, these period 2 orbits are unstable when $C(0)<0$. Based on Eqns. (25)-(28), the quantity $C(0)$ is given as $C(0)=a^2S^2>0$ which implies the existence of stable bifurcating period 2 orbits around the equilibrium state $S_1=(1,0)$.

Figure 4. Creation of a limit cycle around the fixed point S_2 of the discrete-time AMR model (12) using $a=11$, $b=10.65$, $d=3$, $s=0.1$ and $\alpha=0.9$

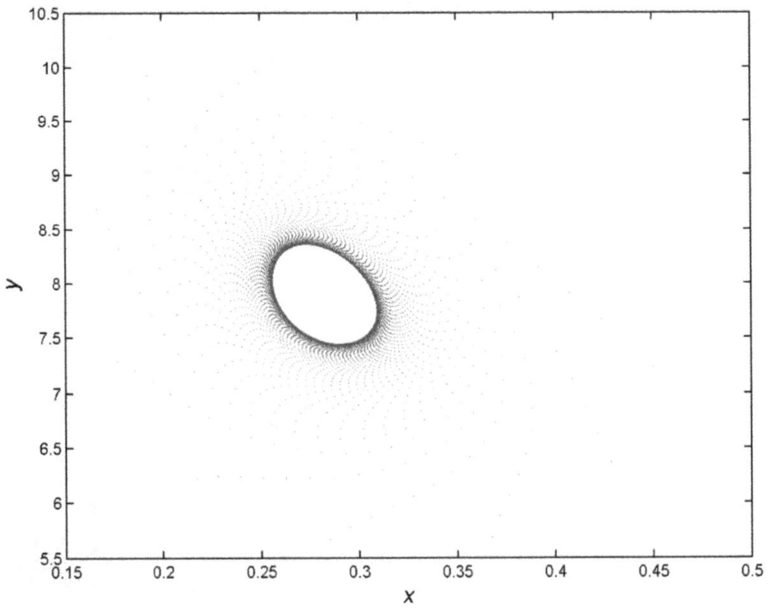

INVESTIGATION OF MAROTTO'S CHAOS IN THE DISCRETE-TIME AMR SYSTEM

In the following, existence of chaos using Marotto's Theorem (Marotto, 2005) is shown. Consider the general form of the discrete-time AMR model system (17), then we have the following definitions and theorem (Marotto, 2005):

Definition 1. Suppose that $f: R^n \rightarrow R^n$ is a differentiable in $B_r(Z)$. The point $Z \in R^n$ is an expanding fixed point of f in $B_r(Z)$, if $f(Z)=Z$ and all eigenvalues of $J(f(X))$ have norms greater than one $\forall X \in B_r(Z)$.

Figure 5. The bifurcation diagram of the discrete-time AMR model (12) as b is varied and fixing a=11, d=3, s=0.1 and α=0.9

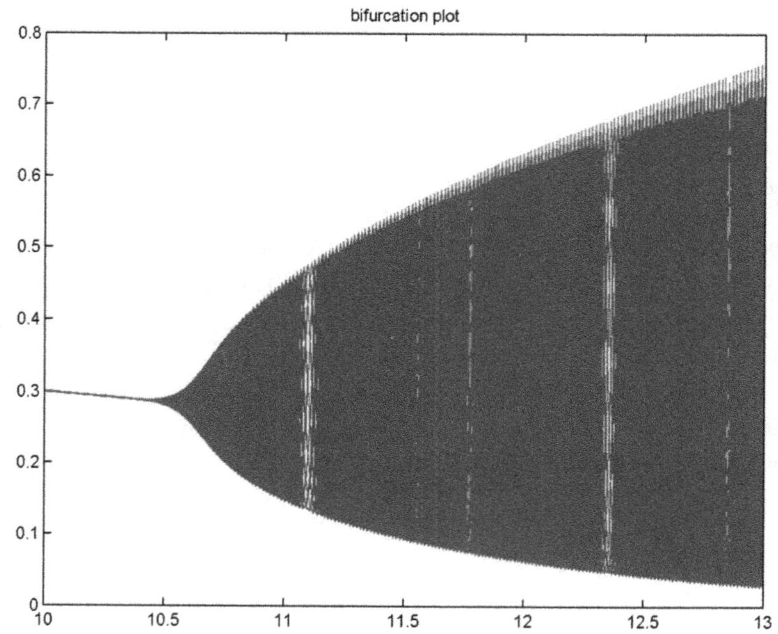

Figure 6. The bifurcation diagram of the discrete-time AMR model (12) as a is varied and fixing b=0.3, d=1, s=0.01 and α=0.1

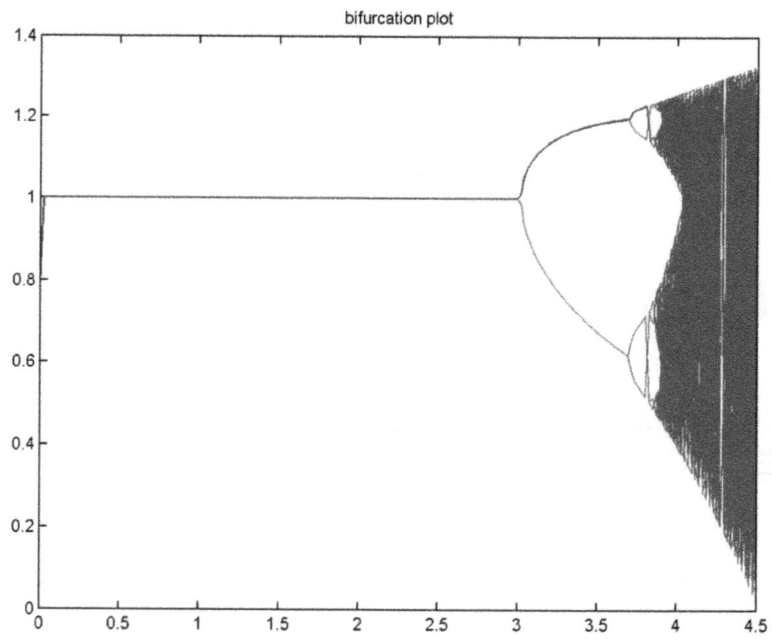

Definition 2. Let Z be an expanding fixed point of f in $B_r(Z)$, for some $r>0$. So, the fixed point Z is called a snap-back repeller of f if

$$\exists X_0 \in B_r(Z) St. X_0 \neq Z, f^l(X_0) = Z$$

and $Jf^l(X_0)\neq 0$ for some positive integer l.

Theorem 1. If the map f: $R^n \rightarrow R^n$ is differentiable and has a snap-back repeller, then the map (12) is chaotic in the sense of Li-Yorke. That is there exists

1. $N\in Z^+$ St. $\forall p\geq N$, the map f has a point of period p;
2. an uncountable set χ that has no periodic points of f, i.e., the so called "scrambled set" of f that has the following properties:

$$f(\chi) \subset \chi,$$

$$\limsup_{k\to\infty} \left\| f^k(X_s) - f^k(Y_s) \right\| > 0 \ \forall X_s; Y_s \in \chi \text{ St. } X_s \neq Y_s,$$

$$\limsup_{k\to\infty} \left\| f^k(X_s) - f^k(Y_p) \right\| > 0 \ \forall X_s \in \chi \text{ and any periodic point } Y_p \text{ of } f;$$

3. an uncountable subset χ_0 of χ St.

$$\limsup_{k\to\infty} \left\| f^k(X_{s_0}) - f^k(Y_{s_0}) \right\| = 0 \ \forall X_{s_0}; Y_{s_0} \in \chi_0.$$

Then, we discuss when a fixed point E^* is an expanding fixed point. The eigenvalues of E^* are solutions of the following equation:

$$\begin{aligned} &\lambda^2 + (-2 - aS + 2aSx_n + Sy_n - bSx_n + Sd)\lambda + 1 + bSx_n + S(a - d - y_n) \\ &+aS^2(bx_n - d) - 2aSx_n + 2aS^2(dx_n - bx_n^2) + dS^2 y_n = 0. \end{aligned} \tag{29}$$

So, our objective is to obtain a pair of complex conjugate solutions of Eq. (29) with modulus greater than one. For this reason, we define the quantity

$$\begin{aligned} &\Theta_1(x_n, y_n) = S^2(4ab + 4a^2 + b^2)x_n^2 + 2S^2(2ay_n - 2ad - by_n - ab - bd - 2a^2)x_n \\ &S^2(y_n^2 + d^2 + 2ad + a^2 - 2ay_n - 2dy_n). \end{aligned} \tag{30}$$

Hence, we need to ensure that $\Theta_1(x_n, y_n)<0$ which is achieved as

$$x_n^1 < x_n < x_n^2,$$
$$x_n^1 = \frac{2y_n(b-2a) + 4ad + 2bd + 4a^2 + 2ab - 4\sqrt{y_n(2abd + ab^2 + db^2 - 2aby_n + 2ba^2)}}{2(4ab + 4a^2 + b^2)}, \tag{31}$$
$$x_n^2 = \frac{2y_n(b-2a) + 4ad + 2bd + 4a^2 + 2ab + 4\sqrt{y_n(2abd + ab^2 + db^2 - 2aby_n + 2ba^2)}}{2(4ab + 4a^2 + b^2)},$$

provided that

$$y_n(2abd + ab^2 + db^2 - 2aby_n + 2ba^2) > 0.$$

Then, define

$$D_1 = \{(x,y) \big| 0 < x_n^1 < x < x_n^2, y > 0\},$$

where x,y refer to x_n, y_n, respectively. If $x \in D_1$ then $\Theta_1(x,y)<0$. Also, define

$$D_2(x,y) = \{(x,y) \big| \; y > \frac{d - a - bx(1 + aS) + adS + 2ax + 2aSx(bx - d)}{dS - 1}, dS \neq 1\}. \tag{32}$$

Thus, the fixed point $E^*=(x^*,y^*)$ is an expanding fixed point in $B_r(E^*)$ if

$$E^*(x^*,y^*) \in B_r(E^*) = \{(x,y) \big| (x,y) \in D_1 \cap D_2 \neq \varphi\}. \tag{33}$$

Now, we need to find a point

$$E_1(x_1,y_1) \in B_r(E^*) St. E_1 \neq E^*, f^l(E_1) = E^*,$$

$|Jf^l(E_1)| \neq 0$ for some positive integer l, where f is defined in the discrete-time AMR map (17). Indeed, we have

$$\begin{aligned} x_1 + S[ax_1(1 - x_1) - x_1y_1] &= x_2, \\ y_1 + S[bx_1y_1 - dy_1] &= y_2. \end{aligned} \tag{34}$$

Also,

$$\begin{aligned} x_2 + S[ax_2(1 - x_2) - x_2y_2] &= x^*, \\ y_2 + S[bx_2y_2 - dy_2] &= y^*. \end{aligned} \tag{35}$$

So, we are able to map $E_1(x_1,y_1)$ to the steady sate $E^*=(x^*,y^*)$ by iterating the map f 2 times if there are solutions for (34) and (35) that differ from $E^*=(x^*,y^*)$. By straightforward computations, the solutions of (35) that differ from $E^*=(x^*,y^*)$ satisfy the following:

$$\begin{aligned} &\gamma_1 x_2^3 + \gamma_2 x_2^2 + \gamma_3 x_2 + \gamma_4 = 0, \\ &y_2 = \frac{y^*}{1 + bSx_1 - dS}, \end{aligned} \tag{36}$$

where

$$\begin{aligned} \gamma_1 &= -abS^2, \\ \gamma_2 &= S(abS + b - a + adS), \\ \gamma_3 &= S(-adS + a - y^* - d - bx^*) + 1, \\ \gamma_4 &= x^*(dS - 1). \end{aligned}$$

Substituting x_2 and y_2 into (34) and solving for x_1 and y_1, we obtain

$$\begin{aligned} &\gamma_1' x_1^3 + \gamma_2' x_1^2 + \gamma_3' x_1 + \gamma_4' = 0, \\ &y_1 = \frac{y_2}{1 + bSx_1 - dS}, \end{aligned} \tag{37}$$

where

$$\begin{aligned} \gamma_1' &= -abS^2, \\ \gamma_2' &= S(abS + b - a + adS), \\ \gamma_3' &= S(-adS + a - y_2 - d - bx_2) + 1, \\ \gamma_4' &= x_2(dS - 1). \end{aligned}$$

Finally, if Eq. (33) is satisfied, the solutions of (34) and (35) agree (36) and (37). Moreover, the solutions

$$E_1(x_1, y_1),\ E_2(x_2, y_2) \neq E^*(x^*, y^*), E_1(x_1, y_1) \in B_r(E^*),$$

and $|Jf^2(E_1)| \neq 0$. So, the fixed point E^* is a snap-back repeller in $B_r(E^*)$. We summarize these results in the following theorem.

Theorem 2. Let the fixed point E^* of the discrete-time AMR model (12) be an expanding fixed point in $B_r(E^*)$; If the solutions $E_1(x_1,y_1)$, $E_2(x_2,y_2)$ of equations (34) and (35) satisfy

$$E_2(x_2, y_2) \neq E^*(x^*, y^*), E_1(x_1, y_1) \in B_r(E^*)$$

and $|Jf^2(E_1)| \neq 0$, then E^* is a snap-back repeller and therefore the discrete-time AMR model (12) is chaotic in the sense of Marotto.

MULTI-DRUG RESISTING BACTERIA

We like to end with the following open problem:

In 2017 Madagascar has been struck by a fast-spreading outbreak of plague. Plague is caused by the bacterium Yersinia pestis. Terrifying as the disease is, it can usually be cured by common antibiotics. Antibiotic-resistant strains have been isolated in Madagascar but are not thought to be a factor in the latest outbreak. The question is WHY? And what can be done if they appear?

Moreover, recalling the following comments on AMR transmission, transmission dynamics between environments are poorly understood. Lack of harmonization between countries, between humans/animals/environment makes data comparison difficult. Differences in health care/animal husbandry systems mean that tailored interventions adapted to each country may be necessary. Knowledge gap in factors determining success of specific clones exists.

Understanding human/animal/food movements and their consequences on AMR is not sufficiently known. Differences regarding human or animal specific clones are yet to be determined. Multidisciplinary in training, education, networking and collaboration is required. It is clear that network models for AMR are needed. This will help us to understand the nature of spreading diseases.

REFERENCES

Ahmad, W. M., & El-Khazali, R. (2007). Fractional-order dynamical models of love. *Chaos, Solitons, and Fractals, 33*(4), 1367–1375. doi:10.1016/j.chaos.2006.01.098

Ahmed, E., & Elgazzar, A. S. (2007). On fractional order differential equations model for nonlocal epidemics. *Physica A, 379*(2), 607–614. doi:10.1016/j.physa.2007.01.010

Al-Khedhairi, A., Askar, S. S., Matouk, A. E., Elsadany, A., & Ghazel, M. (2018). Dynamics, chaos control, and synchronization in a fractional-order Samardzija-Greller population system with order lying in (0,2). *Complexity*.

Al-khedhairi, A., Matouk, A. E., & Askar, S. S. (2019a). Computations of synchronization conditions in some fractional-order chaotic and hyperchaotic systems. *Pramana Journal of Physics, 92*, 72.

Al-khedhairi, A., Matouk, A. E., & Khan, I. (2019b). Chaotic dynamics and chaos control for the fractional-order geomagnetic field model. *Chaos, Solitons, and Fractals, 128*, 390–401. doi:10.1016/j.chaos.2019.07.019

Bagley, R. L., & Calico, R. A. (1991). Fractional order state equations for the control of viscoelastically damped structures. *Journal of Guidance, Control, and Dynamics, 14*(2), 304–311. doi:10.2514/3.20641

Bagley, R. L., & Torvik, P. J. (1983). A theoretical basis for the application of fractional calculus to viscoelasticity. *Journal of Rheology (New York, N.Y.)*, *27*(3), 201–210. doi:10.1122/1.549724

Ben Adda, F. (1997). Geometric interpretation of the fractional derivative. *J. Fract. Calc*, *11*, 21–52.

Blanquart, F. (2019). Evolutionary epidemiology models to predict the dynamics of antibiotic resistance. *Evolutionary Applications*, *12*(3), 365–383. doi:10.1111/eva.12753 PMID:30828361

Caputo, M. (1967). Linear models of dissipation whose Q is almost frequency independent-II. *Geophysical Journal of the Royal Astronomical Society*, *13*(5), 529–539. doi:10.1111/j.1365-246X.1967.tb02303.x

Chen, L., Chai, Y., & Wu, R. (2011). Control and Synchronization of Fractional-Order Financial System Based on Linear Control. *Discrete Dynamics in Nature and Society*, *2011*, 1–21. doi:10.1155/2011/958393

Chen, W. C. (2008). Nonlinear dynamics and chaos in a fractional-order financial system. *Chaos, Solitons, and Fractals*, *36*(5), 1305–1314. doi:10.1016/j.chaos.2006.07.051

Diethelm, K. (1997). An algorithm for the numerical solution of differential equations of fractional order. *Electronic Transactions on Numerical Analysis*, *5*, 1–6.

Diethelm, K., & Ford, N. J. (2002). Analysis of fractional differential equations. *Journal of Mathematical Analysis and Applications*, *265*(2), 229–248. doi:10.1006/jmaa.2000.7194

Diethelm, K., Ford, N. J., & Freed, A. D. (2002). A predictor-corrector approach for the numerical solution of fractional differential equations. *Nonlinear Dynamics*, *29*(1/4), 3–22. doi:10.1023/A:1016592219341

Diethelm, K., & Freed, A. D. (1999). The FracPECE subroutine for the numerical solution of differential equations of fractional order. In S. Heinzel & T. Plesser (Eds.), Forschung und wissenschaftliches Rechnen. Gttingen: Gesellschaft für wissenschaftliche Datenverarbeitung (pp. 57–71). Academic Press.

El-Raheem, Z. F., & Salman, S. M. (2014). On a discretization process of fractional-order logistic differential equation. *J. Egypt. Math. Soc.*, *22*(3), 407–412. doi:10.1016/j.joems.2013.09.001

El-Sayed, A. M. A. (1996). Fractional-order diffusion-wave equation. *International Journal of Theoretical Physics*, *35*(2), 311–322. doi:10.1007/BF02083817

El-Sayed, A. M. A., Elsonbaty, A., Elsadany, A. A., & Matouk, A. E. (2016b). Dynamical analysis and circuit simulation of a new fractional-order hyperchaotic system and its discretization. International Journal of Bifurcation and Chaos, 26.

El-Sayed, A. M. A., Nour, H. M., Elsaid, A., Matouk, A. E., & Elsonbaty, A. (2016a). Dynamical Behaviors, Circuit Realization, Chaos Control and Synchronization of a New Fractional Order Hyperchaotic System. *Applied Mathematical Modelling*, *40*(5-6), 3516–3534. doi:10.1016/j.apm.2015.10.010

El-Sayed, A. M. A., & Salman, S. M. (2013). On a discretization process of fractional order Riccati's differential equation. *J. Fract. Calc. Appl.*, *4*, 251–259.

Elsadany, A. A., & Matouk, A. E. (2015). Dynamical behaviors of fractional-order Lotka– Volterra predator–prey model and its discretization. *Journal of Applied Mathematics and Computing*, *49*(1-2), 269–283. doi:10.100712190-014-0838-6

Heaviside, O. (1971). Electromagnetic theory. New York: Chelsea.

Hegazi, A. S., Ahmed, E., & Matouk, A. E. (2011). The effect of fractional order on synchronization of two fractional order chaotic and hyperchaotic systems. *J. Fract. Calc. Appl.*, *1*(3), 1–15.

Hegazi, A. S., Ahmed, E., & Matouk, A. E. (2013a). On chaos control and synchronization of the commensurate fractional order Liu system. *Communications in Nonlinear Science and Numerical Simulation*, *18*(5), 1193–1202. doi:10.1016/j.cnsns.2012.09.026

Hegazi, A. S., Ahmed, E., & Matouk, A. E. (2013b). On Dynamical Behaviors and Chaos Control of the Fractional-order Financial System. In S. Banerjee (Ed.), *Chaos and Complexity Theory for Management: Nonlinear Dynamics* (pp. 34–49). IGI Global. doi:10.4018/978-1-4666-2509-9.ch002

Hegazi, A. S., & Matouk, A. E. (2011). Dynamical behaviors and synchronization in the fractional order hyperchaotic Chen system. *Applied Mathematics Letters*, *24*(11), 1938–1944. doi:10.1016/j.aml.2011.05.025

Hegazi, A. S., & Matouk, A. E. (2013). Chaos control of the modified Autonomous Van der Pol-Duffing circuits via active control. In S. Banerjee & L. Rondoni (Eds.), *Applications of Chaos and Nonlinear Dynamics in Science and Engineering* (Vol. 3, pp. 185–202). Berlin: Springer. doi:10.1007/978-3-642-34017-8_7

Hilfer, R. (Ed.). (2000). *Applications of fractional calculus in physics*. Upper Saddle River, NJ: World Scientific. doi:10.1142/3779

Ichise, M., Nagayanagi, Y., & Kojima, T. (1971). An analog simulation of noninteger order transfer functions for analysis of electrode process. *Journal of Electroanalytical Chemistry*, *33*(2), 253–265. doi:10.1016/S0022-0728(71)80115-8

Kusnezov, D., Bulgac, A., & Dang, G. D. (1999). Quantum levy processes and fractional kinetics. *Physical Review Letters*, *82*(6), 1136–1139. doi:10.1103/PhysRevLett.82.1136

Kuznetsov, Y. A. (1998). *Elements of applied bifurcation theory* (2nd ed.). New York: Springer.

Laskin, N. (2000). Fractional market dynamics. *Physica A*, *287*(3-4), 482–492. doi:10.1016/S0378-4371(00)00387-3

Latif, M. A., Dragomir, S. S., Matouk, A. E. (2013). New inequalities of Ostrowski type for Co-ordinated s-Convex functions via fractional integrals. *Journal of Fractional Calculus and Applications, 4*(1), 22-36.

Lewnard, J. A., & Grad, Y. H. (2018). Vaccine Waning and Mumps Re-emergence in the United States. *Sci. Transl. Med., 10*, 433.

Magin, R. L. (2006). *Fractional Calculus in Bioengineering*. Begell House.

Mainardi, F. (2010). *Fractional Calculus and Waves in Linear Viscoelasticity: An Introduction to Mathematical Models*. London: Imperial College Press. doi:10.1142/p614

Marotto, F. R. (2005). On redefining a snap-back repeller. *Chaos, Solitons, and Fractals*, *25*(1), 25–28. doi:10.1016/j.chaos.2004.10.003

Matignon, D. (1996). Stability results for fractional differential equations with applications to control processing. In *IEEE-SMC proceedings of the computational engineering in systems and application multiconference* (Vol. 2, pp. 963–968). Lille, France: IMACS.

Matouk, A. E. (2009a). Chaos synchronization between two different fractional systems of Lorenz family. *Mathematical Problems in Engineering*, *2009*, 1–11. doi:10.1155/2009/572724

Matouk, A. E. (2009b). Stability conditions, hyperchaos and control in a novel fractional order hyperchaotic system. *Physics Letters. [Part A]*, *373*(25), 2166–2173. doi:10.1016/j.physleta.2009.04.032

Matouk, A. E. (2010). Dynamical behaviors, linear feedback control and synchronization of the fractional order Liu system. *Journal of Nonlinear Systems and Applications*, *1*(3), 135–140.

Matouk, A. E. (2011a). Chaos, feedback control and synchronization of a fractional-order modified Autonomous Van der Pol-Duffing circuit. *Communications in Nonlinear Science and Numerical Simulation*, *16*(2), 975–986. doi:10.1016/j.cnsns.2010.04.027

Matouk, A. E. (2011b). Synchronization in integer and fractional order chaotic systems. In S. Banerjee (Ed.), *Chaos Synchronization and Cryptography for Secure Communications: Applications for Encryption* (pp. 127–151). IGI Global. doi:10.4018/978-1-61520-737-4.ch007

Matouk, A. E. (2016). Chaos synchronization of a fractional-order modified Van der Pol Duffing system via new linear control, backstepping control and Takagi-Sugeno fuzzy approaches. *Complexity*, *21*(S1), 116–124. doi:10.1002/cplx.21719

Matouk, A. E., & Elsadany, A. A. (2014). Achieving synchronization between the fractional-order hyperchaotic Novel and Chen systems via a new nonlinear control technique. *Applied Mathematics Letters*, *29*, 30–35. doi:10.1016/j.aml.2013.10.010

Matouk, A. E., & Elsadany, A. A. (2016). Dynamical analysis, stabilization and discretization of a chaotic fractional-order GLV model. *Nonlinear Dynamics*, *85*(3), 1597–1612. doi:10.100711071-016-2781-6

Matouk, A. E., Elsadany, A. A., Ahmed, E., & Agiza, H. N. (2015). Dynamical behavior of fractional-order Hastings–Powell food chain model and its discretization. *Communications in Nonlinear Science and Numerical Simulation*, *27*(1-3), 153–167. doi:10.1016/j.cnsns.2015.03.004

Petras, I. (2009). Chaos in the fractional-order Volta's system: Modeling and simulation. *Nonlinear Dynamics*, *57*(1-2), 157–170. doi:10.100711071-008-9429-0

Podlubny, I. (1999). *Fractional differential equations*. New York: Academic Press.

WHO Report, WHO/EMP/IAU/2018.06, 2018.

Sun, H. H., Abdelwahab, A. A., & Onaral, B. (1984). Linear approximation of transfer function with a pole of fractional order. *IEEE Transactions on Automatic Control*, *29*(5), 441–444. doi:10.1109/TAC.1984.1103551

Tavazoei, M. S. (2010). A note on fractional-order derivatives of periodic functions. *Automatica*, *46*(5), 945–948. doi:10.1016/j.automatica.2010.02.023

Tavazoei, M. S., Haeri, M., & Attari, M. (2009). A proof for nonexistence of periodic solutions in time invariant fractional order systems. *Automatica*, *45*(8), 1886–1890. doi:10.1016/j.automatica.2009.04.001

Wen, G. (2005). Criterion to identify Hopf bifurcations in maps of arbitrary dimension. Phys. Rev. E, 72, Article ID 026201.

Zhen, W., Xia, H., & Guodong, S. (2011). Analysis of nonlinear dynamics and chaos in a fractional order financial system with time delay. *Computers & Mathematics with Applications (Oxford, England)*, *62*(3), 1531–1539. doi:10.1016/j.camwa.2011.04.057

Chapter 2
Bifurcation and Chaos in a Discrete Fractional-Order Prey-Predator System Involving Allee Effect

George Maria Selvam A.
https://orcid.org/0000-0003-2004-3537
Sacred Heart College (Autonomous), India

Janagaraj R.
https://orcid.org/0000-0002-9811-078X
Sacred Heart College (Autonomous), India

ABSTRACT

This chapter considers the discrete counterpart of a fractional order prey-predator ODE system involving Allee effect. Several more realistic models were proposed to describe nonlinear interactions between species by introducing different types functional responses and Allee effect. Non-local property of fractional differential equations is useful in modeling population interactions possessing memories. The model under investigation has three steady states, and the positive steady state exists under certain condition. Dynamic nature of the model is discussed through local stability analysis. Enquiry into the qualitative behavior of the model reveals rich and complex dynamics exhibited by the discrete-time model. Moreover, this model undergoes Neimark Sacker bifurcation when the chosen parameter passes through a critical value. The analytical results are strengthened with appropriate numerical examples. The computation of maximal Lyapunov exponents confirms the existence of chaos. Chaos control is achieved by linear feedback control and hybrid control methods.

DOI: 10.4018/978-1-7998-3122-8.ch002

INTRODUCTION

Over the past few centuries the explosion in scientific knowledge and technology has been a contributing factor and also to some extent a direct consequence in advancement of mathematical modeling. The development of more contemporary form of mathematical models has further enhanced to understanding of the world around us. A periodic motivation for those who create or develop mathematical models is the desire to form prediction. Ecology is a branch of biology which deals with the analysis of how species relate with each other and their environment essentially the inter and intra relationships of the species. In particular it encompasses aspects such as population growth, competition, symbiotic ecologic relationships (mutualism), trophic relations, migration and physical environment interactions. Such population dynamics have enthused biologists and mathematicians alike who collectively enhanced the study of these interactions of species by developing mathematical models.

(David Logan, 2011) A set of equations or an equation, that describes some physical phenomenon or problem that has its origin in engineering, science, economics, or some other area are called Mathematical Modeling. The subject of mathematical modeling involves formulation of equations, physical intuition, analysis and solution methods. During the past centuries, many types of models have been developed to investigate the natural and social processes that enlarge over time. Now-a-days, these models are referred as dynamical systems. (Frederick, 2006) Dynamical systems are divided into two general categories, i.e. deterministic models and stochastic models respectively. Deterministic models are employed when the number of quantities involved in the process being modeled is relatively small and all the underlying scientific principles are fairly well understood.

The tools of dynamical systems enable the researchers to better understand and investigate the different nonlinear characteristics, which exhibit new phenomena of the systems from various disciplines. Particularly, the tools of the dynamical systems such as the applied bifurcation theories are successfully employed to investigate the qualitative behaviors of nonlinear systems. This includes the enquiry in to existence of equilibrium and their stability and bifurcations of periodic orbits, quasi-periodic behavior, chaotic attractors, chaos control and synchronization.

There are numerous papers regarding the dynamics of the predator-prey system with and without different kinds of functional responses. In fact, the Allee effect is a phenomenon in biology named after ecologist Warder Clyde Allee (Allee, 1931) who brought attention to the possibility of a positive relationship between aspects of fitness and population size over fifty years ago (Stephens et al, 1999). In 1931, Allee introduced the Allee effect, a reproductive opportunities that cause negative growth rate of a population of biological species at densities below a critical value. It is well known that Allee effect made significant contribution in the stability analysis of equilibrium points of a population dynamics model (Ak Gumus et al, 2012). Biological facts on the Allee function are given below (Kangalgil, 2017):

- When the population density is zero, Allee function is zero.
- The derivatives of the Allee function with respect to independent variables are always strictly positive for all positive values of the population density.
- Allee effect vanishes at high densities.

Many researchers contributed to the study of stability of predator–prey system in the presence/absenceof Allee effect (Pk & Ghorais, 2017). Rarely bifurcation analysis of these systems with the Allee effect has been investigated. Hence, in this chapter, predator–prey system enforced with an Allee effect is considered.(Lakshmanan et al, 2012)Fractional-order systems with order less than three have the ability to exhibit chaos. Hence, investigation of chaos in fractional-order models has become a hot topic of research, and therefore, it has been studied by many biologists and mathematicians.

In recent decades, many authors (Oldham, 1974) has indicted that fractional-order models are more suitable than integer-order models to study biology dynamics due to memory and hereditary properties of fractional derivatives. Hence the use of fractional-order differential equations (FDEs) helps us to have a better understanding of the biological system behaviors. Objective of this work is to propose and discuss the discrete version of a fractional order prey-predator model incorporating Allee effect and to enquire the inevitable consequences of nonlocal properties induced by fractional-order derivatives.

The rest of this chapter is arranged as follows: fractional order model of prey-predator interaction with Allee effect is proposed with its discretization in section 2. Stability of the equilibrium steady states of the discretized system is discussed in section 3. Bifurcation theory is used to study the existence of Flip and Neimark-Sacker bifurcations of the discretized system in section 4. In section 5, two different chaos control strategies are implemented for controlling the chaos due to Flip and Neimark-Sacker bifurcations. Numerical simulations validate the analytical results, including bifurcation diagrams, phase portraits and Lyapunov exponents in section 6. Finally the chapter ends witha brief conclusion.

FRACTIONAL-ORDER SYSTEM WITH DISCRETIZATION

The basic model of two species nonlinear interactions was introduced by Lotka (Lotka, 1925) and Volterra (Volterra, 1926), which had a prominent role in investigating multiple species interactions. In the succeeding year, it has been developed widely to examine the dynamical behavior of the prey-predator model by several researchers from mathematics and ecology. They suggested a system of a pair of first order ODE to determine the dynamics of prey predator population. Consider the following system of nonlinear differential equations in non-dimensional form:

$$\begin{aligned} \frac{dx}{dt} &= rx(1-x) - axy\left[\frac{x}{m+x}\right] \\ \frac{dy}{dt} &= bxy - cy \end{aligned} \tag{1}$$

In system (1), $x(t)$ and $y(t)$ denote the number of prey and predator populations. All the system parameters r,a,b,c,m have positive values that stand for prey intrinsic rate, per capita searching efficiency of the predator, conversion rate, predator death rate and Allee constant imposed on the prey population respectively. From mathematical and biological point of view, this chapter will pay attention to the dynamical behaviors of discrete fractional of (1) in the closed first quadrant R_+^2.

The fractional order systems have risen to prominence due to its distinct features namely they are realistic and practical in their approach, it has non-local property, as it takes into consideration the past and distributed effects of the model which the integer order differential equations lack, and these systems are convenient to model biological systems with memory effect. The non-local property implies that the subsequent status of the model depends not only upon its present stage but also upon all its previous stages. In order to harness the advantage of the properties of fractional order systems, fractional order is introduced in (1) to yield the following system

$$D_t^\alpha(x) = rx(1-x) - axy\left[\frac{x}{m+x}\right]$$
$$D_t^\alpha(y) = bxy - cy \tag{2}$$

with $x(0)=x_0$ and $y(0)=y_0$, where α is the fractional order α∈(0,1] and

$${}_aD_t^\alpha f(t) = \frac{1}{\Gamma(1-\alpha)}\int_a^t \frac{f^{(n)}(\tau)}{(t-\tau)^{\alpha-n+1}}\,d\tau,$$

for $n-1<\alpha<n$ (Caputo, 1967) is the standard Caputo definition of fractional derivatives.

Discretization Process

In applied mathematics, the process of transforming continuous models described by differential equations into its discrete counterpart is termed as discretization (Agarwal et al, 2013). The discrete version of the model is more suitable for implementing numerical methods and computer simulations. The system (2) is discretized with piecewise constant arguments and initial point $x(0)=x_0$ and $y(0)=y_0$. After repeating the discretization process t-times, the discrete fractional order predator prey system involving Allee effect is obtained in the following form:

$$x(t+1) = x(t) + \frac{h^\alpha}{\Gamma(1+\alpha)}\left[rx(t)(1-x(t)) - ax(t)y(t)\left(\frac{x(t)}{m+x(t)}\right)\right]$$
$$y(t+1) = y(t) + \frac{h^\alpha}{\Gamma(1+\alpha)}\left[bx(t)y(t) - cy(t)\right] \tag{3}$$

where α∈(0,1] and $h>0$ is defined as time interval of production.

Existence and Uniqueness of the Solution

Let us consider system (2) in the region $\Omega\times(0,T]$ where

$$\Omega = \left\{(x,y)\in\mathbb{R}^2 : \max(|x|,|y| \le \gamma)\right\}$$

and $\Omega\times(0,T]$ is taken as a region for existence and uniqueness. The approach used in (Hong-LiLi et al, 2016) is utilized. Consider a mapping $H(A)=(H_1(A), H_2(A))$ such that

$$\begin{aligned} H_1(A) &= rx(1-x) - axy\left[\frac{x}{m+x}\right] \\ H_2(A) &= bxy - cy \end{aligned} \tag{4}$$

For any $A, \bar{A} \in \Omega$, it follows from (4) that

$$\begin{aligned} &\left\|H(A) - H(\bar{A})\right\| = \left|H_1(A) - H_1(\bar{A})\right| + \left|H_2(A) - H_2(\bar{A})\right| \\ &= \left|rx(1-x) - axy\left[\frac{x}{m+x}\right] - r\bar{x}(1-\bar{x}) + a\bar{x}\,\bar{y}\left[\frac{\bar{x}}{m+\bar{x}}\right]\right| + \left|bxy - cy - b\bar{x}\,\bar{y} + c\bar{y}\right| \\ &= \left|r(x-\bar{x}) - r\left(x-\bar{x}\right)(x+\bar{x}) + \frac{a}{(m+x)(m+\bar{x})}\left[m\bar{x}^2\bar{y} + \bar{x}^2\bar{y}x - mx^2y - x^2y\bar{x}\right]\right| \\ &+ \left|(bx-c)(y-\bar{y}) + b\bar{y}(x-\bar{x})\right| \le r\left|x-\bar{x}\right| + 2r\gamma\left|x-\bar{x}\right| + \frac{a\eta^2(2m+\gamma)}{(m+\gamma)^2}\left|x-\bar{x}\right| \\ &+ \gamma^2(m+\gamma)\left|y-\bar{y}\right| + (b\gamma+c)\left|y-\bar{y}\right| + b\gamma\left|x-\bar{x}\right| \le \left[r + 2r\gamma + \frac{a\gamma^2(2m+\gamma)}{(m+\gamma)^2} + b\gamma\right]\left|x-\bar{x}\right| \\ &+ \left[\gamma^2(m+\gamma) + b\gamma + c\right]\left|y-\bar{y}\right| \left\|H(A) - H(\bar{A})\right\| \le \Delta\left\|A - \bar{A}\right\| \end{aligned}$$

and

$$\Delta = \max\left\{r + 2r\gamma + \frac{a\eta^2(2m+\gamma)}{(m+\gamma)^2} + b\gamma, \gamma^2(m+\gamma) + b\gamma + c\right\}.$$

For $\Delta<1$, obtain $A=H(A)$ and hence H(A) is a contraction mapping.
The next theorem is proved with the help of above result.

Theorem 1. In the specified region $\Omega\times(0,T]$, sufficient condition for existence and uniqueness of the solution of system (2) is

$$\Delta = \max\left\{r + 2r\gamma + \frac{a\eta^2(2m+\gamma)}{(m+\gamma)^2} + b\gamma, \gamma^2(m+\gamma) + b\gamma + c\right\} < 1.$$

EXISTENCE AND STABILITY OF THE EQUILIBRIUM STEADY STATES OF THE DISCRETIZED SYSTEM

In this section, existence of the equilibrium steady state of the discretized fractional order predator prey model is established. The Jacobian matrix of the system (3) is evaluated and stability of the equilibrium steady state of the model (3) is studied based on the Jury conditions.

Existence of the Equilibrium Steady States of the Discretized System

In this section, equilibrium steady states of the discretized system (3) are computed by solving the system of algebraic equation given by:

$$\begin{aligned} &rx(1-x) - axy\left(\frac{x}{m+x}\right) = 0 \\ &bxy - cy = 0 \end{aligned} \quad . \tag{5}$$

Solving the algebraic system (5), the authors obtain three non-negative equilibrium steady states which are non-negative,

1. $ES_0=(0,0)$
2. $ES_1=(1,0)$
3. $\mathrm{ES}_2 = \left(\frac{c}{b}, \frac{r(c+bm)(b-c)}{abc}\right).$

By simple calculation, the following result for the existence of positive equilibrium steady state of the system (3) is obtained.

Theorem 2. The existence of equilibrium steady states satisfies:

- The trivial and boundary equilibrium steady states ES_0 and ES_1 always exists.
- If $b>c$, then the unique positive interior equilibrium steady state ES_1 exists.

Stability of the Equilibrium Steady States of the Discretized System

This section is devoted to the stability nature of the system (3) around each of the above equilibrium steady state. Stability analysis of the system (3) is discussed by the formulation of the jacobian matrix corresponding to each equilibrium steady state. Jacobian matrix of system (3) evaluated at (x,y) has the form

$$J(x,y) = \begin{bmatrix} 1 + A\left[r(1-2x) - \dfrac{2amxy + ax^2y}{(m+x)^2}\right] & -A\dfrac{ax^2}{m+x} \\ Aby & 1 + A\left[bx - c\right] \end{bmatrix} \tag{6}$$

where $A = \dfrac{h^\alpha}{\Gamma(1+\alpha)}$. Now the characteristic equation of (6) is

$$\Psi(\eta) = \eta^2 - T\eta + D = 0 \tag{7}$$

Where the trace and determinant of the jacobian matrix $J(x,y)$ of (6) are expressed as

$$T = 2 + A[r - c + x(b - 2r)]$$

and

$$D = [1 + A(r - 2rx)][1 + A(bx - c)] + \frac{A^2abx^2y}{m + x}.$$

In the following lemma and theorems, stability of the system (3) is examined through the relations between roots and coefficients of a quadratic equation. For this purpose, following lemma (Leah Edelstein Keshet, 2005) is needed.

Lemma 3. Let us consider equation (7) and η_1, η_2 be the roots of $\Psi(\eta)=0$. Suppose that $\Psi(1)>0$. Then

- $|\eta_1|<1$ and $|\eta_2| < 1 \Leftrightarrow \Psi(-1) > 0$ and $\Psi(0)<1$ (The equilibrium steady state (x^*,y^*) is stable(sink)).
- $|\eta_1|<1$ and $|\eta_2|>1$ (or $|\eta_1|>1$ and $|\eta_2|<1$) $\Leftrightarrow \Psi(-1) < 0$. (The equilibrium steady state (x^*,y^*) is saddle point).
- $|\eta_1|>1$ and $|\eta_2| > 1 \Leftrightarrow \Psi(-1) > 0$ and $\Psi(0)>1$. (The equilibrium steady state (x^*,y^*) is unstable(source)).
- $|\eta_1|=-1$ and $|\eta_2| \neq 1 \Leftrightarrow \Psi(-1) = 0$ and $T\neq0$ and 2.
- η_1 and η_2 are complex and $|\eta_1| = |\eta_2| \Leftrightarrow T^2 - 4D < 0$ and $\Psi(0)=1$.

Theorem 4. The trivial equilibrium steady state ES_0 is a

1. source i.e. unstable for $|\eta_2|>1$ i.e. $h > \left[\frac{2\Gamma(1+\alpha)}{c}\right]^{\frac{1}{\alpha}}$;
2. saddle point for $|\eta_2|<1$, i.e. $0 < h < \left[\frac{2\Gamma(1+\alpha)}{c}\right]^{\frac{1}{\alpha}}$;
3. and non-hyperbolic for $h = \left[\frac{2\Gamma(1+\alpha)}{c}\right]^{\frac{1}{\alpha}}$.

Proof. For ES_0, the Jacobian is written as

$$J_{ES_0} = J(0,0) = \begin{bmatrix} 1 + \frac{h^\alpha}{\Gamma(1+\alpha)}r & 0 \\ 0 & 1 - \frac{h^\alpha}{\Gamma(1+\alpha)}c \end{bmatrix}$$

whose eigenvalues are $\eta_1 = 1 + \frac{h^\alpha}{\Gamma(1+\alpha)}r$ and $\eta_2 = 1 - \frac{h^\alpha}{\Gamma(1+\alpha)}c$. Since $\frac{h^\alpha}{\Gamma(1+\alpha)} > 0$ for $0<\alpha\leq1$.

1. It is obvious that $|m_1|>1$. Then ES_0 is source if $\left|1-\frac{h^\alpha}{\Gamma(1+\alpha)}c\right|>1$ which yields

$$h>\left[\frac{2\Gamma(1+\alpha)}{c}\right]^{\frac{1}{\alpha}}.$$

2. ES_0 is saddle point if $\left|1-\frac{h^\alpha}{\Gamma(1+\alpha)}c\right|<1$ which violates the above condition, i.e.

$$0<h<\left[\frac{2\Gamma(1+\alpha)}{c}\right]^{\frac{1}{\alpha}}.$$

3. Result follows from (a) and (b).

Theorem 5. The boundary equilibrium steady state ES_1 is

1. stable for

$$b<c \text{ and } h<\min\left\{\left[\frac{2\Gamma(1+\alpha)}{r}\right]^{\frac{1}{\alpha}},\left[\frac{2\Gamma(1+\alpha)}{c-b}\right]^{\frac{1}{\alpha}}\right\};$$

2. unstable for

$$b>c \text{ and } h>\max\left\{\left[\frac{2\Gamma(1+\alpha)}{r}\right]^{\frac{1}{\alpha}},\left[\frac{2\Gamma(1+\alpha)}{c-b}\right]^{\frac{1}{\alpha}}\right\};$$

3. and non-hyperbolic for

$$b=c \text{ or } h=\left[\frac{2\Gamma(1+\alpha)}{r}\right]^{\frac{1}{\alpha}} \text{ or } h=\left[\frac{2\Gamma(1+\alpha)}{c-b}\right]^{\frac{1}{\alpha}}.$$

Proof. For ES_1, Jacobian matrix is

$$J_{ES_1}=J(1,0)=\begin{bmatrix}1-\frac{h^\alpha}{\Gamma(1+\alpha)}r & -\frac{h^\alpha}{\Gamma(1+\alpha)}\frac{a}{(1+m)}\\ 0 & 1+\frac{h^\alpha}{\Gamma(1+\alpha)}(b-c)\end{bmatrix}$$

whose eigen values are $\eta_1 = 1 - \frac{h^\alpha}{\Gamma(1+\alpha)} r$ and $\eta_2 = 1 + \frac{h^\alpha}{\Gamma(1+\alpha)}(b-c)$. Since $\frac{h^\alpha}{\Gamma(1+\alpha)} > 0$ for $0<\alpha\leq 1$.

1. ES_1 is stable if

$$\left|1 - \frac{h^\alpha}{\Gamma(1+\alpha)} r\right| < 1 \text{and} \left|1 + \frac{h^\alpha}{\Gamma(1+\alpha)}(b-c)\right| < 1$$

which yields

$$b < c \,\text{and}\, h < \min\left\{\left[\frac{2\Gamma(1+\alpha)}{r}\right]^{\frac{1}{\alpha}}, \left[\frac{2\Gamma(1+\alpha)}{c-b}\right]^{\frac{1}{\alpha}}\right\}.$$

2. ES_1 is unstable if

$$\left|1 - \frac{h^\alpha}{\Gamma(1+\alpha)} r\right| > 1 \text{and} \left|1 + \frac{h^\alpha}{\Gamma(1+\alpha)}(b-c)\right| > 1$$

which violates the above conditions, i.e.

$$b > c \,\text{and}\, h > \max\left\{\left[\frac{2\Gamma(1+\alpha)}{r}\right]^{\frac{1}{\alpha}}, \left[\frac{2\Gamma(1+\alpha)}{c-b}\right]^{\frac{1}{\alpha}}\right\}.$$

3. Result is straight forward from (a) and (b).

Next, stability of positive interior equilibrium state ES_2 is analysed from the jacobian matrix

$$J_{ES_2} = J\left(x^*, y^*\right) = \begin{bmatrix} 1 + Aa_{11} & -Aa_{12} \\ Aa_{21} & 1 \end{bmatrix} \tag{8}$$

The characteristic equation of J_{ES_2} is $\Psi(\eta) = \eta^2 - T\eta + D = 0$, with $T=2+Aa_{11}$ and

$D = 1 + Aa_{11} + A^2 a_{12} a_{21}$,

where

$$A = \frac{h^{\alpha}}{\Gamma(1+\alpha)},$$

$$a_{11} = r - \frac{2rc}{b} - \frac{r(c+bm)(b-c)(2mb+c)}{b(m+c)^2},$$

$$a_{12} = \frac{ac^2}{b(m+c)}$$

and

$$a_{21} = \frac{r(c+bm)(b-c)}{ac}.$$

Simple calculations yield the eigen values

$$\eta_{1,2} = 1 + \frac{AM}{2} \pm \frac{A}{2}\sqrt{M^2 - 4N},$$

here $M=a_{11}$ and $N=a_{12}a_{21}$. Considering Lemma 3, following theorem is stated.

Theorem 6. The equilibrium steady state ES_2 of system (3) has at least four different topological forms for all acceptable values of parameters:

1. ES_2 is asymptotically stable (sink) if any one of the conditions hold:
   a. $S^* \geq 0$ and $h<h_2$,
   b. $S^*<0$ and $h<h_3$,
2. ES_2 is unstable (source) if any one of the conditions hold:
   a. $S^* \geq 0$ and $h>h_1$,
   b. $S^*<0$ and $h>h_3$,
3. ES_2 is unstable (saddle) if any one of the conditions hold:
   a. $S^* \geq 0$ and $h_2<h<h_1$
4. ES_2 is non-hyperbolic if any one of the conditions hold:
   a. $S^*>0$ and $h=h_1$ or $h=h_2$,
   4. b. $S^*<0$ and $h=h_3$,

$S^*=(M^2-4N)$

and

$$h_1 = \left\{\Gamma(1+\alpha)\left[\frac{-M + \sqrt{M^2 - 4N}}{N}\right]\right\}^{\frac{1}{\alpha}},$$

$$h_2 = \left\{\Gamma(1+\alpha)\left[\frac{-M - \sqrt{M^2 - 4N}}{N}\right]\right\}^{\frac{1}{\alpha}},$$

$$h_3 = \left\{\left[\frac{-\Gamma(1+\alpha)M}{N}\right]\right\}^{\frac{1}{\alpha}}$$

BIFURCATIONS OF THE INTERIOR EQUILIBRIUM STEADY STATE OF THE DISCRETIZED FRACTIONAL-ORDER SYSTEM

Bifurcation analysis is an interesting and fruitful topic to investigate the topological nature of the model. Recently, numerous articles have been published on the existence of Flip and Neimark-Sacker bifurcations (Din, 2018). Bifurcation occurs only when a certain parameter value called as a bifurcation parameter value passing through a critical value and the stability of an equilibrium steady state changes. In this current section, Flip and Neimark-Sacker bifurcation behaviors of the interior positive equilibrium steady state ES_2 of system (3) are investigated. Standard bifurcation theory techniques are useful to examine the existence of flip bifurcation (FB) and Neimark-Sacker bifurcation (NSB) for the unique interior positive equilibrium steady state ES_2 of system (3).

Flip bifurcation in a system (3) occurs when a mild change in the parametric value, which leads the system moving to a new behavior with twice the period of the original system. Moreover the system causes an eigen value to pass through -1. Frequently, equilibrium steady state loses stability, and a stable cycle of period 2 appears. Continued parameter changes may result in a cascade of PDBs and onset of chaos. NSB in a system (3) occurs when a mild change in the parametric value, which the system produce invariant circles due to the emergence of NSB. Also the model possesses a pair of complex conjugate eigen values of modulus 1.

Flip Bifurcation of the Discretized Fractional-Order System

First, choose h as the bifurcation parameter for discussing Flip bifurcation. The interior equilibrium steady state ES_2 undergoes Flip bifurcation when one of the eigenvalues of Jacobian matrix is -1 at an equilibrium steady state and another eigenvalue is neither 1 nor -1 (Elaydi, 2008).

The Jacobian matrix J of system (3) at the positive interior equilibrium steady state ES_2 is shown in equation (8). The characteristic polynomial of J is written as

$$\Psi(\eta) = \eta^2 - (2 + AM)\eta + (1 + AM + A^2N).$$

By Theorem 3.4, if $S^*\geq 0$ and $h=h_1$ or h_2, then the eigenvalues of the equilibrium steady state ES_2 are

$$\eta_1 = -1, \eta_2 = \frac{M\sqrt{M^2 - 4N} - M^2 + 3N}{N}.$$

Summarizing the above analysis we consider the following theorem as given below.

Theorem 7. The interior equilibrium steady state ES_2 loses its stability, via a Flip bifurcation when $S^*\geq 0$ and $h=h_1$ or h_2 and

$$\eta_1 = -1, \eta_2 = \frac{M\sqrt{M^2 - 4N} - M^2 + 3N}{N} \neq \pm 1.$$

Neimark-Sacker Bifurcation of the Discretized Fractional-Order System

The present section analyzes Neimark-Sacker bifurcation by using bifurcation theory for the choice of h as a bifurcation parameter. Corresponding to the interior equilibrium steady state ES_2, Neimark-Sacker bifurcation occurs when two eigenvalues of the Jacobian matrix at a fixed point are a pair of complex conjugate numbers with module one (Elaydi, 2008). The Jacobian matrix $J$of system (3) at the positive interior equilibrium steady state ES_2 is shown in equation (8). The characteristic polynomial of J is expressed as

$$\Psi(\eta) = \eta^2 - (2 + AM)\eta + (1 + AM + A^2N).$$

By Theorem 3.4, if $S^*<0$ and $h=h_3$, then the eigenvalues of the interior equilibrium steady state ES_2 are

$$\eta_{1,2} = 1 - \frac{M^2}{2N} \pm i\frac{M}{2N}\sqrt{4N - M^2}.$$

Analyzing the above conditions of Neimark-Sacker bifurcation we consider the following theorem as detailed below.

Theorem 8. The interior equilibrium steady state ES_2 loses its stability and system (3) occurs a Neimark-Sacker bifurcation when $S^*<0$ and $h=h_3$, and

$$\left|\eta_{1,2}\right| = \left|1 - \frac{M^2}{2N} \pm i\frac{M}{2N}\sqrt{4N - M^2}\right| = 1.$$

CHAOS CONTROL

This section presents two control strategies in order to move the unstable fractional periodic orbits or the fractional chaotic orbits towards the stable one. First, linear feedback control method (Din, 2017) is applied to system (3). For this, the controller of fractional-order system (3) is defined by

$$\begin{aligned} x(t+1) &= x(t) + \frac{h^{\alpha}}{\Gamma(1+\alpha)}\left[rx(t)(1-x(t)) - ax(t)y(t)\left(\frac{x(t)}{m+x(t)}\right)\right] + S(t) \\ y(t+1) &= y(t) + \frac{h^{\alpha}}{\Gamma(1+\alpha)}\left[bx(t)y(t) - cy(t)\right] \end{aligned} \tag{9}$$

where

$$S(t) = -s_1(x(t) - x_*) - s_2(y(t) - y_*)$$

is the feedback controlling force, $s_{1,2}$ stands for the feedback gains, and (x^*,y^*) be unique positive interior equilibrium steady state of system (3). The Jacobian matrix of system (9) evaluated at unique positive interior equilibrium steady state(x^*,y^*) is

$$J_1\left(x^*,y^*\right) = \begin{bmatrix} 1 + Aa_{11} - s_1 & -Aa_{12} - s_2 \\ Aa_{21} & 1 \end{bmatrix}, \tag{10}$$

the variables A,a_{11},a_{12},a_{21} have the same meaning as expressed in equation (8). The corresponding characteristic equation of $J_1(x^*,y^*)$ is

$$\eta^2 - (2 + AM - s_1)\eta + (A^2N + AM + 1 + Aa_{21}s_2 - s_1) = 0. \tag{11}$$

If η_1,η_2 are the eigenvalues of (11), then

$$\eta_{1,2} = \frac{(2 + AM - s_1) \pm \sqrt{(2 + AM - s_1)^2 - 4(A^2N + AM + 1 + Aa_{21}s_2 - s_1)}}{2}$$

and

$$\eta_1\eta_2 = Aa_{21}(Aa_{12} + s_2) + 1 + Aa_{11} - s_1 \tag{12}$$

The lines of marginal stability are evaluated by solving equations $\eta_1=\pm 1$ and $\eta_1,\eta_2=1$. These restrictions guarantee that the eigenvalues η_1 and η_2 have absolute value less than 1. Suppose that $\eta_1,\eta_2=1$, then equation (12) implies that

$$l_1 : AM + A^2 N = s_1 - A a_{21} s_2$$

Suppose that $\eta_1=1$ or $\eta_1=-1$, then equation (11) yields

$$l_2 : AN = -a_{21} s_2$$
$$l_3 : 2AM + A^2 N + 4 = 2s_1 - A a_{21} s_2$$

The stable eigenvalues lie within the triangular region bounded by the lines l_1, l_2 and l_3.

Next, in order to control the chaos produced by Neimark-Sacker bifurcation in system (3), the authors introduce hybrid control strategy (Din, 2018). Assuming that system (3) undergoes Neimark-Sacker bifurcation at equilibrium steady state (x^*,y^*), then corresponding fractional-order controlled system is expressed as

$$\begin{aligned} x(t+1) &= \rho x(t) + \frac{\rho h^{\alpha}}{\Gamma(1+\alpha)}\left[rx(t)(1-x(t)) - ax(t)y(t)\left(\frac{x(t)}{m+x(t)}\right)\right] + (1-\rho)x(t) \\ y(t+1) &= \rho y(t) + \frac{\rho h^{\alpha}}{\Gamma(1+\alpha)}\left[bx(t)y(t) - cy(t)\right] + (1-\rho)y(t) \end{aligned} \tag{13}$$

where $0<\rho<1$.Controlling strategy in (13) is a combination of both parameter perturbation and feed back control. Moreover, by proper choice of controlled parameter ρ, the Neimark-Sacker bifurcation of the equilibrium steady state (x^*,y^*) of controlled system (13) can be advanced (delayed) or even completely eliminated. The Jacobian matrix of (13) evaluated at the positive interior equilibrium steady state (x^*,y^*) is

$$J_2\left(x^*, y^*\right) = \begin{bmatrix} 1+\rho A a_{11} & -\rho A a_{12} \\ A\rho a_{21} & 1 \end{bmatrix}. \tag{14}$$

The expressions $A, a_{11}, a_{12}, a_{21}$ have meaning mentioned in (8). Then, the positive equilibrium steady state (x^*,y^*) of the controlled system (13) is locally asymptotically stable if roots of the characteristic equation of (14) lie in an open unit disk.

NUMERICAL EXPERIMENTS

In this section, above theoretical analysis is verified and supported with appropriate examples by considering some special cases of system (4). Numerical simulations manifest clearly interesting rich complex dynamics behaviors. Moreover, linear feedback technique and hybrid control strategy for chaos control are also demonstrated in this section.

Example 9. Take

$\alpha=0.9$; $r=0.95$; $a=1.25$; $m=0.1$; $b=0.99$; $c=0.5$

and $2.4 \leq h \leq 3.4$ in system (3) with initial point $x(0)=0.65$; $y(0)=0.45$. In this example Neimark-Sacker bifurcation is considered. By simple calculation, unique positive interior equilibrium steady state is estimated to be

$$ES_2 = (x^*, y^*) = (0.5051, 0.4506).$$

Moreover, the conditions of Theorem 6are verified as follows:

$M = -0.5557$; $N=0.2347$; $S^* = -0.63$; $h_3=2.4925$.

Eigen values are $\eta_{1,2}=0.3424 \pm i0.7938$ with $|\eta_{1,2}|=1$. By Theorem 6, conditions for Neimark-Sacker bifurcation are obtained near the equilibrium steady state point ES_2 at the bifurcation critical value h_3.

Figure 1(a), 1(b) show Neimark-Sacker bifurcation diagrams in (h,x) and (h,y) planes respectively. Moreover, the corresponding MLEs of Figure 1(a) is displayed in Figure 1(c). Phase portraits of system (3) for different values of h are presented in Figure 2. From Figure 1 and Figure 2, it is observed that unique positive interior equilibrium steady state of system (3) is locally asymptotically stable for $h<h_3=2.4925$, lose its stability at $h=h_3$ and a stable invariant cycle bifurcates from the equilibrium steady state ES_2 for $h>h_3$. Also quasi-periodic orbits on the invariant cycle arise for $h>h_3$ and period orbits emerge in the period-windows, the orbits move towards chaos with the increasing of h. Figure 2 displays how a smooth invariant circle bifurcates from the equilibrium steady state ES_2 when $h>h_3$, then a stable circle disappears and chaotic orbits appear. The MLEs confirm the existence of the chaotic sets in Figure 1(c).

Example 10. This example considers the set of parameter values

$\alpha=0.9$; $r=0.95$; $a=1.25$; $m=0.1$; $b=0.99$; $c=0.5$

with initial point $x(0)=0.65$; $y(0)=0.45$. Example 9 shows that for the system (3), Neimark-Sacker bifurcation occurs as h varies in $h \in [2.4,3.4]$. Moreover, Figure 3 displays a closed invariant circle appearing at $h=2.5$ enclosing unstable positive invariant equilibrium steady state

$$ES_2 = (x^*, y^*) = (0.5051, 0.4506).$$

For these parametric values, the controlled system (13) can be written as

$$\begin{aligned} x(t+1) &= x(t) + \frac{\rho h^\alpha}{\Gamma(1+\alpha)}\left[rx(t)(1-x(t)) - ax(t)y(t)\left(\frac{x(t)}{m+x(t)}\right)\right] \\ y(t+1) &= y(t) + \frac{\rho h^\alpha}{\Gamma(1+\alpha)}\left[bx(t)y(t) - cy(t)\right] \end{aligned} \tag{15}$$

with $\alpha=0.9$; $r=0.95$; $a=1.25$; $m=0.1$; $b=0.99$; $c=0.5$; $h=2.5$ and $0<\rho<1$. Jacobian of controlled system (15) evaluated at ES_2 is

Figure 1. Neimark-Sacker bifurcation diagram in (h,x_1); (h,y_1) planes and Maximum Lyapunov exponents of the system (3)

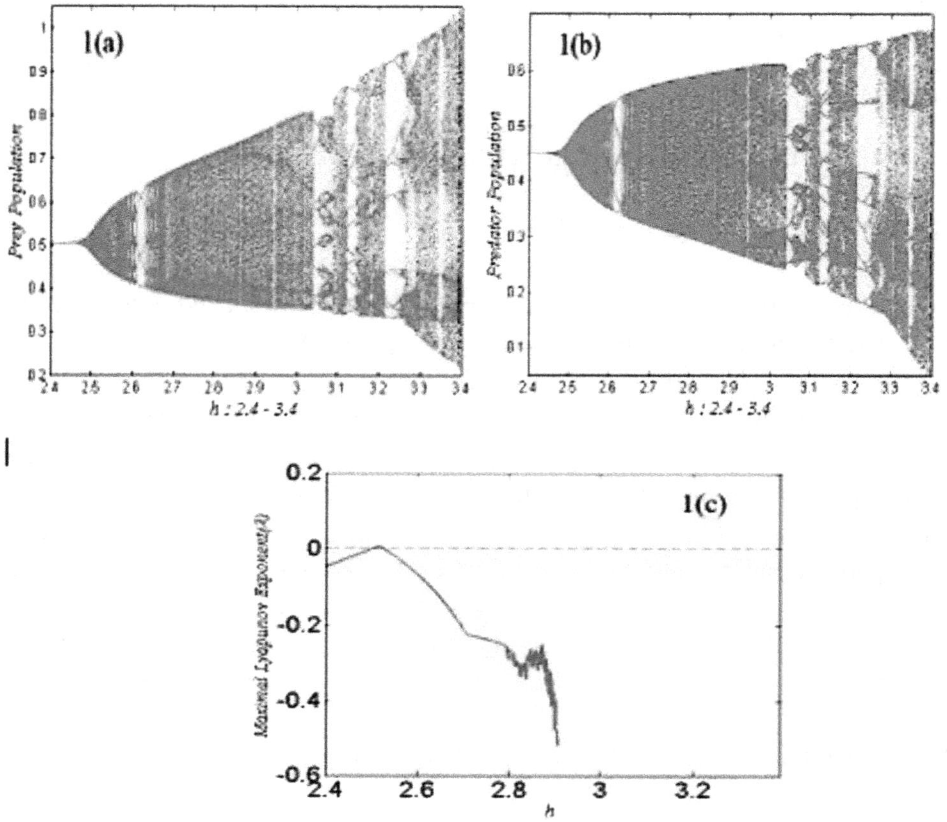

$$J_2\left(x^*,y^*\right)=\begin{pmatrix}1-1.1344\lambda & -1.2444\lambda\\ 0.6594\lambda & 1\end{pmatrix}.$$

The characteristic equation is given by

$$\eta^2-(2-1.1344\lambda)\eta+0.8206\lambda^2-1.1344\lambda+1=0.$$

Then, the roots lie in the unit open disk if and only if $0<\rho<0.9999$. Moreover, the plots for $x(t)$, $y(t)$ of the controlled system (15) are shown in Figure 4 with $\rho=0.995$. From Figure 4(a), 4(b) and 4(c), it is clear that the positive interior equilibrium ES_2 is stable.

CONCLUSION

This chapter examined the qualitative nature of a discrete counterpart for fractional order prey predator system by involving Allee effect. Piecewise constant arguments method was implemented and the discrete fractional order system of the continuous fractional order model was obtained to study the rich

Figure 2. Phase portraits of system (3) for different values of h

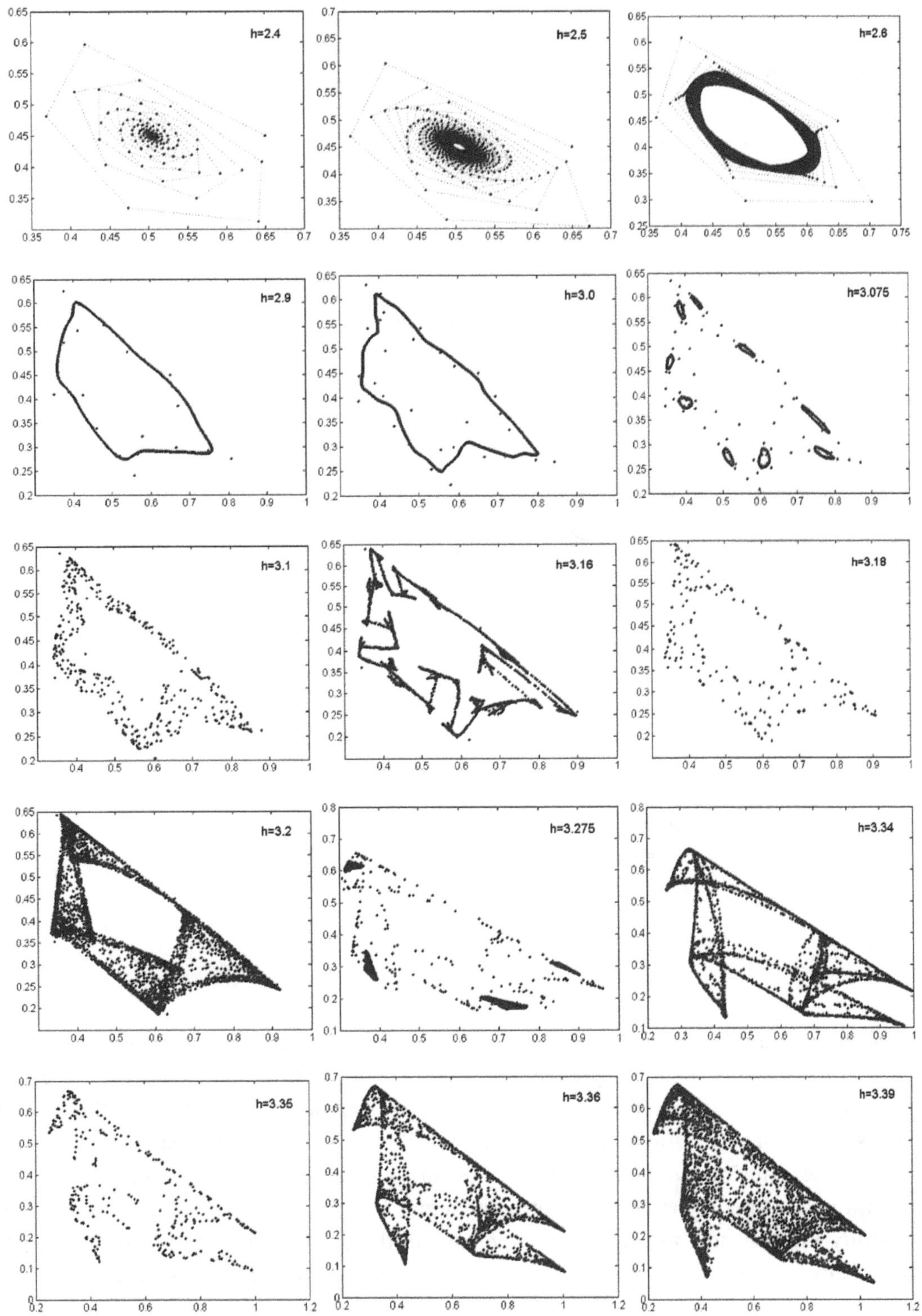

Figure 3. Time series and Phase portrait for the system (3)

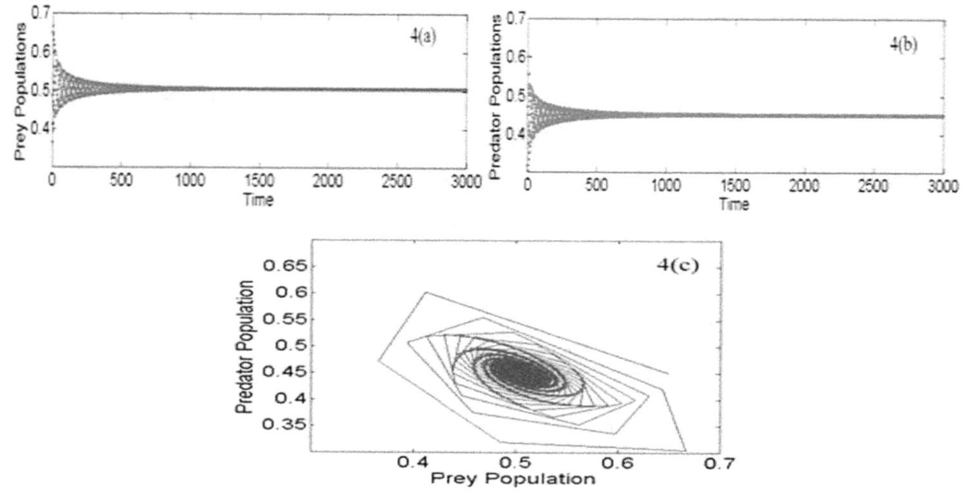

Figure 4. Time series and Phase portrait for the system (13)

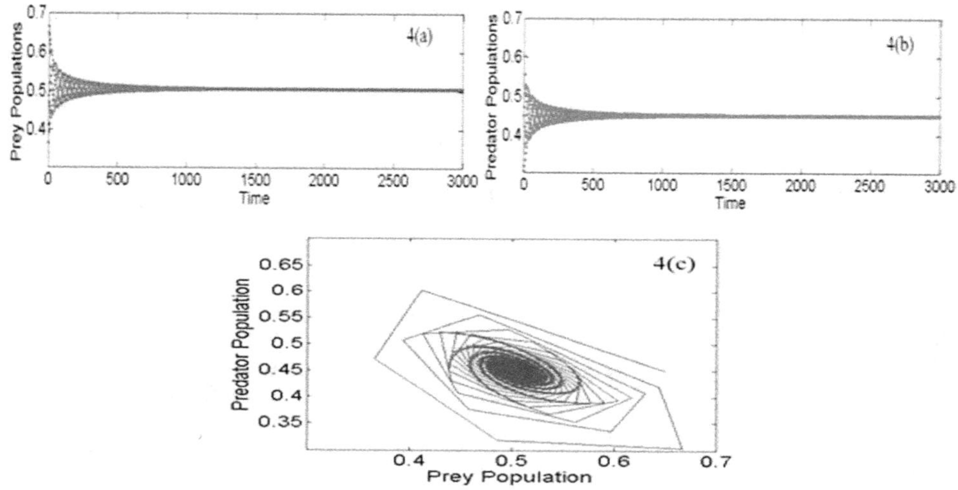

dynamics of the proposed system. A sufficient condition for existence of equilibrium steady states of the fractional order discrete predator-prey system (3) is analyzed. Dynamic behavior of the model is investigated through the local stability analysis of the equilibrium steady states with the help of Jury conditions. Neimark-Sacker bifurcation occurs for a small range of bifurcation parameter h for the system (3). The numerical simulations are shown for distinct parameter values and plotted the time series diagrams with phase lines. Finally, two control strategies like feedback and hybrid control strategies are successfully implemented to control the chaos due to the occurrence of Flip and Neimark-Sacker bifurcations. Numerical examples justify the rich and chaotic dynamics in both species.

REFERENCES

Agarwal, R. P., El-Sayed, A. M. A., & Salman, S. M. (2013). Fractional – Order Chua's system: Discretization, bifurcation and chaos. *Advances in Difference Equations*, *2013*(320).

Ak Gumus, O., & Kose, H. (2012). On the stability of delay population dynamics related with Allee effects. *Math. Comput. Appl.*, *17*(1), 56–67.

Allee, W. C. (1931). *Animal aggressions*. Chicago: University of Chicago press.

Caputo, M. (1967). Linear models of dissipation whose *Q* is almost frequency independent. *Geophysical Journal of the Royal Astronomical Society*, *13*(5), 529–539. doi:10.1111/j.1365-246X.1967.tb02303.x

David Logan, J. (2011). *A First Course in Differential Equations* (2nd ed.). New York: Springer. doi:10.1007/978-1-4419-7592-8

Din, Q. (2017). Complexity and chaos control in a discrete-time prey-predator model. *Communications in Nonlinear Science and Numerical Simulation*, *49*, 113–134. doi:10.1016/j.cnsns.2017.01.025

Din, Q. (2018). Bifurcation analysis and chaos control in discrete-time glycolysis models. *Journal of Mathematical Chemistry*, *56*(3), 904–931. doi:10.100710910-017-0839-4

Din, Q., Elsadany, A. A., & Ibrahim, S. (2018). Bifurcation analysis and chaos control in a second order rational difference equations. *International Journal of Nonlinear Sciences and Numerical Simulation*, *19*(1), 53–68. doi:10.1515/ijnsns-2017-0077

Edelstein Keshet, L. (2005). *Mathematical Models in Biology*. New York: Society for Industrial and Applied Mathematics. doi:10.1137/1.9780898719147

Elabbasy, E. M., Elsadany, A. A., & Zhang, Y. (2014). Bifurcation analysis and chaos in a discrete reduced Lorenz system. *Applied Mathematics and Computation*, *228*, 184–194. doi:10.1016/j.amc.2013.11.088

Elaydi, S. N. (2008). *Discrete Chaos with Applications in Science and Engineering*. Chapman and Hall/ CRC.

Kangalgil, F. (2017). The local stability analysis of a nonlinear discrete-time population model with delay and Allee effect. *Cumhuriyet Sci. J.*, *38*(3), 480–487. doi:10.17776/csj.340491

Lakshmanan, M., & Rajaseekar, S. (2012). *Nonlinear Dynamics Inerrability Chaos and Patterns*. Berlin, Germany: Springer.

Li, H.-L., Long, Z., Cheng, H., Yao-Lin, J., & Zhidong, T. (2016). Dynamical analysis of a fractional-order predator-prey model incorporating a prey refuge. *Journal of Applied Mathematics and Computing*, *54*(1-2), 435–449. doi:10.100712190-016-1017-8

Lotka, A. (1925). *Elements of Physical Biology*. Baltimore: Williams and Wilkins.

Marotto. (2006). *Introduction to Mathematical Modeling Using Discrete Dynamical Systems*. Thomson Brooks/Cole.

Oldham, K. B., & Spanier, J. (1974). *The Fractional Calculus*. London: Academic Press.

Pk, J., & Ghorais, S. (2017). Stability of prey–predator model with holling type response function and selective harvesting. *J. Appl. Computat. Math.*, *6*(03), 358. doi:10.4172/2168-9679.1000358

Stephens, P. A., Sutherland, W. J., & Freckleton, R. P. (1999, October). What is the Allee effect. *Oikos*, *87*(1), 185–190. doi:10.2307/3547011

Volterra, V. (1926). Variazioni e fluttuazioni del numero d'individui in specie animali conviventi. *Mem. Accad. Lincei.*, *6*, 31–113.

ADDITIONAL READING

El-Sayed, A. M. A., & Salman, S. M. (2013). On a discretization process of fractional order Riccati differential equation. *J. Fract. Calc. Appl.*, *4*, 251–259.

Elsadany, A. A., & Matouk, A. E. (2015). Dynamical behaviors of fractional-order Lotka-Volterra predator-prey model and its discretization. *Applied Mathematics and Computation*, *49*, 269–283.

Fowler, M. S., & Ruxton, G. D. (2002). Population dynamics consequences of Allee effects. *Journal of Theoretical Biology*, *215*(1), 39–46. doi:10.1006/jtbi.2001.2486 PMID:12051982

George Maria Selvam, A. & Janagaraj, R. (2018) Numerical Analysis of a Fractional Order Discrete Prey – Predator System with Functional Response. *International Journal of Engineering & Technology,* 7(4.10), 681–684.

George Maria Selvam, A., Janagaraj, R., & Britto Jacob, S. (2017). Stability in a Discrete Fractional Order Prey Predator System with Functional Response. *International Journal of Advanced Research in Computer Science and Software Engineering*, *7*(5), 630–633. doi:10.23956/ijarcsse/SV7I5/0336

George Maria Selvam, A., Janagaraj, R., & Vignesh, D. (2018). Allee effect and Holling type - II response in a discrete fractional order prey - predator model. *IOP Conference Series: Journal of Physics*, 1139, 1-7. 10.1088/1742-6596/1139/1/012003

Kangalgil, F., & Ak Gumus, O. (2016). Allee effect in a new population model and stability analysis. *Gen. Math. Notes.*, *35*(1), 1–6.

Liu, X., & Xiao, D. (2007). Complex dynamic behaviors of a discrete-time predator – prey system. *Chaos, Solitons, and Fractals*, *32*(1), 80–94. doi:10.1016/j.chaos.2005.10.081

Matouk, A. E., & Elsadany, A. A. (2016). Dynamical analysis, stabilization and discretization of a chaotic fractional-order GLV model. *Nonlinear Dynamics*, *85*(3), 1597–1612. doi:10.100711071-016-2781-6

Shi, Y., Ma, Q., & Ding, X. (2018). Dynamical behaviors in a discrete fractional-order predator-prey System. *Filomat*, *32*(17), 5857–5874. doi:10.2298/FIL1817857S

Zhou, S., Liu, Y., & Wang, G. (2005). The stability of predator–prey systems subject to the Allee effects. *Theoretical Population Biology*, *67*(1), 23–31. doi:10.1016/j.tpb.2004.06.007 PMID:15649521

Chapter 3
Application of Fractional Optimal Control Problems on Some Mathematical Bioscience

Ismail Gad Ameen
https://orcid.org/0000-0002-8654-5241
South Valley University, Egypt

Hegagi Mohamed Ali
https://orcid.org/0000-0003-0159-6726
Aswan University, Egypt

ABSTRACT

In this chapter, the authors present a short review of a fractional-order control models, which are described by a system of fractional differential equations. Fractional derivatives describe the behaviour of dynamical systems better than classical calculus, where it can reflect the effect of memory. This survey shows the effect of the control on the fractional models, which represents epidemiological and biomedicine problems. So, the solution to such models is necessary for decision makers in health organizations.

INTRODUCTION

Infectious diseases are the most important problems facing us in the present time. For example, HIV and TB cause 10% of all deaths (Misra et al, 2011; Huo et al, 2015; Hikal & Zahra, 2016). Over recent years, the occurs of infectious diseases outbreaks has increased dramatically which leads to millions of deaths and affecting international economics (Salah et al, 2015; Global Health Policy, 2014). There is an opportunity to collect initial data when a diseases outbreak. These collected data are essential to understand the behaviour of the spreading mechanism, predict the future path and adjust control strategies of diseases. Pathological limitations are one of the main reasons for impossible to collect these initial data (see e.g., Misra et al, 2011; Mukandavire et al, 2009). Consequently, the description of infectious diseases as mathematical models should be used to give a better understanding, analyzing and controlling such diseases.

DOI: 10.4018/978-1-7998-3122-8.ch003

One of the ways in which construct mathematical models is by differential equations of fractional order (see e.g., (Ameen & Novati, 2017; Abderrahim et al, 2018; Khalil et al, 2019). Fractional differential equations (FDEs) have been attracted attention due to their appearance in various applications in biology, physics, epidemiology, fluid mechanics, viscoelasticity, and engineering. Moreover, the fractional-order derivatives (i.e. Caputo fractional derivative) for ordinary or partial differential equations proved that it's a valuable tool for modelling of many various phenomena (see e.g., (Ali & Ameen, 2019; Arafa et al, 2014; Ali, 2019)).

This is due to the fact that the realistic modelling of a phenomenon often does not depend only on the instant time but also on the history of the previous time (memory effects) which can also be successfully achieved by using fractional calculus (where it's non-local operator, for details see Ameen, 2017, p.38-39). More precisely, the modelling of biological and biomedicine by FDEs has more advantages than the integer-order ones, in which memory effects are neglected (Sardar et al, 2015; Du et al, 2013).

Fractional optimal control problems (FOCPs) can be viewed as a generalization of classical optimal control problems (OCPs), in which the dynamics of the control system are characterized by FDEs and might include an objective function given by a fractional integration operator. The motivation to formulate and solve FOCPs depend on the reality that there are a significant number of cases in which FDEs describe the behaviour of the control systems of interest more precisely than the common integer differential equations (Ali et al, 2016).

There have been many efforts in formulating and solving FOCPs to overcome and eliminate certain diseases, for example, Rosa and Torres in (Rosa & Torres, 2018) applied fractional optimal control theory on human respiratory syncytial virus (HRSV) infection. Ding et al (Ding et al, 2011) formulated and addressed a FOCP of HIV-Immune system model and solved it. Sweilam et al (Sweilam et al, 2016) studied a FOCP of a multi-strain tuberculosis model and used numerical methods to find a solution to this problem. Bonyah et al (Bonyah et al, 2018) proposed a FOCP of human African trypanosomiasis model and use preventive and treatment mechanisms to minimize the number of a person infected with trypanosomiasis in the African society. Al Basir and Roy developed and solved a FOCP for HIV-model in (Al Basir & Roy, 2017). There are other some research conducted on diseases which formulated as a FOCPs (see e.g. Elal et al, 2016; Sweilam & AL–Mekhlafi, 2017; Rachah & Torres, 2017; Sweilam et al, 2019; Ali & Ameen, 2020; Ameen et al, 2020).

Throughout this chapter, we will summarize some of the different recent fractional-order control models which describe some diseases and we will show necessary optimality conditions for these FOCPs using Pontryagin's maximum principle.

PRELIMINARIES

In this section, we give some necessary basic concepts of fractional calculus (Podlubny, 1999; Kilbas et al, 2006; Samko et al, 1993).

Definition 1. The left and right Riemann-Liouville fractional integrals of order α>0 of a function $f : \mathbb{R}^+ \to \mathbb{R}$ are, respectively, given by

$$ {}_aI_t^\alpha f(t) = \frac{1}{\Gamma(\alpha)} \int_a^t (t-\tau)^{\alpha-1} f(\tau) d\tau, $$

and

$$ {}_tI_b^\alpha f(t) = \frac{1}{\Gamma(\alpha)} \int_t^b (\tau-t)^{\alpha-1} f(\tau) d\tau, $$

where Γ(.)is the Euler gamma function defined by

$$ \Gamma(z) = \int_0^\infty e^{-t} t^{z-1} dt (\mathrm{Re}(z) > 0). $$

Definition 2. Let a function $f \in AC([a,b])$, where $AC([a,b])$ represents the space of absolutely continuous functions on the interval $[a,b]$, $n-1<\alpha\leq n$ and $n \in \mathbb{N}$. Then, the left and right Riemann-Liouville fractional derivatives are, respectively, defined by

$$ {}_aD_t^\alpha f(t) = \frac{1}{\Gamma(n-\alpha)} \left(\frac{d}{dt}\right)^n \int_a^t (t-\tau)^{n-\alpha-1} f(\tau) d\tau, $$

and

$$ {}_tD_b^\alpha f(t) = \frac{1}{\Gamma(n-\alpha)} \left(-\frac{d}{dt}\right)^n \int_t^b (\tau-t)^{n-\alpha-1} f(\tau) d\tau. $$

Definition 3. The left and right Caputo fractional derivatives of a function $f \in AC([a,b])$, $n-1<\alpha\leq n$ and $n \in \mathbb{N}$ are, respectively, defined by

$$ {}_a^CD_t^\alpha f(t) = \frac{1}{\Gamma(n-\alpha)} \int_a^t (t-\tau)^{n-\alpha-1} f^{(n)}(\tau) d\tau, $$

and

$$ {}_t^CD_b^\alpha f(t) = \frac{(-1)^n}{\Gamma(n-\alpha)} \int_t^b (\tau-t)^{n-\alpha-1} f^{(n)}(\tau) d\tau. $$

Theorem 1. Let $f(\bullet)$ be a differentiable function in interval $[a,b]$ and $g(\cdot) \in L_1([a,b])$. Then, we have the fractional integration by parts as follows

$$\int_a^b g(t)\, {}_a^C D_t^{\alpha} f(t)dt = \int_a^b f(t)\, {}_t D_b^{\alpha} g(t)dt + [{}_t I_b^{1-\alpha} g(t) f(t)]_a^b,$$

and

$$\int_a^b g(t)\, {}_t^C D_b^{\alpha} f(t)dt = \int_a^b f(t)\, {}_a D_t^{\alpha} g(t)dt - [{}_a I_t^{1-\alpha} g(t) f(t)]_a^b.$$

OPTIMAL CONTROL OF FRACTIONAL ORDER RESPIRATORY DISEASES

Respiratory diseases occur when viruses attack the respiratory such as influenza and their types, SARS and Ebola. Most respiratory diseases have no vaccines, such as SARS and Ebola and these diseases spread quickly and maybe re-infected. So, Yaro et al (Yaro et al, 2019) proposed a fractional mathematical model to describe the transmission of respiratory diseases. Moreover, they have incorporated two control measures, namely, quarantine of exposed population and monitoring and treating of the infected population to reduced and eliminating the transmission of these respiratory diseases.

The FOCP in transmission dynamics of respiratory diseases model can be formulated with a given objective function

$$\text{Minimize } J(u_1, u_2) = \int_0^{t_f} (c_1 E + c_2 I + c_3 u_1^2 + c_4 u_2^2) d\xi,$$

subject to the fractional dynamical system

$$\begin{aligned}
{}_a^C D_t^{\sigma} S(\xi) &= b^{\sigma} - \left(\beta \frac{I}{S+E+I}(1-u_1) + \mu_N^{\sigma}\right) S + k^{\sigma} R, \\
{}_a^C D_t^{\sigma} E(\xi) &= \beta \frac{I}{S+E+I}(1-u_1) S - (v^{\sigma}(1-u_2) + \mu_N^{\sigma}) E, \\
{}_a^C D_t^{\sigma} I(\xi) &= v^{\sigma}(1-u_2) E - (\mu_D^{\sigma} + \alpha^{\sigma}) I, \\
{}_a^C D_t^{\sigma} R(\xi) &= \alpha^{\sigma} I - (\mu_N^{\sigma} + k^{\sigma}) R,
\end{aligned} \tag{1}$$

with initial conditions $S(0)=S_0$, $E(0)=E_0$, $I(0)=I_0$, $R(0)=R_0$, where $S(\xi)$, $E(\xi)$, $I(\xi)$, $R(\xi)$ are the susceptible, exposed, infected, recovered populations respectively at time ξ. The parameters b^{σ} represents the rate of new susceptible people entering the populations, β represents the probability of disease transmission, α^{σ} represents the recovery rate, υ^{σ} is the seroconversion rate, μ_N^{σ} is the natural mortality rate, μ_D^{σ} represents the disease induced death rate, and k^{σ} is the rate at which the recovered return to the susceptible population (due to the loss of immunity). The coefficients c_i, i=1,2,3,4 represent the weight constants. While the proposed controls u_1 and u_2 are denoted by quarantine of exposed population groups and monitoring and treatment of infected populations respectively. It is assumed that the exposed population is reduced by the factor $(1-u_1)$ as they are quarantine. Furthermore, the infected population is reduced by a factor of $(1-u_2)$ as they are monitored and treated by health professionals.

The adjoint variables or co-state variables for the state system (1) are given by

$$
\begin{aligned}
{}_tD_{t_f}^{\sigma}\lambda_S &= \lambda_S\left(\frac{\beta I(1-u_1)S}{(S+E+I)^2}-\frac{\beta I(1-u_1)}{S+E+I}-\mu_N^{\sigma}\right)+\lambda_E\left(-\frac{\beta I(1-u_1)S}{(S+E+I)^2}+\frac{\beta I(1-u_1)}{S+E+I}-\mu_N^{\sigma}\right),\\
{}_tD_{t_f}^{\sigma}\lambda_E &= c_1+\frac{\lambda_S\beta I(1-u_1)S}{(S+E+I)^2}+\lambda_E\left(-\frac{\beta I(1-u_1)S}{(S+E+I)^2}-v^{\sigma}(1-u_2)-\mu_N^{\sigma}\right)+\lambda_I v^{\sigma}(1-u_2),\\
{}_tD_{t_f}^{\sigma}\lambda_I &= c_2-\lambda_S\left(\frac{\beta(1-u_1)}{S+E+I}-\frac{\beta I(1-u_1)}{(S+E+I)^2}\right)S+\lambda_E\left(\frac{\beta(1-u_1)}{S+E+I}-\frac{\beta I(1-u_1)}{(S+E+I)^2}\right)S\\
&\quad -\lambda_I(\mu_D^{\sigma}+\alpha^{\sigma})+\lambda_R\alpha^{\sigma},\\
{}_tD_{t_f}^{\sigma}\lambda_R &= \lambda_S k^{\sigma}-\lambda_R(\mu_N^{\sigma}+k^{\sigma}),
\end{aligned}
$$

with the transversality conditions

$$\lambda_S(t_f)=0,\lambda_E(t_f)=0,\lambda_I(t_f)=0,\lambda_R(t_f)=0.$$

The optimal control parameters that minimizes the objective function over admissible control set is given by

$$
\begin{aligned}
u_1^* &= \max\left\{0,\min\left(1,\frac{SI\beta(\lambda_E-\lambda_S)}{2c_3(S+E+I)}\right)\right\},\\
u_2^* &= \max\left\{0,\min\left(1,\frac{Ev^{\sigma}(\lambda_I-\lambda_E)}{2c_4}\right)\right\},
\end{aligned}
$$

to achieves

$$J(u_1^*,u_2^*)=\min_{u_1,u_2}\{J(u_1,u_2)\big|u_1,u_2\in\Omega\},$$

where Ω is admissible control set satisfying

$$\Omega=\left\{(u_1,u_2)\big|u_i:[0,t_f]\to[0,\infty)\,\text{Lebesgue measurable},i=1,2\right\}.$$

After using the Pontryagin's maximum principle to provide the necessary conditions needed for the existence of the optimal solution to this FOCP and doing some numerical simulation, they conclude that if all two control measures u_1 (quarantine of exposed population groups) and u_2 (monitoring and treatment of infected populations) are employed for the same period of time and continue for a considerable period of time, the future of free from transmission of a respiratory disease could be achieved.

FRACTIONAL OPTIMAL CONTROL OF TUBERCULOSIS (TB)

Tuberculosis is a severe infection that is caused by the bacterium Mycobacterium tuberculosis and kills many people each year (for more details see, e.g., (Silva et al, 2017; Silva & Torres, 2013; WHO, 2016)). Many mathematical models have been proposed in order to understand, analyze and control the spread of TB disease (Silva & Torres, 2015; Silva & Torres, 2013; Rodrigues et al, 2014; Denysiuk et al, 2015; Silva & Torres, 2012; Yang et al, 2010). Some of the literature on TB of fractional-order have been published (Sweilam & Al-Mekhlafi, 2016; Wojtak et al, 2018).

In 2019, Silverio and Delfim (Silverio & Delfim, 2019) considered a Caputo fractional-order TB model in (Wojtak et al, 2018), which subject to fractional optimal control theory (Debbouche et al, 2017; Jahanshahi & Torres, 2017). The objective function to minimize the number of infectious individuals and the cost of the control of the disease with the identification and treatment of the patients is given as follows:

$$\text{Minimize}\, J(I(t),u(t)) = \int_0^{t_f} (I(t) + B\rho u^2(t))dt,$$

with $0<B<\infty$, subject to the fractional control system

$$\begin{aligned}
&{}_0^C D_t^\alpha S(t) = \Lambda - \beta I(t)S(t) - \mu S(t),\\
&{}_0^C D_t^\alpha L(t) = \beta I(t)S(t) + (1-k)\delta\, T(t) - (\mu + \varepsilon)\, L(t),\\
&{}_0^C D_t^\alpha I(t) = \varepsilon L(t) + k\delta\, T(t) - (\mu + u(t) + \alpha_1) I(t),\\
&{}_0^C D_t^\alpha T(t) = u(t)\, I(t) - (\mu + \delta + \alpha_2) T(t),
\end{aligned} \tag{2}$$

and the initial conditions are $S(0)$, $L(0)$, $I(0)$, $T(0)\geq 0$, where the population is divided into four classes: $S(t)$ susceptible individuals, $L(t)$ latent individuals, which have been infected but are not infectious and do not show symptoms of the disease, $I(t)$ infectious individuals, which have active TB, may transmit the infection, but are not under treatment and infected individuals in treatment $T(t)$. Λ is the recruitment of the susceptible population, all individuals are exposed to natural death rate μ. The parameter β is the transmission coefficient of susceptible individuals to TB infection individuals and $L(t)$, become infectious at a rate ε, and $I(t)$ start treatment at a rate γ. $T(t)$ leave their compartment at rate δ. The parameter k (where $0\leq k\leq 1$) represents the failure of the treatment. Infectious, $I(t)$, and under treatment individuals, $T(t)$, may suffer TB-induced death at the rates α_1 and α_2, respectively. Finally, in the cost function the maximum number of infectious individuals of the problem without control is ρ.

The co-state variables $\lambda_i(t)$, i=1,…,4 for the state system (2) are given by

$$
\begin{aligned}
{}_tD_{t_f}^{\alpha}\lambda_1(t) &= \mu\lambda_1(t) - \beta I(t)(\lambda_1(t) - \lambda_2(t)),\\
{}_tD_{t_f}^{\alpha}\lambda_2(t) &= (\mu + \varepsilon)\lambda_2(t) - \varepsilon\lambda_3(t),\\
{}_tD_{t_f}^{\alpha}\lambda_3(t) &= -1 + (\mu + u(t) + \alpha_1)\lambda_3(t) - u(t)\lambda_4(t) + \beta S(t)(\lambda_1(t) - \lambda_2(t)),\\
{}_tD_{t_f}^{\alpha}\lambda_4(t) &= (\mu + \delta + \alpha_2)\lambda_4(t) + \delta(k-1)\lambda_2(t) - \delta k\lambda_3(t),
\end{aligned}
$$

satisfying the following transversality conditions:

$$
{}_tD_{t_f}^{\alpha-1}\lambda_i\Big|_{t_f} = 0 \Leftrightarrow {}_tI_{t_f}^{1-\alpha}\lambda_i\Big|_{t_f} = \lambda_i(t_f) = 0, i = 1,\ldots,4.
$$

The optimal control variable $u^*(t)$, which designate the individuals that are put under treatment, is given by

$$
u^*(t) = \min\left\{\max\left(0, \frac{(\lambda_3(t) - \lambda_4(t))\, I(t)}{2B\rho}\right), u_{\max}\right\},
$$

to achieves

$$
J(I, u^*) = \min_u\{J(I,u)\Big| \; u \in \Omega\},
$$

where Ω the set of admissible control functions

$$
\Omega = \left\{u(.) \in L^{\infty}(0,t_f) : 0 \le u(t) \le u_{\max}, \forall\, t \in [0,t_f]\right\}.
$$

After applying the optimal control concept to the fractional TB model, that treatment is effective on the reduction of infected individuals. Also, any change in the value of the fractional derivative order α, corresponds to substantial variations on the solutions of the FOCP.

FRACTIONAL OPTIMAL CONTROL OF AN HIV/AIDS EPIDEMIC MODEL

Over the years, mathematical models which describes the dynamics of HIV/AIDS has played an important role in understanding, analyzing and controlling for the spread of such disease (Pitchaimani & Monica, 2015; Hernandez-Vargasa & Middleton, 2013; Mastroberardino et al, 2015; Okosun et al, 2013; Mukandavire et al, 2009).

Fractional optimal control of HIV/AIDS epidemic in a patchy environment was presented by Kheiri and Jafari in (Kheiri & Jafari, 2018). They discussed the numerical simulation of that model and concluded that one of the best strategies is implementing all the proposed control efforts, which decrease significantly the number of HIV-infected and AIDS people in both patches.

In 2019, Kherir and Jafari have been able to extend the classical model in (Kheiri & Jafari, 2019), which is described a nonlinear mathematical model for the transmission dynamics of HIV/AIDS in Cuba, to the model of fractional order α (where $0<\alpha\leq 1$) with inclusion of constant control parameters of condom use and antiretroviral therapy (ART) treatment. They developed their OCP in order to reduce infection with two control functions, namely that $u_1(t)$ the condom use and $u_2(t)$ the ART treatment. This can be achieved by minimizing the following objective function

$$\text{Minimize } J(u_1, u_2) = \int_0^{t_f} \left(A X + \frac{B_1}{2} u_1^2(t) + \frac{B_2}{2} u_2^2(t) \right) dt,$$

subject to

$$\begin{aligned} {}_0^C D_t^\alpha S(t) &= \Lambda - (\varepsilon_1 u_1(t)\beta + (1 - u_1(t))\beta) X S - \mu S, \\ {}_0^C D_t^\alpha X(t) &= (\varepsilon_1 u_1(t)\beta + (1 - u_1(t))\beta) XS - \kappa XY - (\mu + \gamma + \kappa')X, \\ {}_0^C D_t^\alpha Y(t) &= \kappa XY + \kappa' X - (\mu + (\varepsilon_2 u_2(t)\gamma + (1 - u_2(t))\gamma)) Y, \\ {}_0^C D_t^\alpha Z(t) &= \gamma X + (\varepsilon_2 u_2(t)\gamma + (1 - u_2(t))\gamma)Y - \mu' Z, \end{aligned} \tag{3}$$

with non-negative initial conditions $S(0)=S_0$, $X(0)=X_0$, $Y(0)=Y_0$, $Z(0)=Z_0$, where the variable population size $N(t)$ is divided into four compartment; the susceptible population, $S(t)$, the undiagnosed HIV-infected population, $X(t)$, the diagnosed HIV-infected population, $Y(t)$, and people diagnosed with AIDS, $Z(t)$. The parameter Λ denotes the recruitment rate of $S(t)$, while β the recruitment rate of new members of HIV-infected population infected by sexual transmission with X. The transmission from HIV-infected population to AIDS occurs at the γ rate. κ the rate at which undiagnosed HIV-infected population is diagnosed through contact tracing and κ' the rate at which undiagnosed HIV-infected population is diagnosed through random testing. μ, μ' are the mortality rate of the adult population and the population with AIDS, respectively. $\varepsilon_i \in (0,1)$, $i=1,2$ are two positive constant, which is assumed to measure the effectiveness of the control $u_1(t)$ on β and $u_2(t)$ on γ. Finally, The parameters A, B_1 and B_2 are positive weights and $\frac{B_i}{2} u_i^2(t)$ is the cost of applying control effort u_i, and b is the duration of control programs. In this case, the co-state equations are given in terms of the left fractional derivatives (for more details see, Lemma 3 in (Kheiri & Jafari, 2019)) as follow:

$$
\begin{aligned}
{}_0^C D_t^\alpha \lambda_1(t_f - t) &= (\mu + (\varepsilon_1 u_1(t_f - t)\beta + (1 - u_1(t_f - t))\beta)X(t_f - t))\lambda_1(t_f - t) \\
&\quad + (\varepsilon_1 u_1(t_f - t)\beta + (1 - u_1(t_f - t))\beta)X(t_f - t)\lambda_2(t_f - t), \\
{}_0^C D_t^\alpha \lambda_2(t_f - t) &= A - (\varepsilon_1 u_1(t_f - t)\beta + (1 - u_1(t_f - t))\beta)S(t_f - t)\lambda_1(t_f - t) \\
&\quad + [(\varepsilon_1 u_1(t_f - t)\beta + (1 - u_1(t_f - t))\beta)S(t_f - t) - \kappa Y(t_f - t) \\
&\quad - (\mu + \gamma + \kappa')]\lambda_2(t_f - t) + (\kappa Y(t_f - t) + \kappa')\lambda_3(t_f - t) + \gamma\lambda_4(t_f - t), \\
{}_0^C D_t^\alpha \lambda_3(t_f - t) &= -\kappa X(t_f - t)\lambda_2(t_f - t) + (\varepsilon_2 u_2(t_f - t)\gamma + (1 - u_2(t_f - t))\gamma)\lambda_4(t_f - t) \\
&\quad + [\kappa X(t_f - t) - (\mu + (\varepsilon_2 u_2(t_f - t)\gamma + (1 - u_2(t_f - t))\gamma))]\lambda_3(t_f - t), \\
{}_0^C D_t^\alpha \lambda_4(t_f - t) &= -\mu' \lambda_4(t_f - t),
\end{aligned}
$$

satisfying the following transversality conditions $\lambda_1(t_f) = 0, \lambda_2(t_f) = 0, \lambda_3(t_f) = 0,$ and $\lambda_4(t_f) = 0$.

Furthermore, the optimal controls u_1^* and u_2^* that by implement on the fractional-order model (3), the number of diagnosed and undiagnosed HIV-infected population and AIDS people are significantly reduced, are given by

$$
\begin{aligned}
u_1^* &= \min\left\{1, \max\left(0, \frac{(\lambda_2(t) - \lambda_1(t))(1 - \varepsilon_1)\beta X(t)S(t)}{B_1}\right)\right\}, \\
u_2^* &= \min\left\{1, \max\left(0, \frac{(\lambda_4(t) - \lambda_3(t))(1 - \varepsilon_2)\gamma Y(t)}{B_2}\right)\right\},
\end{aligned}
$$

to achieves

$$
J(u_1^*, u_2^*) = \min_{u_1, u_2}\{J(u_1, u_2) \big| u_1, u_2 \in \Omega\},
$$

where Ω is admissible control set satisfying

$$
\Omega = \{u_i(.) \in L^\infty(0, t_f] \big| 0 \le u_i(t) \le 1, \forall t \in [0, t_f], i = 1,2\}.
$$

FRACTIONAL OPTIMAL CONTROL WITH VECTOR-BORNE DISEASES MODELS

The vector-borne disease is a commonly used term that describes a disease transmitted to a host population, by vectors like mosquitos, insects, and ticks through feeding activity. Being a vector means that it carries a disease from infective one to a healthy host. One of these diseases called West Nile virus (WNV), which is transmitted among mosquitoes, birds, humans, and other animals (Fan et al, 2010). There is no specific treatment for WNV other than supportive therapy (such as hospitalization, intravenous fluids, and respiratory support) for severe cases. Antibiotics will not work because a virus, not bacteria, causes West Nile disease. No vaccine for the virus is currently available. In the absence of effective anti-WNV therapeutic treatment and vaccine, WNV control strategies are based on taking appropriate preventive measures (Garba & Safi, 2013). Sweilam et al (Sweilam et al, 2017) suggested a formulation of the FOCP that describe the transmission dynamics of WNV model.

The main aim of this research is to minimize the exposed and infected human populations, the total number of mosquitoes and the cost of implementing the control which is given in the following FOCP:

$$\text{Minimize}\, J(u_1, u_2) = \int_0^{t_f} (A_1 E(t) + A_2 I(t) + A_3 N_M(t) + B_1 u_1^2 + B_2 u_2^2) dt,$$

subject to a nonlinear system of FDEs

$$\begin{aligned}
{}_a^C D_t^\alpha M_s(t) &= \lambda_M N_M (1 - u_1) - \frac{b_1 \beta_1 M_s B_i}{N_B} - \mu_M M_s - r_0 u_1 M_s, \\
{}_a^C D_t^\alpha M_i(t) &= \frac{b_1 \beta_1 M_s B_i}{N_B} - \mu_M M_i - r_0 u_1 M_i, \\
{}_a^C D_t^\alpha B_s(t) &= \lambda_B + \rho N_B - \frac{b_1 \beta_2 M_i B_s}{N_B} - \Psi_B B_s - \mu_B B_s, \\
{}_a^C D_t^\alpha B_i(t) &= \frac{b_1 \beta_2 M_i B_s}{N_B} - d_B B_i - \Psi_B B_i - \mu_B B_i, \\
{}_a^C D_t^\alpha S(t) &= \lambda_H + \gamma_H N_H - \frac{b_2 \beta_3 M_i S (1 - u_2)}{N_H} - \mu_H S, \\
{}_a^C D_t^\alpha E(t) &= \frac{b_2 \beta_3 M_i S (1 - u_2)}{N_H} - \alpha_w E - \mu_H E, \\
{}_a^C D_t^\alpha I(t) &= \alpha_w E - \gamma I - d_I I - r I - \mu_H I, \\
{}_a^C D_t^\alpha H(t) &= \gamma I - d_H H - \tau H - \mu_H H, \\
{}_a^C D_t^\alpha R(t) &= \tau H + r I - \mu_H R,
\end{aligned} \tag{4}$$

with initial conditions

$$\begin{aligned}
&M_s(0) = M_{s_0}, M_i(0) = M_{i_0}, B_s(0) = B_{s_0}, B_i(0) = B_{i_0}, \\
&S(0) = S_0, E(0) = E_0, I(0) = I_0, H(0) = H_0, R(0) = R_0,
\end{aligned}$$

where the variables of this system defined as follows, $M_s(t)$ and $M_i(t)$ are the populations of susceptible and infected mosquitoes respectively. They considered that the total population of mosquitoes at time t is $N_M(t)=M_s(t)+M_i(t)$. $B_s(t)$, $B_i(t)$ are denoted by susceptible and infected birds respectively and the total population of birds $N_B(t)=B_s(t)+B_i(t)$. While $S(t)$, $E(t)$, $I(t)$, $H(t)$, $R(t)$ are the populations of susceptible, exposed, infected, hospitalized, recovered humans respectively at time t and the total population of humans

$$N_H(t) = S(t) + E(t) + I(t) + H(t) + R(t).$$

Indeed, the interpretation of the parameters is given as, λ_M, λ_H and λ_B are the recruitment rate of mosquitoes, humans and birds respectively. The natural death rate of mosquitoes, birds and humans is denoted by μ_M, μ_B and μ_H respectively. Ψ_B represents the migration rate of birds. The parameter b is the migration rate of birds while b_1 and b_2 are the per capita biting rate of mosquitoes on the primary host (birds) and humans respectively, where $b_1 = \frac{bN_B}{N_B + N_H}$ and $b_2 = \frac{bN_H}{N_B + N_H}$. d_B, d_I are the WNV induced death rate of birds, humans respectively and d_H is the death rate of hospitalized humans. The probability of WNV transmission from infected mosquitoes to a susceptible bird and humans are denoted by β_2 and β_3 respectively while β_1 represents the probability of WNV transmission from an infected bird to a susceptible mosquito. α_w is the rate of development of clinical symptoms of WNV. The natural recovery rate is denoted by r while τ is the treatment-induced recovery rate and γ is the hospitalized rate of infectious humans. The control u_1 represent the measures for the level of larvicide and adulticide used for mosquito control administered at mosquito breeding sites and the control u_2 represent measures for the level of successful prevention (personal protection) efforts. The quantities A_1, A_2 and A_3 represent, respectively, the weight constants of the exposed, infected human and the total mosquito populations. On the other hand, B_1 and B_2 are weight constants for mosquito control and personal protection (prevention of mosquito–human contacts). It is assumed that the reproduction rate of the mosquito population is reduced by a factor of $(1-u_1)$ and the associated force of infection in the human population is reduced by a factor of $(1-u_2)$.

The adjoint variables for the state system (4) are given by

$$
\begin{aligned}
&{}_tD_{t_f}^{\alpha}\lambda_1 = \frac{\partial H_a}{\partial M_s},\ {}_tD_{t_f}^{\alpha}\lambda_2 = \frac{\partial H_a}{\partial M_i},\\
&{}_tD_{t_f}^{\alpha}\lambda_3 = \frac{\partial H_a}{\partial B_s},\ {}_tD_{t_f}^{\alpha}\lambda_4 = \frac{\partial H_a}{\partial B_i},\\
&{}_tD_{t_f}^{\alpha}\lambda_5 = \frac{\partial H_a}{\partial S},\ {}_tD_{t_f}^{\alpha}\lambda_6 = \frac{\partial H_a}{\partial E},\\
&{}_tD_{t_f}^{\alpha}\lambda_7 = \frac{\partial H_a}{\partial I},\ {}_tD_{t_f}^{\alpha}\lambda_8 = \frac{\partial H_a}{\partial H},\\
&{}_tD_{t_f}^{\alpha}\lambda_9 = \frac{\partial H_a}{\partial R},
\end{aligned}
$$

where H_α is Hamiltonian function defined as the following:

$$
\begin{aligned}
H_a &= A_1 E(t) + A_2 I(t) + A_3 N_M(t) + B_1 u_1^2 + B_2 u_2^2 \\
&+\lambda_1 \left[\lambda_M N_M (1-u_1) - \frac{b_1 \beta_1 M_s B_i}{N_B} - \mu_M M_s - r_0 u_1 M_s \right] \\
&+\lambda_2 \left[\frac{b_1 \beta_1 M_s B_i}{N_B} - \mu_M M_i - r_0 u_1 M_i \right] + \lambda_3 \left[\lambda_B + \rho N_B - \frac{b_1 \beta_2 M_i B_s}{N_B} - \Psi_B B_s - \mu_B B_s \right] \\
&+\lambda_4 \left[\frac{b_1 \beta_2 M_i B_s}{N_B} - d_B B_i - \Psi_B B_i - \mu_B B_i \right] + \lambda_5 \left[\lambda_H + \gamma_H N_H - \frac{b_2 \beta_3 M_i S (1-u_2)}{N_H} - \mu_H S \right] \\
&+\lambda_6 \left[\frac{b_2 \beta_3 M_i S (1-u_2)}{N_H} - \alpha_w E - \mu_H E \right] + \lambda_7 \left[\alpha_w E - \gamma I - d_I I - r I - \mu_H I \right] \\
&+\lambda_8 \left[\gamma I - d_H H - \tau H - \mu_H H \right] + \lambda_9 \left[\tau H + r I - \mu_H R \right].
\end{aligned}
$$

The transversality conditions: $\lambda_i(t_f)=0$, $i=1,2,\ldots,9$.

The optimal control parameters that minimizes the objective function over admissible control set are given by

$$
u_1^* = \max\left\{0, \min\left(1, \frac{\lambda_1 (\lambda_M N_M + r_0 M_s) + \lambda_2 r_0 M_i}{2B_1}\right)\right\},
$$

$$
u_2^* = \max\left\{0, \min\left(1, \frac{b \beta_3 M_i S \xi (\lambda_6 - \lambda_5)}{2B_2}\right)\right\},
$$

to achieves

$$
J(u_1^*, u_2^*) = \min_{u_1, u_2} \left\{ J(u_1, u_2) \middle| u_1, u_2 \in \Omega \right\},
$$

where Ω is admissible control set satisfying

$$
\Omega = \{(u_1, u_2) \in (L^\infty(0, t_f]) \,|\, 0 \le u_1, u_2 \le 1, \forall t \in [0, t_f]\}.
$$

CONCLUSION

In this review, we have shown how the FOCPs implemented in some various infectious disease models. Fractional necessary optimality conditions which needed for the existence of the optimal solution for such models are derived using Pontryagin's maximum principle.

Also, the description of these models in the frame of fractional-order distinguishes it from integer-order models. Where the fractional-order models carrying information about the memory of living systems and its associative learning mechanisms which makes it more insight and realistic than integer-order models.

By implementing all the control efforts suggested in each FOCPs, we find that the number of infected individuals is significantly reduced under the controlled conditions.

For this, we can say that the objective of eliminating infectious diseases in a society with restricted resources, it is now a step closer to achieving by using FOCPs.

REFERENCES

Al Basir, F., & Roy, P. K. (2017). Stability analysis and optimal control of a fractional order model for HIV infection. *WSEAS Trans Math, 16*, 152-162. http://www.wseas.org/multimedia/journals/mathematics/2017/a385806-868.php

Ali, H. M. (2019). New approximate solutions to fractional smoking model using the generalized Mittag-Leffler function method. *Progr. Fract. Differ. Appl., 5*(4), 1–9. doi:10.18576/pfda/010101

Ali, H. M., & Ameen, I. (2019). Mittag-Leffler approximation for the solution of nonlinear systems of fractional partial differential equations. *Sylwan, 163*(9), 17–32.

Ali, H. M. & Ameen, I. (2020). Save the pine forests of wilt disease using a fractional optimal control strategy, *Chaos, Solitons & Fractals, 132*, ID 109554. doi: 10.1016/j.chaos.2019.109554

Ali, H. M., Pereira, F. L., & Gama, S. M. (2016). A new approach to the Pontryagin maximum principle for nonlinear fractional optimal control problems. *Mathematical Methods in the Applied Sciences, 39*(13), 3640-3649. doi:10.1002/mma.3811

Ameen, I. (2017). *Fractional calculus: numerical methods and SIR models* (PhD thesis). University of Padova.

Ameen, I., Hidan, M., Mostefaoui, Z. & Ali, H. M. (2020). Fractional optimal control with fish consumption to prevent the risk of coronary heart disease. *Complexity*. . doi:10.1155/2020/9823753

Ameen, I., & Novati, P. (2017). The solution of fractional order epidemic model by implicit Adams methods. *Applied Mathematical Modelling, 43*, 78–84. doi:10.1016/j.apm.2016.10.054

Arafa, A. A. M., Rida, S. Z., & Mohamed, H. (2014). An application of the homotopy analysis method to the transient behaviour of a biochemical reaction model. *Information Sciences Letters, 3*(1), 29.doi:10.12785/isl/030104

Bonyah, E., Gomez-Aguilar, J. F., & Adu, A. (2018). Stability analysis and optimal control of a fractional human African trypanosomiasis model. *Chaos, Solitons & Fractals, 117*, 150-160.

Debbouche, A., Nieto, J. J., & Torres, D. F. M. (2017). Optimal solutions to relaxation in multiple control problems of Sobolev type with nonlocal nonlinear fractional differential equations. *Journal of Optimization Theory and Applications, 174*(1), 7–31. doi:10.100710957-015-0743-7

Denysiuk, R., Silva, C. J., & Torres, D. F. M. (2015). Multiobjective approach to optimal control for a tuberculosis model. *Optimization Methods & Software*, *30*(5), 893–910. doi:10.1080/10556788.2014.994704

Ding, Y., Wang, Z., & Ye, H. (2011). Optimal control of a fractional-order HIV-immune system with memory. *IEEE Transactions on Control Systems Technology*, 20(3), 763-769. doi:10.1109/TCST.2011.2153203

Du, M., Wang, Z., & Hu, H. (2013). Measuring memory with the order of fractional derivative. *Scientific Reports*, *3*(1), 1–3. doi:10.1038rep03431 PMID:24305503

Elal, L., Sweilam, N. H., Nagy, A. M., & Almaghrebi, Y. S. (2016). Computational methods for the fractional optimal control HIV infection. *Journal of Fractional Calculus and Applications*, *7*(2), 121–131.

Fan, G., Liu, J., Van den Driessche, P., Wu, J., & Zhu, H. (2010). The impact of maturation delay if mosquitoes on the transmission of West Nile virus. *Mathematical Biosciences*, *228*(2), 119–126. doi:10.1016/j.mbs.2010.08.010 PMID:20828577

Garba, S. M., & Safi, M. A. (2013). Mathematical analysis of West Nile Virus model with discrete delays. *Acta Mathematica Scientia*, *33B*(5), 1439–1462. doi:10.1016/S0252-9602(13)60095-8

Global Health Policy. (2014). *The U.S. Government & Global Emerging Infectious Disease Preparedness and Response*. http://kff.org/global-healthpolicy/fact-sheet/the-u-s-government-global-emerging-infectious-disease-preparedness-and-response/

Hernandez-Vargasa, E. A., & Middleton, R. H. (2013). Modeling the three stages in HIV infection. *Journal of Theoretical Biology*, *320*, 33–40. doi:10.1016/j.jtbi.2012.11.028 PMID:23238280

Hikal, M. M., & Zahra, W. K. (2016). On fractional model of an HIV/AIDS with treatment and time delay. *Progr. Fract. Differ. Appl*, *2*, 55-66. doi:10.18576/pfda/020106

Huo, J., Zhao, H., & Zhu, L. (2015). The effect of vaccines on backward bifurcation in a fractional order HIV model. *Nonlinear Analysis Real World Applications*, *26*, 289–305. doi:10.1016/j.nonrwa.2015.05.014

Jahanshahi, S., & Torres, D. F. M. (2017). A simple accurate method for solving fractional variational and optimal control problems. *Journal of Optimization Theory and Applications*, *174*(1), 156–175. doi:10.100710957-016-0884-3

Khalil, M., Said, M., Osman, H., Ahmed, B., Ahmed, D., Younis, N., ... Ashmawy, M. (2019). Fractional order models of infectious diseases: A review. *SPC Journal of Environmental Sciences*, *1*, 1–11.

Kheiri, H., & Jafari, M. (2018). Stability analysis of a fractional order model for the HIV/AIDS epidemic in a patchy environment. *Journal of Computational and Applied Mathematics*, *346*, 323–339. doi:10.1016/j.cam.2018.06.055

Kheiri, H., & Jafari, M. (2019). Fractional optimal control of an HIV/AIDS epidemic model with random testing and contact tracing. *Journal of Applied Mathematics and Computing*, *60*(1-2), 387–411. doi:10.100712190-018-01219-w

Kilbas, A. A., Srivastava, H. M., & Trujillo, J. J. (2006). *Theory and Application of Fractional Differential Equations*. New York: Elsevier.

Mastroberardino, A., Cheng, Y., Abdelrazec, A., & Liu, H. (2015). Mathematical modeling of the HIV/AIDS epidemic in Cuba. *International Journal of Biomathematics*, *08*(04), 1550047. doi:10.1142/S1793524515500473

Misra, A. K., Sharma, A., & Shukla, J. B. (2011). Modeling and analysis of effects of awareness programs by media on the spread of infectious diseases. *Mathematical and Computer Modelling*, *53*(5-6), 1221–1228. doi:10.1016/j.mcm.2010.12.005

Mouaouine, A., Boukhouima, A., Hattaf, K., & Yousfi, N. (2018). A fractional order SIR epidemic model with nonlinear incidence rate. *Advances in Difference Equations*, *160*, 1–9. doi:10.118613662-018-1613-z

Mukandavire, Z., Garira, W., & Tchuenche, J. M. (2009). Modelling effects of public health educational campaigns on HIV/AIDS transmission dynamics. *Applied Mathematical Modelling*, *33*(4), 2084–2095. doi:10.1016/j.apm.2008.05.017

Okosun, K. O., Makinde, O. D., & Takaidza, I. (2013). Impact of optimal control on the treatment of HIV/AIDS and screening of unaware infectives. *Applied Mathematical Modelling*, *37*(6), 3802–3820. doi:10.1016/j.apm.2012.08.004

Pitchaimani, M., & Monica, C. (2015). Global stability analysis of HIV-1 infection model with three time delays. *J. Appl. Math. Comput*, *48*(1-2), 293–319. doi:10.100712190-014-0803-4

Podlubny, I. (1999). *Fractional Differential Equations*. New York: Academic Press.

Rachah, A., & Torres, D. F. (2017). Analysis, simulation and optimal control of a SEIR model for Ebola virus with demographic effects. *Commun. Fac. Sci. Univ. Ank. Series A1.*, *67*(1), 179-197. doi:10.1501/Commua1_0000000841

Rodrigues, P., Silva, C. J., & Torres, D. F. M. (2014). Cost-effectiveness analysis of optimal control measures for tuberculosis. *Bulletin of Mathematical Biology*, *76*(10), 2627–2645. doi:10.100711538-014-0028-6 PMID:25245395

Rosa, S., & Torres, D. F. (2018). Optimal control of a fractional order epidemic model with application to human respiratory syncytial virus infection. *Chaos, Solitons & Fractals, 117*, 142-149. Doi:10.1016/j.chaos.2018.10.021

Salah, M., Abdelouahab, & Hamri, N. E. (2015). The Grunwald-Letnikov Fractional-Order Derivative with Fixed Memory Length. *Mediterranean Journal of Mathematics*, 1–16. doi:10.100700009-015-0525-3

Samko, S. G., Kilbas, A. A., & Marichev, O. I. (1993). *Fractional Integrals and Derivatives-Theory and Applications*. Linghorne, PA: Gordon and Breach.

Sardar, T., Rana, S., Bhattacharya, S., Al-Khaled, K., & Chattopadhyay, J. (2015). A generic model for a single strain mosquito-transmitted disease with memory on the host and the vector. *Mathematical Biosciences*, *263*, 18–36. doi:10.1016/j.mbs.2015.01.009 PMID:25645185

Silva, C. J., Maurer, H., & Torres, D. F. M. (2017). Optimal control of a tuberculosis model with state and control delays. *Mathematical Biosciences and Engineering*, *14*(1), 321–337. doi:10.3934/mbe.2017021 PMID:27879136

Silva, C. J., & Torres, D. F. M. (2012). Optimal control strategies for tuberculosis treatment: A case study in Angola. *Numer. Algebra Control Optim*, *2*(3), 601–617. doi:10.3934/naco.2012.2.601

Silva, C. J., & Torres, D. F. M. (2013). Optimal control for a tuberculosis model with reinfection and post-exposure interventions. *Mathematical Biosciences*, *244*(2), 154–164. doi:10.1016/j.mbs.2013.05.005 PMID:23707607

Silva, C. J., & Torres, D. F. M. (2015). Optimal control of tuberculosis: a review, in Dynamics, games and science. *CIM Ser. Math. Sci.*, *1*, 701-722. arXiv:1406.3456

Silverio, S. R., & Torres, D. F. M. (2019). Optimal control and sensitivity analysis of a fractional order TB model. *Statistics Opt. Inform. Comput.*, *7*, 1-9. http://www.IAPress.org

Sweilam, N. H., & Al-Mekhlafi, S. M. (2017). Optimal control for a time delay multi-strain tuberculosis fractional model: a numerical approach. *IMA Journal of Mathematical Control and Information*, *36*(1), 317-340. doi:10.1093/imamci/dnx046

Sweilam, N. H., & Al-Mekhlafi, S. M. (2016). Numerical study for multi-strain tuberculosis (TB) model of variable-order fractional derivatives. *Journal of Advanced Research*, *7*(2), 271–283. doi:10.1016/j.jare.2015.06.004 PMID:26966568

Sweilam, N. H., Saad, O. M., & Mohamed, D. G. (2017). Comparative studies for the fractional optimal control in transmission dynamics of West Nile virus. *International Journal of Biomathematics*, *10*(7). doi:10.1142/S1793524517500954

Sweilam, N. H., Saad, O. M., & Mohamed, D. G. (2019). Fractional optimal control in transmission dynamics of West Nile virus model with state and control time delay: a numerical approach. *Advances in Difference Equations*, *2019*(1), 210. . doi:10.118613662-019-2147-8

WHO. (2016). *Global Tuberculosis Report 2016*. World Health Organization.

Wojtak, W., Silva, C. J., & Torres, D. F. M. (2018). Uniform asymptotic stability of a fractional tuberculosis model. *Math. Model. Nat. Phenom.*, *13*(1), Art. 9. doi:10.1051/mmnp/2018015

Yang, Y., Li, J., Ma, Z., & Liu, L. (2010). Global stability of two models with incomplete treatment for tuberculosis. *Chaos, Solitons, and Fractals*, *43*(1-12), 79–85. doi:10.1016/j.chaos.2010.09.002

Yaro, D., Apeanti, W. O., Akuamoah, S. W., & Lu, D. (2019). Analysis and optimal control of fractional-order transmission of a respiratory epidemic model. *Int. J. Appl. Comput. Math*, *5*(4), 1–21. doi:10.100740819-019-0699-7

Section 2

Chaotic Dynamics of Fractional-Order Systems

Chapter 4
Dynamics of Some Discretized Fractional-Order Differential Equations

Sanaa Moussa Salman
Alexandria University, Egypt

Ahmed M. A. El-Sayed
https://orcid.org/0000-0001-7092-7950
Alexandria University, Egypt

ABSTRACT

This chapter deals with fractional-order differential equations and their discretization. First of all, a discretization process for discretizing ordinary differential equations with piecewise constant arguments is presented. Secondly, a discretization method is proposed for discretizing fractional-order differential equations. Stability of fixed points of the discretized equations are investigated. Numerical simulations are carried out to show the dynamic behavior of the resulting difference equations such as bifurcation and chaos.

INTRODUCTION

The concept of derivative is traditionally associated to an integer; given a function, we can derive it once, twice, three times and so on. It could have an interest to investigate the possibility to derive a function a real number. The main idea is to examine the properties of the ordinary derivative and see where and how is it possible to generalize the concepts.

In 1695, Leibniz posed a question to L'Hospital asking if the meaning of the derivative can be generalized to the form $\frac{d^n y}{dx^n}$ so that $n \in C$? L'Hospital replied by another question what if $n = \frac{1}{2}$? Leibniz replied in a letter (Pertz et al, 1849): It will lead to a paradox, from which one day useful consequences will be drawn.

DOI: 10.4018/978-1-7998-3122-8.ch004

Fractional calculus field is as old as calculus itself but the usefulness of the first in mathematical theory and in pure mathematics has become more evident. The fractional calculus has allowed the operations of integration and differentiation to be applied upon any fractional order. Recently theory of fractional differential equations attracted many scientists and mathematicians to work on (Das, 2007; Hadid et al, 1996; Holmgren, 1994; Ibrahim et al, 2007; Jumarie, 2009], [Yang et al, 2013a; Yang et al, 2013b; Yang et al, Yang et al, 2015a; Yang et al, 2015b). For the existence of solutions for fractional differential equations, one can see [El-Sayed, 1988; El-Sayed, 1992; El-Sayed, 1998; El-Sayed et al, 2007; El-Sayed et al, 2013a; El-Sayed et al, 2013b; Erjaee, 2009]. The results have been obtained by using fixed point theorems like Picard's, Schauder fixed-point theorem and Banach contraction mapping principle. About the development of existence theorems for fractional functional differential equations, many contributions exist [Abbas, 2010; El-Sayed, 1992; Agarwa et al, 2009; Das, 2007]. Many applications of fractional calculus amounting to replace the time derivative in a given evolution equation by a derivative of fractional order. Recently, interesting attempts have been made to give the physical meaning to the initial conditions for fractional differential equations with Riemann-Liouville fractional derivatives which were proposed in (Holmgren, 1994; Ibrahim, et al, 2007). Many equivalent definitions of fractional derivative and fractional integral are introduced and presented by many authors see for example (Furati et al, 2007) and (Jumarie, 2006).

Recently, a new modified Riemann-Liouville derivative is proposed by (Jumarie, 2009). This new definition of fractional derivative has two main advantages: firstly comparing with the Caputo derivative, the function to be differentiated is not necessarily differentiable, secondly different from the Riemann-Liouville derivative the Jumarie's modified Riemann-Liouville derivative of a constant is defined to zero. Moreover, a fractional Taylor series expansion (Jumarie, 2006) was invented by using this new definition.

Dealing with fractional-order differential equations as dynamical systems is somehow new (Erjaee, 2009) and (Yan, 2007). The non-local property of fractional differential equations means that the next state of a system not only depends on its current state but also on its historical states. This property is very close to the real world and thus fractional differential equations have become popular and have been applied to dynamical systems.

BRIEF HISTORICAL BACKGROUND OF DISCRETIZATION METHODS

Generally, there are two discretization methods: direct discretization and indirect discretization. In the direct method one can directly develop discrete-time operators to discretize the fractional-order continuous-time operator. In the indirect discretization method, a rational continuous-time operator should be developed which is then discretized using any of the well-known discretization techniques.

There are several approaches to direct discretization

1. A power series expansion (PSE),
2. Continued fraction expansion (CFE),
3. Matrix approach proposed by (Podlubny, 2000),
4. Taylor/MacLaurin series expansion of generating function suggested by (Duarte et al, 2002; Machado, 1997),
5. Diethelms approach based on solving of FDE in discrete time steps. All mentioned approaches are based on evolution of a generating function. As generating function one can use an Eulers backward

difference, the Tustin rule, the Simpson rule or the Al-Alaoui operator. The simplest method is the direct discretization using finite memory length expansion of Eulers rule from the Grunwald-Letnikov definition of fractional derivative.

On the other hand, the indirect method is based on discretization polynomials (Hennig et all, 2004; Maxim & Kamenetskii, 1997; Tung & Harvey, 1984; Vasumathi & Daniel, 2008). Though the indirect discretization methods are less accurate than the direct ones, it is computationally inexpensive.

To be clear, this chapter is devoted to introduce a new discretization process for discretizing ordinary differential equations with piecewise constant arguments as well as a new method for discretizing fractional-order differential equations.

DISCRETIZATION OF ORDINARY DIFFERENTIAL EQUATIONS

The motivation behind discretization is that it is not possible to solve a differential equation of the form

$$\frac{dx}{dt} = \mathrm{f(x)}, \tag{3.1}$$

analytically, and one turns to numerical or computational methods. A numerical method seeks to approximate the solution to the equation at discrete times where time is subdivided into intervals of length Δt, so that $t_n = n\Delta t$, and then the method approximates the solution at those times, $x_n \approx x(t_n)$.

So, the main goal of any discretization method is to produce a difference equation that resembles the original differential equation in its dynamics, but this is often difficult especially when we are dealing with nonlinear component in a differential equation. Thus, the choice of a discretization scheme that produce a difference equation that resembles the continuous counterpart is a challenging task in numerical analysis.

Euler's Method

One of the oldest ideas for doing this is the Euler method:

$$\frac{dx}{dt} \simeq \frac{\Delta x}{\Delta t},$$

$$\frac{\Delta x}{\Delta t} = \frac{x\left(t_n + \Delta t\right) - x\left(t_n\right)}{\Delta t} = \frac{x_{n+1} - x_n}{\Delta t} \approx \frac{dx}{dt} = f\left(x_n\right).$$

Solving this expression for x_{n+1} we end up with a discrete equation which predicts a future value of x, x_{n+1}, in terms of a past value:

$$x_{n+1} = x_n + \Delta t\, f(x_n).$$

Consider for example the Logistic equation defined by

$$x(t) = \rho x(t)(1 - x(t)), \tag{3.2}$$

with initial value $x(0) = x_0$, where ρ is a positive constant. Applying Euler's method to (3.2) gives

$$x(n + 1) = x(n) + h\,\rho x(n)(1 - x(n)). \tag{3.3}$$

Let

$$y(n) = \frac{h\rho}{1 + h\rho} x(n).$$

So, (3.3) can be written as

$$y(n + 1) = \mu y(n)(1 - y(n)).$$

It is possible for the Euler method (and, in fact, for any numerical approach) to go wrong, particularly when Δt becomes large. In addition, the behavior of dynamics calculated using the Euler approximation generally 'lag' actual system dynamics, however, we should note that Euler's method is an approximation for the derivative.

A Non-Standard Scheme

Consider again the logistic differential equation. Now if we replace $x^2(n)$ in Euler's method by $x(n)x(n + 1)$ we get

$$x(n + 1) = x(n) + h\,\rho x(n) - h\,\rho x(n)x(n + 1).$$

Simplifying we obtain the rational difference equation

$$x(n + 1) = \frac{(1 + h\rho)x(n)}{1 + h\rho x(n)}.$$

Mickens introduced two nonstandard schemes (Mickens, 1994) for asymptotically stable systems. This and some other numerical methods for both differential equations and fractional differential equations are widely used, and the resulting difference equations are dynamically consistent with their continuous counterparts. Mickens formulated a basic set of ad hoc rules for constructing nonstandard schemes for differential equations

$$u^{\cdot} \mapsto \frac{u_{k+1} - \Psi u_k}{\varphi}.$$

where Ψ and ϕ depend on the step size h and other parameters occurring in the differential equations. Functions Ψ and ϕ should satisfy the conditions

$$\Psi = 1 + O(h), \phi = h + O(h^2)$$

Unfortunately, there is no general way to determine Ψ and ϕ. Ψ is usually selected to be $\Psi = 1$, and ϕ is determined in such a way that special solutions of the differential equations have the correct stability properties.

A general choice for ϕ may be

$$\varphi_i\left(h, g_i\right) = \frac{1 - e^{-g_i h}}{g_i},$$

where g_i^0s are determined as

$$g_i = \max\left(\left|\frac{\partial f_i}{\partial u_i}\right|_{u=u_j}\right),$$

where u_j denote the coordinates of the jth fixed point of the continuous system to discretize.

Mickens non-standard discretization method effectively preserves the dynamical behavior of nonlinear ordinary differential equations. For example, the Rössler system (Letellier, 2004)

$$x^{\cdot} = -y - z,$$

$$y^{\cdot} = x + ay,$$

Using $z^{\cdot} = b + xz + cz,$

$$x \mapsto x_k, y \mapsto y_k, z \mapsto z_k, xz \mapsto x_k z_{k+1},$$

Applying Mickens' non-standard scheme with $\Psi = 1$, we obtain the model

$$x_{k+1} = x_k - \varphi_1\left(y_k + z_k\right), y_{k+1} = \left(1 + a\varphi_2\right)y_k + \varphi_2 x_k, z_{k+1} = \frac{z_k\left(1 - c\varphi_3\right) + b\varphi_3}{1 - \varphi_3 x_k},$$

where $\phi_1 = \phi_2 = \phi_3 = sin(h)$.

DIFFERENTIAL EQUATIONS WITH PIECEWISE CONSTANT ARGUMENTS

Consider again equation (3.1). Approximate the R.H.S. of equation (3.1) by

$$f\left(x\left(t\right)\right) \simeq f\left(x\left(\left[\frac{t}{r}\right]r\right)\right), t \in \left(nr, \left(n+1\right)r\right), n = 1, 2, \ldots$$

where [.] represent the greatest integer function. That is

$$\frac{dx\left(t\right)}{dt} = f\left(x\left(\left[\frac{t}{r}\right]r\right)\right), t \in \left(nr, \left(n+1\right)r\right), n = 1, 2, \ldots$$

So, we obtain

$$x_{n+1}(t) = x_n(nr) + rf(x_n(nr)),\ t \in (nr, (n+1)r),\ n = 1,2,\ldots$$

Letting $t \to (n+1)r$, we obtain the discretization of equation (3.1)

$$x_{n+1} = x_n + rf(x_n),\ n = 1,2,3,\ldots$$

Second Order Discretization

Consider the equation given by (4.1)

$$x^0(t) = f_1(x) + f_2(x).$$

We can get a second order discretization of equation (4.2) as (4.2)

$$x_{n+1} = x_n + rf(x_n) + rf(x_n -_1). \tag{4.3}$$

It is worth to mention here that equation (4.3) cannot be obtained by Euler's discretization method.

The Unified System Between Lorenz and Chen Systems

In (El-Sayed & Salman, 2013), a discretization method is proposed to discretize differential equations with piecewise constant arguments.

Consider the unified chaotic system between Lorenz and Chen

$$x^{\cdot} = \left(25\beta + 10\right)\left(y - x\right), t \in \left(0, T\right],$$

$$y^{\cdot} = \left(28 - 35\beta\right)x - xz + \left(29\beta - 1\right)y, \tag{4.4}$$

$$z^{\cdot} = xy - \frac{\beta + 8}{3}z,$$

where $\beta \in [0,1]$.

Meanwhile, consider the corresponding system with piecewise constant arguments given by

$$x\left(t\right) = \left(25\beta + 10\right)\left(y\left(r\left[\frac{t}{r}\right]\right) - x\left(r\left[\frac{t}{r}\right]\right)\right),$$

$$y\left(t\right) = \left(28 - 35\beta\right)x\left(r\left[\frac{t}{r}\right]\right) - x\left(r\left[\frac{t}{r}\right]\right)z\left(r\left[\frac{t}{r}\right]\right) + \left(29\beta - 1\right)y\left(r\left[\frac{t}{r}\right]\right), \tag{4.5}$$

$$z\left(t\right) = x\left(r\left[\frac{t}{r}\right]\right)y\left(r\left[\frac{t}{r}\right]\right) - \frac{\beta + 8}{3}z\left(r\left[\frac{t}{r}\right]\right).$$

The resulting difference equations take the form

$$x_{n+1}\left(t\right) = x_n\left(nr\right) + \left(t - nr\right)\left(\left(25\beta + 10\right)\left(y_n\left(nr\right) - x_n\left(nr\right)\right)\right), t \in \left[nr, \left(n + 1\right)r\right)$$

$$y_{n+1}\left(t\right) = y_n\left(nr\right) + \left(t - nr\right)(\left(28 - 35\beta\right)x_n\left(nr\right) - x_n\left(nr\right)z_n\left(nr\right) + +\left(29\beta - 1\right)y_n\left(nr\right)),$$

$$z_{n+1}\left(t\right) = z_n\left(nr\right) + \left(t - nr\right)\left(x_n\left(nr\right)y_n\left(nr\right) - \frac{\beta + 8}{3}z_n\left(nr\right)\right),$$

as $t \to (n+1)r$,

$$x_{n+1}\left(\left(n + 1\right)r\right) = x_n\left(nr\right) + r\left(\left(25\beta + 10\right)\left(y_n\left(nr\right) - x_n\left(nr\right)\right)\right),$$

$$y_{n+1}\left(\left(n + 1\right)r\right) = y_n\left(nr\right) + r(\left(28 - 35\beta\right)x_n\left(nr\right) - x_n\left(nr\right)z_n\left(nr\right) + +\left(29\beta - 1\right)y_n\left(nr\right)),$$

$$z_{n+1}\left(\left(n + 1\right)r\right) = z_n\left(nr\right) + r\left(x_n\left(nr\right)y_n\left(nr\right) - \frac{\beta + 8}{3}z_n\left(nr\right)\right).$$

That is

$$x_{n+1} = x_n + r\left(\left(25\beta + 10\right)\left(y_n - x_n\right)\right), t \in \left[nr, \left(n+1\right)r\right)$$

$$y_{n+1} = y_n + r\left(\left(28 - 35\beta\right)x_n - x_n z_n + \left(29\beta - 1\right)y_n\right),$$

$$z_{n+1} = z_n + r\left(x_n y_n - \frac{\beta + 8}{3} z_n\right). \tag{4.6}$$

Moreover, if we consider the equations

$$x\left(t\right) = \left(25\beta + 10\right)\left(y\left(r\left[\frac{t-r}{r}\right]\right) - x\left(r\left[\frac{t-r}{r}\right]\right)\right),$$

$$y\left(t\right) = \left(28 - 35\beta\right)x\left(r\left[\frac{t-r}{r}\right]\right) - x\left(r\left[\frac{t-r}{r}\right]\right)z\left(r\left[\frac{t-r}{r}\right]\right) + +\left(29\beta - 1\right)y\left(r\left[\frac{t-r}{r}\right]\right),$$

$$z\left(t\right) = x\left(r\left[\frac{t-r}{r}\right]\right)y\left(r\left[\frac{t-r}{r}\right]\right) - \frac{\beta + 8}{3} z\left(r\left[\frac{t-r}{r}\right]\right).$$

We can apply the same procedure to obtain the discretization which is a second order difference equation

$$x_{n+1} = x_n + r\left(\left(25\beta + 10\right)\left(y_{n-1} - x_{n-1}\right)\right), t \in \left[0, r\right)$$

$$y_{n+1} = y_n + r\left(\left(28 - 35\beta\right)x_{n-1} - x_{n-1}z_{n-1} + \left(29\beta - 1\right)y_{n-1}\right), \tag{4.7}$$

$$z_{n+1} = z_n + r\left(x_{n-1}y_{n-1} - \frac{\beta + 8}{3} z_{n-1}\right).$$

Remark 1. *It should be noticed that the system (4.6) can be obtained also by applying Euler's discretization method. However, Euler's method fails in obtaining system* (4.7).

Fixed Points and Their Asymptotic Stability

Now we study the asymptotic stability of the fixed points of the system (4.6) which has three fixed points if $(\beta + 8)(9 - 2\beta) > 0$:

fix_1=(0,0,0),

$$fix_2 = \left(\sqrt{(\beta+8)(9-2\beta)}, \sqrt{(\beta+8)(9-2\beta)}, 27-6\beta\right),$$

$$fix_3 = \left(-\sqrt{(\beta+8)(9-2\beta)}, -\sqrt{(\beta+8)(9-2\beta)}, 27-6\beta\right).$$

The last two fixed points are symmetrically placed with respect to z−axis.

By considering a Jacobian matrix for one of these fixed points and calculating their eigenvalues, we can investigate the stability of each fixed point based on the roots of the system characteristic equation (Holmgren, 1994).

Linearizing the system (4.6) about fix_1 yields the following characteristic equation

$$P(\lambda) \equiv \lambda^3 + (B + C - 1)\lambda^2 + (BC - B + A)\lambda + (AC - A - D) = 0,$$

where:

$$A = 1 + r(4\beta - 11) - r^2(25\beta + 10)(29\beta - 1)),$$

$$B = -2 - r(-54\beta - 11) + r^2(25\beta + 10)(28 - 35\beta),$$

$$C = r\frac{\beta+8}{3},$$

$$D = r^3\left(25\beta+10\right)\left(28-35\beta\right)\left(\frac{\beta+8}{3}\right) - r^2\left(25\beta+10\right)\left(28-35\beta\right).$$

Now let $a_1 = (B+C-1)$, $a_2 = (BC-B+A)$, $a_3 = (AC-A-D)$. From the Jury test, if $P(1) > 0$, $P(-1) < 0$, and $a_3 < 1$, $|b_3| > b_1$, $c_3 > |c_2|$, where

$$b_3 = 1 - a_3^2,\ b_2 = a_1 - a_3 a_2,\ b_1 = a_2 - a_3 a_1,\ c_3 = b_3^2 - b_1^2,$$

and

$c_2 = b_3 b_2 - b_1 b_2$, then the roots

of $P(\lambda)$ satisfy $\lambda < 1$ and thus fix_1 is a asymptotically stable.

While linearizing the system (4.6) about fix_2 or fix_3 yields the following characteristic equation

$$F(\lambda) \equiv \lambda^3 + a_{11}\lambda^2 + a_{22}\lambda + a_{33} = 0,$$

where:

$a_{11} = 3 - 11r + 4\beta r$, $a_{22} = 25.3\beta r^3 + 31\beta^2 r^2 + 315\beta r^2 + 717.3\beta r + 29.6r - 59.3r^2 - 7.3r^3 - 3$,

$$a_{33} = -r^2\left(25\beta + 10\right)\left(1 - 29\beta\right)(1 - r\left(\frac{\beta + 8}{3}\right) - r^3\left(25\beta + 10\right)\left(\beta + 8\right)\left(9 - 2\beta\right),$$

From the Jury test, if $F(1) > 0$, $F(-1) < 0$, and $a_{33} < 1$, $|b_{33}| > b_{11}$, $c_{33} > |c_{22}|$, where

$$b_{33} = 1 - a_{33}^2,\ b_{22} = a_{11} - a_{33}a_{22},\ b_{11} = a_{22} - a_{33}a_{11},\ c_{33} = b_{33}^2 - b_{11}^2,$$

and $c_{22} = b_{33}b_{22} - b_{11}b_{22}$, then the roots of $F(\lambda)$ satisfy $\lambda < 1$ and thus fix_2 or fix_3 is a asymptotically stable.

Attractors, Bifurcation, and Chaos

In this section we show by numerical experiments that the dynamical behavior of the dynamical systems (4.6) and (4.7) is strongly affected by the change in β. In all simulations we take $r = 0.01$

Take $\beta = -4.2$ in (4.6) (Figure 1).
Take $\beta = -2.5$ in (4.6) (Figure 2).
Take $\beta = 0$ in (4.6) (Figure 3).
Take $\beta = 0.3$ in (4.6) (Figure 4).
Take $\beta = -1.5$ in (4.7) (Figure 5). Take $\beta = -1$ in (4.7) (Figure 6).
Take $\beta = -0.9$ in (4.7) (Figure 7).
Take $\beta = -0.7$ in (4.7) (Figure 8).

Figure 1. Regular attractor of (4.6) with $\beta = -4.2$

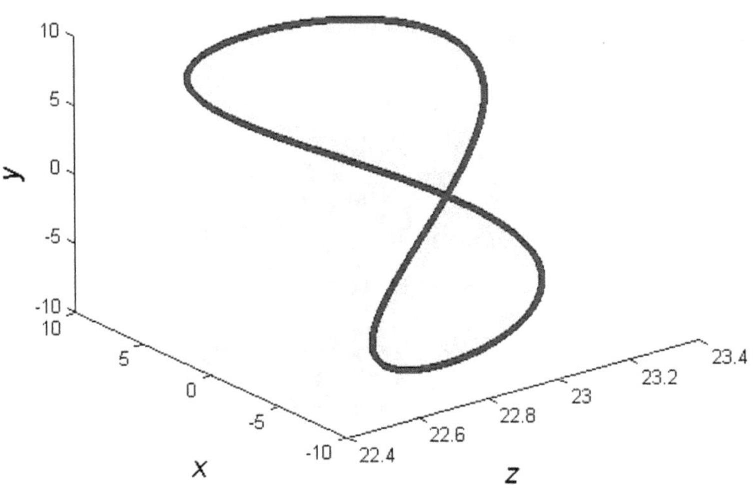

Figure 2. Regular attractor of (4.6) with $\beta = -2.5$

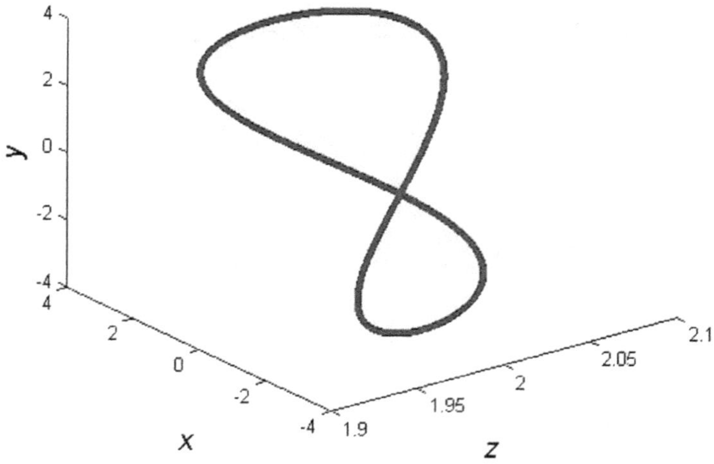

Figure 3. Chaotic attractor of (4.6) with $\beta = 0$

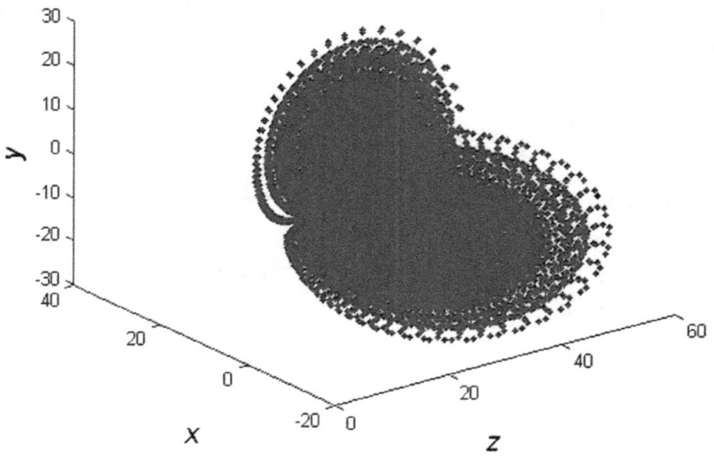

Figure 4. Chaotic attractor of (4.6) with $\beta = 0.3$

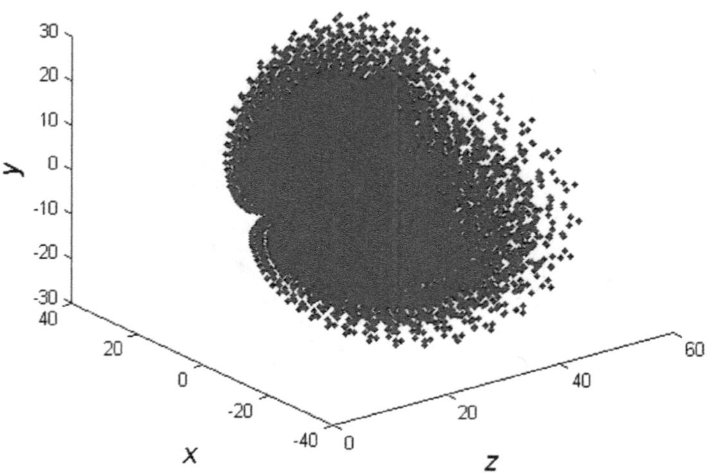

Figure 5. Regular attractor of (4.7) with $\beta = -1.5$

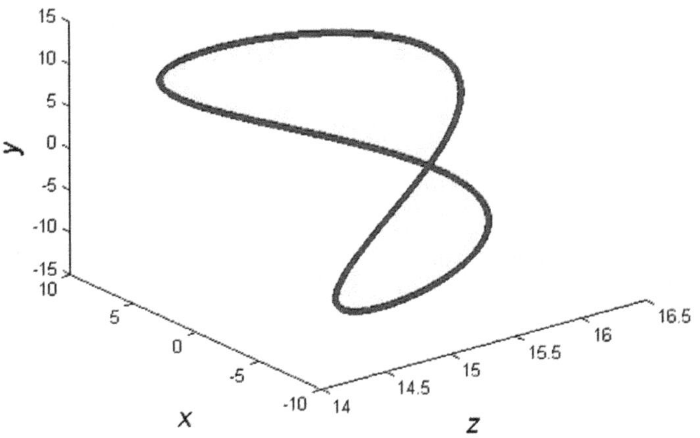

Figure 6. Chaotic attractor of (4.7) with $\beta = -1$

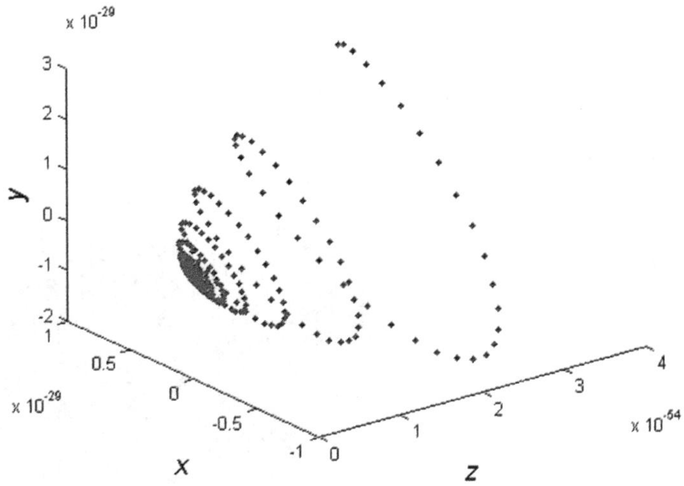

Figure 7. Chaotic attractor of (4.7) with $\beta = -0.9$

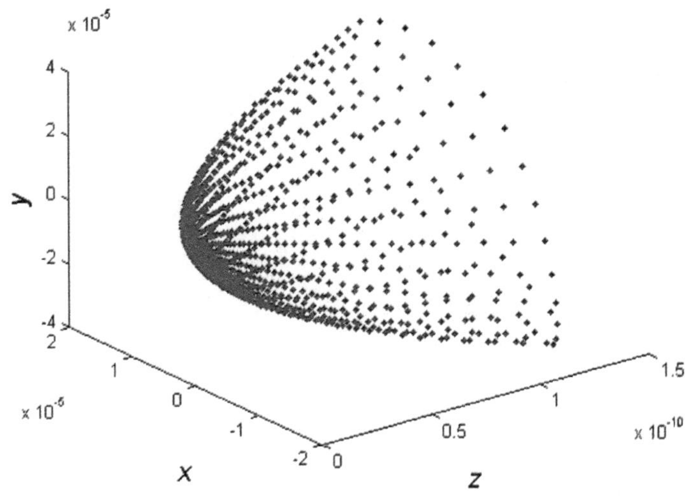

Figure 8. Chaotic attractor of (4.7) with $\beta = -0.7$

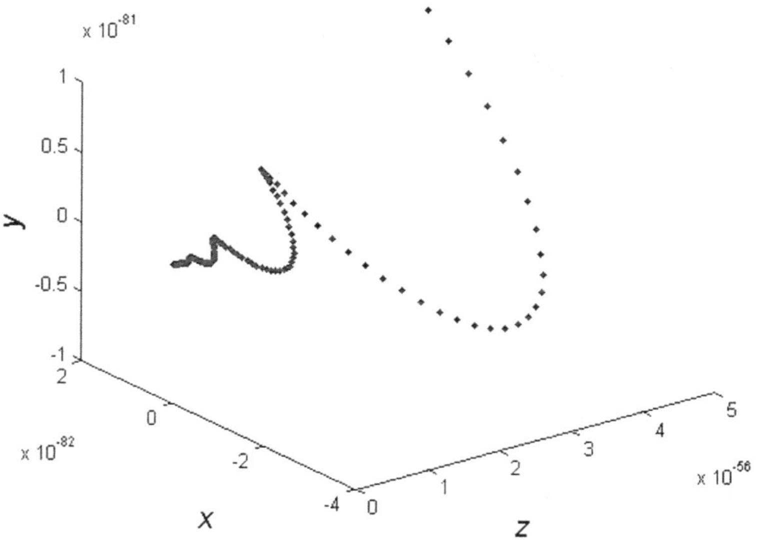

Figure 9. Bifurcation diagram of (4.6) as a function of x

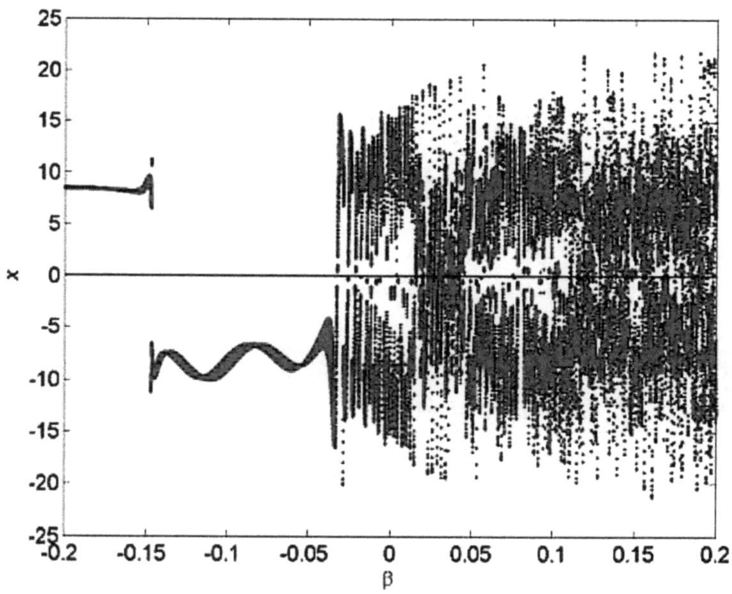

Figure 10. Bifurcation diagram of (4.6) as a function of z

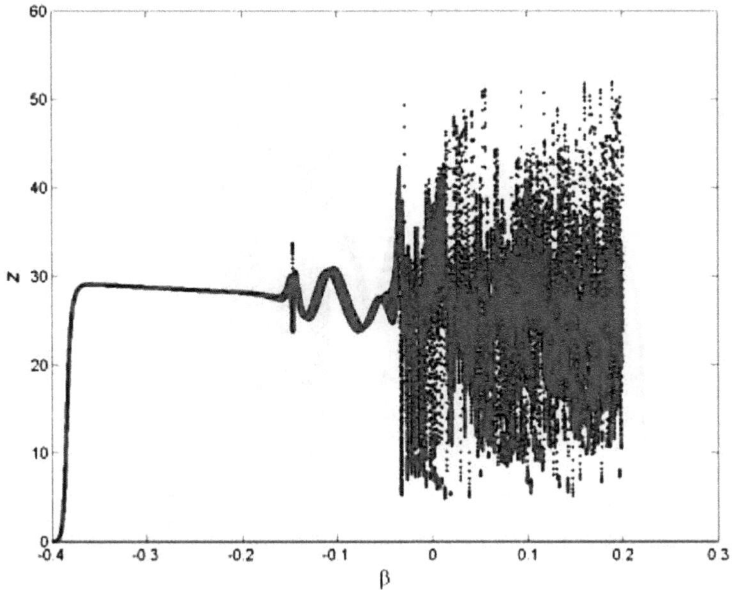

Figure 11. Bifurcation diagram of (4.7) a function of x

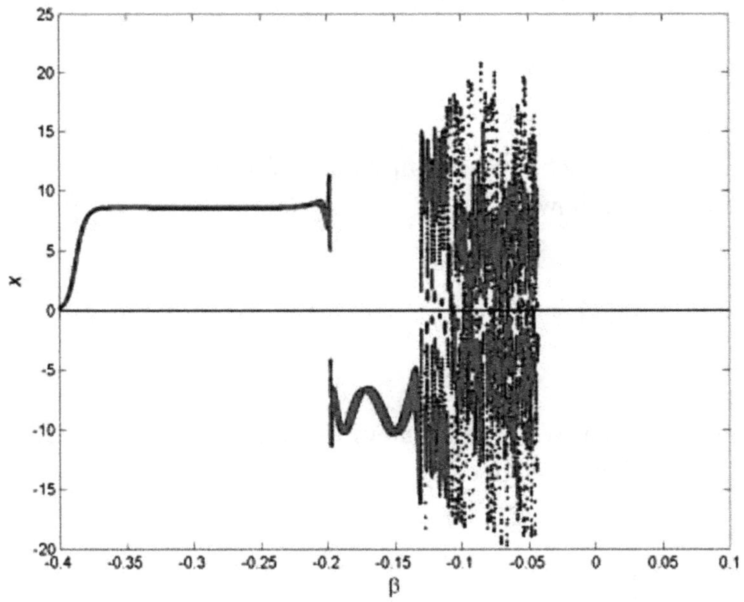

Figure 12. Bifurcation diagram of (4.7) as a function of z

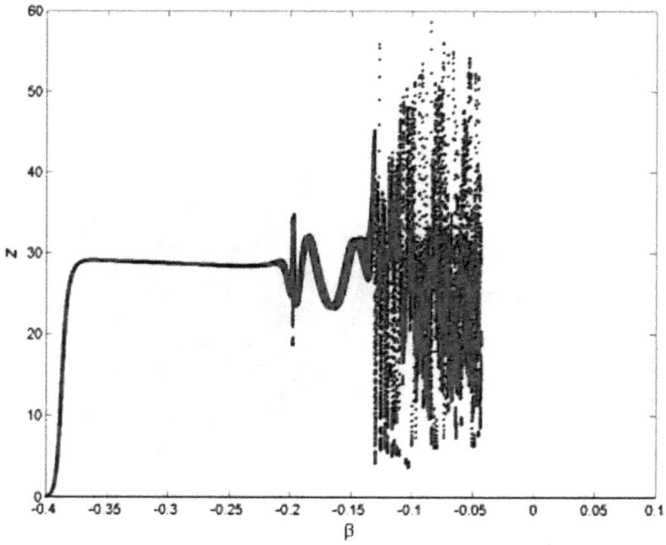

DIFFERENTIAL EQUATIONS WITH FRACTIONAL ORDERS

The Caputo derivative of order α of a function *f*(*t*) is defined as

$$D^{\alpha} = J^{n-\alpha} f^{n}\left(t\right), t > a, \alpha > 0,$$

where *n*=[α] is the first integer which is not less than α, and J^{β} is the β-order Riemann-Liouville integral operator which is described as follows

$$J^{\beta}\varphi\left(t\right) = \frac{1}{\Gamma\left(\beta\right)} \int_{a}^{t} (t-x)^{\beta-1} \varphi\left(x\right) dx,$$

where Γ(.)is the Gamma function. While the Grunwald-Letnikov definition (GL) for the fractional derivative of a continuous function *f*(*t*) is given by

$$D^{\alpha} f\left(t\right) = \lim_{h \to 0} h^{-\alpha} \sum_{j=0}^{[x]} (-1)^{-j} \begin{pmatrix} \alpha \\ j \end{pmatrix} f\left(t - jh\right),$$

where [*x*] is a truncation and $x = \dfrac{t-\alpha}{h}$.

(GL) is a generalization of the derivative. The idea behind is that *h* should approach 0 as n approaches infinity. When *J*=0,1, and α=1, we simply obtain the regular definition of differentiation.

We should point out that the discrete approximations derived from the Grunwald-Letnikov fractional derivative present some limitations. First, they frequently originate unstable numerical methods and henceforth many times a shifted Grunwald- Letnikov formula is used instead. Another disadvantage is that the order of accuracy of such approaches is never higher than one.

On the other hand, a lot of differential equations with Caputo fractional derivative were simulated by the predictor-correctors scheme, such as the fractional Chua system, the fractional Chen system, Lorenz system, and so on.

Consider the IVP

$$D_t^{\alpha} y\left(t\right) = f\left(y\right), 0 \le t \le T, y\left(0\right) = y_0, \alpha \in \left(0,1\right], \tag{1}$$

where f is in general a nonlinear function of its arguments.

The IVP (1) is equivalent to the Volterra integral equation

$$y\left(t\right) = y_0 + \frac{1}{\Gamma\left(\alpha\right)} \int_0^t (t-\tau)^{\alpha-1} f\left(y\left(\tau\right)\right) d\tau.$$

Consider the uniform grid $\{t_n = nh/n = 0,1,\ldots,N\}$ for some integer N and $h := T/N$. Let $y_h(t_n)$ denotes the approximation of $y(t_n)$. So we get by means of (1)

$$y_n\left(t_{n+1}\right) = y_0 + \frac{h^{\alpha}}{\Gamma\left(2+\alpha\right)} f\left(y_n^p\left(t_{n+1}\right)\right) + + \frac{h^{\alpha}}{\Gamma\left(2+\alpha\right)} \sum_{j=0}^{n} a_{j,n+1} f\left(y_n\left(t_j\right)\right).$$

where

$$a_{j,n+1} = \begin{cases} n^{\alpha+1} - \left(n-\alpha\right)(n+1)^{\alpha} & \text{if } j = 0, \\ (n-j+2)^{\alpha+1} + (n-j)^{\alpha+1} - 2(n-j+1)^{\alpha+1} & \text{if } 0 \le j \le n, \\ 1 & \text{if } j = n+1, \end{cases}$$

The preliminary approximation $y_n^p\left(t_{n+1}\right)$ is called predictor and is given by

$$y_n^p\left(t_{n+1}\right) = y_0 + \frac{1}{\Gamma\left(\alpha\right)} \sum_{j=0}^{n} f(y_n\left(t_j\right),$$

where

$$b_{j,n+1} = \frac{h^{\alpha}}{\alpha}((n-1-j)^{\alpha} - \left(n-j\right)^{\alpha}\right).$$

It is worth to mention here that Predictor-Corrector method is an approximation for the fractional order integration. In (El-Sayed & Salman, 2013), a simple discretization scheme is proposed to discretize fractional-order differential equations.

Fractional-Order Riccati Differential Equation

Consider the differential equation of Riccati type

$$\frac{dx}{dt} = 1 - \rho x^2\left(t\right), t \in \left(0, T\right], \tag{5.2}$$

$$x(0) = x_o. \tag{5.3}$$

Its solution is given by

$$x = \frac{1}{\sqrt{\rho}} \tanh\left(t + \tanh^{-1}\sqrt{\rho x_o}\right).$$

On the other hand, consider the corresponding equation with piecewise constant arguments given by

$$\frac{dx}{dt} = 1 - \rho x^2\left(\left[\frac{t}{r}\right]r\right), \tag{5.4}$$

with $x(0) = x_o$.

The discretized form of equations (5.2) and (5.4) is

$x_{n+1}(t) = x_n(nr) + r(1 - \rho x^2_n(nr))$, $t \in [nr,(n + 1)r)$, as $t \to (n + 1)r$, we obtain

$$x_{n+1}((n + 1)r) = x_n(nr) + r(1 - \rho x^2_n(nr)). \tag{5.5}$$

Moreover, if we consider the equation

$$\frac{dx}{dt} = 1 - \rho x\left(\left[\frac{t}{r}\right]r\right)x\left(\left[\frac{t-r}{r}\right]r\right), \tag{5.6}$$

with $x(0) = x_o$.

We can apply our procedure to obtain the discretization of the second order difference equation

$$x_{n+1}((n + 1)r) = x_n(nr) + r(1 - \rho x_n(nr)x_{(n-1)}(nr)). \text{ i.e.} \tag{5.7}$$

$x_{n+1} = x_n + r(1 - \rho x_n x_{n-1})$.

The fractional order differential equation of Riccati type is given by (5.8)

$D^\alpha x(t) = 1 - \rho x^2(t),\ t > 0,$

with the initial condition $x(0) = x_o$.

Consider the counterpart of (5.9) with piecewise constant arguments (5.9)

$$D^\alpha x(t) = 1 - \rho x^2(t), t > 0, \tag{5.10}$$

with the initial condition $x(0) = x_o$.

Applying the proposed discretization process, we get

$$x_{n+1}(t) = x_n(nr) + \frac{(t - nr)^\alpha}{\Gamma(1+\alpha)}\left(1 - \rho x_n^2(nr)\right), t \in \left[nr, (n+1)r\right).$$

Let $t \to (n + 1)r$, we obtain the discretization

$$x_{n+1}((n+1)r) = x_n(nr) + \frac{r^\alpha}{\Gamma(1+\alpha)}\left(1 - \rho x_n^2(nr)\right),$$

we get

$$x_{n+1} = x_n + \frac{r^\alpha}{\Gamma(1+\alpha)}\left(1 - \rho x_n^2\right). \tag{5.11}$$

On a similar manner, consider the corresponding equation of (5.9) with piecewise constant arguments

$$D^\alpha x(t) = 1 - \rho x\left(\left[\frac{t}{r}\right]r\right) x\left(\left[\frac{t-r}{r}\right]r\right), \tag{5.12}$$

with the initial condition $x(0)=x_o$.

So, we obtain the second order difference equation

$$x_{n+1} = x_n + \frac{r^\alpha}{\Gamma(1+\alpha)}\left(1 - \rho x_n x_{n-1}\right). \tag{5.13}$$

It is pretty clear here that our approach is an approximation for the R.H.S.

It is clear that our approach is an approximation for the R.H.S.

Fixed Points and Their Stability

The fixed points of the system (5.11), which has two fixed points namely, $\pm\frac{1}{\sqrt{\rho}}$ are given by solving the equation

$$x = x + \frac{r^{\alpha}}{\Gamma(1+\alpha)}(1-\rho x^{2}).$$

$x_{fix1} = \frac{1}{\sqrt{\rho}}$ is stable if

$$0 < \rho < \frac{(\Gamma(1+\alpha))^{2}}{r^{2\alpha}}, \tag{5.14}$$

and the second fixed point $x_{fix2} = \frac{-1}{\sqrt{\rho}}$ is unstable.

For the stability of fixed points of equation (5.13), we first split it into two equations as follows

$$y_{n+1} = x_{n} \tag{5.15}$$

$$x_{n+1} = x_{n} + \frac{r^{\alpha}}{\Gamma(1+\alpha)}(1-\rho x_{n} y_{n}). \tag{5.16}$$

This system has two fixed points namely, $(x,y)_{fix1} = \left(\frac{1}{\rho},\frac{1}{\rho}\right)$ and $(x,y)_{fix2} = \left(\frac{-1}{\rho},\frac{-1}{\rho}\right)$

If we take for instance $r = 0.2$, $\alpha = 0.85$, and $\rho = 12$, we get $\lambda_1 = 1$ and $\lambda_2 = -0.6021$. This means that this fixed point is unstable. Similarly, if we take $r = 0.3$, $\alpha = 0.95$, and $\rho = 8$, we get $\lambda_1 = 1$ and $\lambda_2 = 0.7616$. This means that this fixed point is unstable.

Bifurcation and Chaos

Numerical experiments show that the dynamic behavior of the dynamical systems (5.11) and (5.13) is affected by the change in both r and α

From the appearance of the above bifurcation diagrams we hinted that the fractional-order Riccati equation exhibits a period-doubling route to chaos.

Figure 13. Bifurcation diagram when r = 0.3, α = 0.85

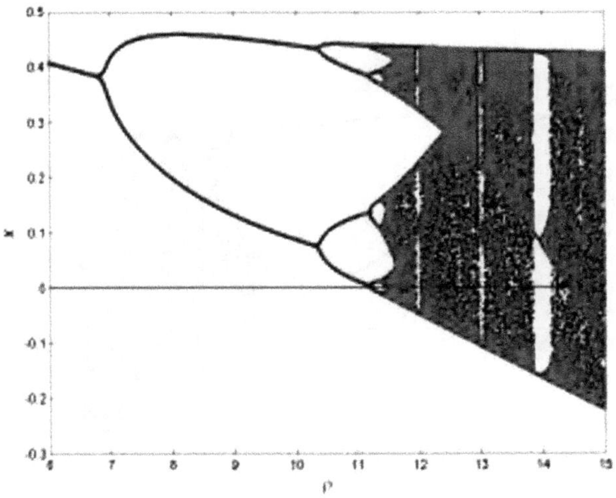

Figure 14. Bifurcation diagram of (5.11) when r = 0.3, α = 0.95

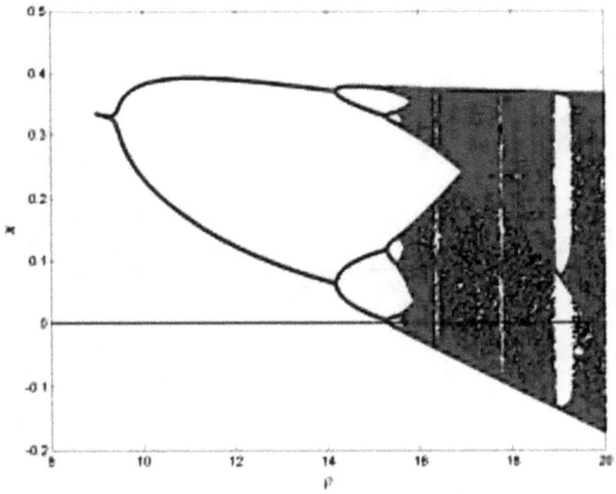

Figure 15. Bifurcation diagram of (5.11) when r = 0.4, α = 0.90

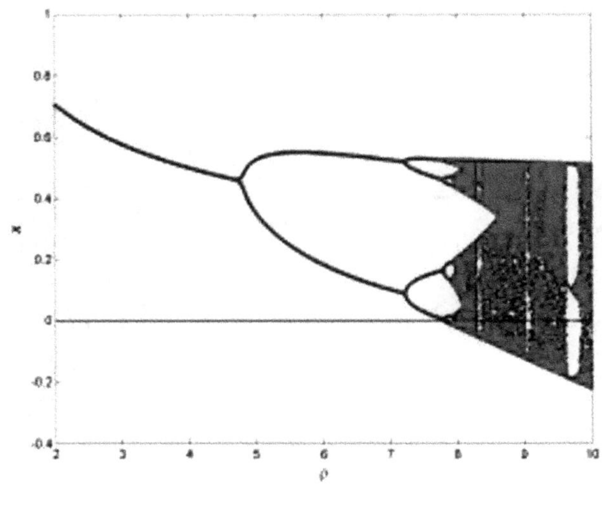

Figure 16. Bifurcation diagram of (5.11) when r = 0.5, α = 0.90

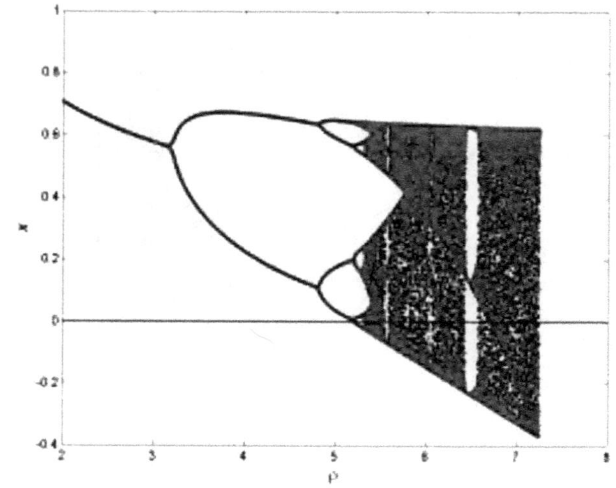

Figure 17. Bifurcation diagram of (5.13) when r = 0.2, α = 0.85

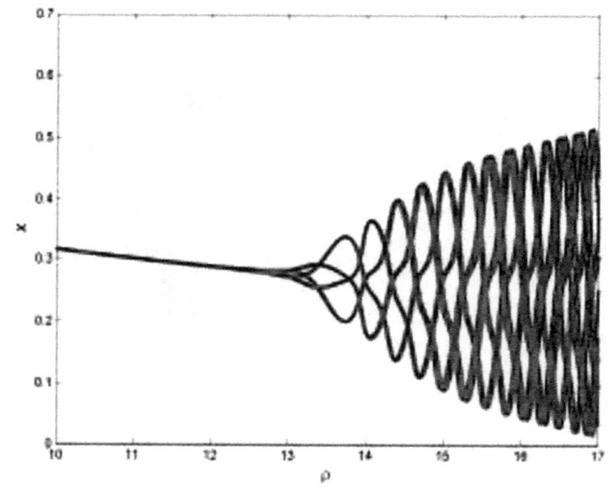

Figure 18. Bifurcation diagram of (5.13) when r = 0.2, α = 0.90

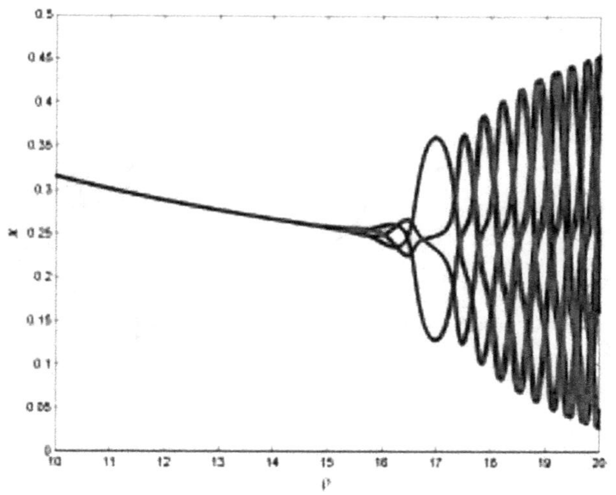

Figure 19. Bifurcation diagram of (5.13) when r = 0.4, α = 0.90

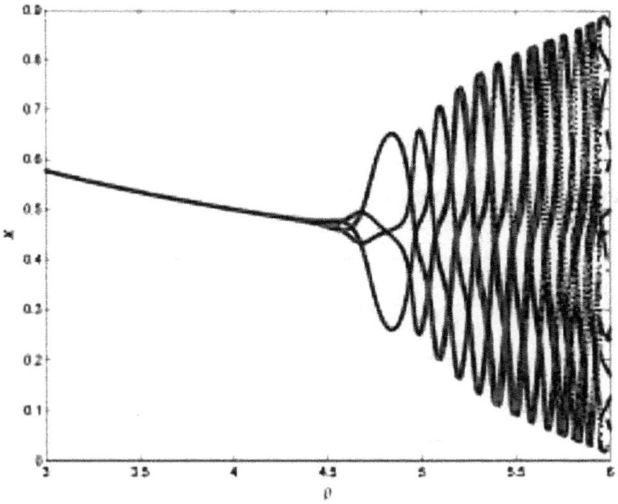

Figure 20. Bifurcation diagram of (5.13) when r = 0.3, α = 0.95

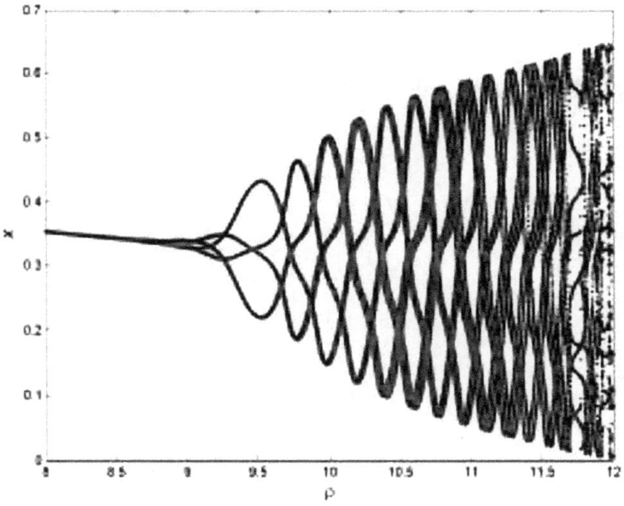

The analysis of system (12) with respect to the second fixed point $\left(\frac{-1}{\sqrt{\rho}}, \frac{-1}{\sqrt{\rho}}\right)$ tells us that a Neimark-Sacker bifurcation exists if

$$4\frac{r^{\alpha}}{\Gamma\left(1+\alpha\right)}\sqrt{\rho} < (1+\frac{r^{\alpha}}{\Gamma\left(1+\alpha\right)}\sqrt{\rho})^{2},$$

That is,

$$2\frac{r^{\alpha}}{\Gamma\left(1+\alpha\right)}\sqrt{\rho}<1+\left(\frac{r^{\alpha}}{\Gamma\left(1+\alpha\right)}\right)^{2}\rho.$$

In this case, a pair of conjugate complex roots

$$\lambda_{1,2}=\frac{\left(1+\frac{r^{\alpha}\sqrt{\rho}}{\Gamma\left(1+\alpha\right)}\right)\pm i\sqrt{4\frac{r^{\alpha}}{\Gamma\left(1+\alpha\right)}\sqrt{\rho}-\left(1+\frac{r^{\alpha}}{\Gamma\left(1+\alpha\right)}\sqrt{\rho}\right)^{2}}}{2}.$$

The eigenvalues of the characteristic equation

$$\lambda^{2}+\left(\frac{r^{\alpha}\sqrt{\rho}}{\Gamma\left(1+\alpha\right)}-1\right)\lambda-\frac{r^{\alpha}\sqrt{\rho}}{\Gamma\left(1+\alpha\right)}=0,$$

are conjugate complex λ(ρ) and $\overline{\lambda\left(\rho\right)}$. The modulus of the eigenvalues is

$$\left|\lambda\left(\rho\right)\right|=\frac{r^{\alpha}\sqrt{\rho}}{\Gamma\left(1+\alpha\right)}\sqrt{\rho}.$$

Thus |λ(ρ)|=1 if and only if

$$\rho=\rho^{*}=\left(\frac{\Gamma\left(1+\alpha\right)}{r^{\alpha}}\right)^{2}.$$

When the parameter ρ passes through such critical value of $\rho^{*}=\left(\frac{\Gamma\left(1+\alpha\right)}{r^{\alpha}}\right)^{2}$, a Neimark-Sacker bifurcation may be expected as shown in figures (17),(18),(19), and(20).

Fractional-Order Logistic Differential Equation

Consider the fractional-order Logistic differential equation given by

$$D^{\alpha}x\left(t\right)=\rho x\left(t\right)\left(1-x\left(t\right)\right),t>0, \tag{5.17}$$

with the initial condition $x(0)=x_o$.

The main purpose of this subsection is to introduce a discretization process to discretize the counterpart of (5.9) with piecewise constant arguments

$$D^{\alpha}x(t) = \rho x\left(\left[\frac{t}{r}\right]r\right)\left(1 - x\left(\left[\frac{t}{r}\right]r\right)\right), \tag{5.18}$$

with the initial condition $x(0)=x_o$. Applying the same procedure exactly as we did in the fractional-order Riccati equation we end up with

$$x_{n+1} = x_n + \frac{r^{\alpha}}{\Gamma(1+\alpha)}\rho x_n(1 - x_n). \tag{5.19}$$

On a similar manner, consider the corresponding equation of (5.17) with piecewise constant arguments

$$D^{\alpha}x(t) = \rho x\left(\left[\frac{t}{r}\right]r\right)\left(1 - x\left(\left[\frac{t-r}{r}\right]r\right)\right), \tag{5.20}$$

with the initial condition $x(0)=x_o$.

So, we obtain the second order discretization

$$x_{n+1} = x_n + \frac{r^{\alpha}}{\Gamma(1+\alpha)}\rho x_n(1 - x_{n-1}). \tag{5.21}$$

Fixed Points and Their Asymptotic Stability

Now we study the stability of the fixed points of the equation (5.19)which has two fixed points namely, 0 and 1 given by solving the equation

$$x = x + \frac{r^{\alpha}}{\Gamma(1+\alpha)}\rho x(1 - x).$$

In case of the first fixed point '0', it is stable if $\left|1 + \frac{r^{\alpha}}{\Gamma(1+\alpha)}\rho\right| < 1$ which is impossible. That is the origin is unstable. For the second fixed point '1', it is stable if

$$0<\rho<\frac{2\left(\Gamma\left(1+\alpha\right)\right)}{r^{\alpha}}. \tag{5.22}$$

On the other hand, to study the stability of the fixed points of equation (5.21) we first split it into two equations as follows

$$y_{n+1}=x_{n}\ x_{n+1}=x_{n}+\frac{r^{\alpha}}{\Gamma\left(1+\alpha\right)}\rho x_{n}\left(1-y_{n}\right). \tag{5.23}$$

This system has two fixed points namely, $(x,y)_{fix1}=\left(0,0\right)$, and $(x,y)_{fix2}=\left(1,1\right)$.

The Jacobian matrix is given by

$$J=\begin{pmatrix} 1 & 0 \\ 1+\dfrac{r^{\alpha}\rho}{\Gamma\left(1+\alpha\right)}\left(1-y\right) & -\dfrac{r^{\alpha}\rho}{\Gamma\left(1+\alpha\right)}x \end{pmatrix}$$

The eigenvalues associated to the Jacobian matrix for the first fixed point are $\lambda_1=0$, and $\lambda_2=1$, that is, this fixed point is unstable.

While the eigenvalues associated to the Jacobian matrix for the second fixed point are

$$\lambda_{1,2}=0.5\left(1-\frac{r^{\alpha}\rho}{\Gamma\left(1+\alpha\right)}\right)\pm 0.5\sqrt{\left(-1+\frac{r^{\alpha}\rho}{\Gamma\left(1+\alpha\right)}\right)^{2}+4\frac{r^{\alpha}\rho}{\Gamma\left(1+\alpha\right)}}.$$

If we take for instance r=0.2, α=0.85, and ρ=3, we get λ_1=1 and λ_2= -0.0539. This means that this fixed point is unstable.

In the next section, the numerical experiments assure our analytical results for different values for r, α, and ρ. It is worth to mention here that Lyapunov exponent for (5.19) is given by

$$Lya.\exp=\lim_{n\to\infty}\Sigma\log_{2}\left|1+\frac{r^{\alpha}}{\Gamma\left(1+\alpha\right)}\rho\left(1-2x\right)\right|.$$

When $\alpha\to 1$ the same Lyapunov exponent for the original discrete system $\left(x_{n+1}=x_{n}+r\rho x_{n}(1-x_{n})\right)$ is obtained. The following figures show the lyapunov exponent for the system (3) for different values of the fractional-order parameter α

Figure 21. Lyapunov exponent for system (5.19) with different values of the fractional-order parameter α.

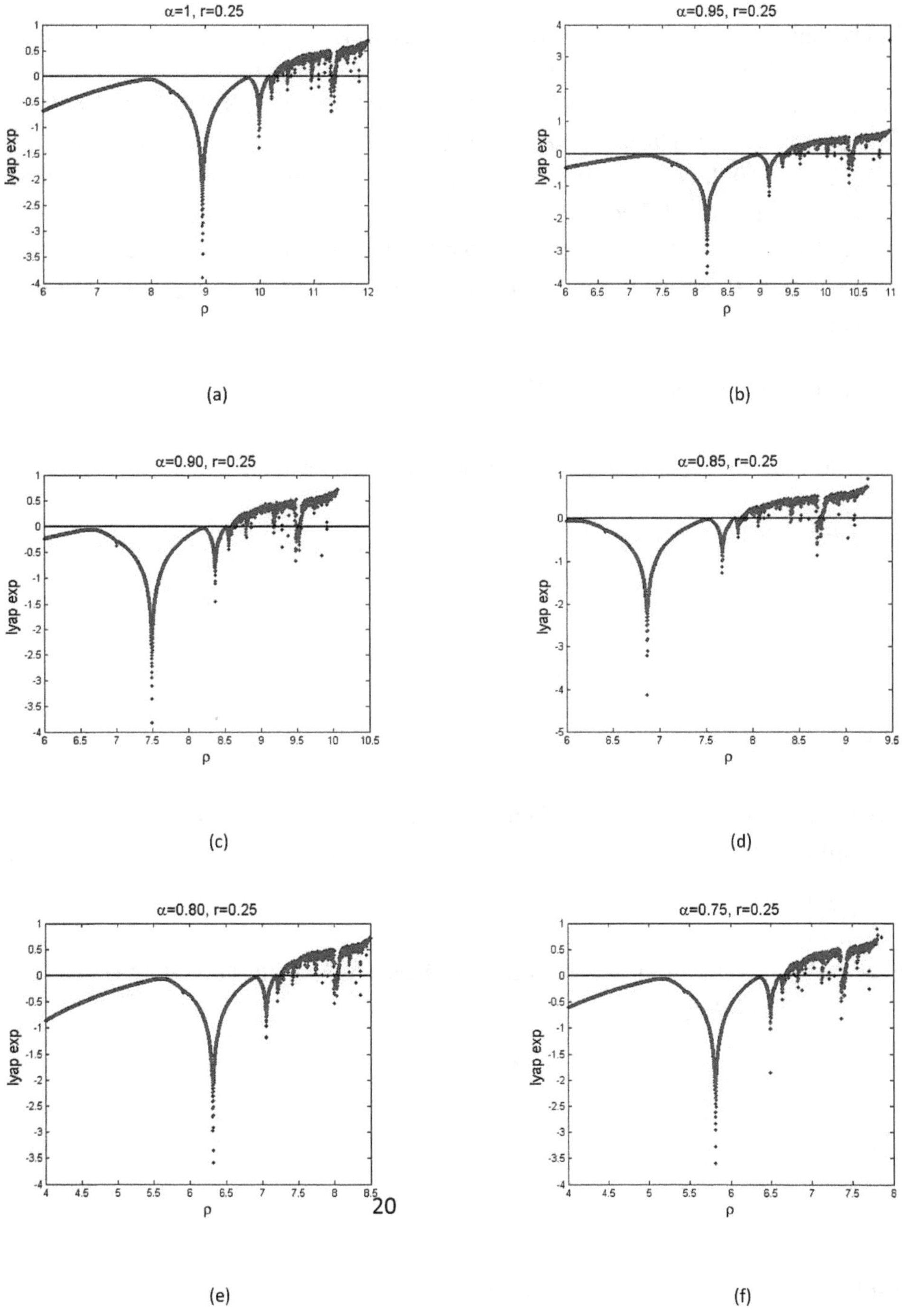

Bifurcation and Chaos

In this section we show by numerical experiments bifurcation and chaos of the dynamical system (5.23) first with respect to the parameter ρ and then with respect to the fractional-order parameter α.

Let $r = 0.25$ be fixed and vary α from 0.70 to 0.95 and ρ from 0 to 8. The initial state of the system (5.23) is $x_0 = 0.1$ and $y_0 = 0.2$. The step size for ρ is 0.001, the resulting bifurcation diagrams are shown in Figure 22 from $(a) - (f)$. It is observed from the figures that increasing the fractional-order parameter α and fixing the parameter r stabilize the chaotic system.

Now vary the fractional-order parameter α from 0.70 to 0.95 but with a fixed system parameter ρ and change the parameter r from 0.15 to 0.30, the resulting bifurcation diagrams are shown in Figure 23 from $(a) - (d)$.

Fractional-Order Nicholson Equation

Consider the fractional-order Nicholson equation given in the form

$$D^{\alpha}x(t) = -\delta x(t) + \rho x(t)e^{-ax(t)},\ t \in (0,T], \tag{5.24}$$

$$x(0) = x_0, \tag{5.25}$$

where $\alpha \in (0,1]$ and δ,ρ,a are positive constants.

Stability of Equilibrium Points

Now we study the equilibrium points and their stability of the fractional-order Nicholson equation. To calculate the equilibrium points, let $D^{\alpha}x(t) = 0$; $\Rightarrow \delta x(t) = \rho x(t)e^{-ax(t)}$. then the equilibrium points are $x = 0$ and $x = \ln\left(\frac{\rho}{\delta}\right)$

To study the stability of the equilibrium points, we have (E. Ahmed et al, 2006; E. Ahmed et al, 2007; El-Sayed et al, 2007; Podlubny, 2002)

$^{-ax}(ax - 1)$.

$f(x(t)) = -\delta - \rho e$

$f'(0) = \rho - \delta,\ f'(1\ \ln(\rho)) = -\delta \ln(\rho)$.

Now the solution of the initial value problem

$$D^{\alpha}\varepsilon(t) = f^{0}(x_{eq})\varepsilon(t) = (\rho - \delta)\varepsilon(t),\ t > 0,$$

and

$$\varepsilon(0) = x_0 - x_{eq},$$

Figure 22. Bifurcation diagram of system (5.23) as a function of ρ with different values of the fractional-order parameter α

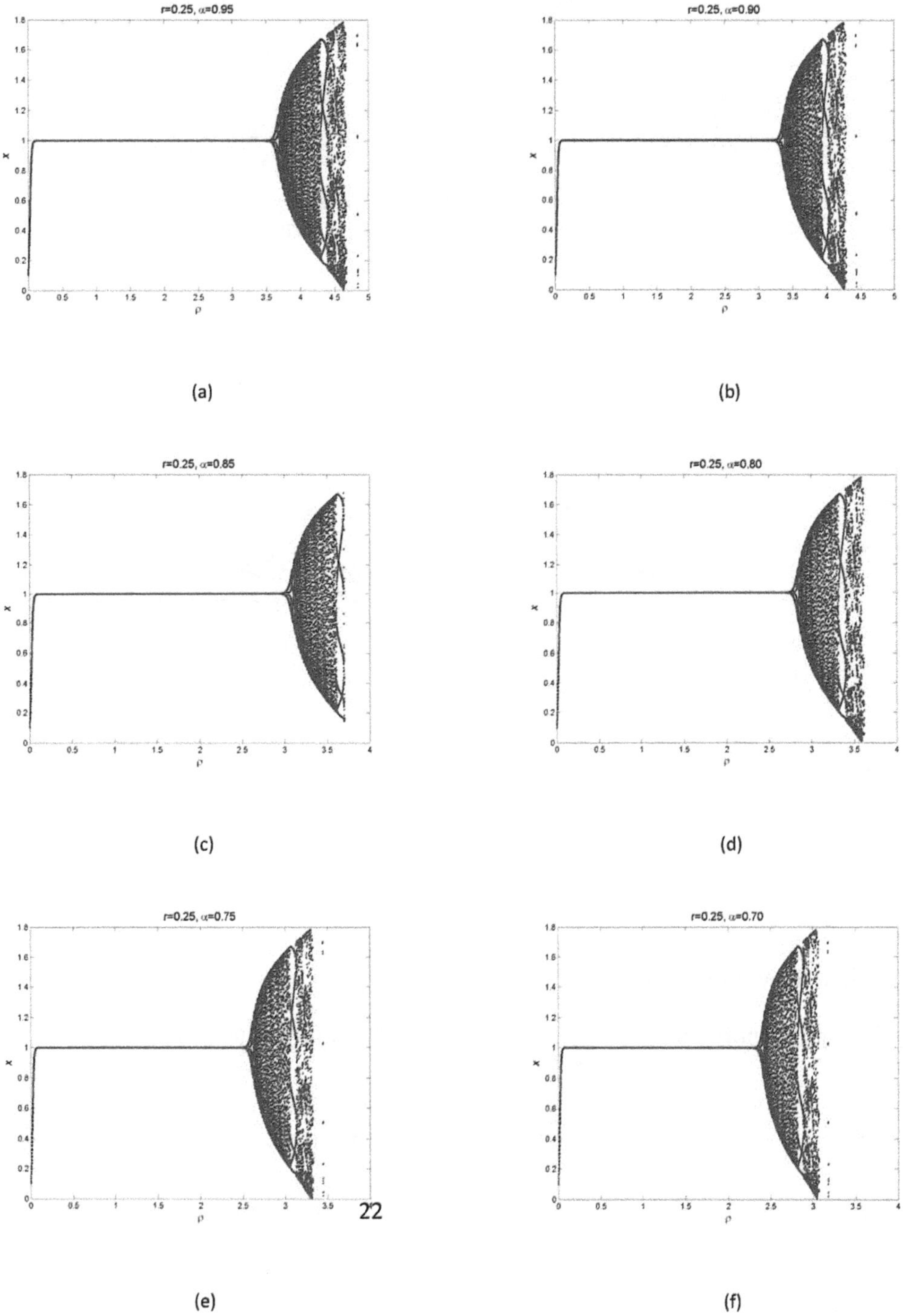

Figure 23. Bifurcation diagram of system (5.23) as a function of the fractionalorder parameter α with different values of the parameters r and ρ

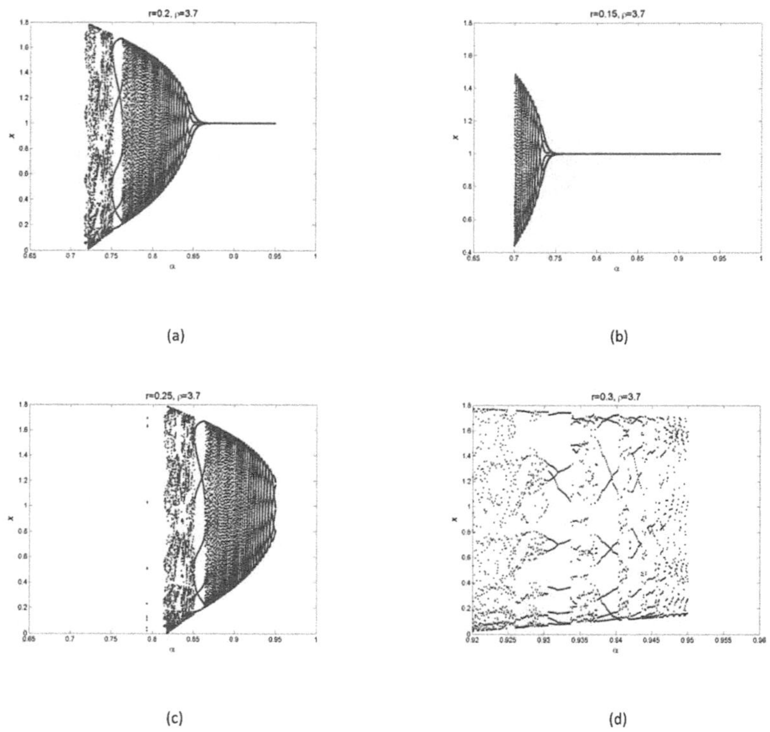

is given by (El-Sayed, 1996).

$$\varepsilon\left(t\right)=\sum_{n=0}^{\infty}\frac{\left(\rho-\delta\right)^{n}t^{\alpha n}}{\Gamma\left(\alpha n+1\right)}x_{0}. \tag{5.26}$$

and for (ρ>δ) then the equilibrium point x=0 is unstable.

Also for the equilibrium point $x=\frac{1}{a}\ln\left(\frac{\rho}{\delta}\right)$ we have the initial value problem

$$D^{\alpha}\varepsilon\left(t\right)=f'\left(x_{eq}\right)=\left(\frac{1}{a}\ln\left(\frac{\rho}{\delta}\right)\right)\epsilon\left(t\right)=-\delta\ln\left(\frac{\rho}{\delta}\right)\varepsilon\left(t\right),$$

t>0 and $\epsilon(0)=x_0-x_{eq}$.

we have the solution

$$\varepsilon\left(t\right)=\sum_{n=0}^{\infty}\frac{\left(-\delta\ln\left(\frac{\rho}{\delta}\right)\right)^{n}t^{\alpha n}}{\Gamma\left(\alpha n+1\right)}x_{0}-\ln\left(\frac{\rho}{\delta}\right)/\ a. \tag{5.27}$$

So, the equilibrium point $x=\frac{1}{a}\ln\left(\frac{\rho}{\delta}\right)$ is asymptotically stable.

Existence and Uniqueness

Let $I=\left[0,T\right],T<\infty$ and $C\left(I\right)$ be the class of all continuous functions defined on I. For the existence of a unique solution $x\in C\left[0,T\right]$ of the problem (5.30)-(5.37), we have the following theorem.

Theorem 1. *If* $\left(\frac{\rho T^{\alpha}}{\Gamma\left(1+\alpha\right)}<1\right)$, then the initial value problem(5.30)-(5.37) has a unique solution $x\in C[0,T]$.

Proof. The proof follows directly from Theorem 1 of (El-Sayed et al, 2011).

Numerical Results

An Adams-type predictor-corrector method has been introduced in (Diethelm & Freed, 1999a) and (Diethelm & Freed, 1999b) and investigated further in (Ahmed et al, 2005), (Diethelm & Ford, 2001; Diethelm, 2002; Diethelm & Ford, 2002; Diethelm & Ford, 2004; El-Sayed et al, 2014; El-Sayed & Salman, 2013; El-Sayed et al, 2013a; El-Sayed et al, 2013b; El-Sayed, 1996) and (Gorenflo & Mainardi, 1997). In this work we apply an Adams-type predictor-corrector method for the numerical solution of a fractional integral equation. The numerical solution of (5.30)-(5.37) is obtained by applying the PECE (Predict, Evaluate, Correct, Evaluate) method and is displayed in Figure 29 for different values of α and we take $\rho = 0.5$, $\delta = 2$, $a = 3$, and $x_0 = 0.85$.

Bifurcation and Chaos

In this part, bifurcation and chaos in the Fractional-order Nicholson system (5.30)(5.37) are numerically investigated, we fix $a = 3$, $\delta = 2$ and $\alpha = 0.90$. The initial state of the Fractional-order Nicholson system is taken as $x(0) = 0.9$. The resulting bifurcation diagram is shown in Figure 30.

Applying the discretization process mentioned in (Chen et al, 2003; El-Sayed et al, 2014; El-Sayed & Salman, 2013) for discretizing fractional order Nicholson equation we obtain

$$x_{n+1}=x_{n}+\frac{(r)^{\alpha}}{\Gamma\left(1+\alpha\right)}\left(-\delta x_{n}+\rho x_{n}e^{-ax_{n}}\right). \tag{5.28}$$

Figure 24. Approximate solutions

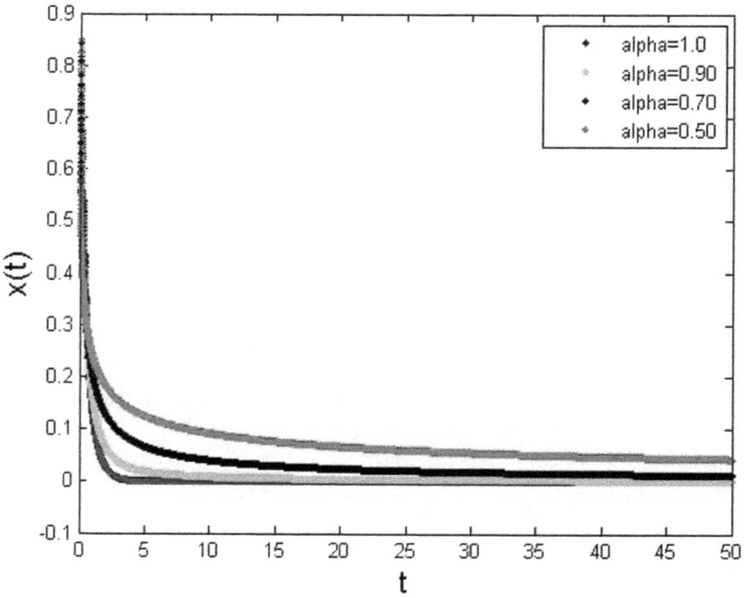

Figure 25. Bifurcation diagram of (5.30) as a function of ρ with α = 0.90.

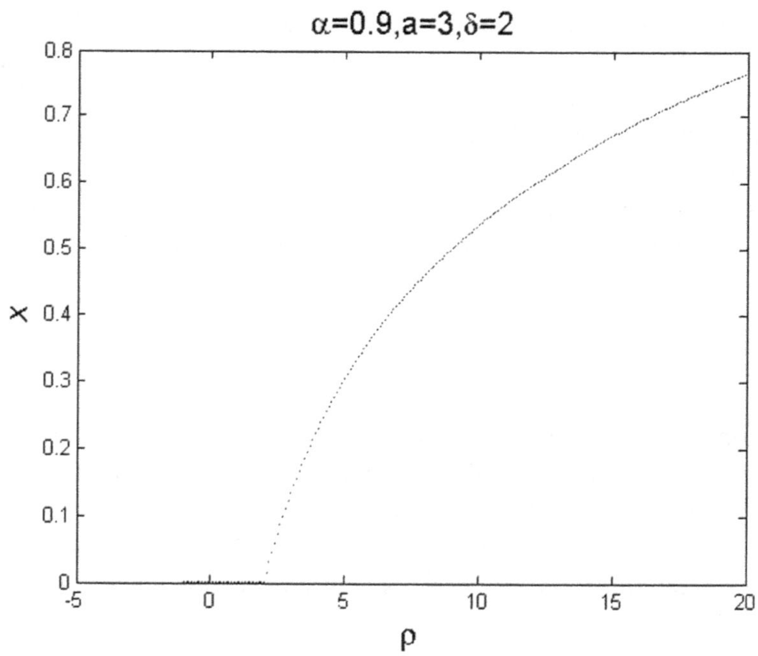

Fixed Points and Stability

Now we study the fixed points of the equation (5.35) which has two fixed points, namely

$$x = 0, x = \frac{1}{a}\ln\left(\frac{\rho}{\delta}\right),$$

which are obtained by solving the equation

$$x = x + \frac{(r)^{\alpha}}{\Gamma\left(1+\alpha\right)}\left(-\delta x + \rho x e^{-ax}\right).$$

To study the stability of these fixed points we rely on the following theorem

Theorem 2. (Elaidy, 2005) Let f be a smooth map on R, and assume that x_0 is a fixed point of f.

1. *If* $|f'(x_0)| < 1$, *then* x_0 *is stable.*
2. *If* $|f'(x_0)| > 1$, *then* x_0 *is unstable.*

In case of the first fixed point $x = 0$ *and* $\rho > \delta$ it is unstable if

$$\left|1 + \frac{(r)^{\alpha}}{\Gamma\left(1+\alpha\right)}\left(\rho - \delta\right)\right| > 1,$$

for the second fixed point $x = \frac{1}{a}\ln\left(\frac{\rho}{\delta}\right)$, it is stable if

$$0 < \rho < \delta e^{\frac{2\Gamma(1+\alpha)}{\delta r^{\alpha}}}.$$

In the next section, the numerical experiments assure our analytical results for different values for r and α. It is worth to mention here that Lyapunov exponent for (5.35) is given by (Udwadia & von Bremen, 2000).

$$Lya.\exp = \lim_{n\to\infty} \Sigma \log_2 \left|1 + \frac{r^{\alpha}}{\Gamma\left(1+\alpha\right)}\left(-\delta - \rho e^{-ax}\left(ax - 1\right)\right)\right|.$$

When ($\alpha = 1$) the same Lyapunov exponent for the original discrete system $x_{n+1} = x_n + r * (-\delta x_n + \rho x_n e^{-ax_n})$ is obtained. The following figures show the Lyapunov exponent for the system (5.35) for different values of the fractional-order parameter α.

Figure 26. Lyapunov exponent for system (5.35) with different values of the fractional-order parameter α and $r = 0.5$.

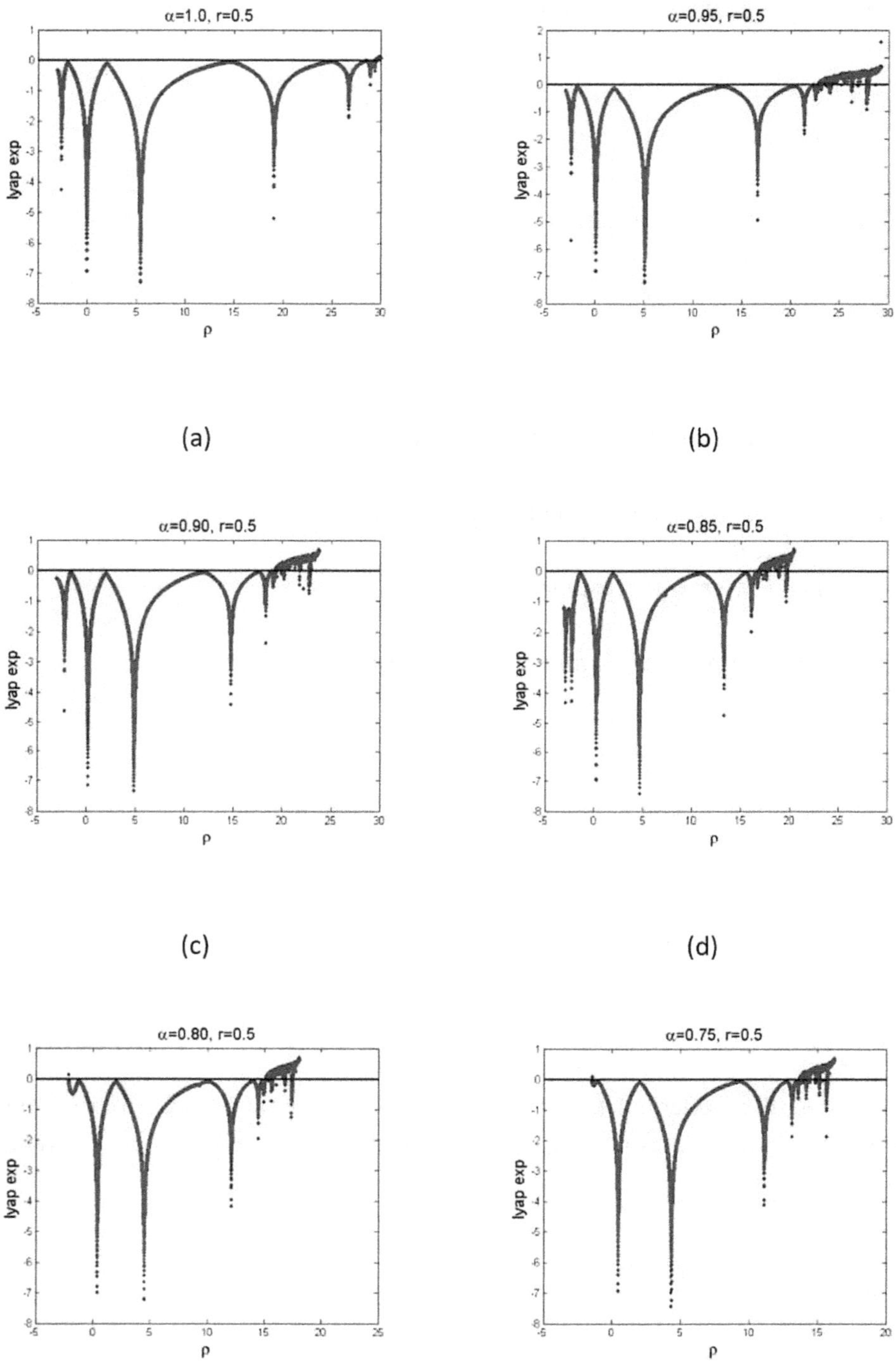

(a) (b) (c) (d) (e) (f)

Bifurcation and Chaos

In this section we study the local bifurcation of the fixed points of system (5.35). In all numerical simulations we take $\delta = 2$, $a = 3$ *and* $r = 0.50$ shows the bifurcation diagrams of system (5.35) as a function of the control parameter ρ with different values of α and $r = 0.50$. If we take, for example, $\alpha = 0.95$, and $r = 0.50$, we can see clearly in Figure 27(a) the bifurcation from a stable fixed point to a stable orbit of period two at $\rho = 13.7$, and then the bifurcation from period two to period four at $\rho \simeq 20.8$. The further period doubling occur at decreasing increments in ρ, and the orbit becomes chaotic for $\rho \simeq 22.5$.

Meanwhile, Figure 33 shows the bifurcation diagrams of system (5.35) as a function of α with different values of ρ.

Figure 27. Bifurcation diagrams for system (5.35) with different values of the fractional-order parameter α and $r = 0.5$.

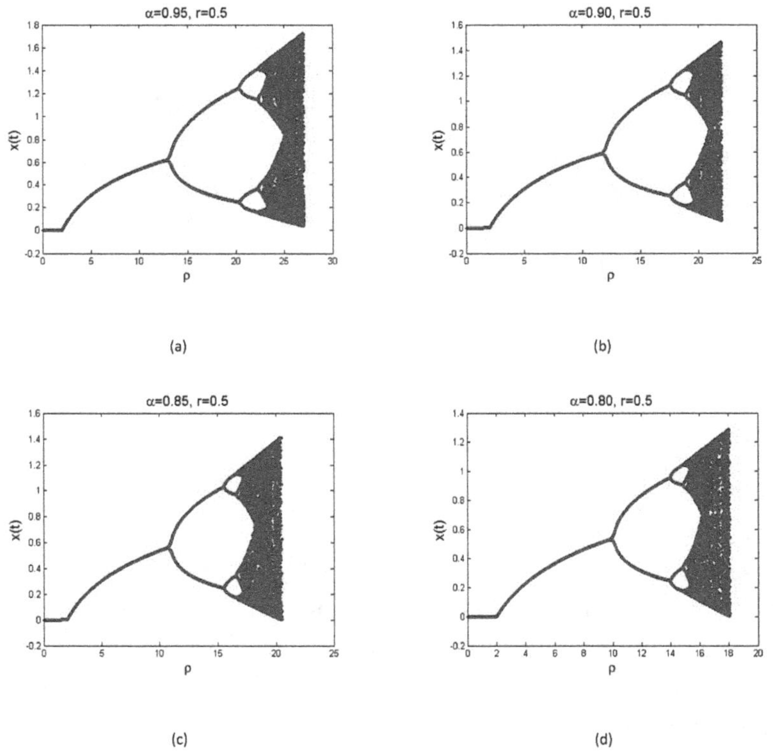

If we compare Figure 30 with Figure 27(b), we can see clearly that the discretized fractional-order Nicholson equation (5.35) has the same tendency as the original fractional-order system (5.30)-(5.37).

Fractional-Order Mackey-Glass Equation

The Mackey-Glass equation is the nonlinear time delay differential equation, which was proposed as a model of hematopoiesis, given by

Figure 28. Bifurcation diagrams for system (5.35) as a function of α with different values of the parameter ρ and $r = 0.5$.

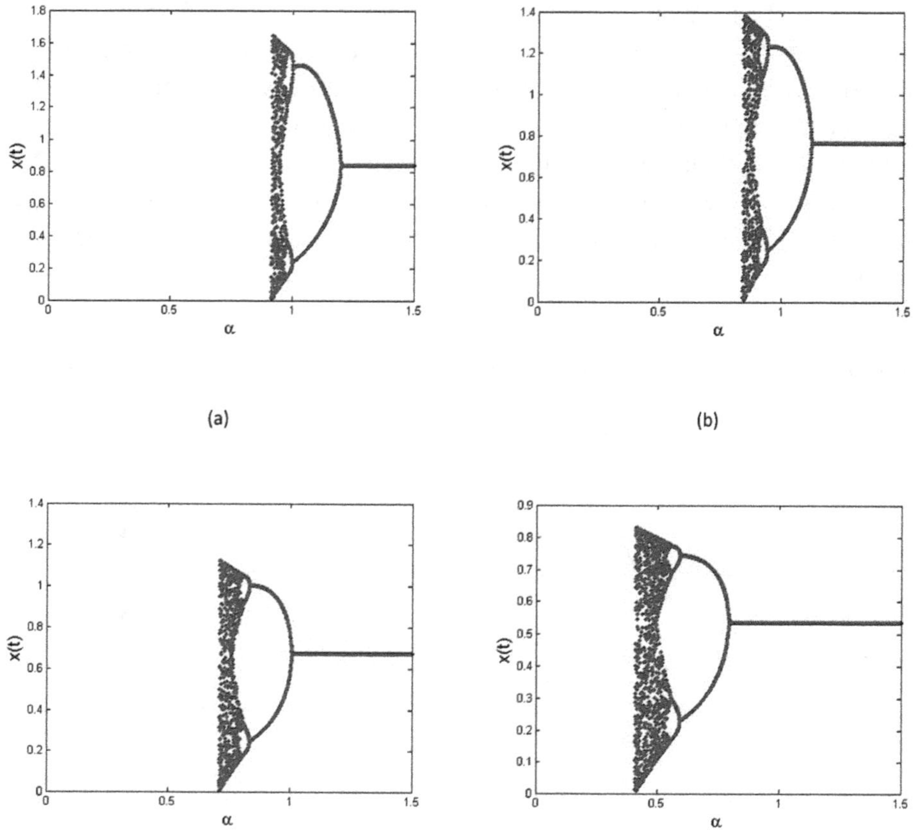

$$\frac{dx}{dt} = \frac{\rho x_\tau}{1 + x_\tau^c} - \gamma x, \gamma, c, \rho > 0 \tag{5.29}$$

where γ, c, ρ, τ are real parameters, and x_τ represents the value of the variable x at time $(t-\tau)$. Depending on the values of the parameters, this equation displays a range of periodic and chaotic dynamics.

In this work, we will show that considering a fractional-order differentiation in equation (5.29) will exhibits more complex and richer dynamics.

Consider the fractional-order Mackey-Glass equation given in the form.

$$D^\alpha x\left(t\right) = \frac{\rho x}{1 + x^c} - \gamma x, t \in \left(0, T\right] \tag{5.30}$$

with the initial condition

$$x(0) = x_0 \tag{5.31}$$

where $\alpha \in (0,1], \gamma = 1$ and $\rho, c > 0$.

Stability of Equilibrium Points

Equilibrium points of the fractional-order Mackey-Glass equation (5.30) can be easily obtained by solving the algebraic equation

$$D^{\alpha}x\left(t\right) = 0, \Rightarrow x = \frac{\rho x}{1 + x^{c}}$$

Indeed, there are only two equilibrium points of equation (5.30) namely,, and $x_1^* = 0$, and $x_2^* = \sqrt[c]{\rho - 1}$. To study the stability of these equilibrium points we calculate the first derivative of the right-hand side of equation (5.30) at each equilibrium point (Ahmed et al, 2006; Ahmed et al, 2007).

$$f'\left(x\right) = \frac{\rho + \rho x^{c}\left(1 - c\right)}{(1 + x^{c})^{2}} - 1.$$

That is, $f'\left(0\right) = \rho - 1$, and $f'\left(\sqrt[c]{\rho - 1}\right) = \frac{c\left(1 - \rho\right)}{\rho}$.

Now the solution of the initial value problem

$$D^{\alpha}\varepsilon\left(t\right) = f'\left(x_1^*\right)\varepsilon\left(t\right) = \left(\rho - 1\right)\varepsilon\left(t\right), t > 0 \text{ and } \varepsilon(0) = x_0,$$

is given by (El-Sayed & Gaafar, 2001).

$$\varepsilon(t) = \sum_{n=0}^{\infty} \frac{\left(\rho - 1\right)^{n} t^{\alpha n}}{\Gamma\left(\alpha n + 1\right)}\left(x_0 - 0\right). \tag{5.32}$$

So, the equilibrium point $x_1^* = 0$ $(if\, \rho > 1)$ is unstable. Similarly, for the equilibrium point $x_2^* = 0$ we have the initial value problem

$$\begin{aligned} D^{\alpha}\varepsilon\left(t\right) &= f'\left(x_2^*\right)\varepsilon\left(t\right) = \frac{c\left(1 - \rho\right)}{\rho}\varepsilon\left(t\right), t > 0, \\ \varepsilon\left(0\right) &= x_0 - x_2^*. \end{aligned} \tag{5.33}$$

and its solution is given by

$$\varepsilon(t)=\sum_{n=0}^{\infty}\frac{\left(\frac{c(1-\rho)}{\rho}\right)^{n}t^{\alpha n}}{\Gamma(\alpha n+1)}\left(x_{0}-\sqrt[c]{\rho-1}\right). \tag{5.34}$$

So, the equilibrium point $x_2^*=\sqrt[c]{\rho-1}$ is asymptotically stable.

Existence and Uniqueness

Let $I=[0,T]$, $T<\infty$ and $C(I)$ be the class of all continuous functions defined on I. For the existence of a unique solution $x\in C[0,T]$ of the problem (5.30)-(5.37), we have the following theorem.

Theorem 3. If $\left(\frac{\rho T^{\alpha}}{2\Gamma(1+\alpha)}<1\right)$, then the initial value problem(5.30)-(5.37) has a unique solution $x\in C[0,T]$.

Proof. The proof follows directly from Theorem 1 of (El-Sayed, 1998).

Numerical Results

An Adams-type predictor-corrector method has been introduced in (Diethelm & Freed, 1999a; Diethelm & Freed, 1999b) and investigated further in (Ahmed et al, 2005; El-Sayed, 1996; Gorenflo & Mainardi, 1997). In this work we apply an Adamstype predictor-corrector method for the numerical solution of a fractional integral equation. The numerical solution of (5.30)-(5.37) is obtained by applying the PECE (Predict, Evaluate, Correct, Evaluate) method and is displayed in Figure 29 for different values of α and we take $\rho=0.5$, $c=6$ and $x_0=0.85$.

Bifurcation and Chaos

In this part, bifurcation and chaos in the Fractional-order Mackey-Glass system (5.30)-(5.37) are numerically investigated. First, we fix $c=6$ and $\alpha=0.90$ and we let ρ varies form -2 to 30. The initial state of the Fractional-order Mackey-Glass system is taken as $x(0)=0.9$. The resulting bifurcation diagram is shown in Figure 30. Second, we fix $c=6$, and $\rho=20$ and we let the fractional-order α varies form 0 to 1. The resulting bifurcation diagram is shown in Figure 31.

Applying a discretization process mentioned in (El-Sayed et al, 2014; El-Sayed & Salman, 2014; El-Sayed et al, 2013) for discretizing fractional-order Mackey-Glass equation(5.30)-(5.37) we obtain

$$x_{n+1}=x_{n}+\frac{r^{\alpha}}{\Gamma(1+\alpha)}\left(\frac{\rho x_{n}}{1+x_{n}^{c}}-x_{n}\right). \tag{5.35}$$

Figure 29. Numerical solution of (5.30)-(5.37) with different α using PECE method.

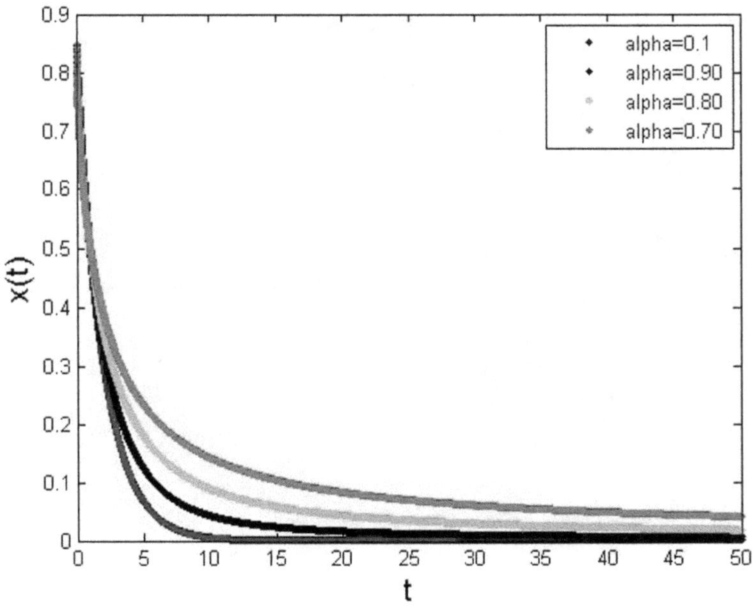

Figure 30. Bifurcation diagram of (5.30)

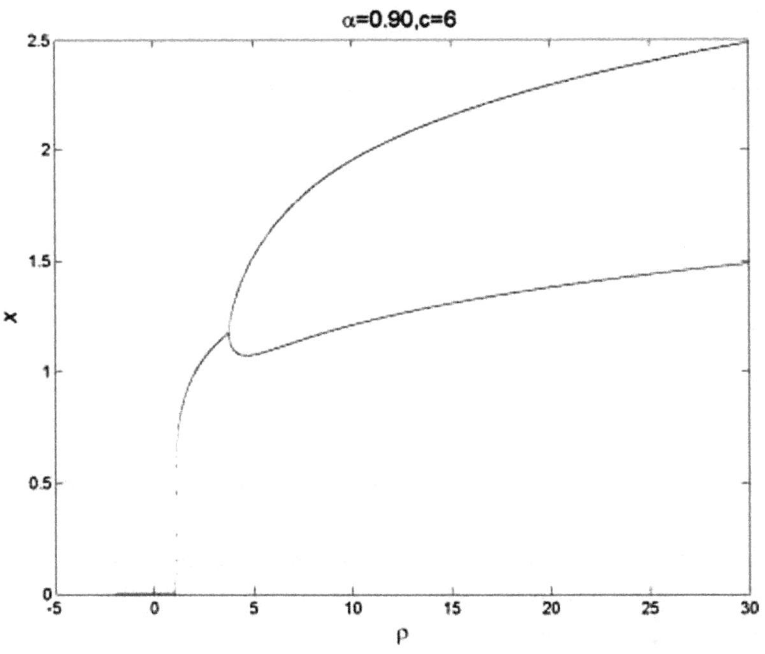

Figure 31. Bifurcation diagram of (5.30) as a function of ρ with α = 0.90. as a function of α with ρ = 20

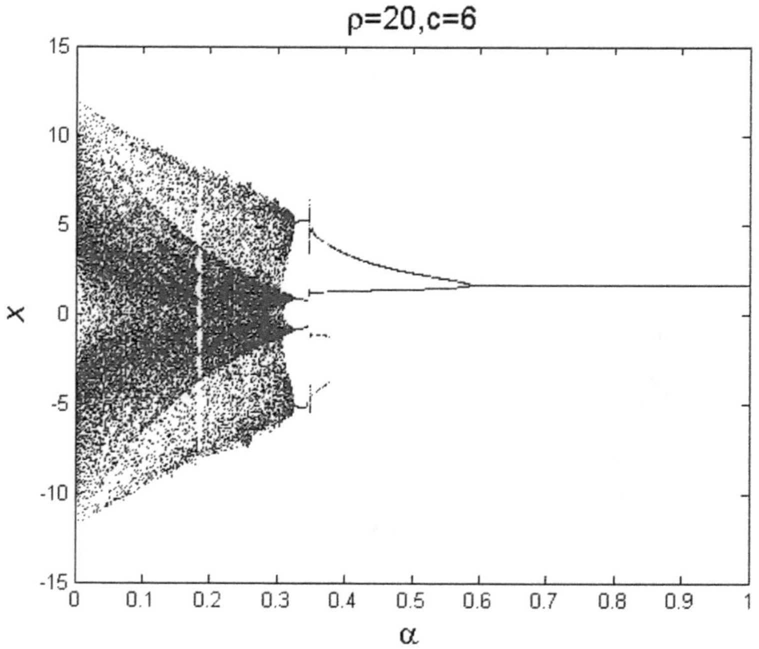

Fixed Points and Stability

Now we study the fixed points of the system (5.35) which has two fixed points, namely

$xfix1 = 0$, and $xfix2 = \sqrt[c]{\rho - 1}$,

which are obtained by solving the equation

$$x = x + \frac{r^{\alpha}}{\Gamma(1+\alpha)}\left(\frac{\rho x}{1+x^{c}} - x\right).$$

To study the stability of these fixed points we rely on the following theorem

Theorem 4. (Elaidy, 2005) Let f be a smooth map on *R*, and assume that x^* is a fixed point of *f*.

1. If $|f'(x_0)| < 1$, then x_0 is stable.
2. If $|f'(x_0)| > 1$, then x_0 is unstable.

That is, x_{fix1} is stable if

$$1 - \frac{2\Gamma(1+\alpha)}{r^{\alpha}} < \rho < 1,$$

while x_{fix2} is stable if

$$1 < \rho < \frac{r^{\alpha} c}{r^{\alpha} c - 2\Gamma\left(1+\alpha\right)}$$

Since the Lyapunov exponent (LE) is a good indicator for existence of chaos and play a key role in the study of nonlinear dynamical systems and they are a measure of the sensitivity of the solutions of a given dynamical system to small changes in the initial conditions. One feature of chaos is the sensitive dependence on initial conditions; for a chaotic dynamical system at least one LE must be positive. Since for non-chaotic systems all LEs are non-positive, the presence of a positive LE has often been used to help determine if a system is chaotic or not.

Indeed, Lyapunov exponent for (5.35) is given by (Udwadia & von Bremen, 2000)

$$Lya.\exp = \lim_{n\to\infty} \Sigma \log_2 \left|1 + \frac{r^{\alpha}}{\Gamma\left(1+\alpha\right)}\left(\frac{\rho + \rho x^{c}\left(1-c\right)}{(1+x^{c})^{2}} - 1\right)\right|.$$

When α=1, the Lyapunov exponent for the discrete system

$$x_{n+1} = x_n + r\left(\frac{\rho x_n}{1+x_n^{c}} - x_n\right),$$

is obtained. Figure 32 shows the Lyapunov exponent for the system (5.35) for different values of the fractional-order parameter α and $r = 0.50$.

Bifurcation and Chaos

In this section we study the local bifurcation of the fixed points of system (5.35). From equation (5.35),

$$f\left(x\right) = x + \frac{r^{\alpha}}{\Gamma\left(1+\alpha\right)}\left(\frac{\rho x}{1+x^{c}} - x\right).$$

So, the eigenvalues associated to f are

$$\lambda_1 = 1 + \frac{r^{\alpha}}{\Gamma\left(1+\alpha\right)}\left(\rho - 1\right),$$

and

$$\lambda_2 = 1 + \frac{r^\alpha c}{\Gamma(1+\alpha)\rho}(1-\rho).$$

When ρ=1, $\lambda_1=\lambda_2=1$, $x_{fix1} = x_{fix2} = 0$. That is, $(x_{fix}, \rho_c) = (0,1)$ is a non-hyperbolic fixed point. In fact, it is a transcritical bifurcation point of *f*. On the other hand, at

$$\rho = \frac{r^\alpha c}{r^\alpha c - 2\Gamma(1+\alpha)},\ \lambda_2 = -1 \text{ and } x_{fix2} = \sqrt[c]{\frac{2\Gamma(1+\alpha)}{r^\alpha c - 2\gamma(1+\alpha)}}.$$

That is,

$$(x_{fix}, \rho_c) = \left(\sqrt[c]{\frac{2\Gamma(1+\alpha)}{r^\alpha c - 2\Gamma(1+\alpha)}}, \frac{r^\alpha c}{r^\alpha c - 2\Gamma(1+\alpha)}\right)$$

is a non-hyperbolic fixed point. In fact, it is a period doubling (supercritical flip) bifurcation point of *f*.

In all numerical simulations we take $c = 6$ and $r = 0.50$. Figure 32 shows the bifurcation diagrams of system (5.35) as a function of the control parameter ρ with different values of α and $r = 0.50$. If we take, for example, $\alpha = 0.95$, $c = 6$, and $r = 0.50$, we can see clearly in Figure 32(a) the bifurcation from a stable fixed point to a stable orbit of period two at $\rho = 2.71$, and then the bifurcation from period two to period four at $\rho \simeq 7.8$. The further period doubling occur at decreasing increments in ρ, and the orbit becomes chaotic for $\rho \simeq 9.4$. Note the intriguing window just beyond $\rho = 13$. It is pretty clear from Figure 32 that when $\alpha \to 0$, the stability region for fixed points $x_{fix1,2}$ is being shrunk. To be clear, let $\alpha = 1$, so x_{fix1} is stable if $-2 < \rho < 1$. While if $\alpha = 0.90$, x_{fix1} is stable if $-1.7234 < \rho < 1$ and so on. The same conclusion can be said to x_{fix2}. That is, when $\alpha = 1$, x_{fix2} is stable if $1 < \rho < 3$. While if $\alpha = 0.90$, x_{fix2} is stable if $1 < \rho < 2.4890$ and so on.

Meanwhile, Figure 33 shows the bifurcation diagrams of system (5.35) as a function of α with different values of ρ.

If we compare Figure 30 with Figure 32(c), we can see clearly that the discretized fractional-order Mackey-Glass (5.35) has the same tendency as the original fractional-order system (5.30)-(5.37). Moreover, if we compare Figure 31 with Figure 33(a), we will also see the same tendency of the dynamic behavior of the two systems (5.30) and (5.35) but not with the same values of α.

Singularly Perturbed Mackey-Glass Equation

Consider the perturbed Mackey-Glass- delay differential equation

$$\epsilon \frac{dx}{dt} = \frac{\rho x_\tau}{1 + x_\tau^c} - x, \gamma = 1, \tag{5.36}$$

Figure 32. Bifurcation diagrams and Lyapunov exponent for system (5.35) with different values of the fractional-order parameter α and r = 0.5

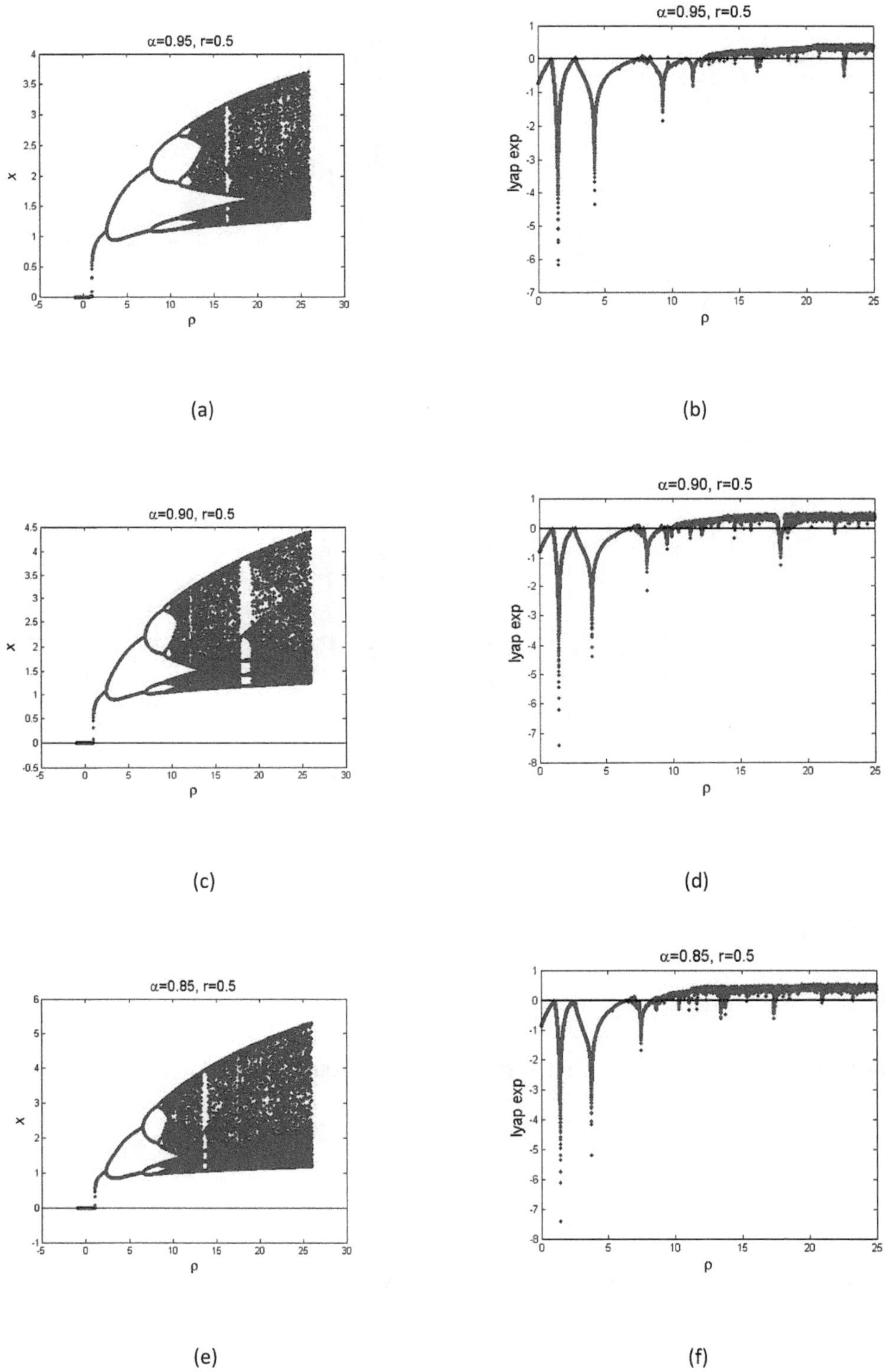

Figure 33. Bifurcation diagrams for system (5.35) as a function of α with different values of the parameter ρ and r = 0.5.

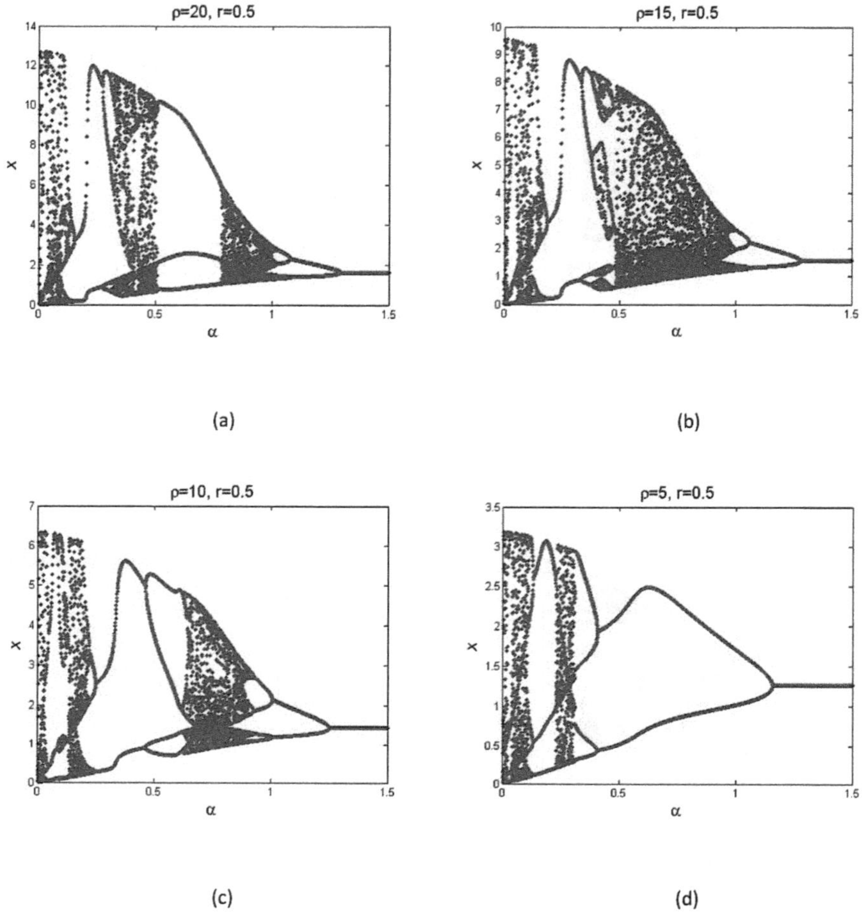

with the initial condition

$x(0)=x_0.$ (5.37)

For ϵ>0 small enough, we investigate the stability of the equilibrium points of the equation

$$\frac{dx}{dt}=\frac{\rho x_{\tau}}{\epsilon\left(1+x_{\tau}^{c}\right)}-\frac{x}{\epsilon}. \tag{5.38}$$

The system (5.38)-(5.37) has two fixed points, namely

$(x_{fix})_1=0,$ which exists for all parameters values,

$(x_{fix})_2=\sqrt[c]{\rho-1},$, which exists only if ρ>1.

To study the stability of the equilibrium points [86] and [82], we have

$$f'\left(x\left(t\right)\right) = \frac{\rho\epsilon + \rho\epsilon x^{c}\left(1-c\right)}{\left(\epsilon + \epsilon x^{c}\right)^{2}} - \frac{1}{\epsilon}.$$

That is, $f'\left(0\right) = \frac{\rho - 1}{\epsilon}$, and $f'\left(\sqrt[c]{\rho - 1}\right) = \frac{c\left(1-\rho\right)}{\rho\epsilon}$.

The characteristic equation reads

$$\lambda - \left(\left(\rho + c - c\rho\right) - \frac{1}{\epsilon}\right)e^{-\lambda\tau} = 0, \tag{5.39}$$

then the equilibrium x^{*} is asymptotically stable for $\rho > \frac{c}{\epsilon + c}$.

In what follows, we show that a Hopf bifurcation may occur at the zero fixed point. Let $i\omega$ be a root of Eq. (5.39) if and only if

$$i\omega = \left(\left(\rho + c - c\rho\right) - \frac{1}{\epsilon}\right)\left(\cos\left(\omega\tau\right) - i\sin\left(\omega\tau\right)\right),$$

Separating the real and imaginary parts, we get

$$0 = \left(\left(\rho + c - c\rho\right) - \frac{1}{\epsilon}\right)\cos\left(\omega\tau\right),$$

$$\omega = \left(\left(\rho + c - c\rho\right) - \frac{1}{\epsilon}\right)\sin\left(\omega\tau\right),$$

Squaring and adding leads to

$$\omega = \left(\left(\rho + c - c\rho\right) - \frac{1}{\epsilon}\right).$$

Let

$$\tau_{k} = \frac{1}{\left(\left(\rho + c - c\rho\right) - \frac{1}{\epsilon}\right)} \times \left[\cos^{-1}\left(\frac{-1}{\left(\left(\rho + c - c\rho\right) - \frac{1}{\epsilon}\right)}\right) + 2\pi k\right], k = 0, 1, \ldots$$

Set

$$\omega_0 = \left((\rho + c - c\rho) - \frac{1}{\epsilon} \right).$$

Let

$$\lambda_k(\tau) = \alpha_k(\tau) + i\omega_k(\tau)$$

be a root of Eq. (5.39) near $\tau = \tau_k$ satisfying

$$\alpha_k(\tau_k) = 0, and\, \omega_k(\tau_k) = 0.$$

We have the following result.

Lemma 1. $\alpha'_k(\tau_k) > 0$.

Proof. Differentiating both sides of Eq.(5.39) with respect to τ, it follows that

$$\frac{d\lambda}{d\tau} = -\left((\rho + c - c\rho) - \frac{1}{\epsilon} \right) e^{-\lambda\tau} \left(\lambda + \tau \frac{d\lambda}{d\tau} \right).$$

That is

$$\frac{d\lambda}{d\tau} = \frac{-\left((\rho + c - c\rho) - \frac{1}{\epsilon} \right) e^{-\lambda\tau} \lambda}{1 + \tau \left(\left((\rho + c - c\rho) - \frac{1}{\epsilon} \right) e^{-\lambda\tau} \right)}.$$

Thus, using Eq. (5.39) we have

$$\frac{d\lambda}{d\tau} = \frac{-\lambda^2}{1 + \tau\lambda},$$

and hence

$$\frac{d\lambda}{d\tau}\Big|_{\tau=\tau_k} = \frac{-\lambda^2}{1 + \tau_k \lambda}.$$

This implies that

$$\alpha'(k) > 0.$$

Lemma 1 implies that the root $\lambda_k(\tau)$ of characteristic equation (5.39) near τ_k crosses the imaginary axis from the left to the right as τ continuously varies from a number less than τ_k to one greater than τ_k by Rouch´e's theorem [87]. This means that Eq. (5.29) undergoes a Hopf bifurcation at the zero fixed point when $\tau = \tau_k$.

The Discretized System

In this section, the discretized analogue of the system (5.38) is obtained via the method of steps as follows (Driver, 1977).

By applying the method of steps then the equation

$$\frac{dx}{dt} = \frac{\rho x_\tau}{(\epsilon)(1 + x_\tau^c)} - \frac{x}{\epsilon}$$

has the solution as follows

1. Let $t \in [0,\tau]$, then

$$x_1(t) = e^{\frac{-1}{\epsilon}t} x_0 + \frac{1}{\epsilon}\int_0^t e^{\frac{-1}{\epsilon}(t-s)} f(x_0)\,ds,$$
$$= e^{\frac{-1}{\epsilon}t} x_0 + f(x_0)\left(1 - e^{\frac{-1}{\epsilon}(t)}\right).$$

Let $(t \to \tau)$, then

$$x_1(\tau) = e^{\frac{-1}{\epsilon}\tau} x_0 + f(x_0)\left(1 - e^{\frac{-1}{\epsilon}(\tau)}\right).$$

2. For $t \in [\tau, 2\tau]$, we have

$$x_2(t) = e^{\frac{-1}{\epsilon}t} x_1 + \frac{1}{\epsilon}\int_{2\tau}^t e^{\frac{-1}{\epsilon}(t-s)} f(x_1)\,ds,$$
$$= e^{\frac{-1}{\epsilon}t} x_1 + f(x_1)\left(1 - e^{\frac{-1}{\epsilon}(t)}\right).$$

Let ($t \to \tau$), then

$$x_2(\tau) = e^{\frac{-1}{\epsilon}\tau} x_1 + f(x_1)\left(1 - e^{\frac{-1}{\epsilon}(\tau)}\right).$$

Repeating the process we can easily deduce that the solution of (5.38)-(5.37) is given by

$$x_{n+1}(t) = e^{\frac{-1}{\epsilon}t} x_n + \frac{1}{\epsilon}\int_{n\tau}^{t} e^{\frac{-1}{\epsilon}(t-s)} f(x_n)\, ds,$$

By letting $t \to (n+1)\tau$, we obtain

$$x_{n+1} = e^{\frac{-1}{\epsilon}n\tau} x_{n\tau} + \left(\frac{\rho x_n}{1 + x_n^c}\right)\left(1 - e^{\frac{-1}{\epsilon}(\tau)}\right). \tag{5.40}$$

Local Stability of Equilibrium Points of the Discretized System

The system (5.40)-(5.37) has two fixed points which are the same as the system (5.38), namely

$(x_{fix})_1 = 0,$

$(x_{fix})_2 = \sqrt[c]{\rho - 1}.$

To study the local stability of these fixed points we rely on the following theorem

Theorem 1. (Elaidy, 2005) Let f be a smooth map on R, and assume that x_0 is a fixed point of f.

1. If $|f'(x_0)| < 1$, then x_0 is stable,
2. If $|f'(x_0)| > 1$, then x_0 is unstable.

Now we have

$$f'(x) = e^{\frac{-1}{\epsilon}} + \left(1 - e^{\frac{-1}{\epsilon}}\right)\left(\frac{\rho + \rho x^c(1-c)}{(1+x^c)^2}\right).$$

Then $(x_{fixed})_1 = 0$ is stable if

$$\left| e^{\frac{-1}{\epsilon}} + \left(1 - e^{\frac{-1}{\epsilon}}\right) \rho \right| < 1,$$

that is, $\rho < 1$.

The second fixed point $(x_{fixed})_2 = \sqrt[c]{\rho - 1}$ is stable if

$$\left| e^{\frac{-1}{\epsilon}} + \left(1 - e^{\frac{-1}{\epsilon}}\right)\left(\frac{\rho - c\rho + c}{\rho}\right) \right| < 1,$$

that is $\rho > 1$.

Mackey-Glass Map in Case $\epsilon \to 0$

In this section we consider the problem of the singularly perturbed differential equation (5.36) and (5.37).

The limiting cases as $\epsilon \to 0$ are special cases of the retarded functional equation

$$x(t) = \frac{\rho x(t - \tau)}{1 + x^c(t - \tau)} t \in (0, T], \tag{5.41}$$

with the initial condition (5.37).

Now we apply the method of steps to the system (5.41) and (5.37).

Step 1: Let $t \in [0,\tau]$, then $t-\tau \in [-\tau,0]$, the solution of (5.41)-(5.37) is given by

$$x(t) = x_1(\tau) = \frac{\rho x_0}{1 + x_0^c}, t \in [0, \tau].$$

Step 2: For $t \in [\tau,2\tau]$, then $t-\tau \in [0,\tau]$, the solution of (5.41)-(5.37) is given by

$$x(t) = x_2(t) = \frac{\rho x_1}{1 + x_1^c}, t \in [\tau, 2\tau].$$

Step n: Repeating the process we can easily deduce that the solution of (5.41)-(5.37) is given by

$$x_{n+1} = \frac{\rho x_n}{1 + x_n^c}, t \in [n\tau, (n + 1)\tau]. \tag{5.42}$$

Fixed Points and Stability of System (5.42)

Exactly like its discrete counterpart, the dynamical system (5.42)-(5.37) has two fixed points which are the solution of the equation

$$x = \frac{\rho x_{\tau}}{1 + x_{\tau}^{c}}.$$

So, we have

$$(x_{fix})_1 = 0,$$

$$(x_{fix})_2 = \sqrt[c]{\rho - 1}.$$

Now since in our case

$$f'(x) = \frac{\rho + \rho x^{c}(1-c)}{(1+x^{c})^{2}},$$

the first fixed point $(x_{fixed})_1 = 0$ is stable if $|\rho| < 1$, that is, $\rho < 1$.

While the second fixed point $(x_{fixed})_2 = \sqrt[c]{\rho - 1}$ is stable if

$$\left|\frac{\rho + c(1-\rho)}{\rho}\right| < 1.$$

So, the second fixed point is stable if $\rho > 1$.

Numerical Results

In this section, we perform some numerical simulations to confirm theoretical analysis obtained. Figure 34 shows how one can branch off at a period doubling to track the period-doubling orbit for Eq. (5.29). A Hopf bifurcation occurs near $\tau = 4.703$, and a sequence of period doubling bifurcations at $\tau = 13.05$, $\tau = 15.78$, and $\tau = 16.31$. Presumably, this is the beginning of a period doubling cascade to chaos. While Figure 35 shows Hopf, first and second period doubling bifurcation of Eq. (5.29) when $\gamma = 1$ and $c = 10$.

Lyapunov exponent is an important method in the stability of system of nonlinear differential equations and it is considered as a good indicator for existence of chaos. The presence of positive Lyapunov exponent used to determine if a system is chaotic or not (Udwadia & von Bremen, 2000).

Figure (36) shows the Lyapunov characteristic exponents (LCEs) and bifurcation diagram for system (5.42)-(5.37). We calculate LCE to indicate chaos. It is clearly seen in this figure the bifurcation from a stable equilibrium point to a stable orbit of period 2 at $\rho = 8$, and then the bifurcation from period two to period four at ρ between 10 and 12. The further period doubling occur at decreasing increments in ρ, and the orbit becomes chaotic for $\rho \simeq 14.5$. The same can be said to Figure 38 which shows the LCEs and bifurcation diagram for system (5.40)-(5.37). We have noticed that in Figure 37 the behavior of the equation MacKey- glass depending on the values of c, as increasing c, shrink the stability region of equilibrium points.

As it can be seen form Figure 36 and Figure 38 there is no difference in the dynamics of Eq. (5.29) and Eq. (5.42) when fixing both the initial condition and the parameter *c*. That is to say, the two equations behave similarly in the long-term behavior. However, when we look at Figure 37 as we change *c*, the dynamics will be different in the limiting case.

Figure 34. Bifurcation diagram for Mackey-glass equation (5.29)

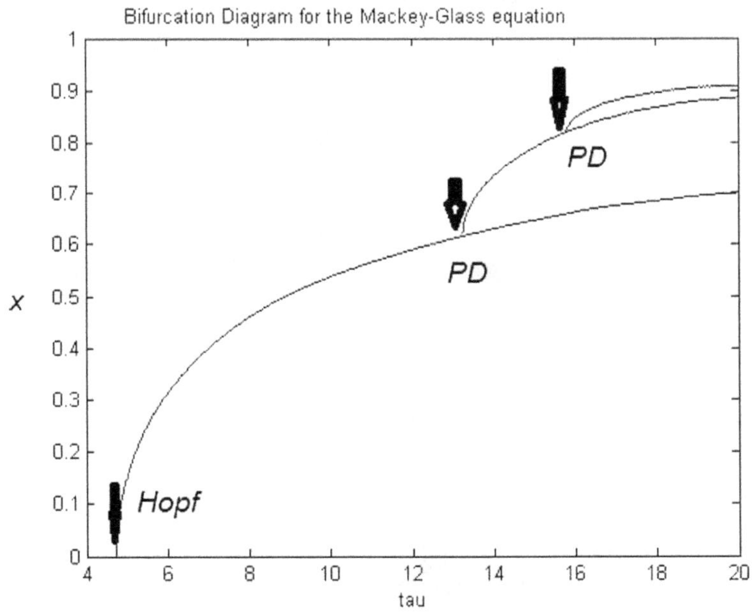

Figure 35. Behavior for Mackey-glass equation (5.29)

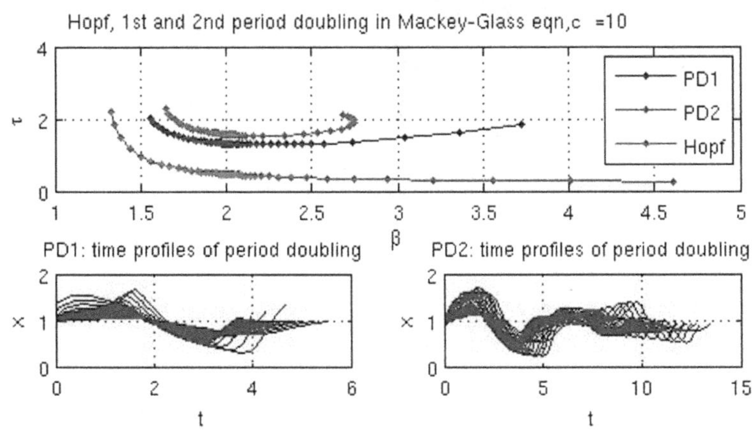

Figure 36. Bifurcation diagram and Lyapunov exponent for system (5.42)

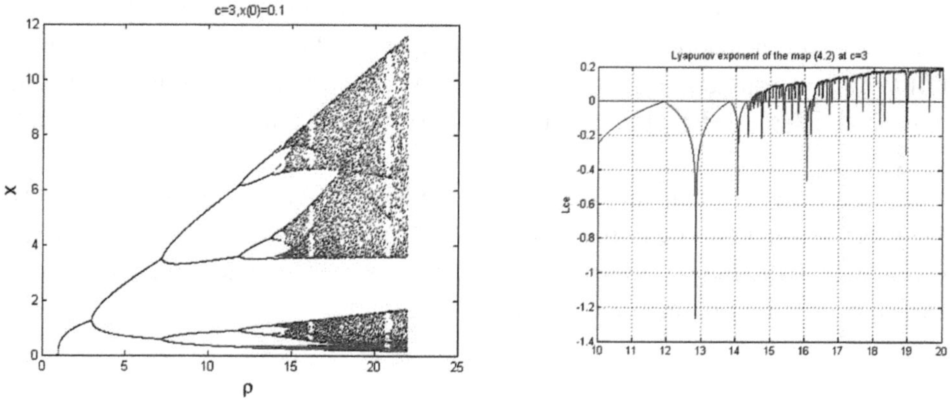

Figure 37. Bifurcation diagram for system (5.42).

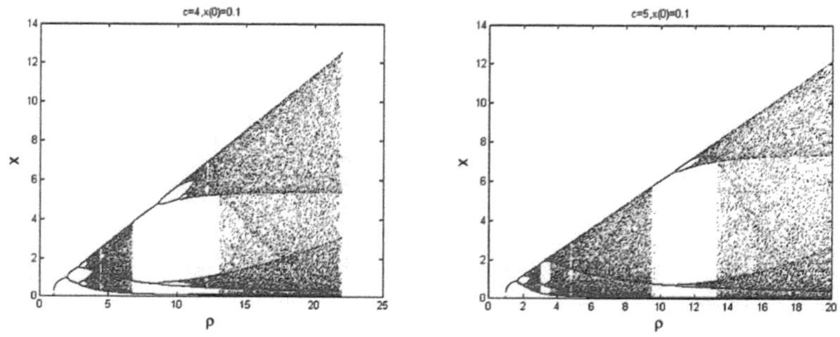

Figure 38. Bifurcation diagram and Lyapunov exponent for system (5.40).

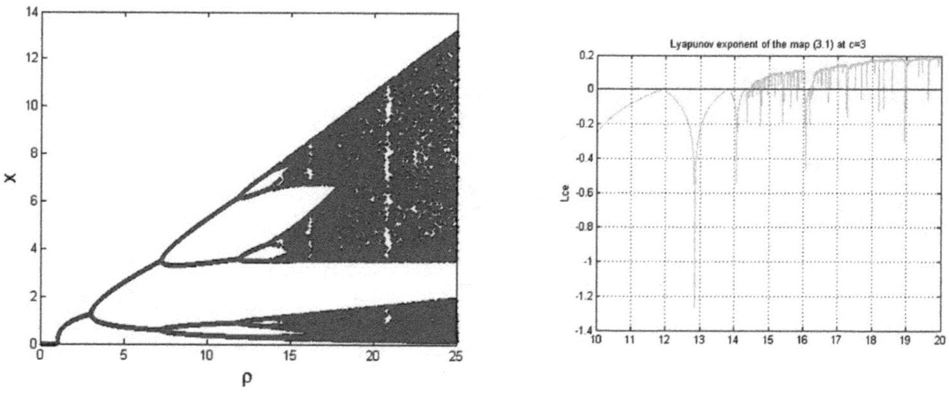

REFERENCES

Abbas, S. (2010). Pseudo almost automorphic solutions of fractional order neutral differential equations. *Semigroup Forum, 81*(3), 393–404. doi:10.100700233-010-9227-0

Agarwal, R. P., Benchohra, M., & Hamani, S. (2009). Boundary value problems for fractional differential equations. *Georgian Mathematical Journal, 16*(3), 401–411.

Aguirregabiria, J. M., & Etxebarria, J. R. (1987). Fractal basin boundaries of a delay differential equation. *Physics Letters. [Part A], 122*(5), 241–244. doi:10.1016/0375-9601(87)90814-0

Ahmed, C., El-Sayed, A. M. A., & El-Saka, H. A. A. (2006). On some RouthHurwitz conditions for fractional order differential equations and their applications in Lorenz, Rossler, Chua and Chen systems. *Physics Letters. [Part A], 358*(1), 1–4. doi:10.1016/j.physleta.2006.04.087

Ahmed, E., El-Sayed, A. M. A., El-Mesiry, E. M., & El-Saka, H. A. A. (2005). *Numerical solution for the fractional replicator equation. Internat. Modern Phys.* doi:10.1142/S0129183105007698

Ahmed, E., El-Sayed, A. M. A., & El-Saka, H. A. A. (2007). Equilibrium points, stability and numerical solutions of fractional-order predator-prey and rabies models. *Journal of Mathematical Analysis and Applications, 325*(1), 542–553. doi:10.1016/j.jmaa.2006.01.087

an der Heiden, U. (1979). Delays in physiological systems. *Journal of Mathematical Biology, 8*(4), 345–364. doi:10.1007/BF00275831 PMID:541577

Chen, Y. Q., Vinagre, B. M., & Podlubny, I. (2003). A new discretization method for fractional-order differentiators via continued fraction expansion. *Proceedings of DETC, 3*, 761–769.

Darus, M., & Ibrahim, R. W. (2011). On the existence of univalent solutions for fractional integral equation of Volterra type in complex plane. *ROMAI J., 7*(1), 77–86.

Das, S. (2007). *Functional Fractional Calculus for System Identification and Controls*. Springer.

Diekmann, O., van Gils, S. A., Verduyn Lunel, S. M., & Walther, H. O. (1955). *Delay Equations: Functional. Complex, and Nonlinear Analysis*. Berlin: Springer.

Diethelm, K., Ford, N. J., & Freed, A. D. (2002a). A predictor-corrector approach for the numerical solution of fractional differential equations. *Nonlinear Dynamics, 29*(1/4), 322. doi:10.1023/A:1016592219341

Diethelm, K., Ford, N. J., & Freed, A. D. (2004b). Detailed error analysis for a fractional Adams method. *Numerical Algorithms, 36*(1), 31–52. doi:10.1023/B:NUMA.0000027736.85078.be

Diethelm, K., & Freed, A. (1999a). On the solution of nonlinear fractional order differential equations used in the modelling of viscoplasticity. In F. Keil, W. Mackens, H. Vo, & J. Werther (Eds.), *Scientific Computing in Chemical Engineering IIComputational Fluid Dynamics. Reaction Engineering, and Molecular Properties* (p. 217224). Heidelberg: Springer.

Diethelm, K., & Freed, A. (1999b). The FracPECE subroutine for the numerical solution of differential equations of fractional order. In S. Heinzel & T. Plesser (Eds.), *Forschung und wissenschaftliches Rechnen 1998* (pp. 57–71). Gottingen: Gesellschaft fur Wisseschaftliche Datenverarbeitung.

Diethelm & Ford. (n.d.). *The numerical solution of linear and non-linear fractional differential equations involving fractional derivatives several of several orders*. Numerical Analysis Report 379. Manchester Center for Numerical Computational Mathematics.

Dieuonn'e, X. (1960). *Foundations of modern analysis*. New York: Academic Press.

Driver, R. D. (1977). Ordinary and Delay Differential Equations. Springer-Verlag.

Duarte, V., & Costa, J. S. (2002). Time domain implementations of non-integer order controllers. Proc. of Controlo 2002, 353-358.

El-Sayed, A. M. A. (1996). Fractional differential difference equations. *J. Fract. Calc.*, *10*, 101–106.

El-Sayed, A.M.A. (1988). Fractional order differential equation. *Kyungpook Math. J., 28*(2), 1822.

El-Sayed, A. M. A. (1992). On the fractional differential equations. *Applied Mathematics and Computation*, *49*(2-3), 205213. doi:10.1016/0096-3003(92)90024-U

El-Sayed, A.M.A. (1998). Nonlinear functional-differential equations of arbitrary orders. *Nonlinear Anal., 33*(2).

El-Sayed, A. M. A. (1998). Nonlinear functional differential equations of arbitrary orders. Nonlinear Anal.: Theory. *Methods Appl.*, *33*, 181–186.

El-Sayed, A. M. A., Ahmed, E., & Herzallah, M. A. E. (2011). On the fractional-order differential Games with non-uniform interaction rate and Asymmetric games. *Journal of Fractional Calculus and Applications*, *1*(1), 1–9.

El-Sayed, A.M.A., El-Mesiry, A.E.M., & El-Saka, H.A.A. (2007). On the fractional-order logistic equation. *Appl. Math. Lett., 20*(7), 817823.

El-Sayed, A. M. A., El-Mesiry, E. M., & El-Saka, H. A. A. (2004). Numerical solution for multiterm fractional (arbitrary) orders differential equations. *Computational & Applied Mathematics*, *23*(1), 33–54. doi:10.1590/S0101-82052004000100002

El-Sayed, A. M. A., El-Mesiry, E. M., & El-Saka, H. A. A. (2007). On the fractional-order logistic equation. *J. Appl. Math. Letters*, *20*(7), 817–823. doi:10.1016/j.aml.2006.08.013

El-Sayed, A.M.A., Gaafar, F.M. (2001). Fractional order differential equations with memory and fractional-order relaxation oscillation model. *Pure Math. Appl., 12*.

El-Sayed, A. M. A., Gaafar, F. M., & Hashem, H. H. (2004). On the mayimal and minimal solutions of arbitrary orders nonlinear functional integral and differential equations. *Math. Sci. Res. J.*, *8*, 336–348.

El-Sayed, A. M. A., & Salman, S. M. (2013, July). The unified system between Lorenz and Chen systems: A discretization process. *Electronic Journal of Mathematical Analysis and Applications*, *1*(2), 318–325.

El-Sayed, A. M. A., & Salman, S. M. (2013). On a discretization process of fractional-order Riccati differential equation. *Journal of Fractional Calculus and Applications*, *4*(2), 251–259.

El-Sayed, A. M. A., & Salman, S. M. (2013, July). On a discretization process of fractional-order Riccati differential equation. *Journal of Fractional Calculus and Applications*, *4*(2), 251–259.

El-Sayed, A. M. A., Agarwal, R. P., & Salman, S. M. (2013). Fractional-order Chua's system: discretization, bifurcation and chaos. *Journal of Advances in Difference Equations, 2013*, 320.

El-Sayed, A. M. A., El-Raheem, Z. F., & Salman, S. M. (2014). Discretization of forced Duffing system with fractional-order damping. *Journal of Advances in Difference Equations, 2014*(66).

Elaidy, S. N. (2005). An introduction to difference equations (3rd ed.). Springer.

Erjaee, G. H. (2009). On analytical justification of phase synchronization in different chaotic systems. *Chaos, Solitons, and Fractals*, *39*(3), 1195–1202. doi:10.1016/j.chaos.2007.06.006

Feng, Q. X., & Yan, J. R. (2002). Global attractivity and oscillation in a kind of Nicholsons blowflies. *J. Biomath.*, *17*, 21–26.

Furati, K. M., & Tatar, N.-E. (2007). Long time behavior for a nonlinear fractional model. *Journal of Mathematical Analysis and Applications*, *332*(1), 441454. doi:10.1016/j.jmaa.2006.10.027

Glass, L., & Mackey, M. C. (1979). Pathological conditions resulting from instabilities in physiological control systems. *Annals of the New York Academy of Sciences*, *316*(1), 214–235. doi:10.1111/j.1749-6632.1979.tb29471.x PMID:288317

Gorenflo, R., & Mainardi, F. (1997). Fractional calculus: Integral and differential equations of fractional order. In A. Carpinteri & F. Mainardi (Eds.), *Fractals and Fractional Calculus in Continuum Mechanics* (pp. 223–276). Wien: Springer. doi:10.1007/978-3-7091-2664-6_5

Gourley, S. A., & Ruan, S. (2000). Dynamics of the diffusive Nicholson's blowflies equation with distributed delay. *Proc. Roy. Soc. Edinburgh Sect. A*, *130*(6), 1275–1291. doi:10.1017/S0308210500000688

Gurney, W. S., Blythe, S. P., & Nisbet, R. M. (1980). Nicholson's blowflies (revisited). *Nature*, *287*(5777), 17–21. doi:10.1038/287017a0

Gurney, W. S. C., Blythe, S. P., & Nisbet, R. M. (1980). blowflies revisited. *Nature*, *287*(5777), 17–21. doi:10.1038/287017a0

Hadid, S.B. (1995). Local and global existence theorems on differential equations of non-integer order. *J. Fract. Calc.*, 101105.

Hadid, S.B., & Momani, S.M. (1996). On the existence of maximal and minimal solutions of differential equations of non-integer order. *J. Fract. Calc.*, 4144.

Hale, J. K., & Verduyn Lunel, S. M. (1993). *Introduction to functional differential equations*. New York, NY: Springer. doi:10.1007/978-1-4612-4342-7

Hayes, N. D. (1950). Roots of the transcendental equation associated with a certain differential difference equations. *J. Lond. Math. Soc.*, *25*, 226–232.

Hennig, D., & Archilla, J. F. R. (2004). Stretching and relaxation dynamics in double stranded DNA. *Physica A*, *331*(34), 579–601. doi:10.1016/j.physa.2003.09.053

Holmgren, R. (1994). *A First Course in Discrete Dynamical Systems*. New York: Springer-Verlag. doi:10.1007/978-1-4684-0222-3

Ibrahim, R. W., & Momani, S. M. (2007). On the existence and uniqueness of solutions of a class of fractional differential equations. *Journal of Mathematical Analysis and Applications, 334*(1), 110. doi:10.1016/j.jmaa.2006.12.036

Diethelm, K. (2002). Predictor-corrector strategies for single- and multi-term fractional differential equations. In *Proceedings of the 5th HellenicEuropean Conference on Computer Mathematics and its Applications*. LEA Press.

Joseph, W.-H., Jianhong, W. U., & Yuanjie, Y. (2000). Numerical steady state and Hopf bifurcation analysis on the diffusive Nicholson's blowflies equation. *Applied Mathematics and Computation, 111*, 33–51.

Jumarie, G. (2006). Modified Riemann-Liouville derivative and fractional Taylor series of nondifferentiable functions further results. *Computers & Mathematics with Applications (Oxford, England), 51*(9-10), 1367–1376. doi:10.1016/j.camwa.2006.02.001

Jumarie, G. (2009). Table of some basic fractional calculus formulae derived from a modified Riemann-Liouville derivative for non-differentiable functions. *Applied Mathematics Letters, 22*(3), 378–385. doi:10.1016/j.aml.2008.06.003

Karakostas, G., Philos, C.G., & Sficas, Y.G. (1991). The dynamics of some discrete population models. *Nonlinear Anal. Theory Methods Appl.,* 1069-1084.

Kocic, L., & Ladas, G. (1990, January). Oscillation and global attractivity in a discrete model of Nicholson's blowflies. *Applicable Analysis, 38*(1-2), 21–31. doi:10.1080/00036819008839952

Kulenovic, M. R. S., & Ladas, G. (1987). Linearized oscillations in population dynamics. *Bulletin of Mathematical Biology, 49*(5), 615–627. doi:10.1016/S0092-8240(87)90005-X PMID:3435801

Kulenovic, M. R. S., Ladas, G., & Sficas, Y. G. (1992). Global attractivity in Nicholson's blowflies. *Applicable Analysis, 43*(1-2), 109124. doi:10.1080/00036819208840055

Letellier, C., Elaydi, S. A., Aguirrec, L., & Alaouid, A. (2004). Difference equations versus differential equations. a possible equivalence for the R¨ossler system? *Physica D. Nonlinear Phenomena, 195*(12), 2949.

Li, J. (1996). Global attractivity in Nicholson's blowflies. *Appl. Math. Ser. B, 11*(4), 425–434. doi:10.1007/BF02662882

Machado, J. A. T. (1997). Analysis and design of fractional-order digital control systems. J. Syst. Anal. *Modeling-Simulation, 27*, 107–122.

Mackey, M. C. (1989). Commodity price actuations: Price dependent delays and non- linearities as explanatory factors. *Journal of Economic Theory, 48*(2), 497–509. doi:10.1016/0022-0531(89)90039-2

Mackey, M. C., & Glass, L. (1977). Oscillation and chaos in physiological control systems. *Science, 197*(4300), 287–289. doi:10.1126cience.267326 PMID:267326

Matignon, D. (1996). Stability results for fractional differential equations with applications to control processing. Computational Engineering in System Application, 2, 963.

Maxim, D., & Kamenetskii, F. (1997). Biophysics of the DNA molecule. *Physics Reports*, *288*(1-6), 13–60. doi:10.1016/S0370-1573(97)00020-3

Mickens, R. E. (1994). Nonstandard finite difference models of differential equations. World Scientific Publishing Co., Inc.

Miller, K. S., & Ross, B. (1993). *An Introduction to The Fractional Calculus and Fractional Differential Equations*. John-Wiley and Sons Inc.

Owa, S. (1978). On the distortion theorems. I. *Kyungpook Math. J.*, *18*(1), 53–59.

Pathak, H., & Shahzad, N. (2007). Fixed point for generalized contractions and applications to control theory. *Nonlinear Anal.* doi.01.041 doi:10.1016/j.na.2007

Pertz, G. H., & Gerhardt, C. J. (1849). Briefwechsel zwischen Leibniz, Hugens van Zulichem und dem Marquis de l'Hospital. Leibnizens gesammelte Werke.

Podlubny, I. (1999). *Fractional Differential Equations*. London: Academic Press.

Podlubny, I. (2000). Matrix approach to discrete fractional calculus. *Fractional Calculus & Applied Analysis*, *3*(4), 359–386.

Podlubny, I. (2002). Geometric and physical interpretation of fractional integration and fractional differffentiation. Dedicated to the 60th anniversary of Prof. Francesco Mainardi. *Fractional Calculus & Applied Analysis*, *5*, 367386.

Podlubny, I., Petras, I., Vinagre, B. M., O'Leary, P., & Dorcak, L. (2002). Analogue realizations of fractional-order controllers. Fractional order calculus and its applications. *Nonlinear Dynamics*, *29*(1/4), 281–296. doi:10.1023/A:1016556604320

Pyragas, V., & Pyragas, K. (2010). Modification of delayed feedback control using ergodicity of chaotic systems. *Lithuanian Journal of Physics*, *50*(3), 305–316. doi:10.3952/lithjphys.50301

Sabatier, J., Agrawal, O. P., & Tenreiro Machado, J. A. (2007). Advances in Fractional Calculus: Theoretical Developments and Applications in Physics and Engineering. Springer.

Saker, S. H., & Agarwal, S. (2002). Oscillation and global attractivity in a periodic Nicholson's blowflies model. *Mathematical and Computer Modelling*, *35*(7-8), 719731. doi:10.1016/S0895-7177(02)00043-2

Smith, H. (2010). An Introduction to Delay Differential Equations with Applications to the Life Sciences. Springer.

So Joseph, W.-H. (2001). Traveling waves for the diffusive Nicholson's blowflies equation. *Applied Mathematics and Computation*, *122*, 385392.

Srivastava, H. M., & Owa, S. (1984). An application of the fractional derivative. *Mathematica Japonica*, *29*(3), 383389.

Tidke, H. L. (2009). Existence of global solutions to nonlinear mixed Volterra-Fredholm integrodifferential equations with nonlocal conditions. *Electronic Journal of Differential Equations*, *55*, 17.

Tung, C. S., & Harvey, S. C. (1984). A molecular mechanical model to predict the helix twists angles of B-DNA. *Nucleic Acids Research*, *12*(7), 3343–3356. doi:10.1093/nar/12.7.3343 PMID:6326059

Udwadia, F. E., & von Bremen, H. (2000). A Note on the Computation of the Largest p-Lyapunov Characteristic Exponents for Nonlinear Dynamical Systems. *J App Math and Comp*, *114*, 205–214. doi:10.1016/S0096-3003(99)00113-7

Vasumathi, V., & Daniel, M. (2008). Perturbed soliton-like molecular excitations in a deformed DNA chain. *Physics Letters. [Part A]*, *373*(1), 76–82. doi:10.1016/j.physleta.2008.10.058

Weng, P., & Liang, M. (1996). Existence and global attractivity of periodic solution of a model in population dynamics. *Acta Math. Appl. Sinica (English Ser.), 12*, 427-434.

Yan, J. P., & Li, C. P. (2007). On chaos synchronization of fractional differential equations. *Chaos, Solitons, and Fractals*, *32*(2), 725735. doi:10.1016/j.chaos.2005.11.062

Yang, X. J. (2013). Approximate solutions for diffusion equations on Cantor spacetime. *Proceedings of the Romanian Academy, Series A*, 127-133.

Yang, X. J. (2013). Mathematical aspects of Heisenberg uncertainty principle within local fractional Fourier analysis. *Boundary Value Problems*, 1–6.

Yang, X. J., & Baleanu, D. (2013). Fractal heat conduction problem solved by local fractional variation iteration method. *Thermal Science*, *17*(2), 625–628. doi:10.2298/TSCI121124216Y

Yang, X. J., Baleanu, D., & Srivastava, H. M. (2015). Local fractional similarity solution for the diffusion equation defined on Cantor sets. *Applied Mathematics Letters*, *47*, 54–60. doi:10.1016/j.aml.2015.02.024

Yang, X. J., Baleanub, D., Lazarevic, M., & Cajic, M. (2015). Fractal boundary value problems for integral and differential equations with local fractional operators. *Thermal Science*, *19*(3), 959–966. doi:10.2298/TSCI130717103Y

Zhang, B. G., & Xu, H. X. (1999). A note on the global attractivity of a discrete model of Nicholsons blowflies. *Discrete Dynamics in Nature and Society*, *3*(1), 51–55. doi:10.1155/S1026022699000072

Chapter 5
On the Influence of Fractional Derivative on Chaos Control of a New Fractional-Order Hyperchaotic System

Ahmed Ezzat Mohamed Matouk
https://orcid.org/0000-0001-5834-4234
Majmaah University, Saudi Arabia

ABSTRACT

The non-local fractional differential operators have potential applications in many fields of science and technology but especially in the field of dynamical systems. This chapter introduces a new hyperchaotic dynamical system involving non-local fractional differential operator with singular kernel (the Caputo type). The system involves three quadratic nonlinearities and also three equilibrium points. Existence of chaotic and hyperchaotic attractors has been illustrated. Based on Matouk's stability theory of four-dimensional fractional-order systems, the influence of the fractional differential operator on stabilizing the proposed system to its three steady states has been shown. Numerical results have been provided to verify the theoretical analysis. This kind of study is expected to add useful applications to chaos-based secure communications and text encryption.

INTRODUCTION

Fractional calculus (FC) has potential applications to science and engineering; For example it has applications to physics (El-Sayed, 1996; Hilfer, 2000; Khan & Kumar, 2018; Saqib et al, 2018a-d, Khan & Kumar, 2019a; Al-Khedhairi et al, 2019a; Al-Khedhairi et al, 2019b); applications to economic (Laskin, 2000; Chen, 2008; Chen et al, 2011; Zhen et al, 2011; Hegazi et al, 2013b; Khan & Kumar, 2016); applications to social systems (Ali, 2019); applications to mathematical biology (Ahmed & Elgazzar, 2007; Matouk et al, 2015; Elsadany & Matouk, 2015; Matouk et al, 2015; Matouk & Elsadany, 2016; Selvam

DOI: 10.4018/978-1-7998-3122-8.ch005

et al, 2017; Selvam et al, 2018a-b; Ameen & Novati, 2017; Al-Khedhairi et al, 2018) and applications to engineering (Sun et al, 1984; Petras, 2009; Matouk, 2009a; Matouk, 2009b; Matouk 2010; Matouk, 2011a; Matouk, 2011b; Hegazi & Matouk, 2011; Hegazi et al, 2011; Hegazi & Matouk, 2013; Hegazi et al, 2013a; Matouk & Elsadany, 2014; Banu & Nasir, 2015; Nasir & Singh, 2015; Matouk, 2016; El-Sayed et al, 2016a; El-Sayed et al, 2016b; Khan & Kumar, 2019b). On the other hand, FC has also important investigations in some interdisciplinary topics of applied mathematics (Matignon, 1996; Ben Adda F, 1997; Tavazoei et al, 2009; Tavazoei, 2010; Abdel Latif, 2011; El-Sayed & Salman, 2013; Latif et al, 2013; Abdel Kader et al, 2017; Abdel Kader et al, 2018; Abdel Latif et al, 2018; Ali & Ameen, 2019; Saqib et al, 2019a-b; Taghavian & Tavazoei, 2017; Taghavian & Tavazoei, 2018; Taghavian & Tavazoei, 2019).

In fact, FC has been employed to describe many real life models whose nature is related to fractional derivatives. However, to discuss the nature of these phenomena in fractional language, many fractional operators have recently been proposed for example the fractional operator given by Caputo (Caputo, 1967)

$$^{C}D^{\alpha}\eta(t) = I^{p-\alpha}\eta^{(p)}(t), \qquad J^{\mu}\xi(t) = \left[\int_{0}^{t}(t-\rho)^{\mu-1}\xi(\rho)d\rho\right] / \Gamma(\mu), \tag{1}$$

and by Riemann-Liouville (Riemann, 1876)

$$^{RL}_{\ *}D^{\alpha}_{a^{+}}\eta(t) = \frac{d^{p}}{dt^{p}}J^{p-\alpha}\eta(t) = \frac{1}{\Gamma(p-\alpha)}\frac{d^{p}}{dt^{p}}\int_{a}^{t}\frac{\eta(\rho)}{(t-\rho)^{\alpha-p+1}}d\rho, \tag{2}$$

where $\alpha\in(0,1)$ is the fractional order, $p=[\alpha]$ and J^{ϑ} is the Riemann–Liouville integral operator of order ϑ. It is evident that each of Caputo and Riemann-Liouville has non-local operator since they are defined by integration. More recently, novel important fractional operators have been appeared e.g. the Caputo-Fabrizio (CF) type (Caputo & Fabrizio, 2015; Atangana & Alqahtani, 2016; Atangana & Baleanu, 2017; Hristov, 2017a), Atangana-Baleanu (AB) type (Atangana, 2018; Atangana & Gómez-Aguilar, 2018a), Yang-Srivastava-Machado (YSM) type (Yang et al, 2016), Yang-Gao-Machado-Baleanu (YGMB) type (Yang et al, 2017; Cattani, 2018) and Yang-Machado (YM) type of fractional operators of variable order (Yang & Machado, 2017).

In fact, non-local differential and integral operators are better candidate to handle complex dynamical behaviors of natural phenomena and are also classified based on their kernels. It is also favorable in real world modeling since the non-locality property implies that the future dynamics of the fractional model also depends on its history states. The Riemann-Liouville and Caputo possess non-local operators with singular kernels, however the Caputo-Fabrizio has non-local operator with non-singular kernel. Moreover, the Atangana and Baleanu fractional derivative has non-local operator with singular kernel. Furthermore, a modified type of Caputo-Fabrizio definition with a smooth kernel has recently been presented by Yépez-Martínez and Gómez-Aguilar (Yépez-Martínez & Gómez-Aguilar, 2019). These definitions have been used for solving various interesting problems in science and engineering. For example, Gómez-Aguilar and Dumitru Baleanu (Gómez-Aguilar & Baleanu, 2014) used the Caputo derivative with order lying in (0,1] to present the fractional transmission line with losses. Also, Gómez-Aguilar et al. utilized Riemann-Liouville, Liouville-Caputo and Caputo-Fabrizio types of fractional derivatives to explore analytical and numerical solutions of fractional models of electrical circuits RC, RL, RLC, power electronic devices and nonlinear loads (Gómez-Aguilar et al, 2016a). Bonyah et al., used Atangana-Baleanu definition to formulate the optimal control problem of a human African trypanosomiasis model (Bonyah et al, 2018). In 2016, Gómez-Aguilar et al. studied analytical solutions of Liénard type system of a pipeline with Caputo-Fabrizio fractional derivatives (Gómez-Aguilar et al, 2016b). Coronel-Escamilla et al. (Coronel-Escamilla et al, 2016) studied non-local dynamics of triple pendulum model with fractional derivatives based on Atangana-Baleanu-Caputo, Caputo-Fabrizio, Riemann-Liouville and Liouville-Caputo fractional derivatives. In addition, Coronel-Escamilla et al. (Coronel-Escamilla et al, 2017) studied the behaviors of fractional Bateman-Feshbach-Tikochinsky and Caldirola-Kanai oscillators based on Liouville-Caputo and Caputo-Fabrizio fractional operators. Yépez-Martínez et al. (Yépez-Martínez et al, 2016) solved the nonlinear mKdV space-time fractional partial differential equation due to Riemann-Liouville operator and the Feng's first integral method. Morales-Delgado et al. (Morales-Delgado, 2017) studied generalized Cauchy problems under diffusion wave by considering fractional order evolution equations based on fractional operators with different kernels such as Caputo-Fabrizio, Liouville-Caputo and Riemann-Liouville operators. In (Atangana & Gómez-Aguilar, 2017), Atangana-Gómez averaging definitions were introduced and applied to examples in real world fractional problems whose operator is local kernel based on the normal distribution. Gómez-Aguilar and Atangana (Gómez-Aguilar & Atangana, 2017) proposed a fractional operator power-law-exponential-Mittag-Leffler kernel with three fractional orders that can be considered as an effective tool to model real world problems involving complex phenomena. Hristov used Caputo-Fabrizio time-fractional derivative to study transient heat diffusion with a non-singular fading memory from the Cattaneo constitutive equation with Jeffrey's kernel (Hristov, 2016a). Jain numerically investigated the fractional diffusion and fractional Buckmaster equations using two-step Laplace Adam-Bashforth method (Jain, 2018). Atangana and Jain used a new numerical approximation to study fractal ordinary differential equation (Atangana & Jain, 2018). For further similar useful applications of fractional derivatives, one can read the articles in (Ichise et al, 1971; Martin et al, 2008; Herzallah, & Baleanu, 2009; Baleanu et al, 2010; Hristov, 2016b; Hristov, 2017b; Laskin, 2017; Atangana & Gómez-Aguilar, 2018b).

It is now shown that there is no unique definition of fractional derivatives since there are several types of fractional derivatives which make difficulties to describe the natural phenomena. Hence, an arising question is: How can we handle this problem? To answer this important question, we recall that the theory of integro-differential operators (an operator which involves both integrals and derivatives of a function) had been appeared before the modern theory of fractional calculus. So, non-local fractional differential derivatives can be classified as integro-differential operators which can be added, subtracted or multiplied by other operators and therefore can be selected to give appropriate description of specific real world phenomena.

In fact, hyperchaotic dynamical systems are better candidate for applications of chaos-based secure communications and text encryption because of their higher unpredictability and more randomness than simple chaotic systems. Furthermore, the fractional parameter enables us to achieve additional degree of freedom for enhancing the process of hiding essential information and encrypting data.

Motivated by the above discussion, we are going to study the influence of a non-local fractional operator (Caputo type) on chaos control of a novel hyperchaotic system proposed by Matouk, known here as Matouk's hyperchaotic system (Matouk, 2019):

$$\begin{aligned}\dot{x}_1 &= ax_4 - ax_2 + hx_1 - x_1x_4,\\ \dot{x}_2 &= bx_1 + x_4 - x_1x_3,\\ \dot{x}_3 &= x_1^2 - cx_3,\\ \dot{x}_4 &= dx_4.\end{aligned} \tag{3}$$

THE MATOUK'S HYPERCHAOTIC SYSTEM WITH CAPUTO FRACTIONAL OPERATOR

The Matouk's hyperchaotic system with Caputo fractional operator is presented as

$$\begin{aligned}{}^{C}D^{\alpha}x_1(t) &= ax_4 - ax_2 + hx_1 - x_1x_4,\\ {}^{C}D^{\alpha}x_2(t) &= bx_1 + x_4 - x_1x_3,\\ {}^{C}D^{\alpha}x_3(t) &= x_1^2 - cx_3,\\ {}^{C}D^{\alpha}x_4(t) &= dx_4.\end{aligned} \tag{4}$$

The system (4) has three equilibrium points (steady states);

$$E_1 = (0,0,0,0), E_2 = \left(\sigma, \frac{\sigma h}{a}, b, 0\right), E_3 = \left(-\sigma, -\frac{\sigma h}{a}, b, 0\right), \sigma = \sqrt{bc},\ a \neq 0, bc \geq 0.$$

The proposed system (4) exhibits hyperchaotic attractors for a= -3, b=15, c=0.6, d= -0.0001, h= -1 and α=0.99 as shown in figure 1. Moreover, a chaotic attractor of system (4) is obtained when a=1, b=45, c=2, d=1, h=1 and α=0.99 as shown in figure 2.

Now, if we consider the eigenvalue equation for arbitrary equilibrium point $\bar{E} = (\bar{x}_1, \bar{x}_2, \bar{x}_3, \bar{x}_4)$ of system (4) as follows:

$$P(\lambda) = \lambda^4 + \gamma_1\lambda^3 + \gamma_2\lambda^2 + \gamma_3\lambda + \gamma_4 = 0, \tag{5}$$

then the stability results of such steady state is determined by the following Matouk's theory:

Theorem 1. (Matouk, 2009b)

1. If D_1, D_2, D_3 are Routh-Hurwitz determinants

$$D_1 = \gamma_1, D_2 = \begin{vmatrix}\gamma_1 & 1\\ \gamma_3 & \gamma_2\end{vmatrix}, D_3 = \begin{vmatrix}\gamma_1 & 1 & 0\\ \gamma_3 & \gamma_2 & \gamma_1\\ \gamma_5 & \gamma_4 & \gamma_3\end{vmatrix},$$

then for α=1, the point $\bar{E}$ is locally asymptotically stable (LAS) iff

Figure 1. Phase portrait of Matouk's fractional-order system (4) with a= -3, b=15, c=0.6, d= -0.0001, h= -1 and α=0.99

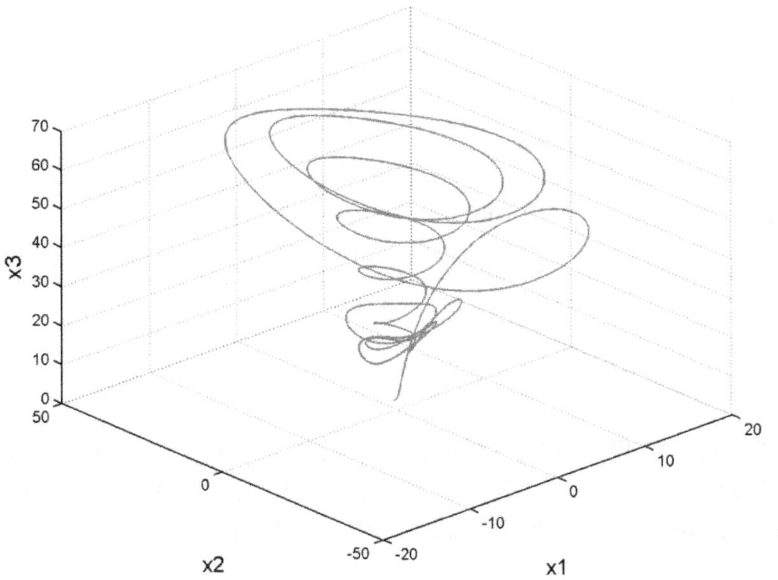

Figure 2. Phase portrait of Matouk's fractional-order system (4) with a=1, b=45, c=2, d=1, h=1 and α=0.99

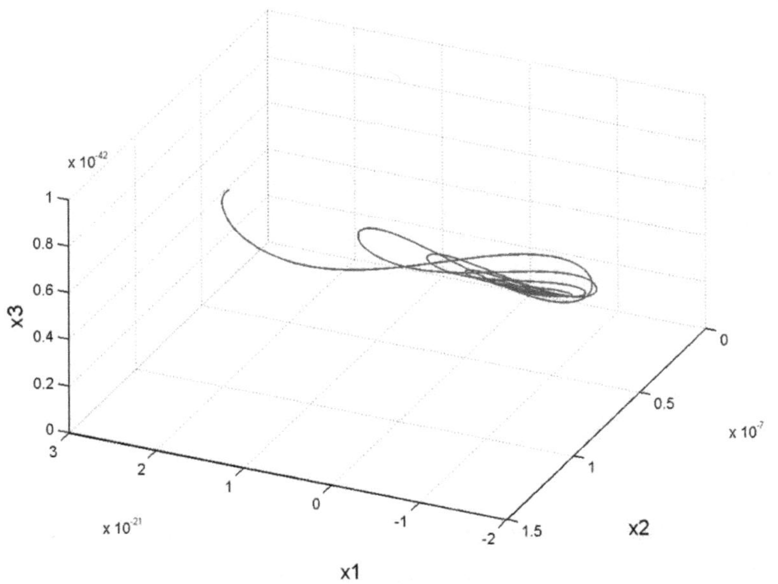

$$D_1 > 0, D_2 > 0, D_3 = 0, \gamma_4 > 0. \tag{6}$$

Also, if the conditions (6) are satisfied with $0 \le \alpha < 1$ then $\bar{E}$ is LAS.

2. If $D(P)>0$, $\gamma_1>0$, $\gamma_2<0$ and $\alpha>2/3$ then $\bar{E}$ is unstable, where $D(P)$ is the discriminant of the polynomial given in Eq. (5).
3. If $D(P)<0$, $\gamma_1>0$, $\gamma_2>0$, $\gamma_3>0$, $\gamma_4>0$ and $\alpha<1/3$, then the point $\bar{E}$ is LAS. Moreover, if $D(P)<0$, $\gamma_1<0$, $\gamma_2>0$, $\gamma_3<0$, $\gamma_4>0$, then the point $\bar{E}$ is unstable.
4. If $D(P)<0$, $\gamma_1>0$, $\gamma_2>0$, $\gamma_3>0$, $\gamma_4>0$ and $\gamma_2 = \frac{\gamma_1\gamma_4}{\gamma_3} + \frac{\gamma_3}{\gamma_1}$, then $\bar{E}$ is LAS for all $0<\alpha<1$.
5. The equilibrium point $\bar{E}$ is LAS only if $\gamma_4>0$.

THE INFLUENCE OF THE FRACTIONAL-ORDER ON CONTROLLING HYPERCHAOS IN MATOUK'S FRACTIONAL-ORDER HYPERCHAOTIC SYSTEM VIA LINEAR FEEDBACK CONTROL TECHNIQUE

The controlled Matouk's fractional-order hyperchaotic system is represented by

$$\begin{aligned}
{}^{C}D^{\alpha}x_1(t) &= ax_4 - ax_2 + hx_1 - x_1x_4 - k_1(x_1 - \bar{x}_1), \\
{}^{C}D^{\alpha}x_2(t) &= bx_1 + x_4 - x_1x_3 - k_2(x_2 - \bar{x}_2), \\
{}^{C}D^{\alpha}x_3(t) &= x_1^2 - cx_3 - k_3(x_3 - \bar{x}_3), \\
{}^{C}D^{\alpha}x_4(t) &= dx_4 - k_4(x_4 - \bar{x}_4),
\end{aligned} \tag{7}$$

where, k_i, $i=1,2,3,4$ all nonnegative feedback gains (FCGs). The Jacobian matrix of the controlled system (7) evaluated at E_1 is described as

$$J(E_1) = \begin{bmatrix} h - k_1 & -a & 0 & a \\ b & -k_2 & 0 & 1 \\ 0 & 0 & -c - k_3 & 0 \\ 0 & 0 & 0 & d - k_4 \end{bmatrix}. \tag{8}$$

To find the suitable feedback gains, we apply the above-mentioned Matouk's stability theory to the coefficients of the eigenvalue equation corresponding to the matrix $J(E_1)$. So, the condition (iv) implies that k_1 equals

$$h - k_2, \quad \frac{k_4(k_4 - h + k_2 - 2d) + (d - k_2)(d + h) + ab}{k_2 + k_4 - d}$$

or

$$-\frac{k_3(k_3-h+k_2+2c)+(c+k_2)(c-h)+ab}{k_2+k_3+c}, \tag{9}$$

which implies that the zero steady state of system (7) is LAS. The controlled system (7) is therefore numerically integrated with the parameters; a=1, b=45, c=2, d=1, h=1 and FCGs $(k_1,k_2,k_3,k_4)=(1,0,2,2)$ which satisfy the condition (iv) of Matouk's stability theory. Figure 3 shows that all state variables of the controlled system (7) with fractional order α=0.99 converge to the steady state E_1. However, this selection of FCGs does not imply the stabilization of system (7) to its steady state E_1 for the integer-order case (See Fig. 4).

On the other hand, the matrices $J(E_2)$ and $J(E_3)$ are calculated as follow:

$$J(E_2)=\begin{bmatrix} h-k_1 & -a & 0 & a-\sqrt{bc} \\ 0 & -k_2 & -\sqrt{bc} & 1 \\ 2\sqrt{bc} & 0 & -c-k_3 & 0 \\ 0 & 0 & 0 & d-k_4 \end{bmatrix} \tag{10}$$

Figure 3. The state variables of the controlled fractional-order system (7) with a=1, b=45, c=2, d=1, h=1, k_1=1, k_2=0, k_3=k_4=2 and α=0.99, are controlled to E_1

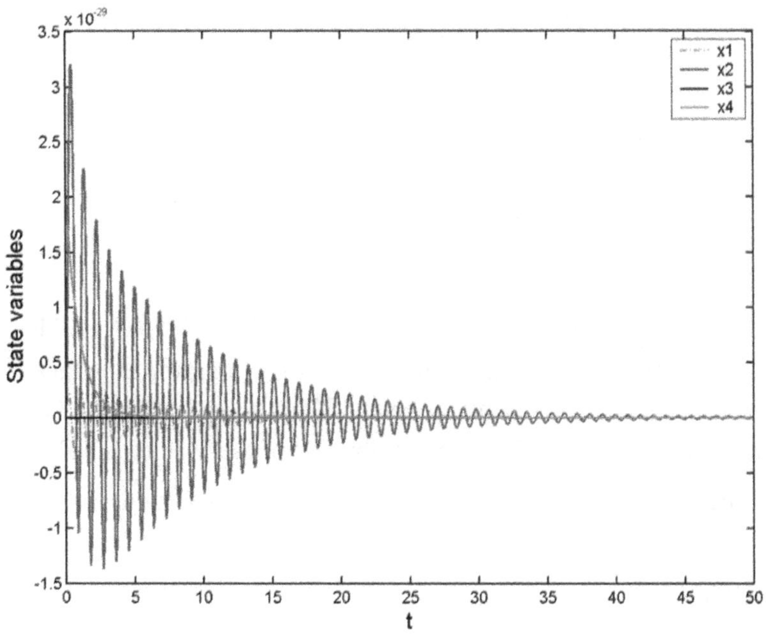

Figure 4. The state variables of the controlled fractional-order system (7) with a=1, b=45, c=2, d=1, h=1, k_1=1, k_2=0, k_3=k_4=2 and α=0.99, are not controlled to E_1

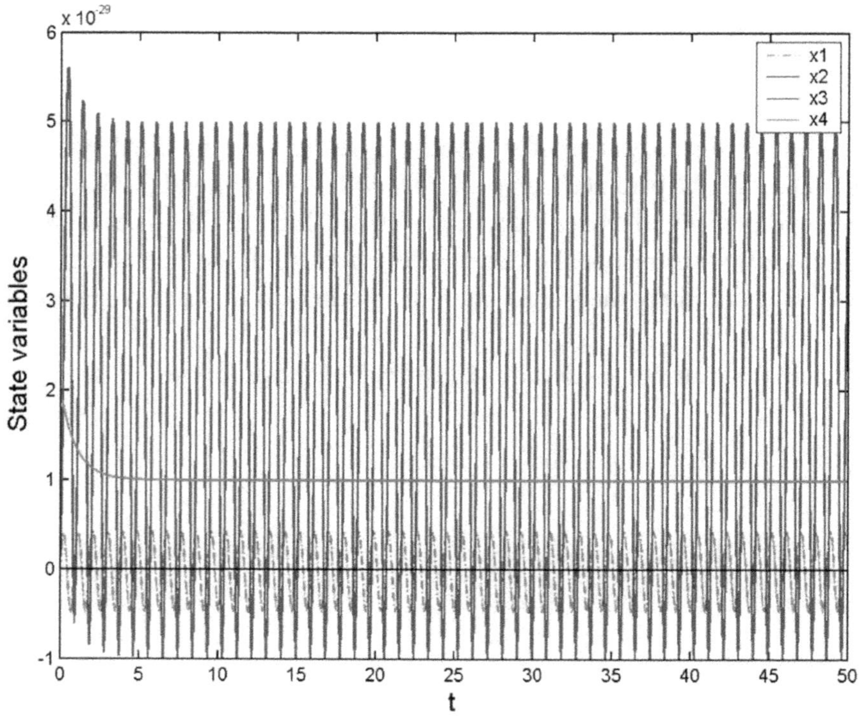

and

$$J(E_3) = \begin{bmatrix} h - k_1 & -a & 0 & a + \sqrt{bc} \\ 0 & -k_2 & \sqrt{bc} & 1 \\ -2\sqrt{bc} & 0 & -c - k_3 & 0 \\ 0 & 0 & 0 & d - k_4 \end{bmatrix}. \tag{11}$$

It is easy to check that each of $J(E_2)$ and $J(E_3)$ yields the same eigenvalue equation $P_{E_2(E_3)}(\lambda)$. So, they have same conditions of stability. Then using the system's parameter values a= -3, b=15, c=0.6, d= -0.0001, h= -1 and α=0.99, we obtain the following FCGs (k_1,k_2,k_3,k_4)=(0.1393502927,2,2,2) which make the coefficients of $P_{E_2(E_3)}(\lambda)$ satisfy the condition (iv) of the Matouk's stability theory. Therefore, all state variables of the controlled system (7) converge to the steady states

$$E_2 = \left(\sigma, \frac{\sigma h}{a}, b, 0\right) \text{ and } E_3 = \left(-\sigma, -\frac{\sigma h}{a}, b, 0\right)$$

as shown in figures 5 and 6, respectively.

Figure 5. The state variables of the controlled fractional-order system (7) with a= -3, b=15, c=0.6, d= -0.0001, h= -1, k_1*=0.1393502927,* k_2*=*k_3*=*k_4*=2 and α=0.99, are controlled to* $E_2 = \left(\sigma, \frac{\sigma h}{a}, b, 0\right)$

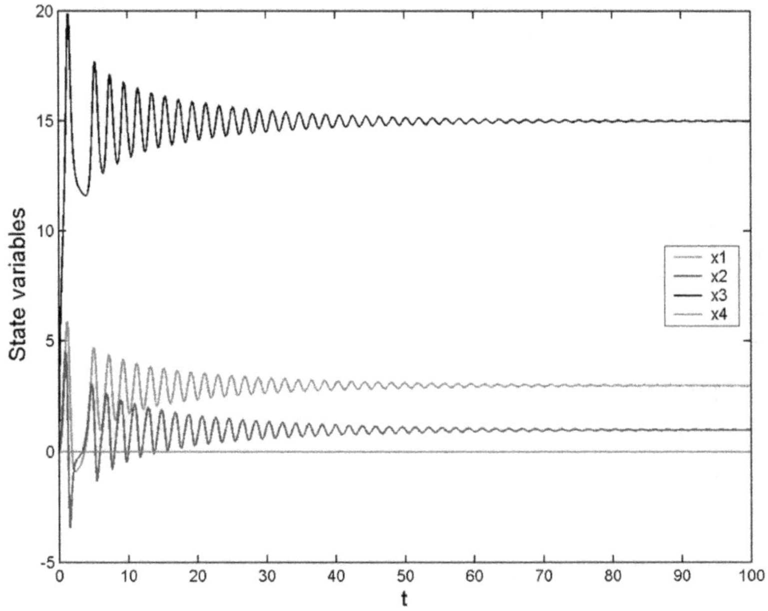

Figure 6. The state variables of the controlled fractional-order system (7) with a= -3, b=15, c=0.6, d= -0.0001, h= -1, k_1*=0.1393502927,* k_2*=*k_3*=*k_4*=2 and α=0.99, are controlled to* $E_3 = \left(-\sigma, -\frac{\sigma h}{a}, b, 0\right)$

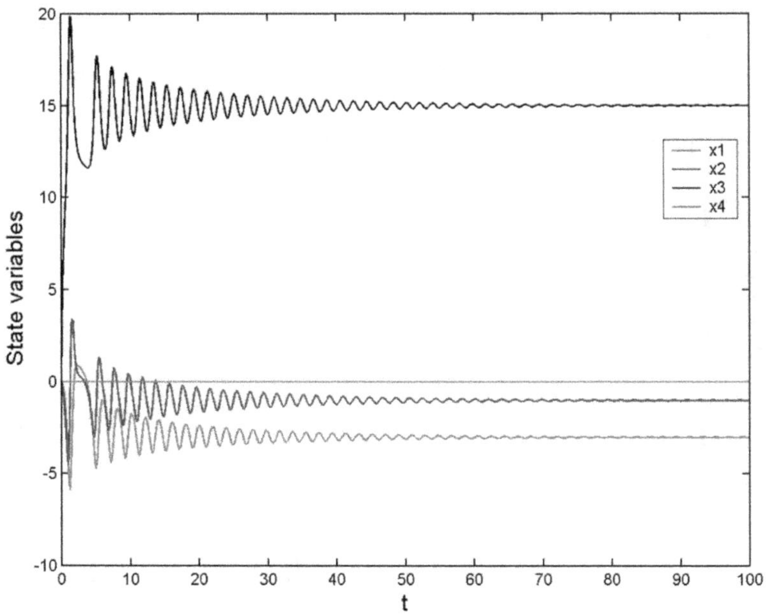

Figure 7. The state variables of the controlled fractional-order system (7) with a= -3, b=15, c=0.6, d= -0.0001, h= -1, k_1*=0.1393502927,* k_2*=*k_3*=*k_4*=2 and α=0.99, are not controlled to* $E_2 = \left(\sigma, \frac{\sigma h}{a}, b, 0\right)$

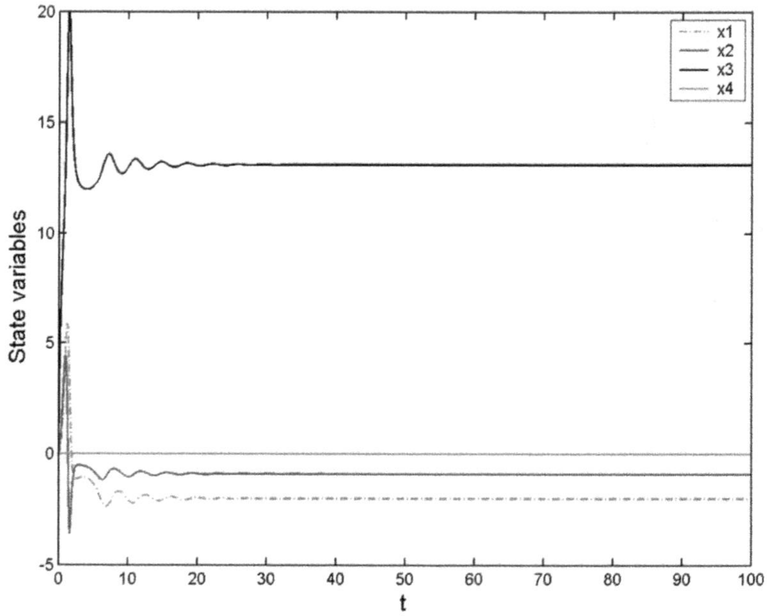

Figure 8. The state variables of the controlled fractional-order system (7) with a= -3, b=15, c=0.6, d= -0.0001, h= -1, k_1*=0.1393502927,* k_2*=*k_3*=*k_4*=2 and α=0.99, are not controlled to* $E_3 = \left(-\sigma, -\frac{\sigma h}{a}, b, 0\right)$

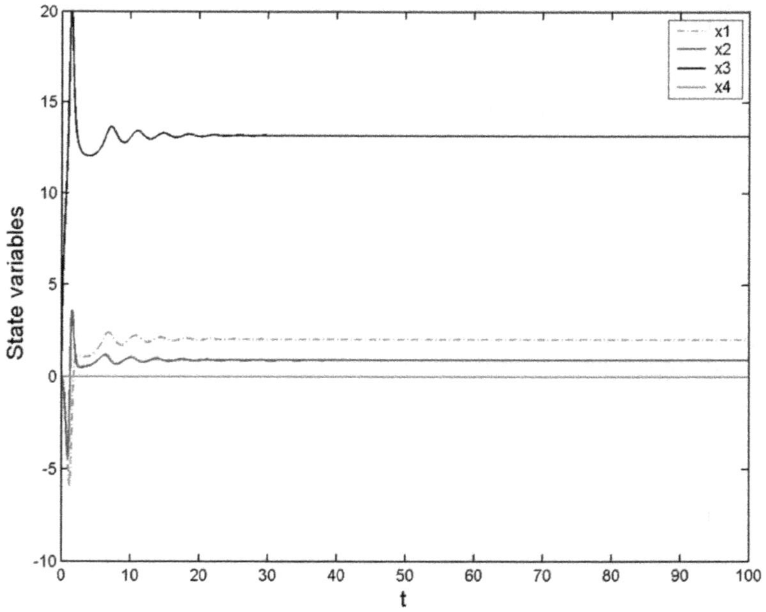

Furthermore, these selections of FCGs and the parameter values do not stabilize system (7) to its steady states $E_2 = \left(\sigma, \frac{\sigma h}{a}, b, 0\right)$ and $E_3 = \left(-\sigma, -\frac{\sigma h}{a}, b, 0\right)$ when α=1 as depicted in Figs. 7 and 8. The results given in Figs. 3-8 show the influence of fractional order α on chaos control of Matouk's system (4).

ACKNOWLEDGMENT

The author thanks Prof. E. Ahmed and Prof. A. M. A. El-Sayed for the useful discussion.

REFERENCES

Abdel Kader, A. H., Abdel Latif, M. S., El Bialy, F., & Elsaid, A. (2018). Symmetry analysis and some new exact solutions of some nonlinear KdV-like equations. *Asian-European Journal of Mathematics, 11*(03), 1850040. doi:10.1142/S1793557118500407

Abdel Kader, A. H., Abdel Latif, M. S., & Nour, H. M. (2017). Some New Exact Solutions of the Modified KdV Equation Using Lie Point Symmetry Method. *International Journal of Applied and Computational Mathematics, 3*(1), 1163–1171. doi:10.100740819-017-0408-3

Abdel Latif, M. S. (2011). Some exact solutions of KdV equation with variable coefficients. *Communications in Nonlinear Science and Numerical Simulation, 16*(4), 1783–1786. doi:10.1016/j.cnsns.2010.07.023

Abdel Latif, M. S., El-Shazly, E., Baleanu, D., Elsaid, A., & Nour, H. M. (2018). Some new soliton-like and doubly periodic-like solutions of Fisher equation with time-dependent coefficients. *Modern Physics Letters B, 32*(33), 1850413. doi:10.1142/S0217984918504134

Ahmed, E., & Elgazzar, A. S. (2007). On fractional order differential equations model for nonlocal epidemics. *Physica A, 379*(2), 607–614. doi:10.1016/j.physa.2007.01.010

Al-Khedhairi, A., Askar, S. S., Matouk, A. E., Elsadany, A., & Ghazel, M. (2018). Dynamics, chaos control, and synchronization in a fractional-order Samardzija-Greller population system with order lying in (0,2). *Complexity.*

Al-khedhairi, A., Matouk, A. E., & Askar, S. S. (2019a). Computations of synchronization conditions in some fractional-order chaotic and hyperchaotic systems. *Pramana Journal of Physics, 92*, 72.

Al-khedhairi, A., Matouk, A. E., & Khan, I. (2019b). Chaotic dynamics and chaos control for the fractional-order geomagnetic field model. *Chaos, Solitons, and Fractals, 128*, 1–12. doi:10.1016/j.chaos.2019.07.019

Ali, H. M. (2019). New approximate solutions to fractional smoking model using the generalized Mittag-Leffler function method. *Progr. Fract. Differ. Appl., 5*(4), 1–9. doi:10.18576/pfda/010101

Ali, H. M., & Ameen, I. (2019). Mittag-Leffler approximation for the solution of nonlinear systems of fractional partial differential equations. *Sylwan, 163*(9), 17–32.

Ameen, I., & Novati, P. (2017). The solution of fractional order epidemic model by implicit Adams methods. *Applied Mathematical Modelling*, *43*, 78–84. doi:10.1016/j.apm.2016.10.054

Atangana, A. (2018). Non validity of index law in fractional calculus: A fractional differential operator with Markovian and non-Markovian properties. *Physica A*, *505*, 688–706. doi:10.1016/j.physa.2018.03.056

Atangana, A., & Alqahtani, R. T. (2016). Numerical approximation of the space-time Caputo-Fabrizio fractional derivative and application to groundwater pollution equation. *Adv. Difference Equations*, *2016*(1), 156. doi:10.118613662-016-0871-x

Atangana, A., & Baleanu, D. (2017). Caputo-Fabrizio derivative applied to groundwater flow within confined aquifer. *Journal of Engineering Mechanics*, *143*(5), D4016005. doi:10.1061/(ASCE)EM.1943-7889.0001091

Atangana, A., & Gómez-Aguilar, J. F. (2017). A new derivative with normal distribution kernel: Theory, methods and applications. *Physica A*, *476*, 1–14. doi:10.1016/j.physa.2017.02.016

Atangana, A., & Gómez-Aguilar, J. F. (2018a). Fractional derivatives with no-index law property: Application to chaos and statistics. *Chaos, Solitons, and Fractals*, *114*, 516–535. doi:10.1016/j.chaos.2018.07.033

Atangana, A., & Gómez-Aguilar, J. F. (2018b). Decolonisation of fractional calculus rules: Breaking commutativity and associativity to capture more natural phenomena. *The European Physical Journal Plus*, *133*(4), 1–22. doi:10.1140/epjp/i2018-12021-3

Atangana, A., & Jain, S. (2018). A new numerical approximation of the fractal ordinary differential equation. *The European Physical Journal Plus*, *133*(2), 37. doi:10.1140/epjp/i2018-11895-1

Baleanu, D., Golmankhaneh, A. K., Nigmatullin, R., & Golmankhaneh, A. K. (2010). Fractional Newtonian mechanics. *Central European Journal of Physics*, *8*, 120–125.

Banu, U. S., & Nasir, A. W. (2015). Design, Analysis and Performance Evaluation of Fractional Order Proportional Integral for Three Interacting Tank Process in Frequency Domain considered as Third Order System. *IFAC-PapersOnLine*, *48*(30), 179–184. doi:10.1016/j.ifacol.2015.12.374

Ben Adda, F. (1997). Geometric interpretation of the fractional derivative. *J. Fract. Calc*, *11*, 21–52.

Bonyah, E., Gómez-Aguilar, J. F., & Adu, A. (2018). Stability analysis and optimal control of a fractional human African trypanosomiasis model. *Chaos, Solitons, and Fractals*, *117*, 150–160. doi:10.1016/j.chaos.2018.10.025

Caputo, M. (1967). Linear models of dissipation whose Q is almost frequency independent-II. *Geophysical Journal of the Royal Astronomical Society*, *13*(5), 529–539. doi:10.1111/j.1365-246X.1967.tb02303.x

Caputo, M., & Fabrizio, M. (2015). A new definition of fractional derivative without singular kernel. *Prog. Fract. Differ. Appl.*, *1*, 73–85.

Cattani, C. (2018). Sinc-fractional operator on Shannon wavelet space. *Frontiers in Physics*, *6*, 118.

Chen, L., Chai, Y., & Wu, R. (2011). Control and Synchronization of Fractional-Order Financial System Based on Linear Control. *Discrete Dynamics in Nature and Society*, *2011*, 1–21. doi:10.1155/2011/958393

Chen, W. C. (2008). Nonlinear dynamics and chaos in a fractional-order financial system. *Chaos, Solitons, and Fractals*, *36*(5), 1305–1314. doi:10.1016/j.chaos.2006.07.051

Coronel-Escamilla, A., Gómez-Aguilar, J. F., Baleanu, D., Córdova-Fraga, T., Escobar-Jiménez, R. F., Olivares-Peregrino, V. H., & Al Qurashi, M. M. (2017). Bateman-Feshbach Tikochinsky and Caldirola-Kanai oscillators with new fractional differentiation. Entropy, 19(2), 55.

Coronel-Escamilla, A., Gómez-Aguilar, J. F., López-López, M. G., Alvarado-Martínez, V. M., & Guerrero-Ramírez, G. V. (2016). Triple pendulum model involving fractional derivatives with different kernels. *Chaos, Solitons, and Fractals*, *91*, 248–261. doi:10.1016/j.chaos.2016.06.007

Diethelm, K. (1997). An algorithm for the numerical solution of differential equations of fractional order. *Electronic Transactions on Numerical Analysis*, *5*, 1–6.

Diethelm, K., & Ford, N. J. (2002). Analysis of fractional differential equations. *Journal of Mathematical Analysis and Applications*, *265*(2), 229–248. doi:10.1006/jmaa.2000.7194

Diethelm, K., Ford, N. J., & Freed, A. D. (2002). A predictor-corrector approach for the numerical solution of fractional differential equations. *Nonlinear Dynamics*, *29*(1/4), 3–22. doi:10.1023/A:1016592219341

Diethelm, K., & Freed, A. D. (1999). The FracPECE subroutine for the numerical solution of differential equations of fractional order. In S. Heinzel & T. Plesser (Eds.), Forschung und wissenschaftliches Rechnen. Gottingen: Gesellschaft für wissenschaftliche Datenverarbeitung (pp. 57–71). Academic Press.

El-Raheem, Z. F., & Salman, S. M. (2014). On a discretization process of fractional-order logistic differential equation. *J. Egypt. Math. Soc.*, *22*(3), 407–412. doi:10.1016/j.joems.2013.09.001

El-Sayed, A. M. A. (1996). Fractional-order diffusion-wave equation. *International Journal of Theoretical Physics*, *35*(2), 311–322. doi:10.1007/BF02083817

El-Sayed, A. M. A., Elsonbaty, A., Elsadany, A. A., & Matouk, A. E. (2016b). Dynamical analysis and circuit simulation of a new fractional-order hyperchaotic system and its discretization. *International Journal of Bifurcation and Chaos in Applied Sciences and Engineering*, 26.

El-Sayed, A. M. A., Nour, H. M., Elsaid, A., Matouk, A. E., & Elsonbaty, A. (2016a). Dynamical Behaviors, Circuit Realization, Chaos Control and Synchronization of a New Fractional Order Hyperchaotic System. *Applied Mathematical Modelling*, *40*(5-6), 3516–3534. doi:10.1016/j.apm.2015.10.010

El-Sayed, A. M. A., & Salman, S. M. (2013). On a discretization process of fractional order Riccati's differential equation. *J. Fract. Calc. Appl.*, *4*, 251–259.

Elsadany, A. A., & Matouk, A. E. (2015). Dynamical behaviors of fractional-order Lotka– Volterra predator–prey model and its discretization. *Journal of Applied Mathematics and Computing*, *49*(1-2), 269–283. doi:10.100712190-014-0838-6

George Maria Selvam, A., & Janagaraj, R. (2018a). Numerical Analysis of a Fractional Order Discrete Prey – Predator System with Functional Response. *International Journal of Engineering & Technology*, *7*(4.10), 681–684.

George Maria Selvam, A., Janagaraj, R., & Britto Jacob, S. (2017). Stability in a Discrete Fractional Order Prey Predator System with Functional Response. *International Journal of Advanced Research in Computer Science and Software Engineering*, *7*(5), 630–633. doi:10.23956/ijarcsse/SV7I5/0336

George Maria Selvam, A., Janagaraj, R., & Vignesh, D. (2018b). Allee effect and Holling type - II response in a discrete fractional order prey - predator model. *IOP Conference Series: Journal of Physics*, *1139*, 1-7. 10.1088/1742-6596/1139/1/012003

Gómez-Aguilar, J. F., & Atangana, A. (2017). New insight in fractional differentiation: Power, exponential decay and Mittag-Leffler laws and applications. *The European Physical Journal Plus*, *132*(13), 1–23. doi:10.1140/epjp/i2017-11293-3

Gómez-Aguilar, J. F., & Baleanu, D. (2014). Fractional transmission line with losses. *Zeitschrift für Naturforschung. Section A. Physical Sciences*, *69*(10-11), 539–546. doi:10.5560/zna.2014-0049

Gómez-Aguilar, J. F., Torres, L., Yépez-Martínez, H., Baleanu, D., Reyes, J. M., & Sosa, I. O. (2016b). Fractional Liénard type model of a pipeline within the fractional derivative without singular kernel. Advances in Difference Equations, 2016(1), 173.

Gómez-Aguilar, J. F., Yépez-Martínez, H., Escobar-Jiménez, R. F., Astorga-Zaragoza, C. M., & Reyes-Reyes, J. (2016a). Analytical and numerical solutions of electrical circuits described by fractional derivatives. *Applied Mathematical Modelling*, *40*(21-22), 9079–9094. doi:10.1016/j.apm.2016.05.041

Hegazi, A. S., Ahmed, E., & Matouk, A. E. (2011). The effect of fractional order on synchronization of two fractional order chaotic and hyperchaotic systems. *J. Fract. Calc. Appl.*, *1*(3), 1–15.

Hegazi, A. S., Ahmed, E., & Matouk, A. E. (2013a). On chaos control and synchronization of the commensurate fractional order Liu system. *Communications in Nonlinear Science and Numerical Simulation*, *18*(5), 1193–1202. doi:10.1016/j.cnsns.2012.09.026

Hegazi, A. S., Ahmed, E., & Matouk, A. E. (2013b). On Dynamical Behaviors and Chaos Control of the Fractional-order Financial System. In S. Banerjee (Ed.), *Chaos and Complexity Theory for Management: Nonlinear Dynamics* (pp. 34–49). IGI Global. doi:10.4018/978-1-4666-2509-9.ch002

Hegazi, A. S., & Matouk, A. E. (2011). Dynamical behaviors and synchronization in the fractional order hyperchaotic Chen system. *Applied Mathematics Letters*, *24*(11), 1938–1944. doi:10.1016/j.aml.2011.05.025

Hegazi, A. S., & Matouk, A. E. (2013). Chaos control of the modified Autonomous Van der Pol-Duffing circuits via active control. In S. Banerjee & L. Rondoni (Eds.), *Applications of Chaos and Nonlinear Dynamics in Science and Engineering* (Vol. 3, pp. 185–202). Berlin: Springer. doi:10.1007/978-3-642-34017-8_7

Herzallah, M. A. E., & Baleanu, D. (2009). Fractional-order Euler-Lagrange equations and formulation of Hamiltonian equations. *Nonlinear Dynamics*, *58*(1-2), 385–391. doi:10.100711071-009-9486-z

Hilfer, R. (Ed.). (2000). *Applications of fractional calculus in physics*. Upper Saddle River, NJ: World Scientific. doi:10.1142/3779

Hristov, J. (2016a). Transient heat diffusion with a non-singular fading memory: From the Cattaneo constitutive equation with Jeffrey's kernel to the Caputo-Fabrizio time-fractional derivative. *Thermal Science*, *20*(2), 757–762. doi:10.2298/TSCI160112019H

Hristov, J. (2016b). Steady-state heat conduction in a medium with spatial non-singular fading memory: Derivation of Caputo-Fabrizio space-fractional derivative with Jeffrey's kernel and analytical solutions. *Thermal Science*, *20*, 115. doi:10.2298/TSCI160112019H

Hristov, J. (2017a). Derivatives with Non-Singular kernels from the Caputo-Fabrizio definition and beyond: Appraising analysis with emphasis on diffusion models. *Frontiers*, *1*, 270–342.

Hristov, J. (2017b). Derivation of the fractional Dodson equation and beyond: Transient diffusion with a non-singular memory and exponentially fading-out diffusivity. *Progr. Fract. Differ. Appl.*, *3*(4), 1–16. doi:10.18576/pfda/030402

Ichise, M., Nagayanagi, Y., & Kojima, T. (1971). An analog simulation of noninteger order transfer functions for analysis of electrode process. *J. Electroanal. Chem.*, *33*(2), 253–265. doi:10.1016/S0022-0728(71)80115-8

Jain, S. (2018). Numerical analysis for the fractional diffusion and fractional Buckmaster equation by the two-step Laplace Adam-Bashforth method. *The European Physical Journal Plus*, *133*(1), 19. doi:10.1140/epjp/i2018-11854-x

Khan, A., & Kumar, S. (2016). Anti-synchronization of chaotic finance systems using fuzzy logic constant controller. *Indian Journal of Industrial and Applied Mathematics*. Doi:10.5958/1945-919X.2016.00013X

Khan, A., & Kumar, S. (2018). Study of chaos in satellite systems. *Pramana Journal of Physics*, *90*(1), 13. doi:10.100712043-017-1502-0

Khan, A & Kumar, S. (2019a). T-S fuzzy modeling and predictive control and synchronization of chaotic satellite systems. *International Journal of Modeling and Simulation, 39*(3).

Khan, A., & Kumar, S. (2019b). Measure of chaos and adaptive synchronization of chaotic satellite systems. *International Journal of Dynamics and Control*, *7*(2), 536–546. doi:10.100740435-018-0481-4

Laskin, N. (2000). Fractional market dynamics. *Physica A*, *287*(3-4), 482–492. doi:10.1016/S0378-4371(00)00387-3

Laskin, N. (2017). Time fractional quantum mechanics. *Chaos, Solitons, and Fractals*, *102*, 16–28. doi:10.1016/j.chaos.2017.04.010

Latif, M. A., Dragomir, S. S., Matouk, A. E. (2013). New inequalities of Ostrowski type for Co-ordinated s-Convex functions via fractional integrals. *Journal of Fractional Calculus and Applications, 4*(1), 22-36.

Martin, R., Quintana, J. J., Ramos, A., & de la Nuez, I. (2008). Modeling of electrochemical double layer capacitors by means of fractional impedance. J. Comput. Nonlinear Dynam., 3, 021303.

Matignon, D. (1996). Stability results for fractional differential equations with applications to control processing. In *IEEE-SMC proceedings of the computational engineering in systems and application multiconference* (Vol. 2, pp. 963–968). Lille, France: IMACS.

Matouk, A. E. (2009a). Chaos synchronization between two different fractional systems of Lorenz family. *Mathematical Problems in Engineering*, *2009*, 1–11. doi:10.1155/2009/572724

Matouk, A. E. (2009b). Stability conditions, hyperchaos and control in a novel fractional order hyperchaotic system. *Physics Letters. [Part A]*, *373*(25), 2166–2173. doi:10.1016/j.physleta.2009.04.032

Matouk, A. E. (2010). Dynamical behaviors, linear feedback control and synchronization of the fractional order Liu system. *Journal of Nonlinear Systems and Applications*, *1*(3), 135–140.

Matouk, A. E. (2011a). Chaos, feedback control and synchronization of a fractional-order modified Autonomous Van der Pol-Duffing circuit. *Communications in Nonlinear Science and Numerical Simulation*, *16*(2), 975–986. doi:10.1016/j.cnsns.2010.04.027

Matouk, A. E. (2011b). Synchronization in integer and fractional order chaotic systems. In S. Banerjee (Ed.), *Chaos Synchronization and Cryptography for Secure Communications: Applications for Encryption* (pp. 127–151). IGI Global. doi:10.4018/978-1-61520-737-4.ch007

Matouk, A. E. (2016). Chaos synchronization of a fractional-order modified Van der Pol Duffing system via new linear control, backstepping control and Takagi-Sugeno fuzzy approaches. *Complexity*, *21*(S1), 116–124. doi:10.1002/cplx.21719

Matouk, A. E. (2019). Dynamics and control in a novel hyperchaotic system. *Int. J. Dyn. Control*, *7*(1), 241–255. doi:10.100740435-018-0439-6

Matouk, A. E., & Elsadany, A. A. (2014). Achieving synchronization between the fractional-order hyperchaotic Novel and Chen systems via a new nonlinear control technique. *Applied Mathematics Letters*, *29*, 30–35. doi:10.1016/j.aml.2013.10.010

Matouk, A. E., & Elsadany, A. A. (2016). Dynamical analysis, stabilization and discretization of a chaotic fractional-order GLV model. *Nonlinear Dynamics*, *85*(3), 1597–1612. doi:10.100711071-016-2781-6

Matouk, A. E., Elsadany, A. A., Ahmed, E., & Agiza, H. N. (2015). Dynamical behavior of fractional-order Hastings–Powell food chain model and its discretization. *Communications in Nonlinear Science and Numerical Simulation*, *27*(1-3), 153–167. doi:10.1016/j.cnsns.2015.03.004

Morales-Delgado, V. F., Taneco-Hernández, M. A., & Gómez-Aguilar, J. F. (2017). On the solutions of fractional order of evolution equations. *The European Physical Journal Plus*, *132*(47), 1–14. doi:10.1140/epjp/i2017-11341-0

Nasir, A. W., & Singh, A. K. (2015). IMC based fractional order controller for non-minimum phase system. In *2015 Annual IEEE India Conference* (pp. 1-6). New Delhi: IEEE. 10.1109/INDICON.2015.7443593

Petras, I. (2009). Chaos in the fractional-order Volta's system: Modeling and simulation. *Nonlinear Dynamics*, *57*(1-2), 157–170. doi:10.100711071-008-9429-0

Podlubny, I. (1999). *Fractional differential equations*. New York: Academic Press.

Riemann, B. (1876). Versuch einer allgemeinen auffassung der integration und differentiation. Gesammelte Werke, 62.

Saqib, M., Ali, F., Khan, I., Sheikh, N. A., & Khan, A. (2018a). Entropy Generation in Different Types of Fractionalized Nanofluids. *Arabian Journal for Science and Engineering*, 1–10.

Saqib, M., Ali, F., Khan, I., Sheikh, N. A., & Shafie, S. B. (2018b). Convection in ethylene glycol-based molybdenum disulfide nanofluid. *Journal of Thermal Analysis and Calorimetry*, 1–10.

Saqib, M., Khan, I., & Shafie, S. (2018c). Application of Atangana–Baleanu fractional derivative to MHD channel flow of CMC-based-CNT's nanofluid through a porous medium. *Chaos, Solitons, and Fractals*, *116*, 79–85. doi:10.1016/j.chaos.2018.09.007

Saqib, M., Khan, I., & Shafie, S. (2018d). Natural convection channel flow of CMC-based CNTs nanofluid. *The European Physical Journal Plus*, *133*(12), 549. doi:10.1140/epjp/i2018-12340-3

Saqib, M., Khan, I., & Shafie, S. (2019a). Application of fractional differential equations to heat transfer in hybrid nanofluid: Modeling and solution via integral transforms. *Advances in Difference Equations*, *2019*(52), 1–18. doi:10.118613662-019-1988-5

Saqib, M., Khan, I., & Shafie, S. (2019b). Generalized magnetic blood flow in a cylindrical tube with magnetite dusty particles. *Journal of Magnetism and Magnetic Materials*, *484*, 490–496. doi:10.1016/j.jmmm.2019.03.032

Sun, H. H., Abdelwahab, A. A., & Onaral, B. (1984). Linear approximation of transfer function with a pole of fractional order. *IEEE Transactions on Automatic Control*, *29*(5), 441–444. doi:10.1109/TAC.1984.1103551

Taghavian, H., & Tavazoei, M. S. (2017). Robust Stability Analysis of Distributed-Order Linear Time-Invariant Systems With Uncertain Order Weight Functions and Uncertain Dynamic Matrices. *Journal of Dynamic Systems, Measurement, and Control*, *139*(12), 121010. doi:10.1115/1.4037268

Taghavian, H., & Tavazoei, M. S. (2018). Exact solution of linear fractional distributed order systems with exponential order weight functions. In *Mathematical Techniques of Fractional Order Systems* (pp. 103–132). Elsevier Publishing. doi:10.1016/B978-0-12-813592-1.00004-0

Taghavian, H., & Tavazoei, M. S. (2019). Algebraic Conditions for Stability Analysis of Linear Time Invariant Distributed Order Dynamic Systems: A Lagrange Inversion Theorem Approach. *Asian Journal of Control*, *21*(2), 879–890. doi:10.1002/asjc.1780

Tavazoei, M. S. (2010). A note on fractional-order derivatives of periodic functions. *Automatica*, *46*(5), 945–948. doi:10.1016/j.automatica.2010.02.023

Tavazoei, M. S., Haeri, M., & Attari, M. (2009). A proof for nonexistence of periodic solutions in time invariant fractional order systems. *Automatica*, *45*(8), 1886–1890. doi:10.1016/j.automatica.2009.04.001

Yang, X.-J., Gao, F., Machado, J. A. T., & Baleanu, D. (2017). A new fractional derivative involving the normalized sinc function without singular kernel. *The European Physical Journal. Special Topics*, *226*(16-18), 3567–3575. doi:10.1140/epjst/e2018-00020-2

Yang, X.-J., & Machado, J. A. T. (2017). A new fractional operator of variable order: Application in the description of anomalous diffusion. *Physica A*, *481*, 276–283. doi:10.1016/j.physa.2017.04.054

Yang, X.-J., Srivastava, H. M., & Machado, J. A. T. (2016). A new fractional derivative without singular kernel: Application to the modelling of the steady heat flow. *Thermal Science*, *20*(2), 753–756. doi:10.2298/TSCI151224222Y

Yépez-Martínez, H., & Gómez-Aguilar, J. F. (2019). A new modified definition of Caputo-Fabrizio fractional-order derivative and their applications to the Multi Step Homotopy Analysis Method (MHAM). *Journal of Computational and Applied Mathematics*, *346*, 247–260. doi:10.1016/j.cam.2018.07.023

Yépez-Martínez, H., Gómez-Aguilar, J. F., Sosa, I. O., Reyes, J. M., & Torres-Jiménez, J. (2016). The Feng's first integral method applied to the nonlinear mKdV space-time fractional partial differential equation. *Revista Mexicana de Física*, *62*(4), 310–316.

Zhen, W., Xia, H., & Guodong, S. (2011). Analysis of nonlinear dynamics and chaos in a fractional order financial system with time delay. *Computers & Mathematics with Applications (Oxford, England)*, *62*(3), 1531–1539. doi:10.1016/j.camwa.2011.04.057

Chapter 6
Synchronization of Fractional-Order Hyperchaotic Finance Systems Using Sliding Mode Control Techniques

Sanjay Kumar
https://orcid.org/0000-0003-2943-4825
Acharya Narendra Dev College, University of Delhi, India & Amity School of Engineering and Technology, Amity University, Patna, India

Ram Pravesh Prasad
Hansraj College, University of Delhi, India

Krishan Pal
Maharaja Agarsen College, University of Delhi, India

Mahendra Pratap Pal
Delhi College of Arts and Commerce, University of Delhi, India

Ajeet Singh
Satyawati College, University of Delhi, India

ABSTRACT

In this chapter, the basic concepts of fractional-order dynamical systems are presented, and the synchronization methodologies of fractional order chaotic dynamical systems are established using slide mode control techniques. Through observation of the different phase portraits and time-series graphs of fractional order finance systems through utilization of the fractional calculus and computer simulation, the authors have obtained that the lowest dimension of fractional order hyper chaotic finance system is 3.90, which is less than 4. Bifurcation diagrams and Lyapunov exponents of fractional order hyper chaotic finance system are calculated to justify the chaos in the systems. Synchronization of two identical fractional-order hyper chaotic finance systems are achieved using sliding mode control techniques.

DOI: 10.4018/978-1-7998-3122-8.ch006

INTRODUCTION

In the last few decades, fractional calculus has established the potential tools to describe in many branches of sciences, engineering, and interdisciplinary fields. Many physical systems can be well defined by fractional-order differential equations which behave chaotically or hyper-chaotically, such as the fractional order Chen system (Li & Chen, 2004), the fractional order Lorenz system (Grigorenko Grigoren, 2003), the fractional-order Duffing system (Gao & Yu, 2005), the fractional order Rossler system (Li & Chen, 2004), the fractional order financial system (Wang, Huang Shi, 2011), the fractional-order hyperchaotic Chen system, the fractional-order hyperchaotic novel system (Matouk, 2009) and so on. The concept of fractional derivative is utilized to model the behavior of real systems in various fields of science and engineering including bioengineering, medicine, biological tissues, cardiac tissue, fluid mechanics, viscoelasticity, material science, etc (Wilkie, Drapaca & Sivaloganathan,2011; Podlubny, 2005; Jun-Guo, 2005; Chen, Liu, Ma & Zhang 2012; Das Gupta, 2012; Odibat, 2010; Kumar et al., 2019). The attributes of fractional order systems, for which they have gained popularity in the investigation of dynamical systems, is that they allow a greater degree of flexibility in the model. However, there are many differences between ordinary differential equation systems (integer-order) and corresponding fractional-order differential equation systems (Faieghi, Delavari & Baleanu, 2013).

Finance system is a part of the economic system. It (finance system) addresses the costs and responsibilities over the allocated time under conditions of certainty and uncertainty. The chaos in the finance systems was first observed in 1985 (Cai & Huang, 2007). Chaos synchronization of chaotic financial system has become an important subject in the application of chaos. Chaotic behaviour in economics system implies the inherent indefiniteness. The main aims of studying the finance system is to achieve stable economic growth. A chaotic finance system is described by the three state variables with time variations as the interest rate, the investment demand and the price index respectively (Khan & Kumar, 2016; Chen, 2008). Many researchers and scientists have worked on the chaotic finance system. Afterwards, a hyper-chaotic finance system from the modified chaotic finance system was presented in 2009 (Ding, Yang & Yao, 2009). The hyper chaotic financial system is described by the four state variables with time variations as the production, money, stock and labor force respectively (Lin, 2018). Its nonlinear dynamical behaviour received great attention by scholars in the past several decades. Recently, Yu, Cai and Li have introduced a new 4D chaotic finance system and achieved its control with the linear feedback and speed feedback controllers. The control of hyper-chaotic finance systems was addressed with the speed feedback, linear feedback and time-delayed feedback controllers by many researchers (Cai & Huang, 2007; Gelberi et al., 2012; Chen, 2008). On the other hand, fractional order hyper chaotic finance systems are not investigated much so far. Therefore, the fractional order hyper-chaotic finance system needs to be further investigated.

Synchronization of fractional-order chaotic systems was first introduced by Deng and Li who have established the synchronization of fractional-order Lu systems (Lu, 2006). After that they have formulated the synchronization of fractional order Chen systems. Since then synchronization of fractional-order chaotic systems have been established by a few researchers and scientists. Gao et al. studied the master-slave synchronization of fractional order chaotic systems (Gao & Yu, 2005). The linear and nonlinear drive-response synchronization methods were applied for synchronizing the fractional-order chaotic Arneodo systems only using a scalar drive signal (Lu, 2005). Also, generalized projective synchronization of two chaotic or hyper-chaotic fractional-order systems have been presented (Hegazi & Matouk, 2011).

Motivated from the above discussion, we study the fractional-order of chaotic behaviour of four dimensional finance systems. We address the system descriptions and sliding mode control techniques of fractional order dynamical systems. In numerical simulations we justify the chaos in fractional order finance system with the lowest dimension. The synchronization of fractional-order hyper chaotic finance systems have been established using the sliding mode control techniques. These studies present new insights to understand the behaviour of fractional order hyper chaotic finance systems with the lowest dimension. Studies can help in macroeconomics analysis, forecasting financial situations, stocks markets and budget allocations.

FRACTIONAL DERIVATIVES AND ITS APPROXIMATION

Fractional calculus is a generalization of concepts of integration and differentiation to a non-integer-order integro-differential operator ${}_aD_t^{\alpha_1}$

It is written as

$$ {}_aD_t^{\alpha_1} = \begin{cases} \dfrac{d^{\alpha_1}}{dt^{\alpha_1}}, & if\ R(\alpha_1) > 0 \\ 1, & if\ R(\alpha_1) = 0 \\ \int\limits_a^t (d\tau)^{-\alpha_1}, & otherwise, i.e\ R(\alpha_1) < 0 \end{cases} \tag{1} $$

The generalized Riemann-Liouville dentition (Podlubny, 1999) is defined by

$$ D^{\alpha_1} g(t) = \frac{d^{\alpha_1}}{dt^{\alpha_1}} J^{n-\alpha_1} g(t), \alpha_1 > 0, \tag{2} $$

where n = $[\alpha_1]$: n is the first integer which is not less than α_1, J^{β_1} is the β_1 -order Riemann-Liouville integral operator which is described as follows

$$ J^{\beta_1} g(t) = \frac{1}{\Gamma(\beta_1)} \int\limits_0^t \frac{g(\tau)}{(t-\tau)^{1-\beta_1}} d\tau \tag{3} $$

for $0<\beta_1\leq 1$ where $\Gamma(.)$ is the gamma function.

The following definition is used as

$$ D^{\alpha_1} g(t) = J^{n-\alpha_1} g^n(t), \alpha_1 > 0, \tag{4} $$

where $n=[\alpha_1]$. It is the common practice to call the operator D_{α_1} the Caputo differential operator of order α_1 because it has apparently first been used for the solution of practical problems by Caputo (Podlubny, 1999; Samko & Kilbas, 1993).

Some Basic Definitions and Properties of Fractional Derivatives and Integrals

Definition 1 (Podlubny, 1999; Samko & Kilbas, 1993; Li, Chen & Podlubny, 2009): A real function $g(t)$, $t>0$ is said to be in Caputo space $C_{\alpha_1}, \alpha_1 \in R$ if there exist a real number $p(>\alpha_1)$ such that $g(t) = t^p g_1(t)$ where $g_1(t) \in C[0,\infty)$

Definition 2 (Podlubny, 1999; Samko & Kilbas, 1993; Li, Chen & Podlubny, 2009): A real function $g(t)$, $t>0$ is said to be in Caputo space $C_{\alpha_1}^m, m \in IN \cup 0$ If $g^{(m)} \in C_{\alpha_1}$

Definition 3 (Podlubny, 1999; Samko & Kilbas, 1993; Li, Chen & Podlubny, 2009): Let $g \in C_{\alpha_1}$ and $\alpha_1 \geq -1$; then the Riemann-Liouville integral of order α_1 ($\alpha_1>0$) is given by

$$I^{\alpha_1} g(t) = \frac{1}{\Gamma(\alpha_1)} \int_0^t (t-\alpha_1)^{\alpha_1 - 1} g(\tau) d\tau, t > 0 \tag{5}$$

Definition 4 (Podlubny, 1999; Samko & Kilbas, 1993; Li, Chen & Podlubny, 2009): The Caputo fractional derivative of $g, g \in C_{-1}^m, m \in IN \cup 0$ is defined as:

$$D^{\alpha_1} g(t) = \frac{d^m}{dt^m} g(t), \alpha_1 = m = I^{m-\alpha_1} \frac{d^m g(t)}{dt^m}, m-1 < \alpha_1 < m, m \in IN \tag{6}$$

Note that for $m-1 < \alpha_1 < m, m \in IN$

$$I^{\alpha_1} D^{\alpha_1} g(t) = g(t) - \sum_{k=0}^{m-1} \frac{d^k g(0)}{dt^k} \frac{t^k}{k}, I^{\alpha_1} t^{\alpha_1} = \frac{\Gamma(v+1)}{\Gamma(\alpha_1 + v + 1)} t^{\alpha_1 + v} \tag{7}$$

Consider the following differential equations as

$$\frac{d^{\alpha_1} x}{dt^{\alpha_1}} = g(t,x), 0 \leq t \leq T \text{ and } x^k(0) = x_0^{(k)}, k = 0,1,2,\ldots,n-1 \tag{8}$$

The differential equation is equivalent to the Volterra integral equation (Diethelm & Ford, 2002; Khan & Kumar, 2018; Khan & Kumar, 2019)

$$x(t) = \sum_{k=0}^{[\alpha_1]-1} x_0^{(k)} \frac{t^k}{k} + \frac{1}{\Gamma(\alpha_1)} \int_0^t \frac{g(\tau,x)}{(t-\tau)^{1-\alpha_1}} d\tau \tag{9}$$

Set

$$h = \frac{T}{N}, t_n = nh(n = 0,1,2,3...,N).$$

Then equation can be discretized as follows:

$$x_h(t_{n+1}) = \sum_{k=0}^{[\alpha_1]-1} x_0^{(k)} \frac{t_{n+1}^k}{k} + \frac{h^{\alpha_1}}{\Gamma(\alpha_1 + 2)} g(t_{n+1}, x_n^p(t_{n+1})) + \frac{h^{\alpha_1}}{\Gamma(\alpha_1 + 2)} \sum a_{j,n+1} g(t_j, x_h(t_j)) \tag{10}$$

where predicted value $x_h^p(t_{n+1})$ is determined by

$$x_h^p(t_{n+1}) = \sum_{k=0}^{[\alpha_1]-1} \frac{t_{n+1}^k}{k} + \frac{1}{\Gamma(\alpha_1)} \sum_{j=0}^{n} b_{j,n+1} g(t_j, x_h(t_j))$$

and

$$a_{j,n+1} = \begin{cases} n^{\alpha_1+1-(n-\alpha_1)(n+1)^{\alpha_1+1}} if\ j = 0 \\ (n-j+2)^{\alpha_1+1} + (n-j^{\alpha_1+1}) - 2(n-j+1)^{\alpha_1+1,} if\ 1 \le j \le n \\ 1, j = n+1 \end{cases}$$

$$b_{j,n+1} = \frac{h^{\alpha_1}}{\alpha_1}((n-j+1)^{\alpha_1} - (n-j)^{\alpha_1}).$$

The estimation error of this approximation is

$$n = Max \mid x(t_j) - x_h(t_j) \mid = o(h^p)(j = 0,1,...,N)$$

in which p=Min(2,1+α_1). Numerical solution of a fractional-order system can be determined by applying the mentioned method.

Now, consider the following fractional-order system:

$$\begin{aligned}
\frac{d^{\alpha_1}x}{dt^{\alpha_1}} &= g_1(x,y,z,w)\\
\frac{d^{\alpha_1}y}{dt^{\alpha_1}} &= g_2(x,y,z,w)\\
\frac{d^{\alpha_1}z}{dt^{\alpha_1}} &= g_3(x,y,z,w)\\
\frac{d^{\alpha_1}w}{dt^{\alpha_1}} &= g_4(x,y,z,w)
\end{aligned} \tag{11}$$

for $0<\alpha_1\leq 1$ and initial condition x_0, y_0, z_0, w_0. System (8) can be written as follows:

$$\begin{aligned}
x_{n+1} &= x_0 + \frac{h^{\alpha_1}}{\Gamma(\alpha_1+2)}[g_1(x_{n+1}^p, y_{n+1}^p, z_{n+1}^p, w_{n+1}^p) + \sum_{j=0}^{n}\gamma_{1,j,n+1}g_1(x_j, w_j, z_j, w_j)],\\
y_{n+1} &= y_0 + \frac{h^{\alpha_1}}{\Gamma(\alpha_1+2)}[g_2(x_{n+1}^p, y_{n+1}^p, z_{n+1}^p, w_{n+1}^p) + \sum_{j=0}^{n}\gamma_{2,j,n+1}g_2(x_j, w_j, z_j, w_j)],\\
z_{n+1} &= z_0 + \frac{h^{\alpha_1}}{\Gamma(\alpha_1+2)}[g_3(x_{n+1}^p, y_{n+1}^p, z_{n+1}^p, w_{n+1}^p) + \sum_{j=0}^{n}\gamma_{3,j,n+1}g_3(x_j, w_j, z_j, w_j)],\\
w_{n+1} &= w_0 + \frac{h^{\alpha_1}}{\Gamma(\alpha_1+2)}[g_4(x_{n+1}^p, y_{n+1}^p, z_{n+1}^p, w_{n+1}^p) + \sum_{j=0}^{n}\gamma_{4,j,n+1}g_4(x_j, w_j, z_j, w_j)],\\
x_{n+1}^p &= x_0 + \frac{1}{\Gamma(\alpha_1)} + \sum_{j=0}^{n}\omega_{1,j,n+1}g_1(x_j, w_j, z_j, w_j)],\\
y_{n+1}^p &= y_0 + \frac{1}{\Gamma(\alpha_1)} + \sum_{j=0}^{n}\omega_{2,j,n+1}g_2(x_j, w_j, z_j, w_j)],\\
z_{n+1}^p &= z_0 + \frac{1}{\Gamma(\alpha_1)} + \sum_{j=0}^{n}\omega_{3,j,n+1}g_3(x_j, w_j, z_j, w_j)],\\
w_{n+1}^p &= w_0 + \frac{1}{\Gamma(\alpha_1)} + \sum_{j=0}^{n}\omega_{4,j,n+1}g_4(x_j, w_j, z_j, w_j)],
\end{aligned}$$

$$\gamma_{i,j,n+1} = \begin{cases} n^{\alpha_1+1} - (n-\alpha_1)(n+1)^{\alpha_1+1}, if\ j=0\\ (n-j+2)^{\alpha_1+1} + (n-j^{\alpha_1+1}) - 2(n-j+1)^{\alpha_1+1}, if\ 1\leq j\leq n\\ 1, j=n+1 \end{cases}$$

$$\omega_{i,j,n+1} = \frac{h^{\alpha_1}}{\alpha_1}((n-j+1)^{\alpha_1} - (n-j)^{\alpha_1})0\leq j\leq n, i=1,2,3,4.$$

SYSTEMS DESCRIPTION AND METHODOLOGY USING SLIDING MODE CONTROL

We consider the chaotic system as

$$\frac{d^{\alpha_1}x(t)}{dt^{\alpha_1}} = Ax(t) + g(x(t)), \tag{12}$$

where $x \in R^n$ is the state vector of the system. A is the $n \times n$ matrix of the system parameters and $g: R^n \rightarrow R^n$ is the nonlinear part of the system. System (12) is treated as the master (drive) system. Again we consider the slave (response) chaotic system as follow:

$$\frac{d^{\alpha_1}y(t)}{dt^{\alpha_1}} = Ay(t) + g(y(t)) + u(t), \tag{13}$$

where $y \in R^n$ is the state vector of the slave system and $u(t) \in R^n$ is the controller to be designed. For that, we define the synchronization error as $e(t)=y(t) - x(t)$.

Then, the error dynamics is defined as

$$\frac{d^{\alpha_1}e(t)}{dt^{\alpha_1}} = Ae(t) + \theta(x, y) + u(t), \tag{14}$$

Here, $\theta(x,y)$ represents the nonlinear terms of the system. Our main aim for chaos synchronization problem is to find the controller $u(t)$ such that

$$\lim_{t \rightarrow \infty} \left\| e(t) \right\| = 0 \text{ for } e(0) \in R^n \tag{15}$$

In this regards, the controller $u(t)$ is defined as

$$u(t) = -\theta(x, y) + Bv(t), \tag{16}$$

where B is a constant gain vector selected such that (A,B) is controllable. $\upsilon(t) \in R^n$ represents the control input. From (14) and (16), the error dynamics is written as

$$\frac{d^{\alpha_1}e(t)}{dt^{\alpha_1}} = Ae(t) + Bv(t), \tag{17}$$

which is the linear time-invariant control system with single input $\upsilon(t)$. The synchronization problem can be reduced by an equivalent problem of stabilizing the zero solution $e(t) \rightarrow 0$ of the system (17) by a suitable choice of the sliding mode control.

In the sliding mode control (SMC), we define the variable (Khan & Kumar, 2018; Khan & Kumar, 2019) as

$$s(e) = Me(t) = m_1 e_1 + m_2 e_2 + ... + m_n e_n \tag{18}$$

where $M=(m_1,m_1,\ldots,m_1)$ is the constant row vector to be determined. We constrain the motion of the system (17) to the sliding manifold which is defined as

$$S = (x \in R^n \text{ such that } s(e(t)) = 0)\,. \tag{19}$$

This is required to be invariant under the flow of the error dynamics (17) which satisfies the following conditions $s(e(t))=0$ in the sliding manifold,

$$\frac{d^{\alpha_1} s(e(t))}{dt^{\alpha_1}} = 0\,. \tag{20}$$

This is the necessary condition for the state trajectory $e(t)$ of (17) to be on the sliding manifold S: From (17) and (18), the equation (20) is rewritten as,

$$\frac{d^{\alpha_1} s(e(t))}{dt^{\alpha_1}} = M(Ae(t)) + Bv(t) = 0\,. \tag{21}$$

Solving (21) for υ, we obtain the equivalent control law

$$v_{eq}(t) = -(MB)^{-1} MAe(t)\,, \tag{22}$$

where M is chosen such that $MB\neq 0$. Putting the equation (22) in the error system (17), we have

$$\frac{d^{\alpha_1} e(t)}{dt^{\alpha_1}} = (I - B(MB)^{-1} M)Ae(t)\,. \tag{23}$$

Here, I is the identity matrix. The row vector, M is selected such that the system matrix of the controlled dynamics $(I-B(MB)^{-1}M)A$ satisfies Hurwitz stability criterion. That is, it has all eigenvalues with negative real parts. Thus, the error system (17) is asymptotically stable. For designing the sliding mode controller for the error dynamics (17), we use the constant plus proportional rate reaching law as

$$\frac{d^{\alpha_1} e(t)}{dt^{\alpha_1}} = \rho sign(s) - \kappa s \tag{24}$$

where *sign*(.)denotes the sign function and gain ρ>0,κ>0 are determined such that sliding condition is satisfied. From (21) and (24), we can obtain the control υ(*t*) as

$$v(t) = -(MB)^{-1}(M(\kappa I + A)e(t) + \rho sign(s)), \tag{25}$$

which gives

$$v(t) = \begin{cases} -(MB)^{-1}(M(\kappa I + A)e(t) + \rho s(e(t))), if\ s(e(t)) > 0, \\ -(MB)^{-1}(M(\kappa I + A)e(t) - \rho s(e(t))), if\ s(e(t)) < 0 \end{cases}. \tag{26}$$

Lemma (Khan & Kumar, 2018; Khan & Kumar, 2019): The motion of sliding mode is asymptotically stable, if the following condition hold:

$$s^T(e(t))\frac{d^{\alpha_1}s(e(t))}{dt^{\alpha_1}} < 0. \tag{27}$$

Theorem (Khan & Kumar, 2018; Khan & Kumar, 2019): The fractional order master system (12) and the fractional order slave system (13) are asymptotically synchronized for the initial condition $x(0), y(0) \in R^n$ by the feedback control law,

$$u(t) = -\theta(x, y) + Bv(t), \tag{28}$$

where υ(*t*) is defined by (26) and *B* is a column vector such that (*A*,*B*) is controllable. Also, the slide mode gain κ and ρ are positive.

Proof: Putting the (27) and (25) into the error dynamics (23), we find the closed-loop error dynamics as

$$\frac{d^{\alpha_1}e(t)}{dt^{\alpha_1}} = Ae(t) - B(MB)^{-1}(M(\kappa I + A)e(t) + \rho sign(s)). \tag{29}$$

To prove the error dynamics(23) is asymptotically stable, we consider the Lyapunov function defined by the equation

$$V(e(t)) = \frac{1}{2}s^2(e(t)). \tag{30}$$

We note that $V(e(t)) \geq 0$ for $e(t) \in R^n$ and

$$V(e(t)) = 0 \Leftrightarrow e(t)) = 0.$$

Thus, it follows that V is a positive definite function on R^n. Differentiating V along the trajectories of (17) of the equivalent dynamics (24), we have

$$\frac{d^{\alpha_1}V(e(t))}{dt^{\alpha_1}} = s^T(e(t))\dot{s}(e(t)) = -\kappa s^2 - \rho sign(s)s, \tag{31}$$

which is the negative definite function of R^n. These calculations show that V is defined positive definite Lyapunov function for the error dynamics (17) and $\frac{d^{\alpha_1}V(e(t))}{dt^{\alpha_1}}$ is defined negative definite function. Thus, by the Lyapunov stability theory (Duan & Yu,2013), the error dynamics (17) is asymptotically stable for the initial condition $e(0)\in R^n$. We have

$$\lim_{t\to\infty} e(t) = 0. \tag{32}$$

Therefore, the fractional order master system (12) and the fractional order slave system (13) are asymptotically synchronized for the initial conditions $x(0),y(0)\in R^n$.

NUMERICAL RESULTS

This is the synchronization of identical fractional order hyper-chaotic finance system using the sliding mode control.

Consider the hyper-chaotic finance system (Lin, 2018)

$$\begin{aligned}
\dot{x}(t) &= z(t) + (y(t) - a)x(t) + w(t),\\
\dot{y}(t) &= 1 - by(t) - x(t)x(t),\\
\dot{z}(t) &= -x(t) - cz(t),\\
\dot{w}(t) &= -dx(t)y(t) - \kappa w(t)
\end{aligned}$$

where $x,y,z,w\in R^n$ is the state vectors; $x(t)$ is the rate of interest, $y(t)$ is the investment demand, $z(t)$ the price exponent. and $w(t)$ is the saving. We take a=0.9, b=0.2, c=1.5, d=0.2 and select κ as a governing parameter. Based on the above descriptions, we consider the new version fractional order hyper-chaotic finance system as follows:

$$\begin{aligned}
\frac{d^{\alpha_1}x}{dt^{\alpha_1}} &= z(t)+(y(t)-a)x(t)+w(t),\\
\frac{d^{\alpha_1}y}{dt^{\alpha_1}} &= 1-by(t)-x(t)x(t),\\
\frac{d^{\alpha_1}z}{dt^{\alpha_1}} &= -x(t)-cz(t),\\
\frac{d^{\alpha_1}w}{dt^{\alpha_1}} &= -dx(t)y(t)-\kappa w(t)
\end{aligned} \tag{33}$$

where α_1 is the fractional order satisfying $0<\alpha_1\leq 1$. Two identical fractional order master (drive) and slave (response) systems for hyper chaotic finance system (33) are rewritten with the subscripts of *x* and *y* respectively as:

Master system:

$$\begin{aligned}
\frac{d^{\alpha_1}x_1}{dt^{\alpha_1}} &= x_3(t)+(x_2(t)-a)x_1(t)+x_4(t),\\
\frac{d^{\alpha_1}x_2}{dt^{\alpha_1}} &= 1-bx_2(t)-x_1(t)x_1(t),\\
\frac{d^{\alpha_1}x_3}{dt^{\alpha_1}} &= -x_1(t)-cx_3(t),\\
\frac{d^{\alpha_1}x_4}{dt^{\alpha_1}} &= -dx_1(t)x_2(t)-\kappa x_4(t).
\end{aligned} \tag{34}$$

Its slave fractional-order hyper chaotic finance system is:

$$\begin{aligned}
\frac{d^{\alpha_1}y_1}{dt^{\alpha_1}} &= y_3(t)+(y_2(t)-a)y_1(t)+y_4(t)+u_1(t),\\
\frac{d^{\alpha_1}y_2}{dt^{\alpha_1}} &= 1-y_2(t)-y_1(t)y_1(t)+u_2(t),\\
\frac{d^{\alpha_1}y_3}{dt^{\alpha_1}} &= -y_1(t)-cy_3(t)+u_3(t),\\
\frac{d^{\alpha_1}y_4}{dt^{\alpha_1}} &= -dy_1(t)y_2(t)-\kappa y_4(t)+u_4(t).
\end{aligned} \tag{35}$$

where, $u(t)=(u_1,u_2,u_3,u_4)$ is the controller. Our aim is to determine the controller $u(t)$ such that the fractionalorder master system synchronize the fractional-order slave system. For this, we define the error signal for (34) and (35) as:

$$
\begin{aligned}
&e(t) = [e_1(t), e_2(t), e_3(t), e_4(t)]^T \\
&e_1(t) = y_1(t) - x_1(t), e_2(t) = y_2(t) - x_2(t), e_3(t) = y_3(t) - x_3(t), \\
&e_4(t) = y_4(t) - x_4(t),
\end{aligned} \tag{36}
$$

The error dynamics is written as

$$
\begin{aligned}
\frac{d^{\alpha_1} e_1(t)}{dt^{\alpha_1}} &= \frac{d^{\alpha_1} y_1(t)}{dt^{\alpha_1}} - \frac{d^{\alpha} x_1(t)}{dt^{\alpha}} = e_3(t) - e_4(t) - ae_1(t) + y_1(t)y_2(t) - x_1(t)x_2(t) + u_1(t), \\
\frac{d^{\alpha_1} e_2(t)}{dt^{\alpha_1}} &= \frac{d^{\alpha_1} y_2(t)}{dt^{\alpha_1}} - \frac{d^{\alpha} x_2(t)}{dt^{\alpha}} = -be_2(t) - y_1(t)y_1(t) - x_1(t)x_1(t) + u_2(t), \\
\frac{d^{\alpha_1} e_3(t)}{dt^{\alpha_1}} &= \frac{d^{\alpha_1} y_3(t)}{dt^{\alpha_1}} - \frac{d^{\alpha} x_3(t)}{dt^{\alpha}} = -be_1(t) - ce_3(t) + u_3(t), \\
\frac{d^{\alpha_1} e_4(t)}{dt^{\alpha_1}} &= \frac{d^{\alpha_1} y_4(t)}{dt^{\alpha_1}} - \frac{d^{\alpha} x_4(t)}{dt^{\alpha}} = -\kappa e_2(t) - dy_1(t)y_1(t)y_2(t) + dx_1(t)x_2(t) + u_4(t).
\end{aligned} \tag{37}
$$

In the matrix notation, the error dynamics is written as

$$
\frac{d^{\alpha_1} e(t)}{dt^{\alpha_1}} = Ae(t) + \theta(x, y) + u(t), \tag{38}
$$

where

$$
A = \begin{bmatrix} -1 & 0 & 1 & 1 \\ 0 & -b & 0 & 0 \\ -1 & 0 & -c & 0 \\ 0 & 0 & 0 & -k \end{bmatrix}; \ \theta(x, y) = \begin{bmatrix} (y_1(t)y_2(t) - x_1(t)x_2(t)) \\ (y_1(t)y_1(t) - x_1(t)x_1(t)) \\ 0 \\ -dy_1(t)y_2(t) - dx_1(t)x_2(t)) \end{bmatrix} \quad u(t) = \begin{bmatrix} u_1(t) \\ u_2(t) \\ u_3(t) \\ u_4(t) \end{bmatrix}.
$$

We set the the sliding mode controller, $u(t)$ as,

$$
u(t) = -\theta(x, y) + Bv(t) \tag{39}
$$

where B is chosen such that (A,B) is controllable. We take B as $B=(1,1,1,1)^T$.

The sliding mode variable is set as

$$
s = Me = [-3, 5, -1, 1]e = -3e_1 + 5e_2 - e_3 + e_4. \tag{40}
$$

Here, we take $M=[-3,5,-1,1]$ such that the all eigenvalues of the matrix $p=[I - B(MB)^{-1}M]*A$ are negative real part or zero. That is,

$$eig(p) = -0.6025 + 1.9704i, -0.6025 - 1.9704i, 0, -0.0949\,.$$

This makes that the dynamical system along the sliding manifold is globally asymptotically stable. We take the sliding mode gain as κ=4, ρ=0.1 and I is the identity matrix of order 4. We observe that the large value of κ can cause chattering (noise) and an appropriate value of ρ is chosen to speed up the time taken to reach the sliding manifold as well as to minimize the system chattering (noise) (Khan & Kumar, 2018; Khan & Kumar, 2019). Using (25), time-invariant control signal with single input, $\upsilon(t)$

$$\begin{aligned} v(t) &= -(MB)^{-1}(M(\kappa I + A)e(t) + \rho sign(s)), \\ &= \frac{1}{8}([-35 - 11]\begin{bmatrix} (4-a) & 0 & -1 & -1 \\ 0 & (4-b) & 0 & 0 \\ -1 & 0 & (4-c) & 0 \\ 0 & 0 & 0 & -\kappa \end{bmatrix} e(t) + \rho sign(s)), \end{aligned} \tag{41}$$

$\upsilon(t) = [4.15, -9.5, 2.0, -0.45]e(t) - 0.05sign(s)$.

Thus, the required sliding mode controller is obtained as $\upsilon(t) = -\theta(x,y) + B\upsilon(t)$. Hence, we can obtain the following result.

Remark 1: The identical fractional order hyper-chaotic finance systems (34) and (35) are asymptotically synchronized for initial conditions with the sliding mode controller $u(t)$ defined in the equation (39).

Simulation

Take the initial conditions for master and slave hyper chaotic finance systems, $x(0)=(1.5,2.5,-0.5,1)^T$ and $y(0)=(-1.5,3.5,1.5,-1)^T$ take parameters values a=0.9, b=0.2, c=1.2, d=0.2, k=0.1 and fractional order α_1=0.90 respectively. Figure 1(a)-(d) is shown the three dimensional phase portrait fractional order hyper chaotic finance systems with different parameters values of fractional orders α_1=0.80, α_1=0.85, α_1=0.88 and α_1=0.90 respectively. Figures 2(a)-(d) is shown as time-series graphs with different axes of hyper chaotic finance systems corresponding to Figure 1(a)-(d). We have observed that Figure 2(a)-(d) are periodic, periodic, periodic and chaotic respectively. We have analysed the figures 1(a)-(d) and 2(a)-(d) and we have observed through computer simulation that the lowest dimension of hyper-chaotic finance system is 3.60. Figure 3(a)-(d) is shown as bifurcation diagrams of hyper chaotic finance systems with respect to the parameters a,b,c,d and k respectively. We have computed the Lyapunov exponent of hyper chaotic finance system at t=300. We have λ_1=0.25449, λ_2=0.054728, λ_3= -0.51089 and λ_4= -0.6213. We observe that out of these four Lyapunov exponent values, two are positive which is required condition for hyper chaotic systems. It represents that four dimensional fractional order finance systems system is hyper chaotic. It is shown in Figure 4. Figures 5(a)-(d) show the tracking the trajectories of slave to master fractional order hyper chaotic finance systems in x_1y_1, x_2y_2, x_3y_3 and x_4y_4 with sliding mode controllers.

Trajectories of master and slave identical fractional order hyper chaotic finance systems are synchronized with sliding mode control techniques in these figures 5(a)-(d). Figure 6 is shown the synchronization of error dynamic of identical fractional order hyper chaotic finance systems in the form of $e_1e_2e_3e_4$ with sliding mode control with respect to time t at the initial condition

$$(e_1(0), e_2(0), e_3(0), e_4(0) = (-3, 1, 2, -2)^T$$

That is,

$$\lim_{t\to\infty} e(t) = 0$$

Figure 1. 3-Dimensional phase portrait of hyper chaotic finance systems (without controller)

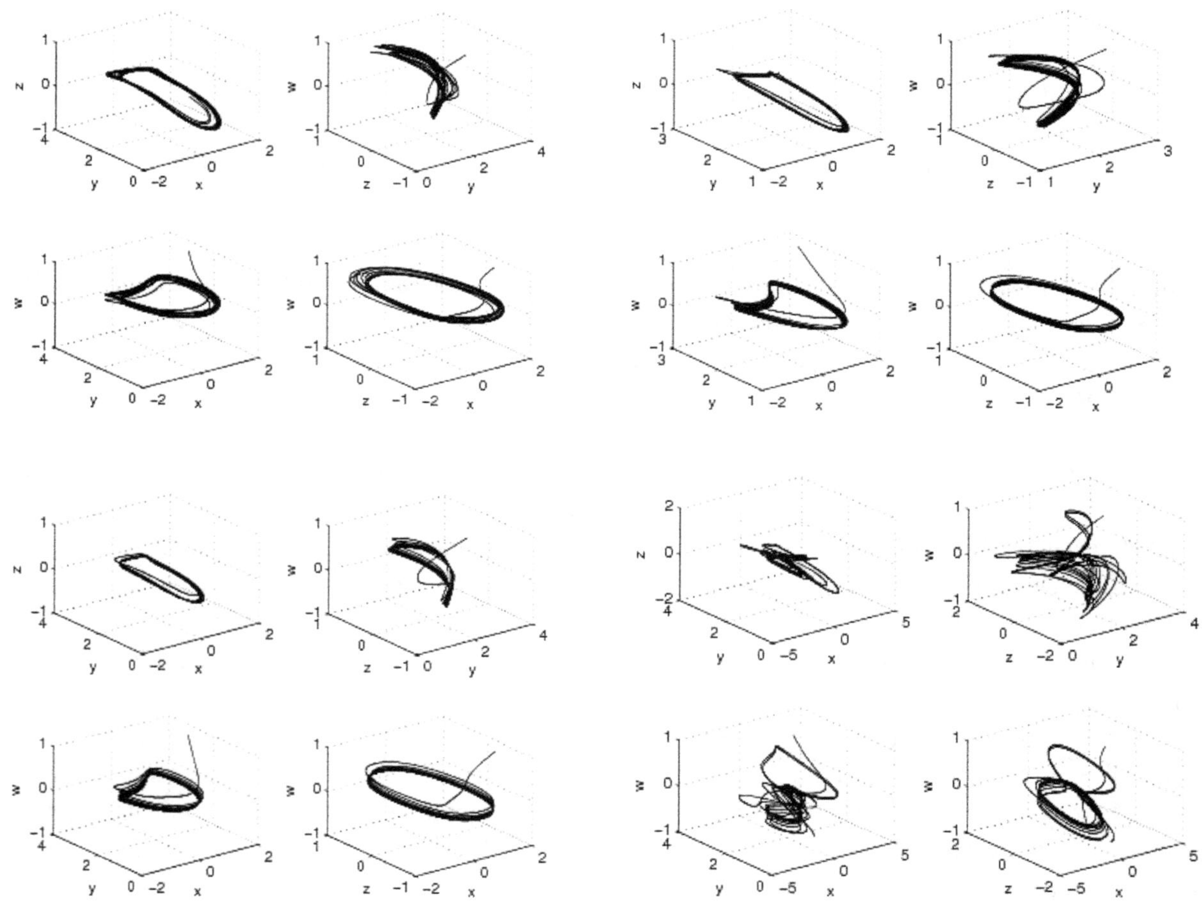

Figure 2. Time series graphs of hyper chaotic finance system (without controller) corresponding to figure 1(a)-1(d).

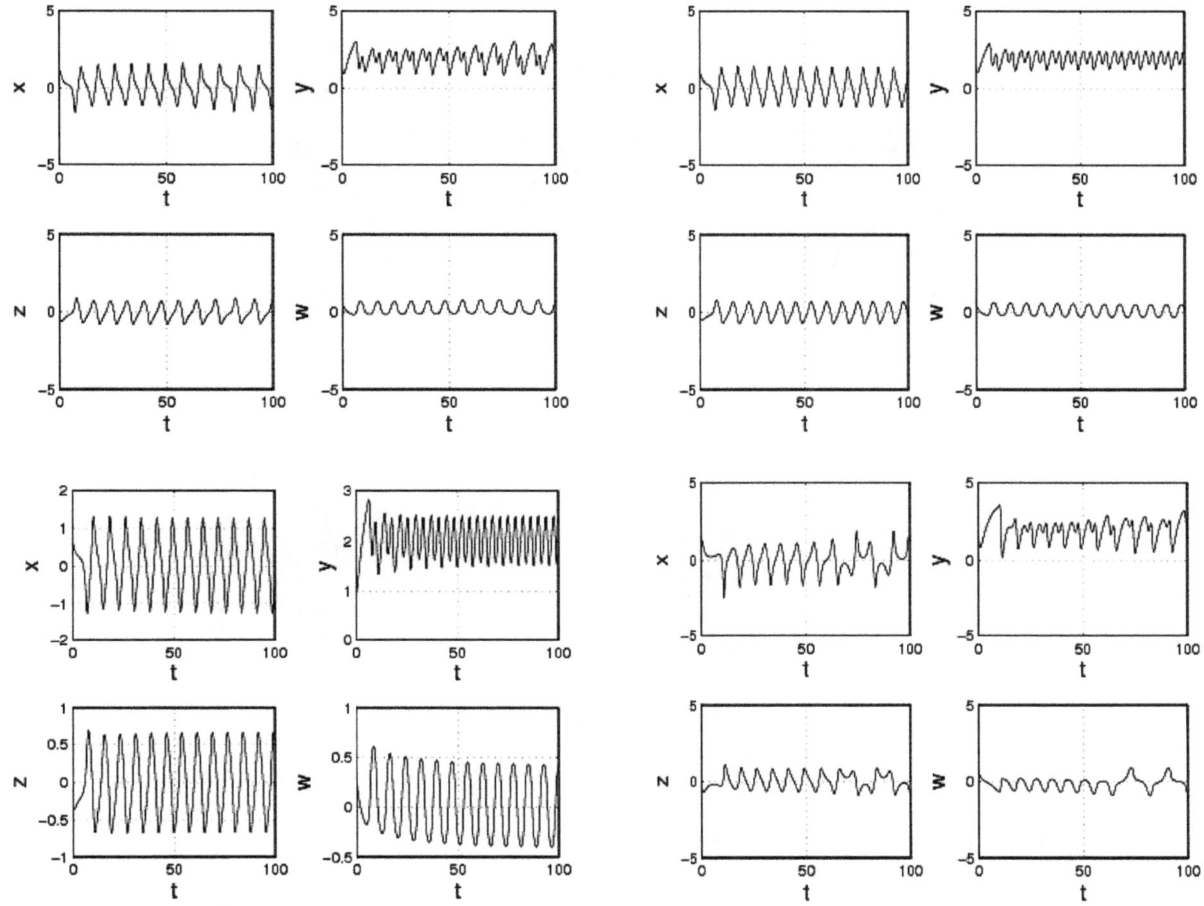

CONCLUSION

In this chapter, we have studied the basic concepts and methods of solving the fractional-order dynamical systems. Synchronization of fractional order dynamical systems has been established using sliding mode control techniques. In numerical results, we have drawn the bifurcation diagrams and Lyapunov exponents of fractional-order hyper chaotic finance systems. We have observed through computer simulation that the lowest dimension of the hyper-chaotic finance system is 3.60. The validity of the results are verified by synchronization of fractional-order systems.

Figure 3. Bifurcation diagrams with the parameter a, b, c, d and k

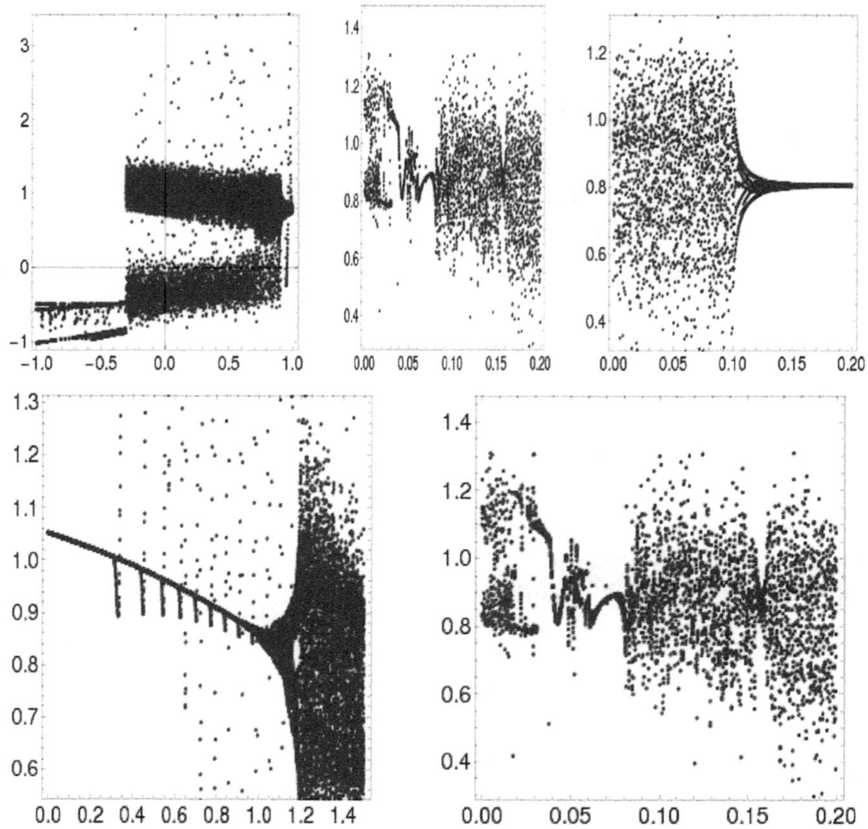

Figure 4. Lyapunov exponent of a hyper chaotic finance system

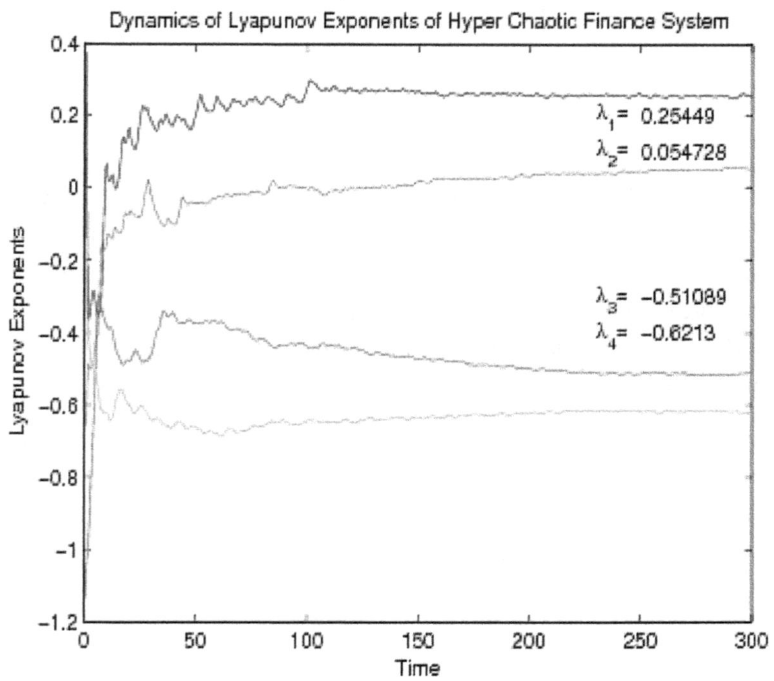

Figure 5. Tracking the trajectory of the master to the salve hyper chaotic finance system systems (with controller)

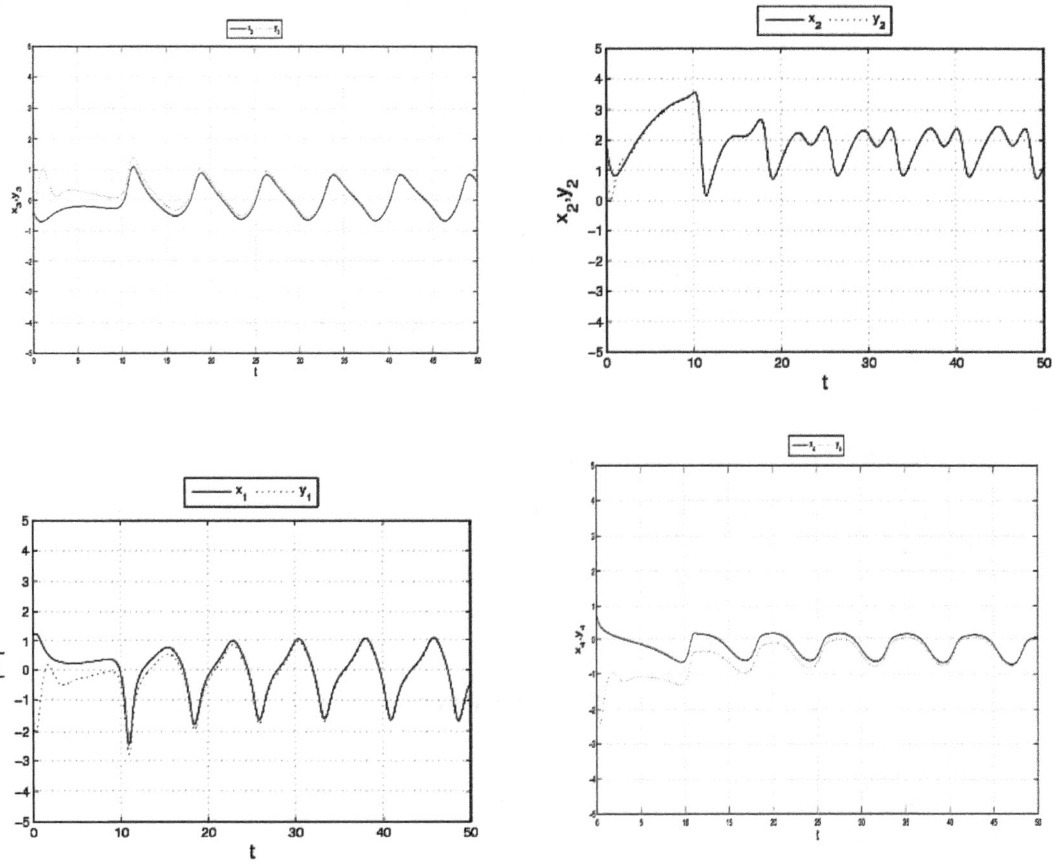

Figure 6. Synchronization of error dynamics of identical hyper chaotic finance systems (with controller)

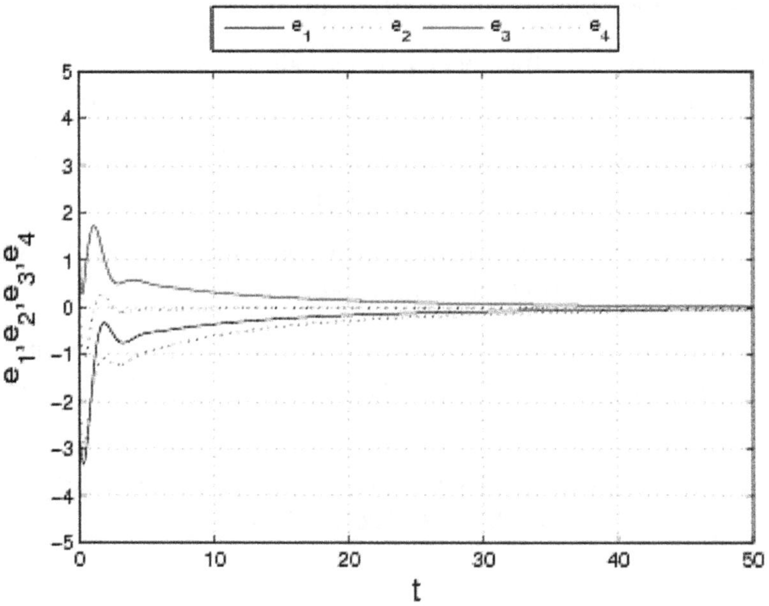

REFERENCES

Cai, G., & Huang, J. (2007). A new finance chaotic attractor. *International Journal of Nonlinear Science, 3*(3), 213–220.

Chen, D. Y., Liu, Y. X., Ma, X. Y., & Zhang, R. F. (2012). Control of a class of fractional-order chaotic systems via sliding mode. *Nonlinear Dynamics, 67*(1), 893901. doi:10.100711071-011-0002-x PMID:22757537

Chen, W. (2008). Dynamics and control of a financial system with time-delayed feed backs. *Chaos, Solitons, and Fractals, 37*(4), 1198–1207. doi:10.1016/j.chaos.2006.10.016

Chen, W. C. (2008). Nonlinear dynamics and chaos in a fractional-order financial system. *Chaos, Solitons, and Fractals, 36*(5), 1305–1314. doi:10.1016/j.chaos.2006.07.051

Daftardar-Gejji, V., & Babakhani, A. (2004). Analysis of a system of fractional differential equations. *J. Math Appl Anal, 293*(2), 511–522. doi:10.1016/j.jmaa.2004.01.013

Das, S., & Gupta, P. A. (2011). mathematical model on fractional Lotka-Volterra equations. *Journal of Theoretical Biology, 277*(1), 16. doi:10.1016/j.jtbi.2011.01.034 PMID:21295596

Diethelm, K., & Ford, N. J. (2002). Analysis of fractional differential equations. *Journal of Mathematical Analysis and Applications, 265*(2), 229–248. doi:10.1006/jmaa.2000.7194

Ding, J., & Yang, W. Y. H. (2009). A new modified hyperchaotic finance system and its control. *International Journal of Nonlinear Science, 8*(1), 59–66.

Duan, G.R. & Yu, H.H. (2013). *LMI in control systems analysis, design and applications.* CRC Press, Taylors and Francis Group.

Faieghi, M. R., Delavari, H., & Baleanu, D. (2013). A note on stability of sliding mode dynamics in suppression of fractional-order chaotic systems. *Computers & Mathematics with Applications (Oxford, England), 66*(5), 832–847. doi:10.1016/j.camwa.2012.11.015

Gao, X., & Yu, J. (2005). Chaos in the fractional order periodically forced complex Duffings oscillators. *Chaos, Solitons, and Fractals, 24*(4), 1097–1104. doi:10.1016/j.chaos.2004.09.090

Gelberi, H. (2012). Time delay feedback control of chaos in a hyper chaotic finance system. In *3rd International Symposium on Sustainable Development.* IEEE.

Grigorenko, I., & Grigorenko, E. (2003). Chaotic dynamics of the fractional Lorenz system. *Physical Review Letters, 91*(3), 034101. doi:10.1103/PhysRevLett.91.034101 PMID:12906418

Hegazi, A. S., & Matouk, A. E. (2011). Dynamical behaviors and synchronization in the fractional order hyperchaotic chen system. *Applied Mathematics Letters, 24*(11), 1938–1944. doi:10.1016/j.aml.2011.05.025

Jun-Guo, L. (2005). Chaotic dynamics and synchronization of fractional-order Genesio-Tesi systems. *Chinese Physics, 14*(8), 15–17. doi:10.1088/1009-1963/14/8/007

Khan, A & Kumar, S. (2016). Anti-synchronization of chaotic finance systems using fuzzy logic constant controller. *Indian Journal of Industrial and Applied Mathematics*. Doi:10.5958/1945-919X.2016.00013X

Khan, A., & Kumar, S. (2018). Study of chaos in satellite systems. *Pramana Journal of Physics*, *90*(1), 13. doi:10.100712043-017-1502-0

Khan, A & Kumar, S. (2018). Analysis and time-delay synchronization of chaotic satellite systems. *Pramana Journal of Physics, 91*, 49. doi:10.100712043-018-1610-5

Khan, A., & Kumar, S. (2018). Measuring chaos and synchronization chaotic satellite systems using sliding mode control. *Optimal Control Applications & Methods*, *39*(5), 1597–1609. doi:10.1002/oca.2428

Khan, A., & Kumar, S. (2018). T–S fuzzy observed based design and synchronization of chaotic and hyper-chaotic dynamical systems. *International Journal of Dynamics and Control*, *6*(3), 1409–1419. doi:10.100740435-017-0358-y

Khan, A. & Kumar, S. (2018). TS Fuzzy modeling and synchronization of chaotic systems. *Journal of Uncertain Systems*, *10*(4), 251-259. Doi:10.1002/oca.2428

Khan, A., & Kumar, S. (2019). T-S fuzzy modeling and predictive control and synchronization of chaotic satellite systems. *International Journal of Modeling and Simulation, 39*(3).

Khan, A., & Kumar, S. (2019). Measure of chaos and adaptive synchronization of chaotic satellite systems. *International Journal of Dynamics and Control*, *7*(2), 536–546. doi:10.100740435-018-0481-4

Kumar, S. (2019). Synchronization of fractional order Rabinovich-Fabrikant systems using sliding mode control techniques. *Archives of Control Sciences*, *29*(2).

Li, C., & Chen, G. (2004). Chaos in the fractional order Chen system and its control. *Chaos, Solitons, and Fractals*, *22*(3), 549–554. doi:10.1016/j.chaos.2004.02.035

Li, C., & Chen, G. (2004). Chaos and hyperchaos in the fractional order Rssler equations. *Physica A*, *341*, 5561. doi:10.1016/j.physa.2004.04.113

Li, Y., Chen, Y., & Podlubny, I. (2009). Mittag-leffler stability of fractional order nonlinear dynamic systems. *Automatica*, *45*(8), 1965–1969. doi:10.1016/j.automatica.2009.04.003

Lin, C. (2018). A Four-Dimensional Hyperchaotic Finance System and Its Control Problems. *Journal of Control Science and Engineering*.

Lu, J. G. (2005). Chaotic dynamics and synchronization of fractional-order Arneodos systems. *Chaos, Solitons, and Fractals*, *26*(4), 1125–1133. doi:10.1016/j.chaos.2005.02.023

Lu, J. G. (2006). Chaotic dynamics of the fractional-order L system and its synchronization. *Physics Letters. [Part A]*, *354*(4), 305–311. doi:10.1016/j.physleta.2006.01.068

Matouk, A. E. (2009). Stability conditions, hyperchaos and control in a novel fractional order hyperchaotic system. *Physics Letters. [Part A]*, *373*(25), 2166–2173. doi:10.1016/j.physleta.2009.04.032

Odibat, Z. M. (2010). Adaptive feedback control and synchronization of non-identical chaotic fractional order systems. *Nonlinear Dynamics*, *60*(4), 479–487. doi:10.100711071-009-9609-6

Podlubny, I. (1999). *Fractional differential equations*. San Diego, CA: Academic Press.

Samko, S. G., Kilbas, A. A., & Marichev, O. I. (1999). *Fractional integrals and derivatives: theory and applications*. Yverdon: Gordon and Breach.

Wang, Z., Huang, X., & Shi, G. D. (2004). Analysis of nonlinear dynamics and chaos in a fractional order financial system with time delay. *Computers & Mathematics with Applications (Oxford, England), 2011*(62), 1531–1539.

Wilkie, K. P., Drapaca, C. S., & Sivaloganathan, S. (2011). A nonlinear viscoelastic fractional derivative model of infant hydrocephalus. *Applied Mathematics and Computation, 217*(21), 8693–8704. doi:10.1016/j.amc.2011.03.115

Section 3

Applications of Fractional Derivatives to Physics

Chapter 7
Generalized Brinkman Type Dusty Fluid Model for Blood Flow

Muhammad Saqib
Universiti Teknologi Malaysia, Malaysia

Sharidan Shafie
Universiti Teknologi Malaysia, Malaysia

Ilyas Khan
College of Engineering, Majmaah University, Saudi Arabia

ABSTRACT

This chapter is dedicated to studying the magnetic blood flow with uniformly distributed magnetite dusty particles (MDP) in a cylindrical tube. For this purpose, the two-phase fractional Brinkman type fluid model is considered. The fractional governing equations are modeled in the cylindrical coordinate system taking into consideration the magnetization of the fluid due to the applied magnetic field. The fractional governing equations are subjected to physical initial and boundary conditions. The joint Laplace and Hankel transform is employed to develop exact analytical solutions. The obtained solutions are computed numerically and plotted in different graphs. It is noticed that for a long time the blood and MDP velocities increase with increasing values of the fractional parameter. In contrast, this effect reverses for a shorter time. In the case of the magnetic parameter, both velocities are decreased with increasing values of the magnetic parameter.

INTRODUCTION

In fluid dynamics, the study of biological fluids (BF) under the influence of a magnetic field is referred to as bio-magnetic fluid dynamics (BFD). This area in fluid dynamics attracted numerous researcher (Carlton *et al.*, 2001; Haik *et al.*, 1999; Liu *et al.*, 2001; Voltairas *et al.*, 2002) due to abundance of

DOI: 10.4018/978-1-7998-3122-8.ch007

application proposed by medical science and bio-engineering which include development of magnetic tracers, targeted transport of drugs using magnetic particles as drug carriers, development of magnetic devices for cell separation, provocation of occlusion of the feeding vessels of cancer tumors, cancer tumor treatment causing magnetic hyperthermia and reduction of bleeding during surgeries (Tzirtzilakis, 2005). The BF exists in all living organisms and their flow is significantly affected by the incidence of the magnetic field. Among different biological fluids, the one electrically conducting fluid is blood. The human blood is composed of plasma and red blood cells (RBC's) (Boyd, 1961). The RBC's consists of a high concentration of hemoglobin, the oxide of iron while the magnetic characteristics of blood are due to the state of oxygenation (Sharma *et al.*, 2015).

To study bio-magnetic fluids particularly blood, the mathematical BFD model developed by Haik *et al.* (1999) is taken into consideration. They suggested that Ferro-hydrodynamics (FHD) is like BFD in which the flow of electrically conducting fluid is influenced by the magnetization of the fluid in the presence of magnetic field and no induce current due to small Reynold number. Hence, in the mathematical model of BFD, the magnetization of the fluid is assumed in the formulation that corresponds to magnetohydrodynamics (MHD) (Haik *et al.*, 2002). Tzirtzilakis (Tzirtzilakis, 2005) considered blood as a Newtonian fluid and develop a mathematical model for blood flow under the effect of magnetic field in a rectangular duct taking into consideration the idea of Haik *et al.* (1999) . Sharma et al. (Sharma *et al.*, 2015) reported a comprehensive study on MHD blood flow with magnetite particles in a cylindrical tube. Ghasemi *et al.* (2015) driven analytical and numerical results for blood flow in the porous artery. The blood is assumed as a non-Newtonian third-grade fluid together with heat transfer and thermophoresis effect. Ellahi *et al.* (2015) considered Prandtl fluid as blood in stenosed arteries with the porous wall. They developed a two-dimensional model for blood flow with slip condition in the cylindrical coordinate system and generated analytical solutions via perturbation technique. Bhatti *et al.* (2016b) analyzed blood with suspension particles considering endoscopic and slip effect in the non-uniform annulus. In another study, Bhatti *et al.* (2016a) extended his work by taking Sisko fluid as blood with magnetite titanium nanoparticles in the uniform tube. A theoretical study on MHD blood flow is carried out by Sinha and Shit (2015) in a capillary with the combined effect of radiative heat transfer.

All the above studies on BMF using MHD based mathematical models are described by partial differential equations with an integer order. But it is proven in the open literature by many investigators (Ali *et al.*, 2016; Khan *et al.*, 2017; Khan *et al.*, 2018; Saqib *et al.*, 2018a; Saqib *et al.*, 2018b; Saqib *et al.*, 2018c; Saqib *et al.*, 2018d) that partial differential equations with fractional order improve the generality and accuracy of the mathematical model. Fractional order tools (derivatives and integration) have extensive applications in engineering and sciences which included electrochemistry, optics, signals processing, biological population models, electromagnetics, viscoelasticity, and fluid mechanics. Further, most physical phenomena of electricity, quantum mechanics, ecological systems, fluid dynamics and many more measured inside their domain of validity via fractional PDE's. Hence, it is even more significant to aware with all the old and newly established methods for the solutions of fractional order PDE's and employment of these methods (Ray *et al.*, 2014).

Motivated by the above-discussed literature, the Brinkman type fluid is considered as blood with uniformly distributed MDP which is present in the hemoglobin of human blood. A fractional mathematical model in Caputo sense is developed in cylindrical coordinates system for blood flow phenomena in a cylindrical tube. The Caputo fractional derivative is used here due to a physical reason which can be illustrated $0<\alpha<1$ by the following formulation (Podlubny, 1998)

$$\left.\begin{array}{l} L\left\{{}^{RL}D_t^{\alpha} f\left(t\right)\right\}\left(q\right) = q^{\alpha}\bar{f}\left(q\right) - {}^{RL}D_t^{\alpha-1} f\left(t\right)\Big|_{t=0} \\ L\left\{{}^{C}D_t^{\alpha} f\left(t\right)\right\}\left(q\right) = q^{\alpha}\bar{f}\left(q\right) - f\left(t\right)\Big|_{t=0} \end{array}\right\}. \tag{1}$$

In the case of the Caputo fractional derivative, the appeared constant term is the usual initial value of the functions. However, in Riemann Liouville case the constant term is not the initial value of the function which has no simple physical interpretation (Naber, 2010). Therefore, the Caputo fractional derivative is more suitable for modeling physical phenomena. The fractional model for the two-phase blood flow under the influence of an applied transverse magnetic field is solved analytically using the joint Laplace and Hankel transforms. The obtained exact analytical results are numerically computed and plotted. It is observed that the blood and MDP velocities exhibit dual behavior for fractional order. Furthermore, the blood and MDP velocities show a decreasing trend for a stronger magnetic field due to resistive type forces.

DESCRIPTION OF THE PROBLEM

Let us consider the blood flow in which MDP are uniformly distributed in a cylindrical tube with a radius r_0 as shown in Figure 1.

Figure 1. The physical configuration of Blood flow

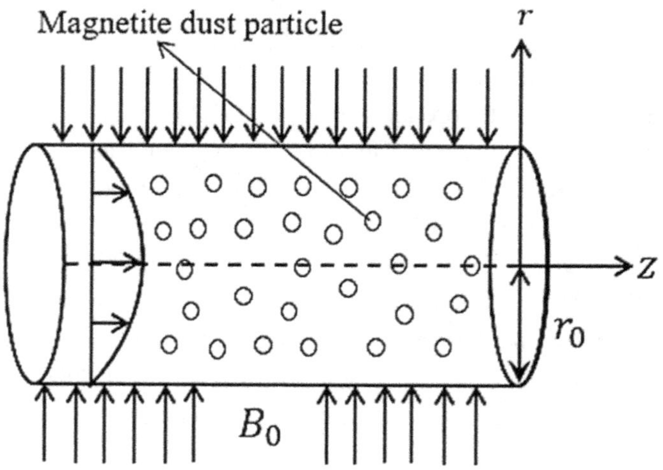

The cylinder is taken along z-axis and r-axis is normal to it. The blood and MDP are moving in z-direction whereas, the magnetic field of strength B_0 is acting transversely on it. It is assumed that the magnetic Reynold number to be very small due to which the induced magnetic field is neglected. Initially, at $t \leq 0$, the blood, MDP, and the cylindrical tube are stationary. The motion of magnetic material under the influence of magnetic field experience the electromotive force which leads to the current flow. The flow of electrically conducting blood generates electromagnetic force due to the application of a transverse magnetic field which depends on speed and electric flux intensity (Tashtoush and Magableh,

2008). The governing equations of the problem involve the Brinkman type fluid equations for the blood flow, Maxwell's relations for the magnetic field and Newton's second law for MDP velocity. Maxwell's set of equations is as follow (Saqib *et al.*, 2019b)

$$\left.\begin{aligned} &\nabla . B = 0, \\ &\nabla \times E = -\frac{\partial B}{\partial t} \\ &\nabla \times B = \mu_0 J, \end{aligned}\right\}. \tag{2}$$

Here B is the magnetic flux intensity, μ_0 is the magnetic permeability, E is electric field intensity and J is the current density described by the generalized Ohm's law as (Abdullah *et al.*, 2018)

$$J=\sigma_0[E+V\times B], \tag{3}$$

where σ_0 is the electrical conductivity and $\vec{V}$ is the velocity field. Moreover, the electromagnetic force is described as (Sheikholeslami, 2018)

$$F_{em} = J \times B = \sigma_0[E + V \times B] \times B = -\sigma B_0^2 u\left(r,t\right) k\,, \tag{4}$$

with k is a unit vector in z-direction, $V = u\left(r,t\right)\vec{k}$ is the axial velocity of the blood. The electromagnetic force F_{em} is incorporated to the momentum equation of unsteady incompressible blood flow in a cylindrical tube of the radius r_0 in the presence of pressure gradient and applied a transverse magnetic field in the cylindrical coordinate system (r,θ,z) takes the following form (Ali *et al.*, 2017; Saqib *et al.*, 2019a)

$$\begin{aligned} &\rho\left(\frac{\partial u\left(r,t\right)}{\partial t} + \beta u\left(r,t\right)\right) = \\ &-\frac{\partial p}{\partial z} + \mu\left(\frac{\partial^2 u\left(r,t\right)}{\partial r^2} + \frac{1}{r}\frac{\partial u\left(r,t\right)}{\partial r}\right) + KN\left(u_p\left(r,t\right) - u\left(r,t\right)\right) - \sigma B_0^{\,2} u\left(r,t\right) \end{aligned}. \tag{5}$$

The pressure gradient $\frac{\partial p}{\partial z}$ is taken of the following oscillating form (Mandal, 2005)

$$\frac{\partial p}{\partial z} = -\lambda_0 - \lambda_1 \cos\omega t\,, \tag{6}$$

where λ_0 and λ_1 are the amplitudes of pressure gradient which give rise to systolic or diastolic pressure. Equation (6) is included in equation (5) which leads to the following form

$$\rho\left(\frac{\partial u(r,t)}{\partial t}+\beta u(r,t)\right)=\lambda_0+\lambda_1\cos\omega t$$
$$+\mu\left(\frac{\partial^2 u(r,t)}{\partial r^2}+\frac{1}{r}\frac{\partial u(r,t)}{\partial r}\right)+KN\left(u_p(r,t)-u(r,t)\right)-\sigma B_0^{\,2}u(r,t), \tag{7}$$

here ρ is the blood density, $u(r,t)$ is the blood velocity in cylindrical coordinates, β is the Brinkman parameter, μ is the blood viscosity, K is the Stocks' constant, and N is the number of MDP per unit volume. The term $KN\left(u_p(r,t)-u(r,t)\right)$ represents the force due to relative motions of blood and MDP. The motion of MDP is governed by Newton's second law of motion as

$$m\frac{\partial u_p(r,t)}{\partial t}=K\left(u(r,t)-u_p(r,t)\right), \tag{8}$$

where m is the average mass of MDP and $u_p(r,t)$ is the velocity of particles. Associated with the flow configuration, the governing equations (7) and (8) are subjected to the following initial and boundary conditions

$$u(r,0)=0, u_p(r,0)=0, \tag{9}$$

$$\left.\begin{aligned} &u(r_0,t)=0, u_p(r_0,t)=0 \\ &\left.\frac{\partial u(r,t)}{\partial r}\right|_{r=0}=0 \end{aligned}\right\}, \tag{10}$$

$$f=\frac{u}{u_0}, g=\frac{u_p}{u_0}, \tau=\frac{t\nu}{r_0^{\,2}}, r_1=\frac{r}{r_0}, \phi_0=\frac{\lambda_0\mu u_0}{r_0^{\,2}}, \phi_1=\frac{\lambda_1\mu u_0}{r_0^{\,2}},$$

into equations (7)-(10) the dimensionless system is obtained as

$$\frac{\partial f(r_1,\tau)}{\partial \tau}=\phi_0+\phi_1\cos\omega t+\frac{\partial^2 f(r_1,\tau)}{\partial r_1^{\,2}}+\frac{1}{r_1}\frac{\partial f(r_1,\tau)}{\partial r_1}+\Phi_c\left(g(r_1,\tau)-f(r_1,\tau)\right)-M_0u(r_1,\tau), \tag{11}$$

$$\Phi_m\frac{\partial g(r_1,\tau)}{\partial \tau}=f(r_1,\tau)-g(r_1,\tau), \tag{12}$$

$$f(r_1,0)=0,\ g(r_1,0)=0, \tag{13}$$

$$\left.\begin{aligned} & f\left(1,\tau\right)=0, g\left(1,\tau\right)=0 \\ & \left.\frac{\partial f\left(r_1,1\right)}{\partial r_1}\right|_{r_1=0}=0 \end{aligned}\right\}, \tag{14}$$

$$\Phi_c=\frac{r_0^{\,2}KN}{\mu}, \Phi_m=\frac{\nu K}{r_0^{\,2}}, M_0=M^2-\beta_1, M=\frac{r_0^{\,2}\sigma B_0^2}{\mu}, \beta_1=\frac{r_0\beta}{\nu},$$

where $f(r_1,\tau)$ is the dimension blood velocity, $g(r_1,\tau)$ is the dimensions MDP velocity, β_1 is the dimensionless Brinkman parameter, Φ_c is the particles concentration parameter, Φ_m is the particles mass parameter and M is the magnetic parameter referred as Hartmann number.

To develop a fractional two-phase blood flow model with MDP, the time derivative on the left-hand side in equations (11) and (12) is replaced by the Caputo time-fractional derivative which yields to

$$D_\tau^{\alpha} f\left(r_1,\tau\right)=\phi_0+\phi_1\cos\omega t+\frac{\partial^2 f\left(r_1,\tau\right)}{\partial r_1^{\,2}}+\frac{1}{r_1}\frac{\partial f\left(r_1,\tau\right)}{\partial r_1}+\Phi_c\left(g\left(r_1,\tau\right)-f\left(r_1,\tau\right)\right)-M_0 f\left(r_1,\tau\right), \tag{15}$$

$$\Phi_m D_\tau^{\alpha} g\left(r_1,\tau\right)=f\left(r_1,\tau\right)-g\left(r_1,\tau\right), \tag{16}$$

where the Caputo fractional derivative is defined as (Podlubny, 1998)

$$D_t^{\alpha} h\left(\eta,t\right)=\begin{cases} \dfrac{1}{\Gamma\left(1-\alpha\right)}\displaystyle\int_0^t \frac{1}{\left(t-s\right)^{\alpha}}\frac{\partial h\left(\eta,s\right)}{\partial s}ds; 0<\alpha<1 \\ \dfrac{\partial h\left(\eta,t\right)}{\partial t}; \alpha=1 \end{cases}. \tag{17}$$

SOLUTIONS OF THE PROBLEM

To solve equations (15) and (17) using the corresponding initial conditions from (13), the Laplace and finite Henkel transforms are used which are defined as

$$\left.\begin{aligned} & L\left\{f\left(r,t\right)\right\}\left(q\right)=\bar{f}\left(q\right)=\int_0^{\infty} f\left(r,t\right)e^{-qt}dt, \\ & H\left\{\bar{f}\left(r,q\right)\right\}\left(r_n\right)=\bar{f}_H\left(r_n,q\right)=\int_0^1 r\bar{f}\left(r,q\right)J_0\left(rr_n\right)dr \end{aligned}\right\}. \tag{18}$$

Here r_n, n=1,2,… are the positive roots of the function $J_0(x)$=0 and J_0 is being the Bessel functions of order zero and first kind. Furthermore, applying the Laplace transform with respect to the time variable τ

$$L\left\{D_\tau^{\ \alpha} f\left(r_1,\tau\right)\right\}(q) = q^\alpha \bar{f}\left(r_1,\tau\right) - f\left(r_1,0\right). \tag{19}$$

Into equations (14) to (18) and (12) yield to

$$q^\alpha \bar{f}\left(r_1,q\right) = \frac{\phi_0}{q} + \frac{q\phi_1}{q^2+\omega^2} + \frac{\partial^2 \bar{f}\left(r_1,q\right)}{\partial r_1^{\ 2}} + \frac{1}{r_1}\frac{\partial \bar{f}\left(r_1,q\right)}{\partial r_1} + \Phi_c\left(\bar{g}\left(r_1,q\right) - \bar{f}\left(r_1,q\right)\right) - M_0 \bar{f}\left(r_1,q\right), \tag{20}$$

$$\bar{g}\left(r_1,q\right) = \frac{\bar{f}\left(r_1,q\right)}{\Phi_m q^\alpha + 1}, \tag{21}$$

$$\bar{f}\left(1,q\right) = 0, \bar{g}\left(1,q\right) = 0. \tag{20}$$

Incorporating equation (21) into equation (20) the blood velocity is obtained in terms of the Laplace transform domain as

$$\frac{\partial^2 \bar{f}\left(r_1,q\right)}{\partial r_1^{\ 2}} + \frac{1}{r_1}\frac{\partial \bar{f}\left(r_1,q\right)}{\partial r_1} = \left[\frac{\Phi_m q^{2\alpha} + \left\{1+\Phi_m\left(\Phi_c + M_0\right)\right\} q^\alpha + M_0}{\Phi_m q^\alpha + 1}\right] \bar{f}\left(r_1,q\right) - \frac{\phi_0}{q} - \frac{q\phi_1}{q^2+\omega^2}, \tag{23}$$

Now, applying the finite Hankel transform on the non-linear term on the left-hand side of equation (23) as

$$\int_0^1 r_1 \left\{\frac{\partial^2 \bar{f}\left(r_1,q\right)}{\partial r_1^{\ 2}} + \frac{1}{r_1}\frac{\partial \bar{f}\left(r_1,\mathrm{q}\right)}{\partial r_1}\right\} J_0\left(r_1 r_{1n}\right) dr_1 = r_{1n}^{\ \ 2} \bar{f}_H\left(r_{1n},q\right), \tag{24}$$

which give

$$r_{1n}^{\ \ 2} \bar{f}_H\left(r_{1n},q\right) = \left[\frac{\Phi_m q^{2\alpha} + \left\{1+\Phi_m\left(\Phi_c + M_0\right)\right\} q^\alpha + M_0}{\Phi_m q^\alpha + 1}\right] \bar{f}_H\left(r_{1n},q\right) - \left(\frac{\phi_0}{q} + \frac{q\phi_1}{q^2+\omega^2}\right)\frac{J_1\left(r_{1n}\right)}{r_{1n}}, \tag{25}$$

and

$$\bar{g}_H\left(r_{1n},q\right)=\frac{1}{\Phi_m}\left(\frac{1}{q^{\alpha}+\dfrac{1}{\Phi_m}}\right)\bar{f}_H\left(r_{1n},q\right), \tag{26}$$

After further simplification, equation (25) takes the following form

$$\bar{f}_H\left(r_{1n},q\right)=\frac{1}{d_{n2}}\left(\frac{\phi_0}{q}+\frac{q\phi_1}{q^2+\omega^2}\right)\left(\frac{1-\Phi_m d_{n3}}{q^{\alpha}+d_{n3}}-\frac{1-\Phi_m d_{n4}}{q^{\alpha}+d_{n4}}\right)\left(\frac{J_1\left(r_{1n}\right)}{r_{1n}}\right), \tag{27}$$

where

$$d_{2n}=\sqrt{d_{1n}^{\ 2}-4\Phi_m\left(r_{1n}^{\ 2}+M_0\right)}, d_{1n}=1+\Phi_m\left(\Phi_c+M_0\right)+\Phi_m r_{1n}^{\ 2},$$
$$d_{3n}=\frac{1}{2\Phi_m}\left(d_{1n}+d_{2n}\right), d_{4n}=\frac{1}{2\Phi_m}\left(d_{1n}-d_{2n}\right).$$

To find the inverse Laplace transform, equation (27) can be written in a more simplified and suitable form as

$$\bar{f}_H\left(r_{1n},q\right)=\left[\begin{array}{l}\dfrac{\phi_0}{d_{2n}}\left\{\left(1-\Phi_m d_{3n}\right)\dfrac{q^{-1}}{q^{\alpha}+d_{3n}}-\left(1-\Phi_m d_{4n}\right)\right\}\\ +\dfrac{\phi_1}{d_{2n}}\dfrac{q}{q^2+\omega^2}\left\{\left(1-\Phi_m d_{3n}\right)\dfrac{1}{q^{\alpha}+d_{3n}}-\left(1-\Phi_m d_{4n}\right)\dfrac{1}{q^{\alpha}+d_{4n}}\right\}\end{array}\right]\left\{\frac{J_1\left(r_{1n}\right)}{r_{1n}}\right\}, \tag{28}$$

Using Lorenzo and Hartleys' and Robotnov and Hartleys' functions which can be respectively defined as (Ali *et al.*, 2017)

$$R_{\alpha,\upsilon}\left(-\eta,t\right)=\sum_{k=0}^{\infty}\frac{\left(-\eta\right)^k t^{(k+1)\alpha-1-\upsilon}}{\Gamma\left\{\left(k+1\right)\alpha-\upsilon\right\}}=L^{-1}\left\{\frac{q^{-\upsilon}}{q^{\alpha}+\eta}\right\}, \tag{29}$$

$$F_{\alpha}\left(-\eta,t\right)=\sum_{k=0}^{\infty}\frac{\left(-\eta\right)^k t^{(k+1)\alpha-1}}{\Gamma\left\{\left(k+1\right)\alpha\right\}}=L^{-1}\left\{\frac{1}{q^{\alpha}+\eta}\right\}, \tag{30}$$

lead equations (26) and (28) to the following form

$$g_H\left(r_{1n},\tau\right)=\frac{1}{\Phi_m}\left\{F_\alpha\left(-\frac{1}{\Phi_m},\tau\right)*f_H\left(r_{1n},\tau\right)\right\}, \tag{31}$$

$$f_H\left(r_{1n},\tau\right)=X\left(\tau\right)\frac{J_1\left(r_{1n}\right)}{r_{1n}}+Y\left(\tau\right)\frac{J_1\left(r_{1n}\right)}{r_{1n}}, \tag{32}$$

where

$$X\left(\tau\right)=\frac{\phi_0}{d_{2n}}\left\{\left(1-\Phi_m d_{3n}\right)R_{\alpha,-1}\left(-d_{3n},\tau\right)-\left(1-\Phi_m d_{4n}\right)R_{\alpha,-1}\left(-d_{4n},\tau\right)\right\}, \tag{33}$$

$$Y\left(\tau\right)=\frac{\phi_1}{d_{2n}}\cos\omega t*\left\{\left(1-\Phi_m d_{3n}\right)F_\alpha\left(-d_{3n},\tau\right)-\left(1-\Phi_m d_{4n}\right)F_\alpha\left(-d_{4n},\tau\right)\right\}. \tag{34}$$

Here * represents the convolution product of the functions. Moreover, upon inverting the Hankel transform

$$\left(H^{-1}\left\{f_H\left(r_{1n},\tau\right)\right\}=f\left(r_1,\tau\right)\right)$$

equations (31) and (32) yield to

$$g\left(r,\tau\right)=\frac{1}{\Phi_m}\int_0^\tau F_\alpha\left(-\frac{1}{\Phi_m},s\right)f\left(r,\tau-s\right)ds, \tag{35}$$

$$f\left(r,\tau\right)=2\sum_{n=1}^{\infty}\frac{J_0\left(r_1 r_{1n}\right)}{r_{1n}J_1\left(r_{1n}\right)}\left\{X\left(\tau\right)+Y\left(\tau\right)\right\}, \tag{36}$$

Limiting Cases

Case I: For α=1

By setting $\alpha\rightarrow1$, the Lorenzo and Hartleys' and Robotnov and Hartleys' functions are respectively reduced to the following form

$$R_{1,-1}\left(-\eta,t\right)=\sum_{k=0}^{\infty}\frac{\left(-\eta\right)^k t^{k+1}}{k!}=L^{-1}\left\{\frac{q^{-1}}{q+\eta}\right\}=\frac{1-e^{\eta t}}{\eta}, \tag{37}$$

$$F_1\left(-\eta,t\right)=\sum_{k=0}^{\infty}\frac{\left(-\eta\right)^k t^k}{k!}=L^{-1}\left\{\frac{1}{q+\eta}\right\}=e^{-\eta t}, \tag{38}$$

that lead $X(\tau)$ to $X_1(\tau)$ and $Y(\tau)$ to $Y_1(\tau)$ as

$$X_1\left(\tau\right)=\frac{\phi_0}{d_{2n}}\left\{\frac{d_{2n}}{r_n^{\,2}-M_0}-\frac{1-\Phi_m}{d_{4n}}e^{-d_{4n}\tau}+\frac{1-\Phi_m}{d_{3n}}e^{-d_{3n}\tau}\right\}, \tag{39}$$

$$Y_1\left(\tau\right)=\frac{\phi_1}{d_{2n}}\left[\begin{aligned}&\frac{1-\Phi_m d_{4n}}{d_{4n}^{\;2}+\omega^2}\left\{d_{4n}\cos\left(\omega t\right)+\omega\sin\left(\omega t\right)-d_{4n}e^{-d_{4n}\tau}\right\}\\&-\frac{1-\Phi_m d_{3n}}{d_{3n}^{\;2}+\omega^2}\left\{d_{3n}\cos\left(\omega t\right)+\omega\sin\left(\omega t\right)-d_{3n}e^{-d_{3n}\tau}\right\}\end{aligned}\right]. \tag{40}$$

Finally, equations (35) and (36) take the following form

$$f\left(r_1,\tau\right)=2\sum_{n=1}^{\infty}\frac{J_0\left(rr_n\right)}{r_nJ_1\left(r_n\right)}\left\{X_1\left(\tau\right)+Y_1\left(\tau\right)\right\}, \tag{41}$$

And

$$g\left(r_1,\tau\right)=\frac{1}{\Phi_m}\int_0^{\tau}e^{-\Phi_m s}f\left(r,\tau-s\right)ds \tag{42}$$

Case II: For Newtonian Viscous Fluid

For $\beta_1\to 0$, equations (41) and (42) can be reduced to the solutions obtained by Zafar *et al.* (Zafar *et al.*, 2018).

NUMERICAL RESULTS AND DISCUSSION

The two-phase flow of Brinkman type dusty fluid under the influence of the transverse magnetic field referred to blood flow in the cylindrical tube is studied. The classical model is transformed into a fractional model by using the Caputo fractional derivative operator. The exact analytical solutions for the non-linear model is developed via the joint Laplace and finite Hankel transforms. As a limiting case, the general solutions for blood velocity and MDP velocity are reduced to classical solutions by setting $\alpha\to 1$. Finally, the obtained solutions are brought into light through numerical computation to explore the impact of the fractional parameter α, dimensionless time τ, Brinkman parameter β_1, Hartmann number M, particles concentration parameter Φ_c and particles mass parameter Φ_m on blood and the MDP velocities.

Figure 2. Variation in velocity profiles of blood and MDP against r_1 due to α for a long time

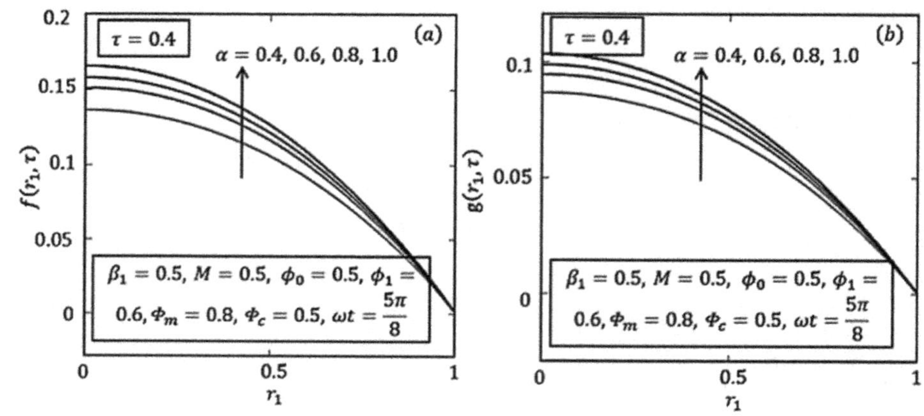

Figures 2 (a) and (b) depict the influence of α on blood and MDP velocities for a long time τ=0.4 respectively. It is found that the blood and MDP velocities are increased with increasing values of α and became maximum at α=1. This is because with increasing values of α, the momentum boundary layer became thicker and thickest at α=1which correspond to an increase in the blood velocity as appeared in the study of Azhar et al. (Azhar *et al.*, 2017). At the same time, the MDP are uniformly suspended in the blood which also shows an increasing trend. In the present case, this behavior of blood and MDP can be observed only for a long time but interestingly for short time as plotted in Figures 3 (a) and (b), the trend of blood and MDP velocities reverses. The same dual effect of α is also reported by Shah *et al.* (2016) and Ali *et al.* (2017). To further explore the influence of α in the present case, Figure 4 (a) and (b) is portraited for blood and MDP velocities against τ. It is detected from these figures that τ=0.33 approximately is the transition point. For τ<0.33 the blood and MDP velocities show an increasing trend for increasing values of α, however, for τ>0.33 the velocities pattern reverses.

Figure 3. Variation in velocity profiles of blood and MDP against r_1 due to α for short time

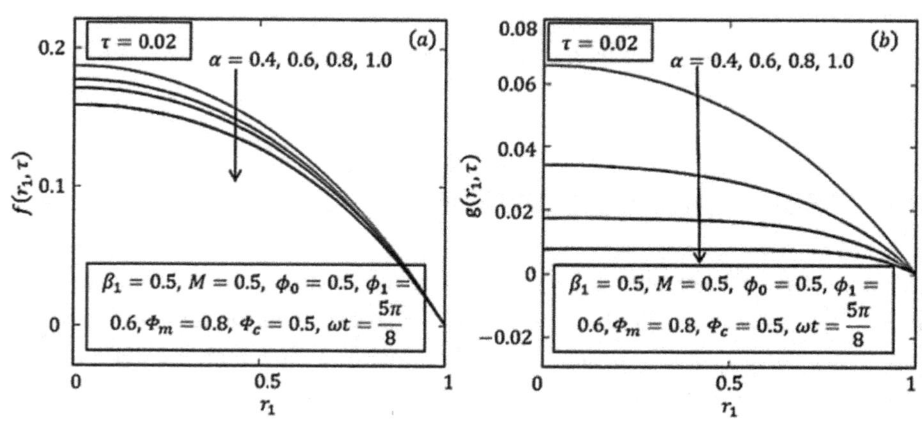

Figure 4. Variation in velocity profiles of blood and MDP against τ due to α

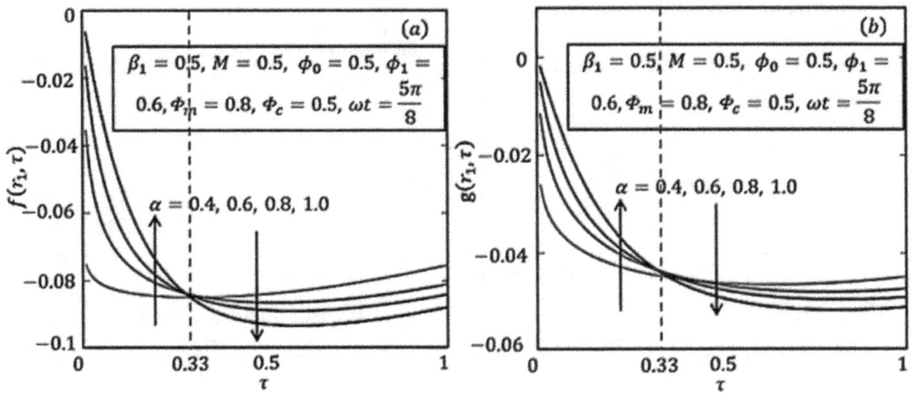

Figures 5 (a) and (b) is plotted to show in the influence of β_1.on blood and MDP velocities. Clearly, both the blood and MDP velocities are decreasing with increasing values of β_1. The greater values of β_1 corresponding to strengthen the drag forces. From equation (6) it can be noticed that the blood velocity is directly related to β_1. Hence, the retardation in the velocities is due to the enhancement in drag forces. This trend of velocity profiles is in strong agreement to that reported by Jan *et al.* (2017).

Figure 5. Variation in velocity profiles of blood and MDP against r_1 due to β_1

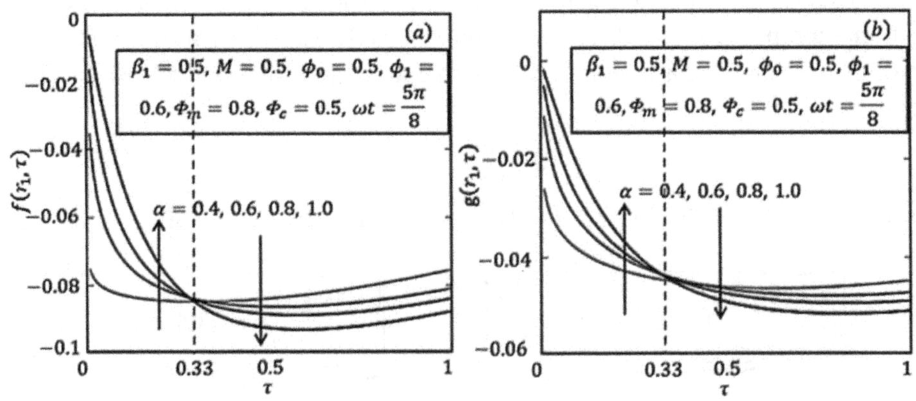

The effect of *M* on the blood and MDP velocities is demonstrated in Figures 6 (a) and (b). It is noticed that the blood and MDP velocities decrease with increase in *M*. This agrees to the Characteristics of the magnetic field that the flow of electrically conducting fluid (blood and MDP) under the influence of applied transverse magnetic field give rise to resistive type forces (Lorentz Forces) which are like drag forces tend to decrease the blood and MDP velocities. The influence of Φ_c and Φ_m on blood and MDP velocities is depicted in Figures 7 (a),(b) and figure 8 (a),(b).respectively. In both the case, the blood and MDP velocities retarded with increasing Φ_c and Φ_m as reported by Nandkeolyar and Das (2014). This is since, when Φ_c and Φ_m are increased the blood became more viscous and denser, as a result, the blood and MDP velocities decelerated.

Figure 6. Variation in velocity profiles of blood and MDP against r_1 due to M

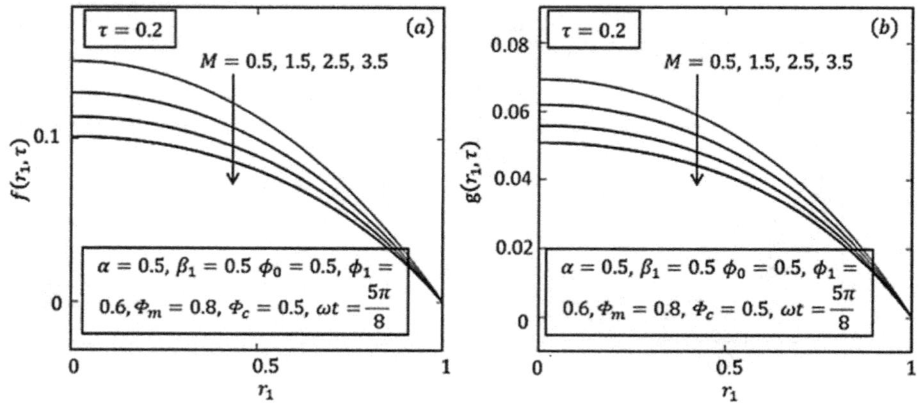

CONCLUSION

The unsteady two-phase blood flow under the influence of a transverse magnetic field in a cylindrical tube is studied. The non-Newtonian Brinkman type fluid is considered for the blood flow, Maxwell's relations for the magnetic field effect, Newton's second law of motion for MDP and the Caputo Fractional derivative for the generalization. The exact analytical solutions for the non-linear system is developed via the joint Laplace and Hankel transforms. The general solutions are reduced to classical solutions by setting $\alpha \to 1$. It is detected that the blood and MDP velocities show a dual effect of α where approximately $\tau=0.33$ is the transition point. For $\tau<0.33$ the blood and MDP velocities show an increasing trend but reverse for $\tau>0.33$. Similarly, retardation occurs in the blood and MPD velocities for greater values of β_1, M, Φ_c and Φ_m.

Figure 7. Variation in velocity profiles of blood and MDP against r_1 due to Φ_c

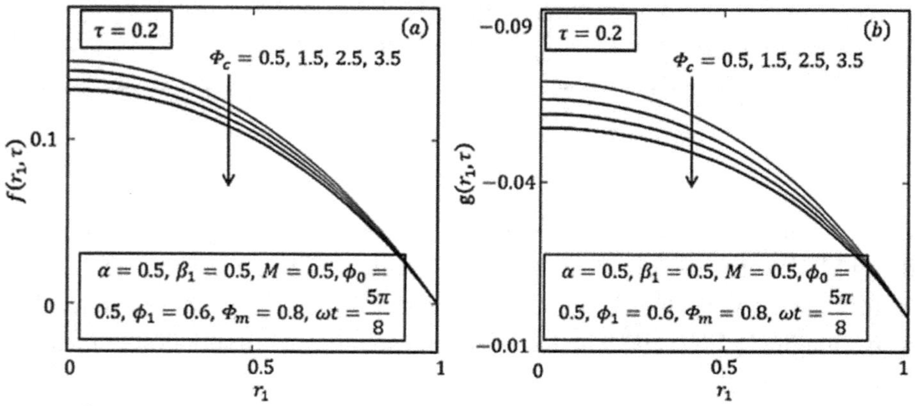

Figure 8. Variation in velocity profiles of blood and MDP against r_1 due to Φ_m

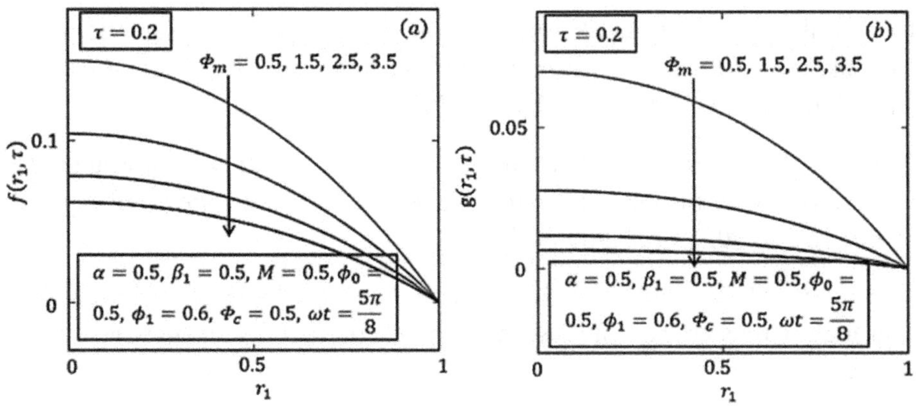

ACKNOWLEDGMENT

The authors would like to acknowledge Ministry of Education (MOE) and Research Management Centre-UTM, Universiti Teknologi Malaysia UTM for the financial support through vote numbers 5F004 and 07G70, 07G72, 07G76 and 07G77 for this research.

REFERENCES

Abdullah, M., Butt, A. R., Raza, N., Alshomrani, A. S., & Alzahrani, A. K. (2018). Analysis of blood flow with nanoparticles induced by uniform magnetic field through a circular cylinder with fractional Caputo derivatives. *Journal of Magnetism and Magnetic Materials*, *446*, 28–36. doi:10.1016/j.jmmm.2017.08.074

Ali, F., Saqib, M., Khan, I., & Sheikh, N. A. (2016). Application of Caputo-Fabrizio derivatives to MHD free convection flow of generalized Walters'-B fluid model. *The European Physical Journal Plus*, *131*(10), 377. doi:10.1140/epjp/i2016-16377-x

Ali, F., Sheikh, N. A., Khan, I., & Saqib, M. (2017). Magnetic field effect on blood flow of Casson fluid in axisymmetric cylindrical tube: A fractional model. *Journal of Magnetism and Magnetic Materials*, *423*, 327–336. doi:10.1016/j.jmmm.2016.09.125

Azhar, W. A., Vieru, D., & Fetecau, C. (2017). Free convection flow of some fractional nanofluids over a moving vertical plate with uniform heat flux and heat source. *Physics of Fluids*, *29*(8), 082001. doi:10.1063/1.4996034

Bhatti, M. M., Zeeshan, A., & Ellahi, R. (2016a). Endoscope analysis on peristaltic blood flow of Sisko fluid with Titanium magneto-nanoparticles. *Computers in Biology and Medicine*, *78*, 29–41. doi:10.1016/j.compbiomed.2016.09.007 PMID:27643464

Bhatti, M. M., Zeeshan, A., & Ijaz, N. (2016b). Slip effects and endoscopy analysis on blood flow of particle-fluid suspension induced by peristaltic wave. *Journal of Molecular Liquids*, *218*, 240–245. doi:10.1016/j.molliq.2016.02.066

Boyd, W. (1961). *Text-book of pathology; structure and function in diseases*. Academic Press.

Carlton, J. M., Yowell, C. A., Sturrock, K. A., & Dame, J. B. (2001). Biomagnetic separation of contaminating host leukocytes from plasmodium-infected erythrocytes. *Experimental Parasitology*, *97*(2), 111–114. doi:10.1006/expr.2001.4594 PMID:11281708

Ellahi, R., Rahman, S. U., Nadeem, S., & Vafai, K. (2015). The blood flow of Prandtl fluid through a tapered stenosed arteries in permeable walls with magnetic field. *Communications in Theoretical Physics*, *63*(3), 353. doi:10.1088/0253-6102/63/3/353

Ghasemi, S. E., Hatami, M., Sarokolaie, A. K., & Ganji, D. D. (2015). Study on blood flow containing nanoparticles through porous arteries in presence of magnetic field using analytical methods. *Physica E, Low-Dimensional Systems and Nanostructures*, *70*, 146–156. doi:10.1016/j.physe.2015.03.002

Haik, Y., Chen, C. J., & Chatterjee, J. (2002). Numerical simulation of biomagnetic fluid in a channel with thrombus. *Journal of Visualization / the Visualization Society of Japan*, *5*(2), 187–195. doi:10.1007/BF03182428

Haik, Y., Pai, V., & Chen, C.-J. (1999). Development of magnetic device for cell separation. *Journal of Magnetism and Magnetic Materials*, *194*(1-3), 254–261. doi:10.1016/S0304-8853(98)00559-9

Jan, S. A. A., Ali, F., Sheikh, N. A., Khan, I., Saqib, M., & Gohar, M. (2017). Engine oil based generalized brinkman-type nano-liquid with molybdenum disulphide nanoparticles of spherical shape: Atangana-Baleanu fractional model. *Numerical Methods for Partial Differential Equations*.

Khan, I., Saqib, M., & Ali, F. (2017). Application of time-fractional derivatives with non-singular kernel to the generalized Convective flow of Casson fluid in a microchannel with constant walls temperature. *The European Physical Journal. Special Topics*, *226*(16-18), 3791–3802. doi:10.1140/epjst/e2018-00097-5

Khan, I., Saqib, M., & Ali, F. (2018). Application of the modern trend of fractional differentiation to the MHD flow of a generalized Casson fluid in a microchannel: Modelling and solution★. *The European Physical Journal Plus*, *133*(7), 262. doi:10.1140/epjp/i2018-11899-9

Liu, J., Flores, G. A., & Sheng, R. (2001). In-vitro investigation of blood embolization in cancer treatment using magnetorheological fluids. *Journal of Magnetism and Magnetic Materials*, *225*(1-2), 209–217. doi:10.1016/S0304-8853(00)01260-9

Mandal, P. K. (2005). An unsteady analysis of non-Newtonian blood flow through tapered arteries with a stenosis. *International Journal of Non-linear Mechanics*, *40*(1), 151–164. doi:10.1016/j.ijnonlinmec.2004.07.007

Naber, M. (2010). Linear fractionally damped oscillator. *International Journal of Differential Equations*.

Nandkeolyar, R., & Das, M. (2014). Unsteady MHD free convection flow of a heat absorbing dusty fluid past a flat plate with ramped wall temperature. *Afrika Matematika*, *25*(3), 779–798. doi:10.100713370-013-0151-9

Podlubny, I. (1998). *Fractional differential equations: an introduction to fractional derivatives, fractional differential equations, to methods of their solution and some of their applications*. Elsevier.

Ray, S. S., Atangana, A., Noutchie, S. C., Kurulay, M., Bildik, N., & Kilicman, A. (2014). Fractional calculus and its applications in applied mathematics and other sciences. *Mathematical Problems in Engineering*, *2014*, 2014. doi:10.1155/2014/849395

Saqib, M., Ali, F., Khan, I., Sheikh, N. A., & Khan, A. (2018a). Entropy Generation in Different Types of Fractionalized Nanofluids. *Arabian Journal for Science and Engineering*, 1–10.

Saqib, M., Ali, F., Khan, I., Sheikh, N. A., & Shafie, S. B. (2018b). Convection in ethylene glycol-based molybdenum disulfide nanofluid. *Journal of Thermal Analysis and Calorimetry*, 1–10.

Saqib, M., Khan, I., & Shafie, S. (2018c). Application of Atangana–Baleanu fractional derivative to MHD channel flow of CMC-based-CNT's nanofluid through a porous medium. *Chaos, Solitons, and Fractals*, *116*, 79–85. doi:10.1016/j.chaos.2018.09.007

Saqib, M., Khan, I., & Shafie, S. (2018d). Natural convection channel flow of CMC-based CNTs nanofluid. *The European Physical Journal Plus*, *133*(12), 549. doi:10.1140/epjp/i2018-12340-3

Saqib, M., Khan, I., & Shafie, S. (2019a). Application of fractional differential equations to heat transfer in hybrid nanofluid: Modeling and solution via integral transforms. *Advances in Difference Equations*, *2019*(52), 1–18. doi:10.118613662-019-1988-5

Saqib, M., Khan, I., & Shafie, S. (2019b). Generalized magnetic blood flow in a cylindrical tube with magnetite dusty particles. *Journal of Magnetism and Magnetic Materials*, *484*, 490–496. doi:10.1016/j.jmmm.2019.03.032

Shah, N. A., Vieru, D., & Fetecau, C. (2016). Effects of the fractional order and magnetic field on the blood flow in cylindrical domains. *Journal of Magnetism and Magnetic Materials*, *409*, 10–19. doi:10.1016/j.jmmm.2016.02.013

Sharma, S., Singh, U., & Katiyar, V. K. (2015). Magnetic field effect on flow parameters of blood along with magnetic particles in a cylindrical tube. *Journal of Magnetism and Magnetic Materials*, *377*, 395–401. doi:10.1016/j.jmmm.2014.10.136

Sheikholeslami, M. (2018). Numerical investigation for CuO-H2O nanofluid flow in a porous channel with magnetic field using mesoscopic method. *Journal of Molecular Liquids*, *249*, 739–746. doi:10.1016/j.molliq.2017.11.069

Sinha, A., & Shit, G. C. (2015). Electromagnetohydrodynamic flow of blood and heat transfer in a capillary with thermal radiation. *Journal of Magnetism and Magnetic Materials*, *378*, 143–151. doi:10.1016/j.jmmm.2014.11.029

Tashtoush, B., & Magableh, A. (2008). Magnetic field effect on heat transfer and fluid flow characteristics of blood flow in multi-stenosis arteries. *Heat and Mass Transfer*, *44*(3), 297–304. doi:10.100700231-007-0251-x

Tzirtzilakis, E. E. (2005). A mathematical model for blood flow in magnetic field. *Physics of Fluids*, *17*(7), 077103. doi:10.1063/1.1978807

Voltairas, P. A., Fotiadis, D. I., & Michalis, L. K. (2002). Hydrodynamics of magnetic drug targeting. *Journal of Biomechanics*, *35*(6), 813–821. doi:10.1016/S0021-9290(02)00034-9 PMID:12021001

Zafar, A. A., Shah, N. A., & Khan, I. (2018). Two Phase Flow of Blood through a Circular Tube with Magnetic Properties. *Journal of Magnetism and Magnetic Materials*.

Chapter 8
Revisiting Anomalous g-Factors for Charged Leptons in a Fractional Coarse-Grained Approach With Axiomatic Local Metric Derivatives

José Weberszpil
https://orcid.org/0000-0002-5016-2329
Universidade Federal Rural do Rio de Janeiro, Brazil

José Abdalla Helayël-Neto
Centro Brasileiro de Pesquisas Físicas, Brazil

ABSTRACT

This contribution sets out to extend the concept of helicity so as to include it in a fractional scenario with a low-level of fractionality. To accomplish this goal, the authors write down the left- and the right-handed Weyl equations from first principles in this extended framework. Next, by coupling the two different fractional Weyl sectors by means of a mass parameter, they arrive at the fractional version of Dirac's equation, which, whenever coupled to an external electromagnetic field and reduced to the non-relativistic regime, yields a fractional Pauli-type equation. From the latter, they are able to present an explicit expression for the gyromagnetic ratio of charged fermions in terms of the fractionality parameter. They then focus their efforts to relate the coarse-grained property of space-time to fractionality and to the (g-2) anomalies of the different leptonic species. To do this, they build up an axiomatic local metric derivative that exhibits the Mittag-Leffler function as eigenfunction and is valid for low-level fractionality, whenever the order parameter is close to 1.

DOI: 10.4018/978-1-7998-3122-8.ch008

INTRODUCTION

The electron gyromagnetic ratio, g_e, is a dimensionless parameter that appears in the spin-orbit interaction and leads to a splitting in the atomic energy levels, giving rise to the so-called fine structure. The g_e-factor appears in the Zeeman effect, by means of the electron's magnetic dipole moment, in the description of the interaction of atoms with external magnetic fields.

It is now known that the anomalous magnetic moment (AMM) of the muon is considered to be one of the most promising observables that can unveil effects of some new Physics beyond the Standard Model (Passera et al, 2008). One also claims that the muon's AMM is one of the most precisely measured quantities in Particle Physics, reaching a precision of 0.54ppm (Jegerlehner, & Nyffeler, 2009). This statement is also reinforced in view of the relatively large value of the muon's mass (de Rafael, 2013). The tau-lepton gyromagnetic ratio is not very well-known experimentally and for this reason it does not provide yet a good test of the Standard Model (SM) and for Physics beyond the Standard Model (de Rafael, 2013)[3].

Relativistic quantum mechanics predicts the remarkable result, g_e=2. Experimentally, the electron's g-ratio has been measured with high precision and the "2010 CODATA value" (Mohr et al, 2012) for this parameter is $g_{e,\exp} = 2.00231930436153\left(53\right)$, which is supported by the theoretical calculations of Quantum Electrodynamics (QED), slightly different from that of Dirac's theoretical prevision in the realm of Relativistic Quantum Mechanics (QM). A good review on the introduction and history of the g-ratios, up to 2007, may be found in (Miller et al, 2007). For a non-perturbative approach, the reader could consult (Brodsky et al, 2004)[6]. An update of the electron's and muo's g-ratios can be found in (de Rafael, 2013).

In this contribution, we revisit the investigation of the fractional coarse-grained aspects of the electron's anomalous magnetic g-ratio (Weberszpil & Helayël-Neto, 2014) with the recent developed axiomatic local deformed derivative (ALDD) (Weberszpil & Helayël-Neto, 2017a), showing the possibility of mapping its experimental value into some theoretical parameter that accounts for fractionality, so that we take the viewpoint that fractionality could be responsible for the deviation from the g_e=2 relativistic quantum-mechanical result. This claim highlights that our motivation to adopt a local deformed operator is more physical than a simple mathematically-motivated extension of classical calculus. This becomes more explicit once we argue that the possible justification for the experimental difference from the theoretical Dirac's value g_e=2 is imputed to the complexity of the interactions of the electrons, taken as a pseudo-particle "dressed" by the interactions and the medium. Here, we look at a dynamical system as an open system that can interact with the environment and we argue that ALDD may become an important tool to study open classical and quantum systems, but considering the limits of low level of fractionality.

The ALDD is a local deformed derivative (LDD) that has the Mittag-Leffler function as an eigen-function and for which deformed Leibniz' and the chain rule hold, similarly to the standard calculus, but whenever a low-level of fractionality is considered. The deformed operators here are local.

Our main goal here is to show that it is possible to build up a local formalism that meets certain similarities to those required by a version of fractional calculus (FC) (Jumarie, 2006; Jumarie, 2009a; Jumarie, 2013), in such a way that reconciles the relevant axioms of the approach outlined here and, additionally, it is an approximation valid within a low fracionality limit. In this way, we try to make clear, in short, that the ALDD operators are in fact local and behave as an approximations but, provided that the order of the derivative is very close to one. That is, being local and considered as approximations.

On the applicability of our approach proposed here, some comments are deserved. Certain systems in Nature seem to be well-described by some fractional parameter in a low-level limit of fractionality. Examples of such low-level systems are the anomalous magnetic g-factor of the charged leptons of the Standard Model, where the low level of fractionality seems to be consistent with the Quantum Electrodynamics corrections (Calcagni, 2012). The possibility of constraining fractal space dimensionality from astrophysics and other areas (Svozil & Zeilinger, 1986), also points to a fractal dimension very close to an integer value, but still noninteger. Thus, the importance of this formalism that it may indicate some higher-order effects that might exist in the systems under study, without the cumbersome calculations of QED.

Several complex systems are described as being dissipative (Calcagni, 2010; Calcagni, 2012). The use of mathematical tools like FC (Podlubny, 1999; Kilbas et al, 2006), that are non-local, or deformed derivatives (DD) (Chen, 2006), local, is also justified here on the basis of our proposition that there exists an intimate relationship between dissipation, coarse-grained medium and a limit energy scale for the interactions. Since we are dealing with open systems, as we already commented, particles should in fact be seen as dressed entities or pseudo-particles that exchange energy with other particles and the environment. Depending on the energy scale, a particular interaction may affect the geometry of space-time. The system composed by particles and their surroundings may be considered nonconservative due to the possible energy exchange. This energy exchange may be the responsible for the resulting non-integer dimension of space-time, giving rise then to an effective coarse-grained medium. This is quite reasonable since, even standard field theory, comes across a granularity in the limit of Planck scale. Also, if we consider a nonstandard geometry, such as the space-time of a cosmological black hole or the space in the surroundings of a region of pair creation, a coarse-grained view of the dynamics of open systems may become a sensible description. Here, we argue that non standard mathematical tools allows us to describe and emulate this kind of dynamics without explicit many-body contributions, dissipation or geometrical terms in the dynamical equations of the system. In some way, these new mathematical approaches may yield an effective theory, with some statistical average without imposing any specific nonstandard statistics, that could describe in a consistent way the connections between coarse-grained medium and dissipation at a certain energy scale.

In recent works, we have developed connections and a variational formalism for deformed or metric derivatives, considering the relevant space-time/phase space as fractal or multifractal (Weberszpil et al, 2015; Weberszpil et al 2016). We emphasized that the paradigm we adopt were different from the standard approach in the generalized statistical mechanics context (Tsallis, 1988; Tsallis, 2009a; Tsallis, 2009b), where the modification of entropy definition leads to the modification of algebra and consequently the derivative concept (Weberszpil et al, 2015; Weberszpil et al 2016). This was structured by mapping to a continuous fractal space (Balankin & Elizarraraz, 2012a; Balankin & Elizarraraz, 2012b; Balankin et al. 2016) which leads naturally to the necessity of modifications in the derivatives, that we called deformed or metric derivatives (Weberszpil et al, 2015; Weberszpil et al 2016; Weberszpil & Chen 2017; Weberszpil & Helayël-Neto, 2017b; Rosa & Weberszpil, 2018; Leopoldino et al, 2019). The modifications of derivatives, accordingly with the metric, bring to a change in the algebra involved, which in turn may conduct to a generalized statistical mechanics with some adequate definition of entropy.

It is worthy to point out here the significant work by Zeilinger and Svozil (Svozil & Zeilinger, 1986) in view of the arguments they adopt to explain the differences between the (g–2) theoretical and experimental results via a non-standard geometry. They propose that such a deviation could also be explained by changes of the dimension of the measure and thus the Hausdorff dimension of space-time, which is closely related to our assumptions here.

Here, we claim that the use of an approach like ALDD is more appropriate to describe the dynamics associated with field theory and particle physics in a coarse-grained scenario for space-time, but for low level of fractionality only.

It seems that a reasonable way to probe the classical framework of physics is to highlight that, in our real world, the generic point is not infinitely small (or thin), it rather has a thickness. In a coarse-grained space, a point is not infinitely thin, and here, this feature is modeled by means of a space in which the generic differential is not dx, but rather $(dx)^\alpha$; likewise for the time variable t. It is noteworthy to emphasize that the notion of fractal space-time was introduced in the 70s with the seminal work by Stillinger (Stillinger,1977), where the axiomatic basis for spaces with non-integer dimension was set up. Later, the concepts associated to a possible non-integer dimension were reinforced with the work by Zeilinger and Svozil (Zeilinger, & Svozil, 1985), where they take into account the intrinsically unavoidable finite resolution of any physical experiment; also the works (Svozil, 1987; Eyink, 1989; Palmer & Stavrinou, 2004) should be quoted here. Along this line, we recommend the work by Nottale (Nottale, 1993), where the notion of a fractal space-time is reassessed. Non-integer differentiability and randomness (Grigolini et al, 1999) are mutually related in their nature, in such a way that studies on fractals on the one hand, and fractional Brownian motion on the other hand, are often parallel in the work (Nottale, 1993). A function which is continuous everywhere, but nowhere integer-differentiable, necessarily exhibits random-like or pseudo-random features, in that various samplings of these functions, on the same given interval, will be different. This may explain the huge amount of literature extending the theory of stochastic differential equations to describe stochastic dynamics driven by fractional Brownian motion (Jumarie, 2007; Jumarie, 2009b; Jumarie, 2009c; Jumarie, 2010). In connection with the anomalous properties of space-time with multifractal structure, we recommend the reader to consult the interesting work by Gianluca (Calcagni et al, 2013) and references therein. Also, we draw special attention to the efforts in building up a solid foundation for the construction of a suitable geometry and a consistent field theory in fractional spaces, multi-fractals (Calcagni, 2012) and multi-scale (Calcagni et al, 2014) space-times.

The great majority of actual classical systems is nonconservative but, in spite of that, the most advanced formalisms of classical mechanics deal only with conservative systems (Riewe, 1996). Dissipation (Godinho et al, 2014), for example, is present even at the microscopic level. There is dissipation in every non-equilibrium or fluctuating process, including dissipative tunneling (Caldeira & Leggett, 1981; Caldeira & Leggett, 1983; Ankerhold et al, 1995) and electromagnetic cavity radiation (Senitzky, 1960; Stevens, 1958), for instance. In (Weberszpil et al, 2015; Weberszpil et al 2016; Weberszpil & Chen 2017; Weberszpil & Helayël-Neto, 2017b; Rosa & Weberszpil, 2018; Leopoldino et al, 2019), we adopt that a new way to suitably treat nonconservative systems is through DD.

Field theory aspects of non-linear dynamics are today an important subject of study in different sub-areas of Physics and Mathematics, but the real success and a radically new understanding of non-linear processes have acquired body over the past 40 years. This understanding has been inspired by the discovery and insights of chaotic dynamics, where the randomness of physical processes is considered; more precisely, when particle trajectories are indistinguishable for random process (Zaslavsky, & Zaslavskij, 2005).

In (Godinho et al, 2012) we have argued that the modeling of TeV-physics may demand an approach based on fractal operators and FC, and we claim that, in the realm of complexity, nonlocal theories and memory effects are connected to complexity. We also adopt that FC and the nondifferentiable nature of microscopic dynamics may be connected with time scales.

Here, to achieve our goals, we carefully build up a fractional Dirac equation from underlying fractional Weyl equations written down in terms of the helicity operator. We begin by discussing the fractional angular momentum algebra in a coarse-grained scenario in order to understand the fractional spin operator used to set up the helicity operator. Also, a gauge-covariant fractional derivative is proposed so that, from the gauge transformation of charged matter fields, the gauge transformation of the fractional vector potential can be read off. Then, minimal coupling to the electromagnetic field is naturally achieved by writing down fractional Lagrangian densities in terms of fractional gauge-covariant derivatives of the matter fields. It is important to stress that we are not proposing the matter transformation from the very start (Herrmann, 2008); we rather get it as an output, from the matter-gauge coupling together with the gauge field transformation. After all, gauge symmetry is actually intrinsic to vector bosons from which we need to decouple spurious degrees of freedom. Matter fields should feel gauge symmetry indirectly, through their coupling to the gauge fields. This is the line of thought we advocate here. Finally, by working out the non-relativistic approximation of the fractional Dirac's equation, the fractional version of the Pauli's equation is read off and this yields an explicit expression for the fractional g-ratio if an external magnetic field is switched on. We point out that fractional Dirac's equation has been the object of interest of several authors over the past decade (Plyushchay, De Traubenberg, 2000; Raspini, 2001; Závada, 2002).

AXIOMATIC DEFORMED DERIVATIVE

In this Section, we present a version of DD that has the Mittag-Leffler functions as eigenfunctions (Weberszpil & Helayël-Neto, 2017a) (but is valid only for low level of fractionality). This is due to the fact that the appearance of Mittag-Leffler function is very present in the description of dynamics for some complex systems like soil, porous media or some viscous systems. Here, we consider that neither the space nor the time are infinitely thin but have instead some "thickness". As the use of certain calculation rules are essential to our approach, we briefly comment on this point, before presenting these rules.

Axiom 1. Linearity

$$D^{\alpha}[c_1 f(x) + c_2 g(x)] = c_1 D^{\alpha} f(x) + c_2 D^{\alpha} g(x). \tag{2.1}$$

Axiom 2. Leibniz Rule

$$D^{\alpha}(fg) = (D^{\alpha} f)g + f(D^{\alpha} g). \tag{2.2}$$

Axiom 3. Derivative of a Power Function

$$D^{\alpha}(t-a)^{\nu} = \frac{\Gamma(1+\nu)}{\Gamma(1+\nu-\alpha)}(t-a)^{\nu-\alpha}, \; (\alpha,\nu,x>0), \tag{2.3}$$

Axiom 4. Low-Level Fractionality

Here, we consider α very close to 1. This is what we call low-level fractionality.

Clearly, $D^\alpha 1=0$, since the Leibniz rule is assumed.

We consider axiom 4 to give consistency to the the simultaneous validity of the axioms 2 and 3. But, we emphasize that we are dealing here with local operators.

With axioms 3 and $D^\alpha 1=0$, we show that the Mittag-Lefffler functions is an eigenfunction of this local metric derivative: $D^\alpha E_\alpha(\lambda x^\alpha) = \lambda E_\alpha(\lambda x^\alpha)$.

Of course, if we do not impose axiom 2 for Leibniz rule and substitute the axiom by another one that attributes the derivative of a constant to be zero, $D^\alpha 1=0$, the result above would be valid too. But, our intention is to permit the use o Leibniz rule, even if it would result in an approximation.

From the Leibniz Rule to the Chain Rule for Local Deformed Derivative

We are restricting ourselves only to this type of LDD and not fractional derivatives which are nonlocal and do not allow no-violated Leibniz rules. But other type of local DD could be uses instead.

Consider a deformed or metric derivative D^α of order α that satisfies the Leibniz rule (axiom 2), whose domain includes all locally Hölder-continuous functions of order α. Using the Hadamard's representation theorem for $f \in C^2$, around an arbitrary point yields a fractional chain rule for such D^α applied to $f¡w$, where w is locally Hölder of exponent α, accordingly to axiom 2. Remember axiom 4. Then the following statement holds:

$$D_x^\alpha (f \circ w)(x_0) = f'(w(x_0))(D_x^\alpha w)(x_0). \tag{2.4}$$

Or in a simple form:

$$\frac{d^\alpha}{dx^\alpha} f[u(x)] = \frac{df}{du}\frac{d^\alpha u}{dx^\alpha}. \tag{2.5}$$

This kind of chain rule holds for every kind of DD (which are local operators).

In a recent article (Weberszpil, 2014), the authors show that the Leibniz rule for fractional derivatives in a coarse-grained medium treated as a Hölder-Space yields a modified chain rule.

Here, another comment is pertinent and concerns certain definitions called "local fractional derivatives", as the ones introduced by Kolwankar and Gandal (Kolwankar & Gangal, 1996; Kolwankar & Gangal, 1998; Kolwankar & Gangal, 1999), with several works related to this approach; for example, the works (Babakhani & Daftardar-Gejji 2002; Chen et al, 2010; Carpinteri & Sapora, 2010) or the approaches with a Hausdorff derivative and also the treatment with the so-called fractal derivative (Chen, 2006; Chen et al, 2010a). All of the mentioned approaches seem to be applicable to power-law phenomena. There is also the some other recently-developed deformed derivatives (Chen, 2006; Kobelev, 2006; Rosa & Weberszpil, 2019). All of them deserve to be more deeply investigated, under a mathematical and a physical point of view (Chen, 2010; Budini, & Cáceres, 2004; Bolonha et al, 1999).

Now that we have set up these fundamental expressions, we are ready to carry out the calculations of our main concern.

FRACTIONAL ANGULAR MOMENTUM ALGEBRA

Here, we shall derive the commutation algebra for spin-1/2 particles in a coarse-grained medium.

We follow the steps in (Weberszpil & Helayël-Neto, 2014).

Since in the ALDD approach the chain and Leibniz rules hold, it is not difficult to obtain the commutation relation between position and momentum operators and to define the fractional angular momentum components (Herrmann, 2007; Herrmann, 2008), we can write

$$\hat{L}_z^\alpha := \left[\hat{x}^\alpha, \widehat{p_y}^\alpha\right], \hat{L}_y^\alpha := \left[\hat{z}^\alpha, \widehat{p_x}^\alpha\right], \hat{L}_x^\alpha := \left[\hat{y}^\alpha, \widehat{p_z}^\alpha\right], \tag{3.1}$$

where $\widehat{x^\alpha}$, $\widehat{y^\alpha}$, $\widehat{z^\alpha}$ stands for the position operators whereas $\widehat{p_x^\alpha}$, $\widehat{p_y^\alpha}$ and $\widehat{p_z^\alpha}$ refer to the corresponding momentum operators. An important warning is worthy at this stage: we should confuse the α-parameter that is simply a label, as in the set of equations above, with the case where it is a true fractional power, as when it appears in some constants like in the factor $\hbar^\alpha$.

Again, one can obtain the commutation relations using the definitions given above and the Leibniz (2.2) and chain rules (2.5) rules for ALDD as

$$\left[\hat{L}_x^\alpha, \hat{L}_y^\alpha\right] = i\Gamma(\alpha+1)\hbar^\alpha M_\alpha \hat{L}_z^\alpha, \left[\hat{L}_z^\alpha, \hat{L}_x^\alpha\right] = i\Gamma(\alpha+1)\hbar^\alpha M_\alpha \hat{L}_y^\alpha, \left[\hat{L}_y^\alpha, \hat{L}_z^\alpha\right] = i\Gamma(\alpha+1)\hbar^\alpha M_\alpha \hat{L}_x^\alpha, \tag{3.2}$$

where

$$\sigma_\mu = (\sigma_0, \sigma_1, \sigma_2, \sigma_3) = (I_2, \sigma_x, \sigma_y, \sigma_z)$$

and the factor M_α is not an operator; it is simply a multiplicative factor introduced for the sake of ensuring the right dimensionality of these commutators.

Let us now draw the reader's attention to the different character of the α-parameter. In $\hbar^\alpha$, α is a true power. In the factor M_α, α it is simply a label. In the position, linear momentum and angular momentum operators, α stands for a label that denotes the fractionality. To build up the algebra, we define a set of operators, as in the sequel.

The square of fractional angular moment operator is defined as in (Herrmann, 2007):

$$\left(\hat{L}^\alpha\right)^2 = \hat{L}^\alpha \hat{L}^\alpha = \left(\hat{L}_x^\alpha\right)^2 + \left(\hat{L}_y^\alpha\right)^2 + \left(\hat{L}_z^\alpha\right)^2. \tag{3.3}$$

We also define the fractional angular momentum operator $\vec{L}^\alpha$ and the fractional raising and lowering operators, L_+^α and their hermitian conjugates L_-^α , respectively as

$$\vec{L}^{\alpha} \equiv \hat{L}_x^{\alpha}\hat{i} + \hat{L}_y^{\alpha}\hat{j} + \hat{L}_z^{\alpha}\hat{k} \text{ and } \hat{L}_+^{\alpha} \equiv \hat{L}_x^{\alpha} + i\hat{L}_y^{\alpha},\ L_-^{\alpha} \equiv \hat{L}_x^{\alpha} - i\hat{L}_y^{\alpha}. \tag{3.4}$$

The commutation relations then follow as below:

$$\left[\hat{L}_+^{\alpha}, \hat{L}_-^{\alpha}\right] = 2\Gamma(\alpha+1)\hbar^{\alpha}M_{\alpha}\hat{L}_z^{\alpha}, \left[\hat{L}_z^{\alpha}, \hat{L}_+^{\alpha}\right] = \Gamma(\alpha+1)\hbar^{\alpha}M_{\alpha}\hat{L}_+^{\alpha}, \left[\hat{L}_z^{\alpha}, \hat{L}_-^{\alpha}\right] = -\Gamma(\alpha+1)\hbar^{\alpha}M_{\alpha}\hat{L}_-^{\alpha}, \tag{3.5}$$

$$\left[\hat{L}^{\alpha}\hat{L}^{\alpha}, \hat{L}_+^{\alpha}\right] = \left[\hat{L}^{\alpha}\hat{L}^{\alpha}, \hat{L}_-^{\alpha}\right] = \left[\hat{L}^{\alpha}\hat{L}^{\alpha}, \hat{L}_x^{\alpha}\right] = \left[\hat{L}^{\alpha}\hat{L}^{\alpha}, \hat{L}_y^{\alpha}\right] = \left[\hat{L}^{\alpha}\hat{L}^{\alpha}, \hat{L}_z^{\alpha}\right] = 0. \tag{3.6}$$

The commutation relations written above indicate that the ordinary integer angular momentum algebra does not change. This implies that the raising (lowering) operator, acting on an eigenstate $|j,m\rangle$ of the $\hat{L}_z^{\alpha}$-operator, raises (lowers) the eigenvalue of the latter by the amount $\Gamma(\alpha+1)\hbar^{\alpha}M_{\alpha}$. That is, upon the action of the raising (lowering) operator, the eigenvalue of $\hat{L}_z^{\alpha}$ is raised (lowered) by $\Gamma(\alpha+1)\hbar^{\alpha}M_{\alpha}$.

Another important conclusion is that, for the ordinary Pauli spin matrices, the representation for the basis of $\left(\hat{L}^{\alpha}\right)^2$ is the same as the integer case. This allows us to rewrite all usual relations from ordinary quantum mechanics spin algebra in a coarse-grained scenario. The only difference is that, instead of $\hbar$, we have to replace it in all relations by an effective factor $\hbar_{eff} = \Gamma(\alpha+1)\hbar^{\alpha}M_{\alpha}$. There is no fractional number of particles but, there exists an effective Planck constant.

FRACTIONAL FIELD EQUATIONS - WEYL AND DIRAC FERMIONS

This section sets out to show our proposal for building fractional versions of Weyl and Dirac equations on the basis of fundamental principles. Rather than adopting a fractionalized version for these equations from the very beginning, we propose to extend the concept of helicity to account for fractionality and, once the suitable helicity operators are written down, we adopt them to propose the fractional Weyl equation for left- and the right-handed fermions. Next, by coupling the different fractional Weyl equations by means of a mass parameter, we shall end up with what we propose to be the fractional Dirac equation.

Weyl's equation is a relativistic wave equation to describe massless spin-1/2 particles. Recalling that helicity is the projection of the spin onto the direction of momentum, we proceed to write down Weyl's equations as statements on the helicity of the left- and right-handed fermions, since, in the case of massless fermions, chirality and helicity are equivalent.

The projection of the spin onto the linear momentum is written as follows:

$$\lambda_{\alpha} = \overrightarrow{S^{\alpha}} \circ \frac{\overrightarrow{p^{\alpha}}}{\left|\overrightarrow{p^{\alpha}}\right|}, \lambda_{\alpha} = \pm 1, \tag{4.1}$$

with the spin vector, $\overrightarrow{S_{\alpha}}$, obtained from the angular momentum algebra, which reads:

$$\overrightarrow{S^{\alpha}} = \frac{\hbar^{\alpha}}{2}\Gamma(\alpha+1)M_{\alpha}\vec{\sigma}. \tag{4.2}$$

Here, the Pauli matrices are the same since the structure of the algebra is not modified, as shown in Section 3.

For a massless particle, the relativistic relation between energy and momentum reads as (Weberszpil et al, 2012): $E^{\alpha} = \left|\overrightarrow{p^{\alpha}}\right| c^{\alpha}$, where, in the c^{α} factor, the α-parameter plays the role of a power; it is not just a superscript. As before, the α appearing in $\overrightarrow{p^{\alpha}}$ and in the energy operator E^{α} corresponds to a label.

Let us consider a 2×1 component fermionic field $\chi_{L,\alpha}$ that transforms under a spinor representation of the space-time group. Taking into account the equations above, we can write

$$\frac{\hbar^{\alpha}}{2}\Gamma(\alpha+1)M_{\alpha}\vec{\sigma}^{\circ}\overrightarrow{p^{\alpha}}\chi_{L,\alpha} = \frac{\hbar^{\alpha}}{2}\Gamma(\alpha+1)M_{\alpha}\chi_{L,\alpha} \tag{4.3}$$

or

$$c^{\alpha}\vec{\sigma}^{\circ}\overrightarrow{p^{\alpha}}\chi_{L} = \left|\overrightarrow{p^{\alpha}}\right|\chi_{L,\alpha} = E^{\alpha}\chi_{L,\alpha}. \tag{4.4}$$

To get a quantum wave equation, we invoke the correspondence principle and we propose the fractional operators that represent energy and momentum to be given by the equations that follow:

$$\hat{E}^{\beta} = i\left(\hbar\right)^{\beta}\frac{\partial^{\beta}}{\partial t^{\beta}}, \tag{4.5}$$

$$\hat{p}^{\alpha} = -i\left(\hbar\right)^{\alpha}M_{\alpha}\frac{\partial^{\alpha}}{\partial x^{\alpha}}; \tag{4.6}$$

where the constant M_{α} is included for dimensional reasons, as before. We remark here that, if in the definition of the fractional moment operator, eq. (4.6), the complex i is replaced by an i^{α}, the canonical commutation relations change.

By virtue of the correspondence principle, eq. (4.4) can be rewritten as

$$i\left(\hbar\right)^{\alpha}\vec{\sigma}^{\circ}M_{\alpha}\frac{\partial^{\alpha}}{\partial x^{\alpha}}\chi_{L,\alpha} + i\left(\hbar\right)^{\alpha}\frac{M_{t,\alpha}}{c^{\alpha}}\frac{\partial^{\alpha}}{\partial t^{\alpha}}\chi_{L,\alpha} = 0 \tag{4.7}$$

$M_{t,\alpha}$ is included for dimensional reasons, as already stated above. Here, we are considering space and time with the same fractionality.

The Weyl equation above can be cast in a covariant form as it follows below:

$$i\left(\hbar\right)^{\alpha}\sigma^{\mu}\partial_{\mu}^{(\alpha)}\chi_{L,\alpha}=0. \tag{4.8}$$

Here,

$$\sigma^{\mu}=(\sigma^{0},\sigma^{i})=(\sigma^{0},\sigma^{1},\sigma^{2},\sigma^{3})=(I_{2},\sigma_{x},\sigma_{y},\sigma_{z})$$

are the usual Pauli-spin matrices and we define the space-time fractional derivative in Minkowsky space as

$$\partial_{\mu}^{(\alpha)}=\left(\frac{M_{t,\alpha}}{c^{\alpha}}\frac{\partial^{\alpha}}{\partial t^{\alpha}};M_{\alpha}\frac{\partial^{\alpha}}{\partial x^{\alpha}}\right). \tag{4.9}$$

Also, defining the conjugated Pauli-spin matrices as $\overline{\sigma}^{\mu}=(\sigma^{0},-\sigma^{i})$, the following properties hold:

$$tr(\sigma^{\mu}\overline{\sigma}^{\nu})=tr(\overline{\sigma}^{\mu}\sigma^{\nu})=2\eta^{\mu\nu}, \tag{4.10}$$

or

$$\sigma^{\mu}\overline{\sigma}^{\nu}+\sigma^{\nu}\overline{\sigma}^{\mu}=2\eta^{\mu\nu}\mathbf{1}, \tag{4.11}$$

$$\overline{\sigma}^{\mu}\sigma^{\nu}+\overline{\sigma}^{\nu}\sigma^{\mu}=2\eta^{\mu\nu}\mathbf{1}. \tag{4.12}$$

Inserting $\overline{\sigma}^{\nu}\partial_{\nu}^{(\alpha)}$ in the eq. (4.8), we obtain

$$i\left(\hbar\right)^{\alpha}\overline{\sigma}^{\nu}\partial_{\nu}^{(\alpha)}\sigma^{\mu}\partial_{\mu}^{(\alpha)}\chi_{L,\alpha}=0. \tag{4.13}$$

Next, we symmetrize in (u, v) and make use of eq. (4.12).
The last equation can be rewritten, with the help of the properties of σ^{μ} as

$$i\left(\hbar\right)^{\alpha}\eta^{\mu\nu}\partial_{\nu}^{(\alpha)}\partial_{\mu}^{(\alpha)}\chi_{L,\alpha}=0$$

which yields the propagating fractional wave equation for left-helicity fermion $i\left(\hbar\right)^{\alpha}\Box^{(\alpha)}\chi_{L,\alpha}=0$. The notation for the box symbol it is not to be confused with the fractional power operator in distribution theory. Here, the box symbol is defined as

$$\Box^{(\alpha)}\equiv\frac{M_{t,\alpha}}{c^{\alpha}}\frac{\partial^{\alpha}}{\partial t^{\alpha}}\frac{\partial^{\alpha}}{\partial t^{\alpha}}-M_{\alpha}^{2}\frac{\partial^{\alpha}}{\partial x^{\alpha}}\frac{\partial^{\alpha}}{\partial x^{\alpha}}.$$

Right-Helicity

A similar procedure can be used to build up a projection for right-chirality fermion by taking in eq.(4.1) $\lambda_\alpha = -1$.

Now, the spinor field is noted as ξ_R and belongs to a group representation such as $\xi_R \in (0;1/2)$. We are then ready to write down Weyl's equation for the right-handed sector:

$$i\left(\hbar\right)^{\alpha} \bar{\sigma}^{\mu}\partial_{\mu}^{(\alpha)}\xi_{R,\alpha} = 0, \tag{4.14}$$

that also conduct us to the equation

$$i\left(\hbar\right)^{\alpha}\square^{(\alpha)}\xi_{R,\alpha} = 0. \tag{4.15}$$

In compact notation, the fractional derivative is given by $\partial_{\mu}^{\alpha} = \left(\frac{M_{t,\alpha}}{c^{\alpha}}\partial_{t}^{\alpha}; M_{\alpha}\nabla^{\alpha}\right)$.

The two Weyl equations can be written in a more compact form in terms of a four-dimensional spinorial field Ψ_α as

$$i\hbar^{\alpha}\begin{pmatrix} 0 & \bar{\sigma}^{\mu}\partial_{\mu}^{\alpha} \\ \sigma^{\mu}\partial_{\mu}^{\alpha} & 0 \end{pmatrix}\begin{pmatrix} \chi_{L,\alpha} \\ \xi_{R,\alpha} \end{pmatrix} = i\hbar^{\alpha}\gamma^{\mu}\partial_{\mu}^{\alpha}\Psi_{\alpha} = 0, \tag{4.16}$$

where γ^μ are the usual Dirac gamma matrices.

We go ahead now by introducing a mass parameter that mixes up the two chiral components to obtain an equation that describes the dynamics of a charged massive fermion. To achieve that, we present the two fractional Weyl equations as

$$i\left(\hbar\right)^{\alpha} \sigma^{\mu}\partial_{\mu}^{(\alpha)}\chi_{L,\alpha} + u^{\alpha}c^{\alpha}\xi_{R,\alpha} = 0,$$

$$i\left(\hbar\right)^{\alpha} \bar{\sigma}^{\mu}\partial_{\mu}^{(\alpha)}\xi_{R,\alpha} + \tilde{u}^{\alpha}c^{\alpha}\chi_{L,\alpha} = 0, \tag{4.17}$$

where u^α and $\tilde{u}^\alpha$ are parameters which, as we shall show below, will combine to give the mass of the charged fermion. Imposing that the equations above be compatible with the fractional energy-momentum relationship (Weberszpil et al, 2012), we are led to

$$E^{2\alpha} = p^{2\alpha}c^{2\alpha} + m^{2\alpha}c^{4\alpha}. \tag{4.18}$$

From the second of eq. (4.17), we obtain

$$\chi_{L,\alpha} = -\frac{i\hbar^{\alpha}}{\tilde{u}^{\alpha}c^{\alpha}}\bar{\sigma}^{\mu}\partial_{\mu}^{(\alpha)}\xi_{R,\alpha}, \tag{4.19}$$

that, once replaced in the first of those equations, yields

$$\hbar^{2\alpha}\sigma^{\mu}\partial_{\mu}^{(\alpha)}\bar{\sigma}^{\nu}\partial_{\nu}^{(\alpha)}\xi_{R,\alpha} + u^{\alpha}\tilde{u}^{\alpha}c^{2\alpha}\xi_{R,\alpha} = 0\,, \tag{4.20}$$

or

$$\Box^{(\alpha)}\xi_{R,\alpha} + \frac{u^{\alpha}\tilde{u}^{\alpha}c^{2\alpha}}{\hbar^{2\alpha}}\xi_{R,\alpha} = 0\,.\ \text{or} \tag{4.21}$$

This equation indicates, upon comparison with a fractional Klein-Gordon (Weberszpil et al, 2012) equation, that $u^{\alpha} = \tilde{u}^{\alpha} = -m^{\alpha}$. With these observations, the low lever fractionality fractional Dirac equation can finally be written, from eq. (4.17), in the general case $\alpha \neq 1$ (but close to 1) as

$$(i\hbar^{\alpha}\gamma^{\mu}\partial_{\mu}^{\alpha} - 1m^{\alpha}c^{\alpha})\Psi_{\alpha} = 0\,. \tag{4.22}$$

Field Transformation, Minimal Coupling and the Gauge-Covariant Fractional Derivative

We start off with the conjugated of fractional Dirac's equation:

$$\bar{\Psi}_{\alpha}(-i\hbar^{\alpha}\gamma^{\mu}\partial_{\mu}^{\alpha} - 1m^{\alpha}c^{\alpha}) = 0. \tag{4.23}$$

Now, multiplying eq. (4.22)[eq:Fractional Dirac] by $\bar{\Psi}$ and eq. (4.23)[eq:Conjug Fractional Dirac] by Ψ and then subtracting them, we obtain

$$\partial_{\mu}^{\alpha}\bar{\Psi}_{\alpha}\Psi_{\alpha}\gamma^{\mu} + \bar{\Psi}_{\alpha}\gamma^{\mu}\partial_{\mu}^{\alpha}\Psi_{\alpha} = \partial_{\mu}^{\alpha}(\bar{\Psi}_{\alpha}\gamma^{\mu}\Psi_{\alpha}) = 0. \tag{4.24}$$

The quantity in the bracket can be identified as a conserved current and this equation is interpreted as a fractional continuity equation. From standard electromagnetism coupling with matter, we now take that j_{μ}, the electromagnetic current density, is coupled to the electromagnetic field, A^{μ}, where j_{μ} is of the form $j_{\mu} = e_{\alpha}\bar{\Psi}\gamma^{\mu}\Psi, e_{\alpha}$ is the electron charge measured in the coarse-grained medium and α is here simply a label.

For the fractional case, we can think of a minimal coupling term as $e_{\alpha}\bar{\Psi}_{\alpha}\gamma^{\mu}\Psi_{\alpha}A_{\mu}^{\alpha}$, where $A_{\mu}^{\alpha} = (\phi^{\alpha}; -\vec{A}^{\alpha})$ is the fractional potential tensor transform under Gauge transform as $(A_{\mu}^{\alpha})' = A_{\mu}^{\alpha} + \partial_{\mu}^{\alpha}\chi$.

Following the Lagrangian for the model with integer dimensions, we can write down the Lagrangian with electromagnetic coupling in the fractional case as given by

$$\mathcal{L} = \bar{\Psi}_{\alpha}(i\hbar^{\alpha}\gamma^{\mu}\partial_{\mu}^{\alpha} - 1m^{\alpha}c^{\alpha})\Psi_{\alpha} - e_{\alpha}\bar{\Psi}_{\alpha}\gamma^{\mu}\Psi_{\alpha}A_{\mu}^{\alpha}, \tag{4.25}$$

that can be re written as

$$\mathcal{L} = \bar{\Psi}_{\alpha}i\hbar^{\alpha}\gamma^{\mu}\left(\partial_{\mu}^{\alpha} + \frac{ie_{\alpha}}{\hbar^{\alpha}}A_{\mu}^{\alpha}\right)\Psi_{\alpha} - 1m^{\alpha}c^{\alpha}\bar{\Psi}_{\alpha}\Psi_{\alpha}. \tag{4.26}$$

In order that the theory remain covariant, we take the spinor field transformation as follows:

$$\Psi_{\alpha}^{'} = R(\chi)\Psi_{\alpha}, \tag{4.27}$$

where $R(\chi)$ is taken to be unitary, but its explicit form shall be determined in the sequel. The Lagrangian density with the electromagnetic coupling term suggests the definition of the fractional gauge-covariant derivative to be of the form:

$$D_{\mu}^{\alpha} = \partial_{\mu}^{\alpha} + \frac{ie_{\alpha}}{c^{\alpha}\hbar^{\alpha}}kA_{\mu}^{\alpha}, \tag{4.28}$$

where the tensor field A_{μ}^{α} is considered to have a fractional Gage symmetry and transforms by

$$(A_{\mu}^{\alpha})^{'} = A_{\mu}^{\alpha} + \partial_{\mu}^{\alpha}\chi. \tag{4.29}$$

The fractional covariant derivative of the field Ψ_{α} obeys, as usually done in the approach of gauge theories, the same transform rule as the matter field:

$$(D_{\mu}^{\alpha}\Psi_{\alpha})^{'} = R(\chi)D^{\alpha}\Psi_{\alpha}. \tag{4.30}$$

In more details,

$$\partial_{\mu}^{\alpha}\Psi_{\alpha}^{'} + \frac{ie_{\alpha}}{c^{\alpha}\hbar^{\alpha}}kA_{\mu}^{'\alpha}\Psi_{\alpha}^{'} = R(\chi)\partial_{\mu}^{\alpha}\Psi_{\alpha} + \frac{ie_{\alpha}}{c^{\alpha}\hbar^{\alpha}}kA_{\mu}^{\alpha}\Psi_{\alpha}, \tag{4.31}$$

which turn into a fractional differential equation of the form

$$\frac{\partial_{\mu}^{\alpha}R}{R} = -\frac{ie_{\alpha}}{c^{\alpha}\hbar^{\alpha}}\partial_{\mu}^{\alpha}\chi. \tag{4.32}$$

The solution to the equation above reads:

$$R = \exp\left(-\frac{ie_{\alpha}}{c^{\alpha}\hbar^{\alpha}}\chi\right). \tag{4.33}$$

This can be readily proven by means of the following the steps: fractionally deriving the above equation, with the use of eq. (2.5), we get that

$$\partial^{\alpha}_{\mu}R = -\frac{ie_{\alpha}}{c^{\alpha}\hbar^{\alpha}}\exp\left(-\frac{ie_{\alpha}}{c^{\alpha}\hbar^{\alpha}}\chi\right)\partial^{\alpha}_{\mu}\chi = -\frac{ie_{\alpha}}{c^{\alpha}\hbar^{\alpha}}R\partial^{\alpha}_{\mu}\chi\,, \tag{4.34}$$

and this proves our assertion.

The low-level fractionality fractional Dirac equation, in a coarse-grained scenario, may now be written with the minimal electromagnetic coupling as given below:

$$i\hbar^{\alpha}\gamma^{\mu}\left(\partial^{\alpha}_{\mu} + \frac{ie_{\alpha}}{c^{\alpha}\hbar^{\alpha}}A^{\alpha}_{\mu}\right)\Psi_{\alpha} - 1m^{\alpha}c^{\alpha}\Psi_{\alpha} = 0\,. \tag{4.35}$$

Splitting the space and time components, multiplying by γ^0 and using the properties of the Dirac gamma-matrices, we obtain

$$\begin{aligned} &i\hbar^{\alpha}\gamma^{0}\gamma^{0}\frac{1}{c^{\alpha}}\partial^{\alpha}_{t}\Psi_{\alpha} + i\hbar^{\alpha}\gamma^{0}\gamma^{i}M_{x,\alpha}\partial^{\alpha}_{x}\Psi_{\alpha} + i\hbar^{\alpha}\gamma^{0}\gamma^{0}\frac{ie_{\alpha}}{c^{\alpha}\hbar^{\alpha}}\phi^{\alpha}\Psi_{\alpha} + \\ &+ i\hbar^{\alpha}\gamma^{0}\gamma^{i}\frac{ie_{\alpha}}{c^{\alpha}\hbar^{\alpha}}(-\overrightarrow{A^{\alpha}})\Psi_{\alpha} - \gamma^{0}m^{\alpha}c^{\alpha}\Psi_{\alpha} = 0 \end{aligned} \tag{4.36}$$

or

$$i\hbar^{\alpha}1\partial^{\alpha}_{t}\Psi_{\alpha} = c^{\alpha}\gamma^{0}\gamma^{i}(-i\hbar^{\alpha}M_{x,\alpha}\partial^{\alpha}_{x}\Psi_{\alpha} - \frac{e_{\alpha}}{c^{\alpha}}\vec{A}^{\alpha}\Psi_{\alpha}) + \gamma^{0}\gamma^{0}e_{\alpha}\phi^{\alpha}\Psi_{\alpha} + \gamma^{0}m^{\alpha}c^{2\alpha}\Psi_{\alpha}, \tag{4.37}$$

$$i\hbar^{\alpha}1\partial^{\alpha}_{t}\Psi_{\alpha} = c^{\alpha}\gamma^{0}\gamma^{i}(-i\hbar^{\alpha}M_{x,\alpha}\partial^{\alpha}_{x}\Psi_{\alpha} - \frac{e_{\alpha}}{c^{\alpha}}\overrightarrow{A^{\alpha}}\Psi_{\alpha}) + 1e_{\alpha}\phi^{\alpha}\Psi_{\alpha} + \gamma^{0}m^{\alpha}c^{2\alpha}\Psi_{\alpha}. \tag{4.38}$$

Using the correspondence principle, we have

$$i\hbar^{\alpha}1\partial^{\alpha}_{t}\Psi_{\alpha} = c^{\alpha}\vec{\alpha}^{\circ}(\overrightarrow{p^{\alpha}} - \frac{e_{\alpha}}{c^{\alpha}}\overrightarrow{A^{\alpha}})\Psi_{\alpha} + 1e_{\alpha}\phi^{\alpha}\Psi_{\alpha} + \gamma^{0}m^{\alpha}c^{2\alpha}\Psi_{\alpha}. \tag{4.39}$$

Now, we write the Dirac spinor as a column-vector in the coarse-grained scenario as $\Psi_{\alpha} = \begin{pmatrix} \psi_{\alpha,s} \\ \psi_{\alpha,w} \end{pmatrix}$, where we have named the strong(s) and weak(w) components as the bi-spinors: $\psi_{\alpha,s} = \begin{pmatrix} \psi_{\alpha,1} \\ \psi_{\alpha,2} \end{pmatrix}$, $\psi_{\alpha,w} = \begin{pmatrix} \psi_{\alpha,3} \\ \psi_{\alpha,4} \end{pmatrix}$, respectively. In a symplectic form, we can write

$$\psi_{\alpha,s} = \psi_{\alpha,1} + i\psi_{\alpha,2}, \tag{4.40}$$

$$\psi_{\alpha,w} = \psi_{\alpha,3} - i\psi_{\alpha,4}. \tag{4.41}$$

The Non-Relativistic Regime of Fractional Dirac's Equation and the Fractional g-ratio

In order to work out the non-relativistic limit of the fractional Dirac's Equation, we have to consider that the dominant term in the Hamiltonian is the rest energy, given by $m^{\alpha}c^{2\alpha}$. We then propose an ansatz for the solution to the fractional Dirac's Equation as

$$\Psi'_{\alpha} = e^{\left(-i\frac{E^{\alpha}}{\hbar^{\alpha}}t^{\alpha}\right)} \begin{pmatrix} \psi_{\alpha,s} \\ \psi_{\alpha,w} \end{pmatrix}, E^{\alpha} \cong m^{\alpha}c^{2\alpha}. \tag{4.42}$$

Inserting this ansatz into the fractional equation, eq. (4.22), with the help of the chain rule (2.5) and the Leibniz rule (2.2), we obtain

$$i\hbar^{\alpha}1\partial_t^{\alpha}\begin{pmatrix} \psi_{\alpha,s} \\ \psi_{\alpha,w} \end{pmatrix} = \begin{pmatrix} c^{\alpha}\vec{\sigma}^{\circ}(\overrightarrow{p^{\alpha}} - \frac{e_{\alpha}}{c^{\alpha}}\overrightarrow{A^{\alpha}}) \\ c^{\alpha}\vec{\sigma}^{\circ}(\overrightarrow{p^{\alpha}} - \frac{e_{\alpha}}{c^{\alpha}}\overrightarrow{A^{\alpha}}) \end{pmatrix}\begin{pmatrix} \psi_{\alpha,w} \\ \psi_{\alpha,s} \end{pmatrix} + e_{\alpha}\phi^{\alpha}\begin{pmatrix} \psi_{\alpha,s} \\ \psi_{\alpha,w} \end{pmatrix} + m^{\alpha}c^{2\alpha}\begin{pmatrix} \psi_{\alpha,s} \\ -\psi_{\alpha,w} \end{pmatrix} - m^{\alpha}c^{2\alpha}\Gamma(\alpha+1)\begin{pmatrix} \psi_{\alpha,s} \\ \psi_{\alpha,w} \end{pmatrix}, \tag{4.43}$$

which yields two equations as cast below:

$$i\hbar^{\alpha}1\partial_t^{\alpha}\psi_{\alpha,b} = c^{\alpha}\vec{\sigma}^{\circ}(\overrightarrow{p^{\alpha}} - \frac{e_{\alpha}}{c^{\alpha}}\overrightarrow{A^{\alpha}})\psi_{\alpha,w} + e_{\alpha}\phi^{\alpha}\psi_{\alpha,s} + (1-\Gamma(\alpha+1))m^{\alpha}c^{2\alpha}\psi_{\alpha,s}, \tag{4.44}$$

$$i\hbar^{\alpha}1\partial_t^{\alpha}\psi_{\alpha,w} = c^{\alpha}\vec{\sigma}^{\circ}(\overrightarrow{p^{\alpha}} - \frac{e_{\alpha}}{c^{\alpha}}\overrightarrow{A^{\alpha}})\psi_{\alpha,s} + e_{\alpha}\phi^{\alpha}\psi_{\alpha,w} - (1+\Gamma(\alpha+1))m^{\alpha}c^{2\alpha}\psi_{\alpha,w}. \tag{4.45}$$

We now consider that the mass term is dominant over the electrostatic interaction energy, that is,

$$e_{\alpha}\phi^{\alpha} \lll m^{\alpha}c^{2\alpha}(1+\Gamma(\alpha+1)),$$

and that the weakly-relativistic component $\psi_{\alpha,w}$ has slow evolution whenever compared to the rest energy,

$$i\hbar^{\alpha}1\partial_t^{\alpha}\psi_{\alpha,w} < m^{\alpha}c^{2\alpha}(1+\Gamma(\alpha+1))\psi_{\alpha,w}.$$

With these approximations, we can write that the second equation takes the form

$$0 \cong c^{\alpha}\vec{\sigma}^{\circ}(\overrightarrow{p^{\alpha}} - \frac{e_{\alpha}}{c^{\alpha}}\overrightarrow{A^{\alpha}})\psi_{\alpha,s} + e_{\alpha}\phi^{\alpha}\psi_{\alpha,w} - (1+\Gamma(\alpha+1))m^{\alpha}c^{2\alpha}\psi_{\alpha,w}, \tag{4.46}$$

which leads to the relation between the relativistic weak and strong components as

$$\psi_{\alpha,w} \cong \frac{c^{\alpha}\vec{\sigma}^{\circ}(\overrightarrow{p^{\alpha}} - \frac{e_{\alpha}}{c^{\alpha}}\overrightarrow{A^{\alpha}})}{(1+\Gamma(\alpha+1))m^{\alpha}c^{2\alpha}}\psi_{\alpha,s}. \tag{4.47}$$

The expression above readily gives that the weak component is, within the approximations carried out above, suppressed with respect to the strong bi-spinor, $\psi_{\alpha,w} \lll \psi_{\alpha,s}$.

Now, inserting this result into the first eq. (4.44), the equation for the strong component turns out to be

$$i\hbar^{\alpha}1\partial_t^{\alpha}\psi_{\alpha,s} = \frac{\vec{\sigma}^{\circ}(\overrightarrow{p^{\alpha}} - \frac{e_{\alpha}}{c^{\alpha}}\overrightarrow{A^{\alpha}})\vec{\sigma}^{\circ}(\overrightarrow{p^{\alpha}} - \frac{e_{\alpha}}{c^{\alpha}}\overrightarrow{A^{\alpha}})}{(1+\Gamma(\alpha+1))m^{\alpha}}\psi_{\alpha,s} + e_{\alpha}\phi^{\alpha}\psi_{\alpha,s} + (1-\Gamma(\alpha+1))m^{\alpha}c^{2\alpha}\psi_{\alpha,s}. \tag{4.48}$$

We define the fractional momentum operator as $\overrightarrow{\pi^{\alpha}} \equiv \overrightarrow{p^{\alpha}} - \frac{e_{\alpha}}{c^{\alpha}}\overrightarrow{A^{\alpha}}$, we use the well-known propriety of Pauli matrices

$$(\overrightarrow{\sigma^{\circ}\pi^{\alpha}})(\overrightarrow{\sigma^{\circ}\pi^{\alpha}}) = \pi^{\alpha\circ}\pi^{\alpha} + i\vec{\sigma}^{\circ}(\pi^{\alpha}\wedge\pi^{\alpha})$$

and also that

$$(\pi^{\alpha}\wedge\pi^{\alpha}) = -\frac{e_{\alpha}}{c^{\alpha}}\overrightarrow{p^{\alpha}}\wedge\overrightarrow{A^{\alpha}}.$$

Now, with the definition

$$\overrightarrow{p^{\alpha}} = i\hbar^{\alpha} M_{x,\alpha} \vec{\nabla}^{\alpha}, \vec{\nabla}^{\alpha} \equiv \hat{i}\frac{\partial^{\alpha}}{\partial x^{\alpha}} + \hat{j}\frac{\partial^{\alpha}}{\partial y^{\alpha}} + \hat{k}\frac{\partial^{\alpha}}{\partial z^{\alpha}}$$

and in analogy with the integer case, $\vec{\nabla}^{\alpha} \wedge \overrightarrow{A^{\alpha}} = \overrightarrow{B^{\alpha}}$, we can write down

$$i\hbar^{\alpha} 1\partial_t^{\alpha}\psi_{\alpha,b} = \frac{\frac{(\overrightarrow{p^{\alpha}} - \frac{e_{\alpha}}{c^{\alpha}}\overrightarrow{A^{\alpha}})^2}{}}{(1+\Gamma(\alpha+1))m^{\alpha}}\psi_{\alpha,b} - \frac{\hbar^{\alpha} M_{x,\alpha}\frac{e_{\alpha}}{c^{\alpha}}\vec{\sigma}\circ\overrightarrow{B^{\alpha}}}{(1+\Gamma(\alpha+1))m^{\alpha}}\psi_{\alpha,b} + \\ + e_{\alpha}\phi^{\alpha}\psi_{\alpha,b} + (1-\Gamma(\alpha+1))m^{\alpha}c^{2\alpha}\psi_{\alpha,b}. \tag{4.49}$$

The eq. (4.49) is the low level fractionality fractional version of the Pauli equation in a coarse-grained scenario.

Recalling that the spin term is $\overrightarrow{S_{\alpha}} = \frac{\hbar^{\alpha}}{2}\Gamma(\alpha+1)M_{x,\alpha}\vec{\sigma}$, the second term in the r.h.s. of the above equation can be written as

$$-\frac{e_{\alpha}}{2m^{\alpha}c^{\alpha}}\left(\frac{4}{(1+\Gamma(\alpha+1))\Gamma(\alpha+1)}\right)\overrightarrow{S_{\alpha}}\circ\overrightarrow{B^{\alpha}}\psi_{\alpha,s}. \tag{4.50}$$

The factor into the bracket can be identified as the fractional gyromagnetic ratio or g-factor g_{free},

$$g_{frac} = \frac{4}{(1+\Gamma(\alpha+1))\Gamma(\alpha+1)}. \tag{4.51}$$

Notice that eq. (4.51) has an universal form in that it is the same for a general α. However, as we shall be discussing in Section 6, α is species-dependent: electron, muon and the tau will exhibit different fractionality, that is, different α-parameters. In this sense, we may say that eq. (4.51) is species-dependent. This remark will be very important for a relationship that will be proposed in Section 6 between the mean life time of the lepton and the α-parameter. Notice that, whenever α=1, g_{free}=g=2, the equation turns into the usual Pauli equation.

We can map a fractional parameter with the CODATA (Mohr et al, 2012) known value of g_{exp} for electrons and muons, and according to (Eidelman et al, 2007) for the lepton tau. The mapping can be achieved by numerically solving the equation

$$g_{frac} = \frac{4}{(1+\Gamma(\alpha+1))\Gamma(\alpha+1)} = g_{exp}. \tag{4.52}$$

Since we expect the α-parameter to be very close to 1 (low fractionality), we could write that α=1-ε, ε being a deviation from 1 as an effect of the fractionality; then, we could write $g_{frac} = 2 + \delta(\varepsilon)$, where δ is a small deviation. Then, by expanding the gamma functions in the denominator of eq. (4.52) around ε=0, δ(ε) it could be readily computed. However, our option is to avoid the expansion in ε in order to have a result that includes all orders in ε so that it can compared with a better precision with the experimental value of g, written here as g_{exp}. In our Discussion and Conclusions (Section 6), we shall elaborate more on this result and we shall show how we make use of it to fit the g-ratios of the charged leptonic particles.

We believe that, for the sake of completeness, it would be instructive to present the Gordon decomposition for the electronic current in its fractional form. The Gordon decomposition readily manifests the magnetic dipole piece of the relativistic current. So, after we have studied the effect of the fractionality on the g-factor, it is worthy to show explicitly that the spin contribution couples to the electron momentum transfer. We shall do that in the next Section and we choose to work it out in the configuration space, rather than in momentum space, as one usually does.

THE GORDON DECOMPOSITION AND THE FRACTIONAL DIRAC'S EQUATION: A FRACTIONAL SPIN CURRENT

From eq. (4.24)[eq:Fractional Continuity], we can define the fractional current density as

$$j_{\alpha}^{\mu} = e_{\alpha}\bar{\Psi}\gamma^{\mu}\Psi_{\alpha}, \tag{5.1}$$

where α here does not stand for the spinor index, as it might appear; α refers to fractionality, as in the previous sections.

The above equation may be rewritten as

$$j_{\alpha}^{\mu} = \frac{1}{2}e_{\alpha}\left(\bar{\Psi}_{\alpha}\gamma^{\mu}\Psi_{\alpha} + \bar{\Psi}_{\alpha}\gamma^{\mu}\Psi_{\alpha}\right). \tag{5.2}$$

From the Dirac equation eq. (4.22) and its Dirac-conjugated version, (4.23) we may write, respectively,

$$\Psi_{\alpha} = \frac{i\hbar^{\alpha}}{m^{\alpha}c^{\alpha}}\gamma^{\mu}\partial_{\mu}^{\alpha}\Psi_{\alpha}, \bar{\Psi}_{\alpha} = -\frac{i\hbar^{\alpha}}{m^{\alpha}c^{\alpha}}\partial_{\mu}^{\alpha}\bar{\Psi}_{\alpha}\gamma^{\mu}. \tag{5.3}$$

Now, inserting these results in the first and second terms of the eq. (5.2) [eq:Current Symmetric], respectively, we can write for the fractional current density

$$j_{\alpha}^{\mu} = \frac{i\hbar^{\alpha}e_{\alpha}}{m^{\alpha}c^{\alpha}}\left(\left(\bar{\Psi}_{\alpha}\gamma^{\mu}\gamma^{\nu}\right)\left(\partial_{\nu}^{\alpha}\Psi_{\alpha}\right) - \left(\partial_{\nu}^{\alpha}\bar{\Psi}_{\alpha}\right)\left(\gamma^{\nu}\gamma^{\mu}\Psi_{\alpha}\right)\right). \tag{5.4}$$

For the gamma-matrices, we can write that

$$\gamma^{\mu}\gamma^{\nu} = \frac{1}{2}\left(\gamma^{\mu}\gamma^{\nu} + \gamma^{\nu}\gamma^{\mu}\right) + \frac{1}{2}\left(\gamma^{\mu}\gamma^{\nu} - \gamma^{\nu}\gamma^{\mu}\right) = \eta^{\mu\nu}1 + \Sigma^{\mu\nu}, \tag{5.5}$$

and

$$\gamma^{\nu}\gamma^{\mu} = \eta^{\mu\nu}1 + \Sigma^{\nu\mu} = \eta^{\mu\nu}1 - \Sigma^{\mu\nu}, \tag{5.6}$$

where

$$\Sigma^{\mu\nu} = \frac{1}{2}\left(\gamma^{\mu}\gamma^{\nu} - \gamma^{\nu}\gamma^{\mu}\right) = \frac{1}{2}\left[\gamma^{\mu}, \gamma^{\nu}\right]$$

and $\eta^{\mu\nu}$ is the metric tensor.

Using the definitions above and inserting them into eq. (5.4), we obtain for the current a suitable form, where we split the contribution due to the spin:

$$j^{\mu}_{\alpha} = \frac{i\hbar^{\alpha} e_{\alpha}}{2m^{\alpha}c^{\alpha}}\left(\bar{\Psi}_{\alpha}\left(\partial^{\alpha}_{\nu}\Psi_{\alpha}\right) - \left(\partial^{\alpha}_{\nu}\bar{\Psi}_{\alpha}\right)\Psi_{\alpha}\right) + \frac{i\hbar^{\alpha} e_{\alpha}}{2m^{\alpha}c^{\alpha}}\left(\bar{\Psi}_{\alpha}\Sigma^{\mu\nu}\left(\partial^{\alpha}_{\nu}\Psi_{\alpha}\right) + \left(\partial^{\alpha}_{\nu}\bar{\Psi}_{\alpha}\right)\Sigma^{\mu\nu}\Psi_{\alpha}\right). \tag{5.7}$$

Defining $\sigma^{\mu\nu} = i\Sigma^{\mu\nu}$, the second term in the equation above may be identified with the spin contribution to the fractional current and reads as

$$j^{\mu}_{\alpha,spin} = \frac{\hbar^{\alpha} e_{\alpha}}{2m^{\alpha}c^{\alpha}}\partial^{\alpha}_{\nu}\left(\bar{\Psi}_{\alpha}\sigma^{\mu\nu}\Psi_{\alpha}\right). \tag{5.8}$$

Therefore, we are here giving the Gordon decomposition carried out in configuration space rather then presenting it in Fourier space. We choose to do so because we are not dealing with generalized functions, but with non-differentiable functions.

CONCLUSION

With the help of the eq. (4.52), we are able to compare the the fractional α-parameter for electrons, muons and taus. The numerical results show that the fractionality, α, exhibit a hierarchy, that is α(*electron*)>α(*muon*)>α(*tau*).

(In what follows g_{e-exp} is the g-factor for electrons given by experimental CODATA reference. Analogously for $g_{\mu-exp}$. The $\alpha_{\mu-exp}$ parameter is the corresponding one for the fermion species)

For the electron (CODATA), we have that:

$$g_{e-exp} = 2.00231930436146.$$

This yields, in our approach,

$$\alpha_{e-exp} = 0.9981697906061296726\,,$$

indicating a system with low fractionality (α near 1).

Analogously, for the muon (CODATA):

$$g_{\mu-exp} = 2.00233184182\,,$$

which gives

$$\alpha_{\mu-exp} = 0.9981598882409105161\,;$$

for the tau-lepton (Data from QED, see e.g. (Eidelman et al, 2007): g_τ=2.00235442, such that our

α_τ=0.998142055249517567.

Notice the hierarchy for the g-factors: $g_\tau > g_{\mu-Exp} > g_{e-Exp}$: an inverted hierarchy with respect to the values of α-parameters.

We can interpret these results as follows: the higher is the g-ratio, the farther is the α -parameter from α=1 (g-factor is a decreasing function with α, as it can be seen numerically from eq. (4.51). A larger α -parameter means obviously fractionality closer to 1. Moreover, if this parameter deviates significantly from α=1, it may indicate that the system is more sparse, or equivalently, roughness is larger. This may suggest that there is a smaller variety of different types of interactions (particles and fields) in the surroundings of the pseudo-particle. In other words, if a particle interacts less or has a lower mean life than another particle, its corresponding α -parameter would be a little more distant (downwards) from and which indicates a lower complexity of interactions and also that the medium "seen" by the particle is more rugged. We understand that a value of alpha closer to 1 suggests an enhancement in the complexity of the interactions. By this we mean that the particle interacts more significantly with more structures in its neighborhood. An α -parameter equal to 1 means one type of total mixing memory (providing an idealized integer model, which is not natural), with the pseudo-particles feeling the interactions of all kinds, so that, in average, these fluctuations are vanishing (a statistical average). We can speculate that the medium seen by this kind of particle is less rugged or closer to the continuum.

Now, in terms of mean life, muon decays in $10^{-6}s$, tau decays in $10^{-15}s$, while the electron is stable. The relation between mean life, the α -parameter and the coarse-grained character can be understood as follows: The lower is the mean life, this means that the particle does not have sufficient time to interact with the surrounding environment and with other particles, implying that the α -parameter departs from the value 1. The overall interaction felt by those particles is limited to the closer interacting particles and the neighbor environment. Thinking this way, the medium appears more rugged for this particle. Stable particles, like electrons, can experience the entire environment, therefore it interacts more, justifying a fractional parameter closer to 1. So, this medium could be considered less rugged for the electrons than for the other leptonic species. The time scale of the interaction expressed by the mean life can give clues on the fractionality.

From QED calculations[4, 3], we also have for electron, muon and tau (In what follows, g_{e-QED} is the g-factor for electrons given by QED. Analogously for $g_{\mu-QED}$ and $g_{\mu-QED}$. The $\alpha_{\mu-QED}$-parameter is the corresponding one for the fermion species):

$$g_{e-QED} = 2.00231930436364, \alpha_{e-QED} = 0.99816979060440786 81;$$

$$g_{\mu-QED} = 2.002331694362, \alpha_{\mu-QED} = 0.9981600047070900225.$$

Comparing the parameters above from QED with those from the measured anomalies [4], we can see that

$$g_{e-QED} = 2.00231930436364 > g_{e-exp} = 2.00231930436146.$$

This leads to

$$\alpha_{e-QED} = 0.9981697906044078681 < \alpha_{e-exp} = 0.9981697906061296726.$$

We can see the opposite behavior for muon:

$$g_{\mu-QED} = 2.002331694362 < g_{\mu-exp} = 2.00233184182,$$

what leads to

$$\alpha_{\mu-QED} = 0.9981600047070900225 > \alpha_{\mu-exp} = 0.9981598882409105161.$$

We can interpret these results as follows: for the electron, the results may be indicating that the complexity of the interactions taken into account in the QED calculation may be lower than it would be in the reality seen by the experiment. That is, the electron, as a pseudo-particle, keeps hidden other interactions or even structures that are not well described or are incomplete in the SM description. Thus, this may justify the deviation from the QED calculations as compared to the experimental results. In other words, in the experimental reality for electrons, we may think that the complexity is larger than that considered in the QED calculations based on the SM and the fractionality should indicate that the SM, although very good, may not be providing all information necessary to describe the interactions in a more complete view, either at high energies or in granular or fractal space-time. Thus, the FC may gives evidences that the SM could require corrections (or higher order calculations by QED).

For the muon, an opposite behavior in terms of fractional parameter could be observed and indicates that the QED calculations might be taken into account with more interactions and a consequent complexity than the particle really experience, suggesting that the SM may not be complete to describe the whole interaction scenario.

In summary, in the present work, we have built up a fractional Dirac equation in a coarse-grained scenario by taking into account a fractional Weyl equation, a fractional angular momentum algebra, by introducing a mass parameter and imposing that the equations be compatible with the fractional energy-momentum relation. Considering then a fractional gauge-covariant derivative, we could minimally couple the charged fermion to an external electromagnetic field. The step further consisted in working out the non-relativistic limit of the Dirac's equation to obtain a fractional version of the Pauli's equation. We have then investigated the anomalous magnetic moment for the charged leptons: electrons, muons and taus. With the fractional approach to a coarse-grained scenario, we were able to get the fractionality associated to each lepton species.

We have shown that a mapping of the anomalous magnetic g-ratio is possible in terms of a fractional parameter that could be used to give some insight about the precision of the QED calculations necessary to better emulate reality.

Defining a fractional current density, we have also performed a fractional Gordon decomposition and identified the spin contribution to the fractional electromagnetic current density.

We also suggested that the understanding of the results comparatively for electrons, muons e taus may be thought in the realm of complexity, mean life, coarse-grainedness and pseudo-particle concepts.

We have finally shown that the fractionalities exhibit a hierarchy $\alpha(electron) > \alpha(muon) > \alpha(tau)$.

We have established a connection between mean life of the particle, fractionality and roughness, showing that a small mean life can lead to a fractional parameter more distant from $\alpha=1$ if compared with more stable particles. Consequently, a small mean life means less interaction, less complexity related to interactions and to the effective medium felt by the particle. This means more roughness.

In this study, we have proposed an axiomatic version of metric derivative that has the Mittag-Leffler function as an eigenfunction and is only valid for low-level fractionality systems.

The structure of the DD generated shows similarities with some versions claimed to hold or not hold in the literature, but here it is axiomatically set up and the consistency are given in the context of low level of fractionality,i.e., is valid within a low-fractionality limit and it is in essence a local formalism.

A final remark can be made. We could have used different types of DD. The choice for the axiomatic one intended to revisit our article in ref. Weberszpil & Helayël-Neto, 2014) with a similar formalism (Weberszpil & Helayël-Neto, 2017), but making here clear that it is only valid for low level fractionality.

ACKNOWLEDGMENT

The authors would like to thank Dr. A. E. Matouk, for the invitation and to Monica G. Weberszpil for the help with the bibliographic organization.

REFERENCES

Ankerhold, J., Grabert, H., & Ingold, G. L. (1995). Dissipative quantum systems with a potential barrier: General theory and the parabolic barrier. *Physical Review. E*, *51*(5), 4267–4281. doi:10.1103/PhysRevE.51.4267 PMID:9963138

Babakhani, A., & Daftardar-Gejji, V. (2002). On calculus of local fractional derivatives. *Journal of Mathematical Analysis and Applications*, *270*(1), 66–79. doi:10.1016/S0022-247X(02)00048-3

Balankin, A. S., Bory-Reyes, J., & Shapiro, M. (2016). Towards a physics on fractals: Differential vector calculus in three-dimensional continuum with fractal metric. *Physica A*, *444*, 345–359. doi:10.1016/j.physa.2015.10.035

Balankin, A. S., & Elizarraraz, B. E. (2012a). Map of fluid flow in fractal porous medium into fractal continuum flow. *Physical Review. E*, *85*(5), 056314. doi:10.1103/PhysRevE.85.056314 PMID:23004869

Balankin, A. S., & Elizarraraz, B. E. (2012b). Hydrodynamics of fractal continuum flow. *Physical Review. E*, *85*(2), 025302. doi:10.1103/PhysRevE.85.025302 PMID:22463270

Bologna, M., Grigolini, P., & Riccardi, J. (1999). Lévy diffusion as an effect of sporadic randomness. *Physical Review. E*, *60*(6), 6435–6442. doi:10.1103/PhysRevE.60.6435 PMID:11970559

Brodsky, S. J., Franke, V. A., Hiller, J. R., McCartor, G., Paston, S. A., & Prokhvatilov, E. V. (2004). A nonperturbative calculation of the electron's magnetic moment. *Nuclear Physics B*, *703*(1-2), 333–362. doi:10.1016/j.nuclphysb.2004.10.027

Budini, A. A., & Cáceres, M. O. (2004). Functional characterization of linear delay Langevin equations. *Physical Review. E*, *70*(4), 046104. doi:10.1103/PhysRevE.70.046104 PMID:15600457

Calcagni, G. (2010). Fractal universe and quantum gravity. *Physical Review Letters*, *104*(25), 251301. doi:10.1103/PhysRevLett.104.251301 PMID:20867360

Calcagni, G. (2012a). Geometry and field theory in multi-fractional spacetime. *Journal of High Energy Physics*, *2012*(1), 65. doi:10.1007/JHEP01(2012)065

Calcagni, G. (2012b). Geometry of fractional spaces. *Advances in Theoretical and Mathematical Physics*, *16*(2), 549–644. doi:10.4310/ATMP.2012.v16.n2.a5

Calcagni, G., Eichhorn, A., & Saueressig, F. (2013). Probing the quantum nature of spacetime by diffusion. *Physical Review. D*, *87*(12), 124028. doi:10.1103/PhysRevD.87.124028

Calcagni, G., Magueijo, J., & Fernández, D. R. (2014). Varying electric charge in multiscale spacetimes. *Physical Review. D*, *89*(2), 024021. doi:10.1103/PhysRevD.89.024021

Caldeira, A. O., & Leggett, A. J. (1981). Influence of dissipation on quantum tunneling in macroscopic systems. *Physical Review Letters*, *46*(4), 211–214. doi:10.1103/PhysRevLett.46.211

Caldeira, A. O., & Leggett, A. J. (1983). Quantum tunnelling in a dissipative system. *Annals of Physics*, *149*(2), 374–456. doi:10.1016/0003-4916(83)90202-6

Carpinteri, A., & Sapora, A. (2010). Diffusion problems in fractal media defined on Cantor sets. *ZAMM-Journal of Applied Mathematics and Mechanics/Zeitschrift für Angewandte Mathematik und Mechanik. Applied Mathematics and Mechanics*, *90*(3), 203–210.

Chen, W. (2006). Time–space fabric underlying anomalous diffusion. *Chaos, Solitons, and Fractals*, *28*(4), 923–929. doi:10.1016/j.chaos.2005.08.199

Chen, W., Sun, H., Zhang, X., & Korošak, D. (2010b). Anomalous diffusion modeling by fractal and fractional derivatives. *Computers & Mathematics with Applications (Oxford, England)*, *59*(5), 1754–1758. doi:10.1016/j.camwa.2009.08.020

Chen, Y., Yan, Y., & Zhang, K. (2010a). On the local fractional derivative. *Journal of Mathematical Analysis and Applications*, *362*(1), 17–33. doi:10.1016/j.jmaa.2009.08.014

de Rafael, E. (2013). Update of the Electron and Muon g-Factors. *Nuclear Physics B - Proceedings Supplement*, *234*, 193–198. doi:10.1016/j.nuclphysbps.2012.12.012

Eidelman, S., Giacomini, M., Ignatov, F. V., & Passera, M. (2007). The tau lepton anomalous magnetic moment. *Nuclear Physics B - Proceedings Supplements, 169*, 226-231. arXiv preprint hep-ph/0702026

Eyink, G. (1989). Quantum field-theory models on fractal spacetime. *Communications in Mathematical Physics*, *125*(4), 613–636. doi:10.1007/BF01228344

Godinho, C. F., Weberszpil, J., & Helayël-Neto, J. A. (2012). Extending the d'Alembert solution to space–time modified Riemann–Liouville fractional wave equations. *Chaos, Solitons, and Fractals*, *45*(6), 765–771. doi:10.1016/j.chaos.2012.02.008

Godinho, C. F., Weberszpil, J., & Neto, J. H. (2014). Fractional canonical quantization: A parallel with noncommutativity. *International Journal of Theoretical Physics*, *53*(7), 2379–2395. doi:10.100710773-014-2037-5

Grigolini, P., Rocco, A., & West, B. J. (1999). Fractional calculus as a macroscopic manifestation of randomness. *Physical Review. E*, *59*(3), 2603–2613. doi:10.1103/PhysRevE.59.2603

Herrmann, R. (2007). The fractional symmetric rigid rotor. *Journal of Physics. G, Nuclear and Particle Physics*, *34*(4), 607–625. doi:10.1088/0954-3899/34/4/001

Herrmann, R. (2008). Gauge invariance in fractional field theories. *Physics Letters. [Part A]*, *372*(34), 5515–5522. doi:10.1016/j.physleta.2008.06.063

Jegerlehner, F., & Nyffeler, A. (2009). The muon g− 2. *Physics Reports*, *477*(1-3), 1–110. doi:10.1016/j.physrep.2009.04.003

Jumarie, G. (2006). Modified Riemann-Liouville derivative and fractional Taylor series of nondifferentiable functions further results. *Computers & Mathematics with Applications (Oxford, England)*, *51*(9-10), 1367–1376. doi:10.1016/j.camwa.2006.02.001

Jumarie, G. (2007). Fractional partial differential equations and modified Riemann-Liouville derivative new methods for solution. *Journal of Applied Mathematics and Computing*, *24*(1-2), 31–48. doi:10.1007/BF02832299

Jumarie, G. (2009a). Table of some basic fractional calculus formulae derived from a modified Riemann–Liouville derivative for non-differentiable functions. *Applied Mathematics Letters*, *22*(3), 378–385. doi:10.1016/j.aml.2008.06.003

Jumarie, G. (2009b). From Lagrangian mechanics fractal in space to space fractal Schrödinger's equation via fractional Taylor's series. *Chaos, Solitons, and Fractals*, *41*(4), 1590–1604. doi:10.1016/j.chaos.2008.06.027

Jumarie, G. (2010). Derivation and solutions of some fractional Black–Scholes equations in coarse-grained space and time. Application to Merton's optimal portfolio. *Computers & Mathematics with Applications (Oxford, England)*, *59*(3), 1142–1164. doi:10.1016/j.camwa.2009.05.015

Jumarie, G. (2013). On the derivative chain-rules in fractional calculus via fractional difference and their application to systems modelling. *Open Physics*, *11*(6), 617–633. doi:10.247811534-013-0256-7

Kilbas, A. A. A., Srivastava, H. M., & Trujillo, J. J. (2006). *Theory and applications of fractional differential equations* (Vol. 204). Elsevier Science Limited. doi:10.1016/S0304-0208(06)80001-0

Kobelev, V. V. (2006). The variant of post-Newtonian mechanics with generalized fractional derivatives. *Chaos (Woodbury, N.Y.)*, *16*(4), 043117. doi:10.1063/1.2384864 PMID:17199395

Kolwankar, K. M., & Gangal, A. D. (1996). Fractional differentiability of nowhere differentiable functions and dimensions. *Chaos (Woodbury, N.Y.)*, *6*(4), 505–513. doi:10.1063/1.166197 PMID:12780280

Kolwankar, K. M., & Gangal, A. D. (1998). Local fractional Fokker-Planck equation. *Physical Review Letters*, *80*(2), 214–217. doi:10.1103/PhysRevLett.80.214

Kolwankar, K. M., & Gangal, A. D. (1999). Local fractional calculus: a calculus for fractal space-time. In *Fractals* (pp. 171–181). London: Springer. doi:10.1007/978-1-4471-0873-3_12

Leopoldino, A. P., Weberszpil, J., Godinho, C. F., & Helayël-Neto, J. A. (2019). Discussing the extension and applications of a variational approach with deformed derivatives. *Journal of Mathematical Physics*, *60*(8), 083507. doi:10.1063/1.5099046

Metzler, R., & Klafter, J. (2004). The restaurant at the end of the random walk: Recent developments in the description of anomalous transport by fractional dynamics. *Journal of Physics. A, Mathematical and General*, *37*(31), R161–R208. doi:10.1088/0305-4470/37/31/R01

Miller, J. P., de Rafael, E., & Roberts, B. L. (2007). Muon (g− 2): Experiment and theory. *Reports on Progress in Physics*, *70*(5), 795–881. doi:10.1088/0034-4885/70/5/R03

Mohr, P. J., Taylor, B. N., & Newell, D. B. (2012). CODATA recommended values of the fundamental physical constants: 2010. *Journal of Physical and Chemical Reference Data*, *84*(4), 1527.

Nottale, L. (1993). *Fractal space-time and microphysics: towards a theory of scale relativity*. World Scientific. doi:10.1142/1579

Palmer, C., & Stavrinou, P. N. (2004). Equations of motion in a non-integer-dimensional space. *Journal of Physics. A, Mathematical and General*, *37*(27), 6987–7003. doi:10.1088/0305-4470/37/27/009

Passera, M., Martiano, W. J., & Sirlin, A. (2008, November). The muon g-2 discrepancy: errors or new physics? In. AIP Conference Proceedings: Vol. 1078. *No. 1* (pp. 378–381). AIP. doi:10.1063/1.3051967

Plyushchay, M. S., & De Traubenberg, M. R. (2000). Cubic root of Klein-Gordon equation. *Physics Letters. [Part B]*, *477*(1-3), 276–284. doi:10.1016/S0370-2693(00)00190-8

Podlubny, I. (1999). Fractional differential equations. Academic Press.

Raspini, A. (2001). Simple solutions of the fractional Dirac equation of order 2/3. *Physica Scripta, 64*(1), 20–22. doi:10.1238/Physica.Regular.064a00020

Riewe, F. (1996). Nonconservative lagrangian and hamiltonian mechanics. *Physical Review. E, 53*(2), 1890.

Riewe, F. (1997). Mechanics with fractional derivatives. *Physical Review. E, 55*(3), 3581. doi:10.1103/PhysRevE.55.3581

Rosa, W., & Weberszpil, J. (2018). Dual conformable derivative: Definition, simple properties and perspectives for applications. *Chaos, Solitons, and Fractals, 117*, 137–141. doi:10.1016/j.chaos.2018.10.019

Senitzky, I. R. (1960). Dissipation in quantum mechanics. The harmonic oscillator. *Physical Review, 119*(2), 670–679. doi:10.1103/PhysRev.119.670

Stevens, K. W. H. (1958). The wave mechanical damped harmonic oscillator. *Proceedings of the Physical Society, 72*(6), 1027–1036. doi:10.1088/0370-1328/72/6/311

Stillinger, F. H. (1977). Axiomatic basis for spaces with noninteger dimension. *Journal of Mathematical Physics, 18*(6), 1224–1234. doi:10.1063/1.523395

Svozil, K. (1987). Quantum field theory on fractal spacetime: A new regularisation method. *Journal of Physics. A, Mathematical and General, 20*(12), 3861–3875. doi:10.1088/0305-4470/20/12/033

Svozil, K., & Zeilinger, A. (1986). Dimension of space-time. *International Journal of Modern Physics A, 1*(04), 971–990. doi:10.1142/S0217751X86000368

Tsallis, C. (1988). Possible generalization of Boltzmann-Gibbs statistics. *Journal of Statistical Physics, 52*(1-2), 479–487. doi:10.1007/BF01016429

Tsallis, C. (2009a). Nonadditive entropy and nonextensive statistical mechanics-an overview after 20 years. *Brazilian Journal of Physics, 39*(2a2A), 337–356. doi:10.1590/S0103-97332009000400002

Tsallis, C. (2009b). *Introduction to nonextensive statistical mechanics: approaching a complex world.* Springer Science & Business Media.

Weberszpil, J. (2014). *Validity of the fractional Leibniz rule on a coarse-grained medium yields a modified fractional chain rule.* arXiv preprint arXiv:1405.4581

Weberszpil, J., & Chen, W. (2017). Generalized Maxwell relations in thermodynamics with metric derivatives. *Entropy (Basel, Switzerland), 19*(8), 407. doi:10.3390/e19080407

Weberszpil, J., Godinho, C. F. L., Cherman, A., & Helayël-Neto, J. A. (2012). Aspects of the Coarse-Grained-Based Approach to a Low-Relativistic Fractional Schrodinger Equation. *Mathematical Methods in Physics - ICMP 2012, 2012, Rio de Janeiro. Proceedings of Science (PoS).* arXiv preprint arXiv:1206.2513

Weberszpil, J., & Helayël-Neto, J. A. (2014). Anomalous-Factors for Charged Leptons in a Fractional Coarse-Grained Approach. *Advances in High Energy Physics.*

Weberszpil, J., & Helayël-Neto, J. A. (2016). Variational approach and deformed derivatives. *Physica A, 450*, 217–227. doi:10.1016/j.physa.2015.12.145

Weberszpil, J., & Helayël-Neto, J. A. (2017). Structural scale q-derivative and the LLG equation in a scenario with fractionality. *EPL, 117*(5), 50006. doi:10.1209/0295-5075/117/50006

Weberszpil, J., & Helayel-Neto, J. A. (2017a). Axiomatic Local Metric Derivatives for Low-Level Fractionality with Mittag-Leffler Eigenfunctions. *Journal of Advances in Physics, 13*(3), 4751–4755. doi:10.24297/jap.v13i3.5943

Weberszpil, J., & Helayël-Neto, J. A. (2017b). Structural scale q-derivative and the LLG equation in a scenario with fractionality. *EPL, 117*(5), 50006. doi:10.1209/0295-5075/117/50006

Weberszpil, J., Lazo, M. J., & Helayël-Neto, J. A. (2015). On a connection between a class of q-deformed algebras and the Hausdorff derivative in a medium with fractal metric. *Physica A, 436*, 399–404. doi:10.1016/j.physa.2015.05.063

Zaslavsky, G. M., & Zaslavskij, G. M. (2005). *Hamiltonian chaos and fractional dynamics*. Oxford University Press on Demand.

Závada, P. (2002). Relativistic wave equations with fractional derivatives and pseudodifferential operators. *Journal of Applied Mathematics, 2*(4), 163–197. doi:10.1155/S1110757X02110102

Zeilinger, A., & Svozil, K. (1985). Measuring the dimension of space-time. *Physical Review Letters, 54*(24), 2553–2555. doi:10.1103/PhysRevLett.54.2553 PMID:10031375

Chapter 9
Properties of Spatial-Rotation Transformations of Fractional Derivatives

Ehab Malkawi
https://orcid.org/0000-0002-6461-3741
Physics Department, United Arab Emirates University, UAE

ABSTRACT

The transformation properties of the fractional derivatives under spatial rotation in two-dimensional space and for both the Riemann-Liouville and Caputo definitions are investigated and derived in their general form. In particular, the transformation properties of the fractional derivatives acting on scalar fields are studied and discussed. The study of the transformation properties of fractional derivatives is an essential step for the formulation of fractional calculus in multi-dimensional space. The inclusion of fractional calculus in the Lagrangian and Hamiltonian dynamical formulation relies on such transformation. Specific examples on the transformation of the fractional derivatives of scalar fields are discussed.

INTRODUCTION

Fractional calculus deals with differentiation and integration to arbitrary real or complex orders. The idea could be traced to the time of the integer-order calculus. For extensive mathematical discussion of fractional calculus, the reader can consult (Samko et al., 1993; Miller & Ross, 1993; Oldham & Spanier, 1974; Nishimoto, 1989; Podlubny, 1999) and references therein. The techniques of fractional calculus have been applied to wide range of fields, such as physics, engineering, chemistry, biology, economics, control theory, signal image processing, groundwater problems, and many others.

Physics applications of fractional calculus span a wide range of topics and problems. For a review of selected physics applications, see (Herrman, 2011; Hilfer, 2000; Carpinteri & Mainardi, 1997; West et al., 2003; Tarasov, 2013; Zaslavsky, 2005) and references therein. Generalizing fractional calculus to several variables, multidimensional space, and generalization of fractional vector calculus has been

DOI: 10.4018/978-1-7998-3122-8.ch009

reported (Hanyga, 2002; Adda, 2001; Engheta, 1998; Naqvi & Abbas, 2004; Meerschaert et al., 2006; Cottrill-Shepard & Naber, 2001; Tarasov, 2005; Yong et al., 2003). In addition, progress has been reported on generalization of Lagrangian and Hamiltonian systems (Riewe, 1996; Herzalla et al., 2011; Golmankhaneh et al., 2012; Muslih & Baleanu, 2005; Klimek, 2002; Agrawal, 2002).

Despite the significant progress in applying fractional calculus to a wide range of physical problems, there is still a lack of satisfactory geometric and physical interpretation, in comparison with the simple interpretations of their integer-order counterparts (see (Podlubny, 2002) and references therein). Effort has been devoted to relate fractional calculus and fractal geometry (Calcgni, 2012; Moshrefi-Torbati & Hammond, 1998; Nigmatullin, 1992; Rutman, 1995; Machado, 2003). A different approach to geometric interpretation of fractional calculus is based on the idea of the contact of α-th order (Adda, 1997). However, a satisfactory interpretation is still missing (Podlubny, 2002).

A study of the symmetry of physical systems described by fractional calculus and investigating the transformation properties of fractional derivative operators under specific groups can shed light on the nature of such operators. For example, studying the transformation of the fractional operators under point transformation (or change of variables) can be found in (Buckwar & Luchko, 1998; Gazizov et al., 2009; Gazizov et al., 2011), even though the class of variables substitution preserving the form of the fractional derivative is found to be very narrow. Generalizing the definition of fractional derivative of a function with respect to another function can be used to study general transformation of point transformation (Gazizov et al., 2012).

The question that this work is trying to address is basically the following: In two dimensional space, given a quantity that is described by a fractional derivative acting on some function in a specific coordinate system (x,y), how could this quantity be expressed in a rotated coordinate system $\left(x', y'\right)$? Thus, this work tries to relate the same fractional quantity when expressed in two equivalent coordinate systems. This current work only considers the Cartesian space coordinates in two dimensions and adheres to the notion of time as invariant in all coordinate systems (nonrelativistic). In other words, this work is looking into the transformation of the spatial fractional derivative operators and their effect on physical quantities under the rotation group $SO(2)$. Currently this work is not interested in constructing the irreducible representations of $SO(2)$, as done in the standard integer-order differentiation case (Jeffreys, 1931; Arfken et al., 2012; Baker, 2003; Helgason, 2001), rather a focus on the orthonormal basis of the two dimensional Cartesian space of R^2 is pursued. Intuitively, quantities expressed through fractional derivative operators are expected to behave differently from scalar, vector, and tensor quantities. An understanding of such transformations should shed some light on the physical nature of fractional derivatives and therefore provide a guide in the construction of physical systems when expressed through fractional calculus.

In addition, this work investigates the transformation properties of the fractional derivative and its action on invariant scalar fields, under space rotation in two dimensions, and using both the Riemann-Liouville and Caputo definitions. A comparison is pursued between the Riemann-Liouville and Caputo definitions in relating the same physical quantity under spatial rotation to check for major differences in the transformation properties between the two definitions. This work is built upon several published works (Malkawi, 2015; Malkawi, 2018; Malkawi; 2016; Malkawi & Rousan, 2004). In the initial work (Malkawi & Rousan, 2004), the transformation properties of the Riemann-Liouville fractional derivative of a scalar field under infinitesimal transformation of $SO(2)$ were derived using elementary series expansion. The current presented work is more general and can be applied to any functional dependence without being bounded to infinitesimal transformation. In addition, this work reiterates the case of the Riemann-Liouville fractional derivative as applied to a scalar field in a similar manner to (Malkawi & Rousan, 2004) and shows that the result agrees with the derived general treatment.

The results derived in this work are considered important because they could be used to write interaction terms that are consistent with particular transformations under spatial rotation. For example, in writing the Lagrangian terms of a particular interaction, one would seek to determine all the scalar invariant (Lorentz) interaction terms. Therefore, the study of fractional derivatives transformation under spatial rotation becomes important in the construction. In (Riewe, 1996) a method is suggested to include nonconservative forces such as friction in the Lagrangian and Hamiltonian formalism containing derivatives of fractional order. The current work could be used to help in constructing such formalism. Also, the results presented in this work could be used to study the transformation properties of fractional differential equations and their solutions. This work could shed some light on the nature and behavior of fractional derivatives, which is important in the development of fractional calculus applications to physical problems.

In section 2, a brief introduction to fractional calculus is given to lay out relevant notation used in this work. Section 3 lays out the *SO*(2) spatial rotational transformation properties of scalar and vector fields and their integer derivatives expressed in Cartesian coordinates. Section 4 applies the previous construction to the Riemann-Liouville and Caputo fractional derivatives and derive their corresponding transformations under *SO*(2). Section 4 also discusses the transformation properties of the fractional derivatives as applied to a scalar field. In section 5, another basic treatment to the Riemann-Liouville fractional derivative as applied to a scalar field is derived and equivalency to the general treatment is shown. Finally, in section 6 a brief discussion and conclusions are highlighted.

FRACTIONAL CALCULUS

Several definitions of the fractional differentiation and integration exist in literature. In this work, the focus will be on the Riemann-Liouville and the Caputo derivatives. From a mathematical point of view, all definitions are well founded. However, from physics point of view this may not be the case. For example, the Riemann-Liouville derivative of a constant function is not zero while Caputo's derivative of a constant function is zero. For mathematical properties of fractional derivatives and integrals one can consult (Samko et al., 1993; Miller & Ross, 1993; Oldham & Spanier, 1974; Nishimoto, 1989; Podlubny, 1999) and references therein. In this section, the notation used in the rest of this work is outlined. This work considers real analytic functions on R^2 and only discusses the Riemann-Liouville and Caputo definitions of the fractional derivative.

Let $f(x,y)$ to be a real analytic function in a specific domain in the Euclidian space R^2; f: $R^2 \rightarrow R$. The x partial fractional derivative of order $\alpha>0$ (keeping y constant) and where a is the lower limit of x is written as ${}_aD_x^\alpha f(x,y)$. Similarly for the y partial fractional derivative of order α (keeping x constant) and where b is the lower limit of y is written as ${}_bD_y^\alpha f(x,y)$. Since $f(x,y)$ is analytic then the partial fractional derivatives ${}_aD_x^\alpha f(x,y)$ and ${}_aD_x^\alpha f(x,y)$ are assumed to commute, i.e., $\left[{}_aD_x^\alpha, {}_bD_y^\alpha\right] = 0$.

Definition 1. The Cauchy's repeated integration formula of the nth-order integration of the function $f(x,y)$ along x, keeping y constant, can be written as

$$ {}_{a}D_{x}^{n}f\left(x,y\right)=\int_{a}^{x}dx_{n-1}\int_{a}^{x_{n-1}}dx_{n-2}\cdots\int_{a}^{x_{1}}f(x_{0},y)dx_{0}=\frac{1}{\left(n-1\right)!}\int_{a}^{x}\frac{f\left(u,y\right)du}{\left(x-u\right)^{1-n}}. \tag{2.1} $$

A similar relation for the nth-order integration of the function $f(x,y)$ along y, keeping x constant,

$$ {}_{b}D_{y}^{n}f\left(x,y\right)=\int_{b}^{y}dx_{n-1}\int_{b}^{y_{n-1}}dx_{n-2}\cdots\int_{b}^{y_{1}}f(x,y_{0})dy_{0}=\frac{1}{\left(n-1\right)!}\int_{b}^{y}\frac{f\left(x,u\right)du}{\left(y-u\right)^{1-n}}. \tag{2.2} $$

Definition 2. The Riemann-Liouville fractional integration of order $\alpha<0$ and along x, keeping y constant, is defined as

$$ {}_{a}D_{x}^{\alpha}f\left(x,y\right)=\frac{1}{\Gamma\left(-\alpha\right)}\int_{a}^{x}\frac{f\left(u,y\right)du}{\left(x-u\right)^{1+\alpha}}, \tag{2.3} $$

and along y, keeping x constant, is

$$ {}_{b}D_{y}^{\alpha}f\left(x,y\right)=\frac{1}{\Gamma\left(-\alpha\right)}\int_{b}^{y}\frac{f\left(x,u\right)du}{\left(y-u\right)^{1+\alpha}}, \tag{2.4} $$

where $\Gamma(.)$ is the Gamma function. Throughout this work, the choice $a=b=0$ is taken and dropped from the notation henceforth.

Definition 3. The Riemann-Liouville partial fractional derivatives of the order $\alpha>0$, where $n-1<\alpha<n$ and $n\in N$, are defined as

$$ D_{x}^{\alpha}f\left(x,y\right)=D_{x}^{n}D_{x}^{\alpha-n}f\left(x,y\right)=\frac{1}{\Gamma\left(n-\alpha\right)}D_{x}^{n}\int_{a}^{x}\frac{f\left(u,y\right)du}{\left(x-u\right)^{\alpha-n+1}}, \tag{2.5} $$

and

$$ D_{y}^{\alpha}f\left(x,y\right)=D_{y}^{n}D_{y}^{\alpha-n}f\left(x,y\right)=\frac{1}{\Gamma\left(n-\alpha\right)}D_{y}^{n}\int_{a}^{x}\frac{f\left(x,u\right)du}{\left(y-u\right)^{\alpha-n+1}}, \tag{2.6} $$

where the unified notation

$$D_x^n \equiv \frac{\partial^n}{\partial x^n}, D_y^n \equiv \frac{\partial^n}{\partial y^n}$$

is used throughout this work.

The Leibniz rule applied to the Riemann-Liouville partial fractional derivatives can be written as,

$$D_x^\alpha\left(f\left(x,y\right)g\left(x,y\right)\right) = \sum_{k=0}^{\infty}\frac{\Gamma\left(\alpha+1\right)}{\Gamma\left(k+1\right)\Gamma\left(\alpha-k+1\right)}D_x^{\alpha-k}f\left(x,y\right)D_x^k g\left(x,y\right), \tag{2.7}$$

$$D_y^\alpha\left(f\left(x,y\right)g\left(x,y\right)\right) = \sum_{k=0}^{\infty}\frac{\Gamma\left(\alpha+1\right)}{\Gamma\left(k+1\right)\Gamma\left(\alpha-k+1\right)}D_y^{\alpha-k}f\left(x,y\right)D_y^k g\left(x,y\right). \tag{2.8}$$

Definition 4. The Caputo partial fractional derivatives of order $\alpha>0$, where $n-1<\alpha<n$ and $n\in N$, are defined as

$${}^C D_x^\alpha f\left(x,y\right) = \frac{1}{\Gamma\left(n-\alpha\right)}\int_a^x \frac{D_u^n f\left(u,y\right)du}{\left(x-u\right)^{\alpha-n+1}}, \tag{2.9}$$

$${}^C D_y^\alpha f\left(x,y\right) = \frac{1}{\Gamma\left(n-\alpha\right)}\int_a^x \frac{D_u^n f\left(x,u\right)du}{\left(y-u\right)^{\alpha-n+1}}. \tag{2.10}$$

The Riemann-Liouville and Caputo definitions of the fractional derivative are related as follows:

$${}^C D_x^\alpha f\left(x,y\right) = D_x^\alpha f\left(x,y\right) - \sum_{k=0}^{\infty}\frac{x^{k-\alpha}}{\Gamma\left(k+1-\alpha\right)}D_x^k f\left(x,y\right)\Big|_{x=0}, \tag{2.11}$$

$${}^C D_y^\alpha f\left(x,y\right) = D_y^\alpha f\left(x,y\right) - \sum_{k=0}^{\infty}\frac{x^{k-\alpha}}{\Gamma\left(k+1-\alpha\right)}D_y^k f\left(x,y\right)\Big|_{y=0}. \tag{2.12}$$

As an example, for the function $f(x,y)=x^\beta y^\lambda$, where $\beta,\lambda> -1$ one finds

$$D_x^\alpha\left(x^\beta y^\lambda\right) = \frac{\Gamma\left(\beta+1\right)}{\Gamma\left(\beta-\alpha+1\right)}\left(x^{\beta-\alpha}y^\lambda\right),$$

$$D_y^{\alpha}\left(x^{\beta}y^{\lambda}\right)=\frac{\Gamma\left(\lambda+1\right)}{\Gamma\left(\lambda-\alpha+1\right)}\left(x^{\beta}y^{\lambda-\alpha}\right),$$

$$D_x^{\alpha}D_y^{\alpha}\left(x^{\beta}y^{\lambda}\right)=\frac{\Gamma\left(\beta+1\right)\Gamma\left(\lambda+1\right)}{\Gamma\left(\beta-\alpha+1\right)\left(\lambda-\alpha+1\right)}\left(x^{\beta-\alpha}y^{\lambda-\alpha}\right). \tag{2.13}$$

The Caputo partial fractional derivatives give the same result for the quantity $(x^{\beta}y^{\lambda})$ except for $\beta,\lambda\in N<n$ because in this case the Caputo partial fractional derivatives vanish, as expected from (2.9) and (2.10).

TRANSFORMATION OF INTEGER-ORDER DERIVATIVES UNDER SPATIAL ROTATION

Any rotation is considered a motion of a certain space that leaves at least one point fixed. For example, rotation describes the motion of a rigid body around a fixed point. A rotation is different from translation, which has no fixed point. Spatial rotations specify how physical quantities transform in rotated coordinate systems. Spatial rotation and the transformations of scalar, vector, and tensor quantities are discussed in standard mathematics and physics textbooks such as (Jeffreys, 1931; Arfken et al., 2012) and references therein. The invariance of physical quantities under spatial rotations underlines the existence of spatial symmetry of the physical system.

Mathematically, a rotation is a mapping. All rotations about a fixed point form a Lie group under composition called the rotation group of a particular space (Baker, 2003; Helgason, 2001). Matrices are often used to represent transformations, especially when a large number of points are being transformed, as they are a direct representation of the linear operator. Spatial rotation is classified as either proper or improper rotation. The former can be represented by square matrices with unit determinant (+1). The latter is represented by square matrices with determinant -1 and consist of a proper rotation combined with a spatial reflection (inversion). In physics, spatial rotation is frequently understood as a coordinate transformation.

There are two types of proper rotations called active and passive transformations. In physics, an active transformation leads to an actual change to the physical position of a system, which makes sense even in the absence of a coordinate system whereas a passive transformation is a change in the coordinate description of the physical system (change of basis). Put differently, a passive transformation refers to observation of the same event from two different coordinate frames. In two dimensions, only a single angle is needed to specify a rotation about the origin, the angle of rotation that specifies an element of the circle group (also known as $U(1)$). The rotation group is acting to rotate an object counterclockwise through an angle about the origin. Two-dimensional rotations about the same point commute. Rotations about different points, in general, do not commute.

Rotations in three-dimensional space differ from those in two dimensions in a number of important ways. Rotations in three dimensions are generally not commutative, so the order in which rotations are applied is important even about the same point. Rotations about the origin have three degrees of freedom, the same as the number of dimensions. A three-dimensional rotation can be specified in a number of ways such as Euler angles (for example see (Wigner, 1959; Arfken at al. 2012)). Any rotation about the origin can be represented as the composition of three rotations obtained by changing one of the Euler angles while leaving the other two constant.

In this section, an elegant formalism of realizing spatial rotations using differential operators rather than matrices is outlined. The new formalism can unify the transformation of all physics quantities into a general form of transformation. In Subsection 3.1, the new approach is outlined in two-dimensional space, while in Subsection 3.2, a generalization to three dimensional space is carried out.

Spatial Rotation in Two-Dimensional Space

Consider the Euclidian space R^2 and work with orthonormal basis of the Cartesian coordinate system. The distinction between covariant and contravariant quantities disappears and therefore, notation will not distinguish between. Consider the two Cartesian coordinate systems (x,y) and (x',y') sharing the same origin (0,0). The two dimensional (x',y') coordinates are related to the (x,y) coordinates though a counter-clockwise rotation by an angle θ about their identical axis of rotation $z = z' = 0$. For θ=0, the two systems of coordinates are overlapping. This work considers the case of passive transformation of the coordinate system, i.e., the observation of the same quantity from two different rotated coordinate frames, without actually rotating physical quantities. The two coordinate systems are related by the transformation matrix

$$R(\theta) = \begin{bmatrix} \cos\theta & \sin\theta \\ -\sin\theta & \cos\theta \end{bmatrix}. \tag{3.1}$$

For the active transformation case the angle θ could be replaced by -θ. For passive transformation, the coordinates (x',y') are related to the coordinates (x,y) through the transformation

$$\begin{bmatrix} x' \\ y' \end{bmatrix} = \begin{bmatrix} \cos\theta & \sin\theta \\ -\sin\theta & \cos\theta \end{bmatrix} \begin{bmatrix} x \\ y \end{bmatrix}. \tag{3.2}$$

Consider the generator L of the spatial rotation in two dimensions, which is also the angular momentum generator, $L=y\partial_x-x\partial_y$ where $\partial_x \equiv \frac{\partial}{\partial x}$ and $\partial_y \equiv \frac{\partial}{\partial y}$. The rotation elements of $SO(2)$ are written as $e^{\theta L}$, where the exponential expression is defined as an infinite sum of differential operators, explicitly $e^{\theta L} = \sum_{n=0}^{\infty} \frac{\theta^n}{n!} L^n$. Since $Lx=y$ and $Ly= -x$, it is easy to find the action of the element $e^{\theta L}$ on the coordinates x and y, explicitly

$$e^{\theta L} x = \cos\theta x + \sin\theta y = x,$$

$$e^{\theta L}y = -\sin\theta x + \cos\theta y = y'. \tag{3.3}$$

Thus we retrieve the standard result in (3.2). Furthermore, one can show that

$$e^{\theta L}x^2 = x'^2, e^{\theta L}y^2 = y'^2, e^{\theta L}xy = x'y'.$$

Generally, one writes the transformation

$$x'^n y'^m = \left(e^{\theta L}x^n\right)\left(e^{\theta L}y^m\right) = e^{\theta L}\left(x^n y^m\right), \tag{3.4}$$

where $n,m \in N$. This result emerges from proving the following two Lemmas:

Lemma I. Given any two real functions $A(x,y)$ and $B(x,y)$ that are analytic (of Class C^∞) in their common domain, then

$$L^n\left(AB\right) = \sum_{k=0}^{\infty}\frac{n!}{k!\left(n-k\right)!}\left(L^k A\right)\left(L^{n-k}B\right). \tag{3.5}$$

This is just the general Leibniz rule applied to the differential operator L and the proof is similar using mathematical induction. The second important Lemma is

Lemma II. Given any two functions $A(x,y)$ and $B(x,y)$ that are analytic (of Class C^∞) in their common domain, then

$$e^{\theta L}\left(A\left(x,y\right)B\left(x,y\right)\right) = \left(e^{\theta L}A\left(x,y\right)\right)\left(e^{\theta L}B\left(x,y\right)\right). \tag{3.6}$$

To prove this, one first writes

$$e^{\theta L}\left(A\left(x,y\right)B\left(x,y\right)\right) = \sum_{n=0}^{\infty}\frac{\theta^n}{n!}L^n\left(A\left(x,y\right)B\left(x,y\right)\right) = \sum_{n=0}^{\infty}\frac{\theta^n}{n!}\sum_{k=0}^{\infty}\frac{n!}{k!\left(n-k\right)!}\left(L^k A\right)\left(L^{n-k}B\right).$$

Noting that $1/(n-k)!$ vanishes for $k>n$ one rewrites the above results as

$$e^{\theta L}\left(A\left(x,y\right)B\left(x,y\right)\right) = \sum_{n=0}^{\infty}\sum_{k=0}^{\infty}\frac{\theta^n}{k!\left(n-k\right)!}\left(L^k A\right)\left(L^{n-k}B\right).$$

Shifting the order of the sum and then shifting the index $n \to n+k$, one writes

$$e^{\theta L}\left(A(x,y)B(x,y)\right)=\sum_{n=0}^{\infty}\sum_{k=0}^{\infty}\frac{\theta^{n+k}}{k!n!}\left(L^{k}A\right)\left(L^{n}B\right).$$

Noting that $1/n!$ vanishes for $n= -k,-k+1,\ldots,-1$, one reaches the conclusion

$$e^{\theta L}\left(A(x,y)B(x,y)\right)=\sum_{k=0}^{\infty}\frac{\theta^{k}}{k!}\left(L^{k}A\right)\sum_{n=0}^{\infty}\frac{\theta^{n}}{n!}\left(L^{n}B\right)=\left(e^{\theta L}A(x,y)\right)\left(e^{\theta L}B(x,y)\right).$$

Thus, (3.6) follows.

Based on (3.6) one can discuss the transformation of functions under spatial rotation. A function $\varnothing(x,y)$ that is analytic at the origin (0,0), can be expanded in a Taylor series as

$$\varnothing(x,y)=\sum_{n,m}^{\infty}\frac{a_{n,m}}{n!m!}x^{n}y^{m}.$$

In the transformed coordinates frame, the transformed function $\varnothing'(x',y')$ can be also expanded in a Taylor series

$$\varnothing'(x',y')=\sum_{n,m}^{\infty}\frac{a'_{n,m}}{n!m!}x'^{n}y'^{m}=e^{\theta L}\sum_{n,m}^{\infty}\frac{a'_{n,m}}{n!m!}x^{n}y^{m}.$$

If the constants $a_{n,m}$ are independent of the coordinates, then one concludes

$$\varnothing'(x',y')=e^{\theta L}\varnothing(x,y).$$

For example, the function $\varnothing(x,y)=\sin(x+y)$ transforms in the rotated coordinates as the function

$$\varnothing'(x',y')=\sin(x'+y')=e^{\theta L}\sin(x+y).$$

If the constants $a_{n,m}$ depend on the coordinate system, their transformation will be more involved. A scalar function $\varnothing(x,y)$ is invariant under spatial rotation, i.e., $\varnothing'(x',y')=\varnothing(x,y)$. Therefore, one defines a scalar field as follows:

Definition 3.1. A function $\varnothing(x,y)$ is considered to be a scalar function if it satisfies the following relation

$$L\varnothing(x,y)=x\partial_{y}\varnothing(x,y)-y\partial_{x}\varnothing(x,y)=0, \tag{3.7}$$

assuming the dependence on the rotation angle θ occurs only through its arguments (*x,y*). In general, if the constants $a_{n,m}$ depend on the angle of rotation θ then, for a scalar function $\varnothing\left(x,y\right)$.

$$\varnothing'\left(x',y'\right)=e^{\theta L}\sum_{n,m}^{\infty}\frac{a'_{n,m}}{n!m!}x^{n}y^{m}=\varnothing\left(x,y\right),$$

where the constants $a_{n,m}$ transform as

$$a'_{n,m}=\partial_{x}^{n}\partial_{y}^{m}\left(e^{-\theta L}\varnothing\left(x,y\right)\right)\Big|_{(0,0)}.$$

For example, consider the function

$$\varnothing\left(x,y\right)=b_{1}x+b_{2}y,$$

where b_1 and b_2 are constants in the coordinate frame (*x,y*). If $\varnothing\left(x,y\right)$ is a scalar function then in the rotated frame $\left(x',y'\right)$ the constants b_1 and b_2 must transform as

$$b_{1}'=\partial_{x}\left(e^{-\theta L}\varnothing\left(x,y\right)\right)\Big|_{(0,0)}=b_{1}\cos\theta+b_{2}\sin\theta,$$

$$b_{2}'=\partial_{y}\left(e^{-\theta L}\varnothing\left(x,y\right)\right)\Big|_{(0,0)}=-b_{1}\sin\theta+b_{2}\cos\theta.$$

Thus and must transform as the components of a vector. Another example to consider is the following analytic function written in the expanded Taylor series form; It is straightforward to prove that is a scalar function satisfying (3.7). It is easy to spot that is a scalar function once one write.

A vector field can be defined through appropriate transformation in the rotated coordinates. In this formalism a vector-field transformation is obtained by generalizing the transformation of the position vector and noting that *Lx*=*y* and *Ly*= -*x*.

Definition 3.2. Given two functions $A_1(x,y)$ and $A_2(x,y)$, that are analytic (of Class C^{∞}) in their common domain, $A_1(x,y)$ and $A_2(x,y)$ are said to be the components of a vector field $A(x,y)$ if they satisfy the two relations

$$A_{1}\left(x,y\right)=-LA_{2}\left(x,y\right),A_{2}\left(x,y\right)=LA_{1}\left(x,y\right).\qquad(3.8)$$

Note that one can also write

$$A_1(x,y) = -L^2 A_1(x,y), A_2(x,y) = -L^2 A_2(x,y),$$

or equivalently $(1+L^2)A_{1,2}(x,y) = 0$.

For example, the two functions

$$A_1(x,y) = \sum_{n,m}^{\infty} \frac{x^{2n+1}y^{2m}}{n!m!}, A_2(x,y) = \sum_{n,m}^{\infty} \frac{x^{2n}y^{2m+1}}{n!m!},$$

satisfy (3.8) and thus form the components of a vector field. This is clear once one writes the functions as

$A_1(x,y) = xe^{x^2+y^2}$ and $A_2(x,y) = ye^{x^2+y^2}$.

Next, to derive the transformation of derivatives, one notes the following

$$\partial_{x'} x'^n = nx'^{n-1} = ne^{\theta L}x^{n-1} = e^{\theta L}\partial_x e^{-\theta L}e^{\theta L}x^n = e^{\theta L}\partial_x e^{-\theta L}x'^n. \tag{3.9}$$

Thus one derives the important result $\partial_{x'} = e^{\theta L}\partial_x e^{-\theta L}$ and $\partial_{y'} = e^{\theta L}\partial_y e^{-\theta L}$..

The following two commutation relations are easy to derive $[\partial_x, L] = -\partial_y$ and $[\partial_y, L] = \partial_x$. Therefore, it is straightforward to show that

$$\partial_{x'} = e^{\theta L}\partial_x e^{-\theta L} = \cos\theta\partial_x + \sin\theta\partial_y,$$

$$\partial_{y'} = e^{\theta L}\partial_y e^{-\theta L} = -\sin\theta\partial_x + \cos\theta\partial_y.$$

Transformation of higher derivatives can be easily derived, for example

$$\partial_{x'}^2 = e^{\theta L}\partial_x e^{-\theta L}e^{\theta L}\partial_x e^{-\theta L} = e^{\theta L}\partial_x^2 e^{-\theta L}.$$

In general, one concludes

$$\partial_{x'}^n = e^{\theta L}\partial_x^n e^{-\theta L},$$

$$\partial_{x'}^n\partial_{y'}^m = e^{\theta L}\partial_x^n\partial_y^m e^{-\theta L}. \tag{3.10}$$

One can easily show that

$$\left[\partial_x^2, L\right] = -2\partial_{x,y}^2,$$

$$\left[\partial_y^2, L\right] = +2\partial_{x,y}^2.$$

Thus, one concludes that

$$\left[\partial_x^2 + \partial_y^2, L\right] = 0.$$

Indicating that the Laplacian operator $\partial_x^n + \partial_y^n$ is a scalar operator, as expected. An important remark is to check the transformation of the operator L itself under spatial rotation. One can easily show that

$$L' = e^{\theta L} L e^{-\theta L} = L.$$

Thus, the operator L is a scalar itself.

Spatial Rotation in Three Dimensional Space

In three dimensions, the rotation matrix contains three angles that represent three consecutive rotations in a specific order. There exist several conventions in literature (Arfken, 2012). Here the adopted convention is chosen to be $R(\alpha,\beta,\delta)=R_x(\alpha)R_y(\beta)R_z(\delta)$ where

$$R_x(\alpha) = \begin{bmatrix} 1 & 0 & 0 \\ 0 & \cos\alpha & \sin\alpha \\ 0 & -\sin\alpha & \cos\alpha \end{bmatrix},$$

$$R_y(\beta) = \begin{bmatrix} \cos\beta & 0 & -\sin\beta \\ 0 & 1 & 0 \\ \sin\beta & 0 & \cos\beta \end{bmatrix},$$

$$R_z(\delta) = \begin{bmatrix} \cos\delta & \sin\delta & 0 \\ -\sin\delta & \cos\delta & 0 \\ 0 & 0 & 1 \end{bmatrix}.$$

The $SO(3)$ generators of the Lie algebra of the spatial rotation are as follows:

$$L_x = z\partial_y - y\partial_z,\ L_y = x\partial_z - z\partial_x,\ L_z = y\partial_x - x\partial_y.$$

With the commutation rules

$$\left[L_x, L_y\right] = L_z, \left[L_y, L_z\right] = L_x, \left[L_z, L_x\right] = L_y.$$

Based on the selected convention of rotation, the group elements are defined in the following order

$$e^{\alpha L_x} e^{\beta L_y} e^{\delta L_z}.$$

The coordinates (x,y,z) transform under spatial rotation as expected, explicitly one can show

$$z' = e^{\alpha L_x} e^{\beta L_y} e^{\delta L_z} z = x \sin\beta - y \sin\alpha \cos\beta + z \cos\beta \cos\alpha$$

Similarly, one can generate the correct transformation of (x,y) and all other products. In general, one writes the important result

$$x'^n y'^m z'^k = e^{\alpha L_x} e^{\beta L_y} e^{\delta L_z} \left(x^n y^m z^k\right).$$

The proof of this result comes directly from **Lemma II** of the last subsection, Similar to the result from last section, the partial derivatives transform as

$$\partial^n_{x'} \partial^m_{y'} \partial^k_{z'} = e^{\alpha L_x} e^{\beta L_y} e^{\delta L_z} \left(\partial_x^{\,n} \partial_y^{\,m} \partial_z^{\,k}\right) e^{-\delta L_z} e^{-\beta L_y} e^{-\alpha L_x}.$$

Given any analytic function $\varnothing\left(x,y,z\right)$, where all higher derivatives exist and its dependence on the angles of rotation come only through its arguments (x,y,z), then

$$\varnothing'\left(x',y',z'\right) = e^{\alpha L_x} e^{\beta L_y} e^{\delta L_z} \varnothing\left(x,y,z\right).$$

For example, the function $\sin(x+y+z)$ transforms as

$$\sin\left(x'+y'+z'\right) = e^{\alpha L_x} e^{\beta L_y} e^{\delta L_z} \sin\left(x+y+z\right).$$

If $\varnothing\left(x,y,z\right)$, is a scalar field then we conclude that it must satisfy the conditions

$$L_x \varnothing\left(x,y,z\right) = L_y \varnothing\left(x,y,z\right) = L_z \varnothing\left(x,y,z\right) = 0.$$

If the function $\varnothing\left(x,y,z\right)$, is analytic at the origin (0,0,0) and higher derivatives exist to all orders, then it can be expanded as

$$\varnothing\left(x,y,z\right)=\sum_{n,m,k}\frac{a_{n,m,k}}{n!m!k!}x^{n}y^{m}z^{k}.$$

If the constants $a_{n,m,k}$ are dependent on the angles of rotation then we get

$$\varnothing'\left(x',y',z'\right)=e^{\alpha L_{z}}e^{\beta L_{y}}e^{\delta L_{z}}\sum_{n,m,k}\frac{a'_{n,m,k}}{n!m!k!}x^{n}y^{m}z^{k}.$$

If the function $\varnothing\left(x,y,z\right)$ is a scalar field then

$$a'_{n,m,k}=\partial_{x}^{\ n}\partial_{y}^{\ m}\partial_{z}^{\ k}\,e^{-\delta L_{z}}\,e^{-\beta L_{y}}e^{-\alpha L_{z}}\varnothing\left(x,y,z\right)\Big|_{(0,0)}.$$

Defining vector fields and tensors in three dimensions using this formalism could be done similar to the two dimensional case, but we leave this to future work. The new formalism could be used to define new physical quantities that have specific constraints on their general spatial transformation.

TRANSFORMATION OF FRACTIONAL DERIVATIVES UNDER SPATIAL ROTATION IN TWO DIMENSIONAL SPACE

In general, one can generalize the results of the previous section to discuss fractional derivatives, in particular the Riemann-Liouville and Caputo definitions of the fractional derivatives. This work seeks to derive the spatial transformation properties of fractional derivatives and their operation on scalar fields. This work considers only real functions and real values of the fractional order $0<\alpha$. This restriction limits the allowed domain space (x,y). This is clear if one considers the transformation of x^{β} and y^{β}, where $\beta\in R$.

$$e^{\theta L}x^{\beta}=x'^{\beta},e^{\theta L}y^{\beta}=y'^{\beta}. \tag{4.1}$$

Therefore, one requires $x,y>0$ and $x',y'>0,$. which is equivalent to only consider the allowed domain:

$$\left[\left(x,y\right):x,y>0\cap y\,/\,x>\tan\theta\right].$$

Despite the restriction on the domain, one can still study the transformation of the fractional derivatives and deduce their properties. An extension to the whole domain can be pursued in the complex plane once a clear understanding of such transformation is acquired in the restricted domain.

Based on the results of the previous section, one makes the following remarks:

Remark 4.1. In the domain

$$\left[\left(x,y\right):x,y>0\cap y\,/\,x>\tan\theta\right]$$

the quantity $x^{\beta}y^{\lambda}$ transforms as

$$x'^{\beta}y'^{\lambda}=\left(e^{\theta L}x^{\beta}\right)\left(e^{\theta L}y^{\lambda}\right)=e^{\theta L}\left(x^{\beta}y^{\lambda}\right),\qquad(4.2)$$

where $\beta,\lambda\in R$. Next one looks at the transformation of the partial fractional derivatives. Using the Riemann-Liouville and Caputo definition of the fractional derivative, one notes the following

$$D_{x'}^{\alpha}\left(x'^{\beta}\right)=\frac{\Gamma\left(\beta+1\right)}{\Gamma\left(\beta-\alpha+1\right)}\left(x'^{\beta-\alpha}\right)=\frac{\Gamma\left(\beta+1\right)}{\Gamma\left(\beta-\alpha+1\right)}\left(e^{\theta L}x^{\beta-\alpha}\right)=e^{\theta L}D_{x}^{\alpha}x^{\beta}$$
$$=e^{\theta L}D_{x}^{\alpha}e^{-\theta L}e^{\theta L}x^{\beta}=e^{\theta L}D_{x}^{\alpha}e^{-\theta L}x'^{\beta}.$$

Remark 4.2. The partial fractional derivatives according to the Riemann-Liouville and Caputo definitions transform as

$$D_{x'}^{\alpha}=e^{\theta L}D_{x}^{\alpha}e^{-\theta L},$$

$$D_{y'}^{\alpha}=e^{\theta L}D_{y}^{\alpha}e^{-\theta L}.\qquad(4.3)$$

Considering an infinitesimal angle transformation, $\theta\ll1$, and keeping only linear terms of θ, one writes

$$D_{x'}^{\alpha}\approx D_{x}^{\alpha}+\theta\left[L,D_{x}^{\alpha}\right],$$

$$D_{y'}^{\alpha}\approx D_{y}^{\alpha}+\theta\left[L,D_{y}^{\alpha}\right].$$

Remark 4.3. Given an analytic function $f(x,y)$, its partial fractional derivatives transform as

$$D_{x'}^{\alpha}f'\left(x',y'\right)=e^{\theta L}D_{x}^{\alpha}f\left(x,y\right),$$

$$D_{y'}^{\alpha}f'\left(x',y'\right)=e^{\theta L}D_{y}^{\alpha}f\left(x,y\right).\qquad(4.4)$$

Thus, one derives the transformation of the partial fractional derivatives according to the Riemann-Liouville and Caputo definitions. The result in (4.4) is valid for general finite rotation in two-dimensional space. The result can be applied to any function of definite transformation such as scalar or vector fields. Considering the infinitesimal angle approximation, $\theta\ll1$, and keeping only linear terms of θ, one writes

$$D_{x'}^{\alpha} f'(x',y') \approx D_x^{\alpha} f(x,y) + \theta L D_x^{\alpha} f(x,y),$$

$$D_{y'}^{\alpha} f'(x',y') \approx D_y^{\alpha} f(x,y) + \theta L D_y^{\alpha} f(x,y). \tag{4.5}$$

Next, one considers the transformation of the partial fractional derivatives as applied to a scalar field $\Psi(x,y)$ and using both Riemann-Liouville and Caputo definitions in two-dimensional space.

Transformation of the Riemann-Liouville Partial Fractional Derivatives of a Scalar Field

Consider a scalar function $\Psi(x,y)$, that is invariant under spatial rotation, i.e., $L\Psi(x,y)=0$, then, under infinitesimal angle approximation, $\theta \ll 1$ and according to (4.5)

$$D_{x'}^{\alpha} \Psi'(x',y') \approx D_x^{\alpha} \Psi(x,y) + \theta L D_x^{\alpha} \Psi(x,y), \tag{4.6}$$

where

$$L D_x^{\alpha} \Psi(x,y) = y D_x^1 D_x^{\alpha} \Psi(x,y) - x D_y^1 D_x^{\alpha} \Psi(x,y). \tag{4.7}$$

Since y and D_y^1 commute with D_x^{α} one can write the above equation as

$$L D_x^{\alpha} \Psi(x,y) = D_x^1 D_x^{\alpha} \left(y \Psi(x,y)\right) - x D_x^{\alpha} D_y^1 \Psi(x,y). \tag{4.8}$$

Rewriting the above equation using commutation relations, one finds

$$L D_x^{\alpha} \Psi(x,y) = \left[D_x^1, D_x^{a}\right]\left(y \Psi(x,y)\right) - \left[x, D_x^{a}\right]\left(D_y^1 \Psi(x,y)\right) + D_x^{a}\left(y D_x^1 - x D_y^1\right) \Psi(x,y).$$

The last term vanishes since $L\Psi(x,y)=0$, thus one arrives at the following result

$$L D_x^{\alpha} \Psi(x,y) = \left[D_x^1, D_x^{a}\right]\left(y \Psi(x,y)\right) - \left[x, D_x^{a}\right]\left(D_y^1 \Psi(x,y)\right).$$

Note that from the properties of the fractional derivative, $D_x^m D_x^a = D_x^{m+\alpha}$, while $\left[D_x^1, D_x^a\right] \neq 0$ except for $\alpha \in Z^+$. Using the known relation below (Samko et al., 1993; Miller & Ross, 1993; Oldham & Spanier, 1974; Nishimoto, 1989; Podlubny, 1999)

$$D_x^{a} D_x^1 \Psi(x,y) = D_x^{a+1} \Psi(x,y) - \frac{x^{-\alpha-1} \Psi(0,y)}{\Gamma(-\alpha)}, \tag{4.9}$$

one concludes

$$\left[D_{x}^{1}, D_{x}^{a}\right]\left(y\Psi\left(x,y\right)\right) = +\frac{x^{-\alpha-1}\Psi\left(0,y\right)y}{\Gamma\left(-\alpha\right)}. \tag{4.10}$$

Using the Leibniz rule (2.7), substituting $f=\Psi(x,y)$ and $g(x)=x$, one can show that

$$\left[x, D_{x}^{a}\right]\left(D_{y}^{1}\Psi\left(x,y\right)\right) = -\alpha D_{x}^{a-1}D_{y}^{1}\Psi\left(x,y\right).$$

Thus, one concludes that

$$D_{x'}^{\alpha}\Psi'\left(x',y'\right) \approx D_{x}^{\alpha}\Psi\left(x,y\right) + \theta\left(\alpha D_{x}^{a-1}D_{y}^{1}\Psi\left(x,y\right) + \frac{x^{-\alpha-1}\Psi\left(0,y\right)y}{\Gamma\left(-\alpha\right)}\right).. \tag{4.11}$$

A similar relation holds for y,

$$D_{y'}^{\alpha}\Psi'\left(x',y'\right) \approx D_{y}^{\alpha}\Psi\left(x,y\right) - \theta\left(\alpha D_{y}^{a-1}D_{x}^{1}\Psi\left(x,y\right) + \frac{y^{-\alpha-1}\Psi\left(x,0\right)x}{\Gamma\left(-\alpha\right)}\right). \tag{4.12}$$

Note that for $\alpha \in Z^{+}$, the last term in the above result vanishes. Also, note that for $0<\alpha<1$, the term D_{y}^{a-1} represents a fractional integral. One notices the following two remarks:

Remark 4.1.1. For $\alpha=1$, one recovers the standard transformation of the gradient operator,

$$D_{x'}^{1}\Psi'\left(x',y'\right) = D_{x}^{1}\Psi\left(x,y\right) + \theta D_{y}^{1}\Psi\left(x,y\right),$$

$$D_{y'}^{1}\Psi'\left(x',y'\right) = D_{y}^{1}\Psi\left(x,y\right) - \theta D_{x}^{1}\Psi\left(x,y\right).$$

Remark 4.1.2. For $\alpha=2$, one recovers the known scalar invariance of the Laplacian operator,

$$\left(D_{x'}^{2} + D_{y'}^{2}\right)\Psi'\left(x',y'\right) = \left(D_{x}^{2} + D_{y}^{2}\right)\Psi\left(x,y\right).$$

Transformation of the Caputo Partial Fractional Derivatives of a Scalar Field

To derive the transformation of the Caputo partial fractional derivatives one considers $0<\alpha<1$, and uses the relation between the Riemann-Liouville and Caputo definitions, as expressed in (2.11) and (2.12). In addition, one uses the results of (4.9) and (4.10) to show that

$$\left[x, {}^{C}D_{x}^{a}\right]\left(D_{y}^{1}\Psi\left(x,y\right)\right) = \left[x, D_{x}^{\alpha}\right]D_{y}^{1}\Psi\left(x,y\right) - \frac{x^{1-\alpha}D_{y}^{1}\Psi\left(0,y\right)}{\Gamma\left(1-\alpha\right)}, \tag{4.13}$$

$$\left[D_{x}^{1}, {}^{C}D_{x}^{a}\right]\left(D_{y}^{1}\Psi\left(x,y\right)\right) = \left[D_{x}^{1}, D_{x}^{\alpha}\right]\left(y\Psi\left(x,y\right)\right) - \frac{x^{-\alpha-1}\Psi\left(0,y\right)y}{\Gamma\left(-\alpha\right)}. \tag{4.14}$$

Therefore, the Caputo partial fractional derivative transforms, under infinitesimal angle θ≪1, as

$${}^{C}D_{x'}^{\alpha}\Psi'\left(x',y'\right) \approx {}^{C}D_{x}^{a}\Psi\left(x,y\right) + \theta\left(\alpha D_{x}^{\alpha-1}D_{y}^{1}\Psi\left(x,y\right) + \frac{x^{-\alpha+1}D_{y}^{1}\Psi\left(0,y\right)}{\Gamma\left(1-\alpha\right)}\right), \tag{4.15}$$

$${}^{C}D_{y'}^{\alpha}\Psi'\left(x',y'\right) \approx {}^{C}D_{y}^{a}\Psi\left(x,y\right) - \theta\left(\alpha D_{y}^{\alpha-1}D_{x}^{1}\Psi\left(x,y\right) + \frac{y^{-\alpha+1}D_{x}^{1}\Psi\left(x,0\right)}{\Gamma\left(1-\alpha\right)}\right). \tag{4.16}$$

Note that α<1 and thus $D_{x}^{\alpha-1}$ and $D_{y}^{\alpha-1}$ represent fractional integrals.

The observed difference between the Riemann-Liouville and Caputo transformation of the fractional derivative of a scalar field is that in the Riemann-Liouville transformation, the last terms in (4.11) and (4.12) depend on the value of the scalar function at the boundary. While, for the Caputo transformation, the last terms in (4.15) and (4.16) depend on the normal derivative of the scalar function at the boundary. In the case of a constant scalar field, the normal derivatives vanish at the boundary and thus the mentioned terms vanish in the case of Caputo fractional derivative transformations, while they survive in the case of the Riemann-Liouville fractional derivative transformations. The derived transformation properties of the fractional derivatives can be used to construct Lorentz invariant terms, as discussed in section 6.

ELEMENTARY METHOD TO DRIVE THE TRANSFORMATION OF THE FRACTIONAL DERIVATIVE OF A SCALAR FIELD UNDER SPACE ROTATION

The authors in (Malkawi & Rousan, 2004) derived the infinitesimal transformation of the Riemann-Liouville partial-fractional derivatives of a scalar field, by using the function's Taylor expansion and then applying the Riemann-Liouville fractional derivative to each term of the expansion. This method was only applied for scalar fields and for infinitesimal rotations. The method could in principle be used for any analytic field and for infinitesimal rotations. However, the method is tedious and cannot be used for finite rotations. The result presented in (Malkawi & Rousan, 2004) agrees with the general result derived in the previous section. In this section, a summary of the method used is summarized and equivalency is highlighted.

Consider a scalar field Ψ(*x*,*y*), which is required to be analytic at the origin (0,0). One can expand Ψ(*x*,*y*) in a Taylor series written as

$$\Psi\left(x,y\right)=\sum_{n,m}^{\infty}\frac{a_{n,m}}{n!m!}x^{n}y^{m}, \tag{5.1}$$

where

$$\left(a_{n,m}=D_{x}^{n}D_{y}^{m}\Psi\left(x,y\right)\Big|_{(0,0)}\right).$$

Under an infinitesimal rotation θ≪1 in the (*x*,*y*) plane, the coordinates transform according to (3.3), as

$$x'\approx x+\theta y,$$

$$y'\approx -\theta x+y. \tag{5.2}$$

The scalar function Ψ(*x*,*y*) is invariant under space rotation, i.e., $\Psi'\left(x',y'\right)=\Psi\left(x,y\right)$. The first-order partial derivatives of the scalar function transform, under an infinitesimal rotation θ≪1, as

$$D_{x'}^{1}\approx D_{x}^{1}+\theta D_{y}^{1},$$

$$D_{y'}^{1}\approx -\theta D_{x}^{1}+D_{y}^{1}. \tag{5.3}$$

The transformations of the first-order partial derivatives indicate that the gradient field ∇Ψ(*x*,*y*) (directional derivative) is a vector field.

One could start by investigating the transformation property of the Riemann-Liouville fractional-partial derivative $D_{x}^{\alpha}\Psi\left(x,y\right)$ first. Starting from (5.1) and using the result in (2.13), one writes the fractional derivative as

$$D_{x}^{\alpha}\Psi\left(x,y\right)=\sum_{n,m}^{\infty}a_{n,m}\frac{x^{n-\alpha}y^{m}}{\Gamma\left(n-\alpha+1\right)!m!}. \tag{5.4}$$

A similar relation holds for $D_{y}^{\alpha}\Psi\left(x,y\right)$

$$D_{y}^{\alpha}\Psi\left(x,y\right)=\sum_{n,m}^{\infty}a_{n,m}\frac{x^{n}y^{m-\alpha}}{n!\Gamma\left(m-\alpha+1\right)!}. \tag{5.5}$$

The transformed fractional-partial derivatives $D_{x'}^{\alpha}\Psi'\left(x',y'\right)$ and $D_{y'}^{\alpha}\Psi'\left(x',y'\right)$ can be written as

$$D_{x'}^{\alpha}\Psi'\left(x',y'\right)=\sum_{n,m}^{\infty}a'_{n,m}\frac{x'^{n-\alpha}y'^{m}}{\Gamma\left(n-\alpha+1\right)!m!},$$

$$D_{y'}^{\alpha}\Psi'\left(x',y'\right)=\sum_{n,m}^{\infty}a'_{n,m}\frac{x'^{n}y'^{m-\alpha}}{n!\Gamma\left(m-\alpha+1\right)},\qquad(5.6)$$

where $a'_{n,m}$ are the transformed quantities of $a_{n,m}$ For the rest of this section, one considers infinitesimal spatial rotation $\theta \ll 1$. Using the transformations in (5.2), one can write

$$x'^{n-\alpha}\approx\left(x+\theta y\right)^{n-\alpha}\approx x^{n-\alpha}+\left(n-\alpha\right)\theta x^{n-\alpha-1}y,$$

$$y'^{m}\approx\left(y-\theta x\right)^{m}\approx y^{m}-m\theta xy^{m-1}.$$

Thus one can show that, to first order in $\theta \ll 1$

$$a'_{n,m}=D_{x'}^{n}D_{y'}^{m}\Psi'\left(x',y'\right)_{(0,0)}\approx\left(D_{x}^{n}D_{y}^{m}+n\theta D_{x}^{n-1}D_{y}^{m+1}-m\theta D_{x}^{n+1}D_{y}^{m-1}\right)\Psi\left(x,y\right)\Big|_{(0,0)}.$$

Substituting into (5.6) one finds

$$D_{x'}^{\alpha}\Psi'\left(x',y'\right)\approx\sum_{n,m}^{\infty}\frac{1}{\Gamma\left(n-\alpha+1\right)!m!}\left[\left(D_{x}^{n}D_{y}^{m}+n\theta D_{x}^{n-1}D_{y}^{m+1}-m\theta D_{x}^{n+1}D_{y}^{m-1}\right)\Psi\left(x,y\right)\Big|_{(0,0)}\right]$$
$$\left[x^{n-\alpha}y^{m}+\left(n-\alpha\right)\theta x^{n-\alpha-1}y^{m+1}-m\theta x^{n-\alpha+1}y^{m-1}\right]$$

Expanding the expression above and keeping terms to first order in θ one gets

$$D_{x'}^{\alpha}\Psi'\left(x',y'\right)\approx\sum_{n,m}^{\infty}\frac{1}{\Gamma\left(n-\alpha+1\right)!m!}$$
$$\left[\begin{pmatrix}D_{x}^{n}D_{y}^{m}\Psi\left(0,0\right)x^{n-\alpha}y^{m}+\left(n-\alpha\right)\theta D_{x}^{n}D_{y}^{m}\Psi\left(0,0\right)x^{n-\alpha-1}y^{m+1}-m\theta D_{x}^{n}D_{y}^{m}\Psi\left(0,0\right)x^{n-\alpha+1}y^{m-1}\\+n\theta D_{x}^{n-1}D_{y}^{m+1}\Psi\left(0,0\right)x^{n-\alpha}y^{m}-m\theta D_{x}^{n+1}D_{y}^{m-1}\Psi\left(0,0\right)x^{n-\alpha}y^{m}\end{pmatrix}\right]$$

The above equation can be rewritten as

$$D_{x'}^{\alpha}\Psi'\left(x',y'\right)\approx D_{x}^{\alpha}\Psi\left(x,y\right)+\theta\sum_{n,m}^{\infty}\frac{1}{\Gamma\left(n-\alpha+1\right)!}$$
$$\left[\begin{pmatrix}\frac{\left(n-\alpha\right)}{m!}D_{x}^{n}D_{y}^{m}\Psi\left(0,0\right)x^{n-\alpha-1}y^{m+1}-\frac{1}{\left(m-1\right)!}D_{x}^{n}D_{y}^{m}\Psi\left(0,0\right)x^{n-\alpha+1}y^{m-1}\\+\frac{n}{m!}D_{x}^{n-1}D_{y}^{m+1}\Psi\left(0,0\right)x^{n-\alpha}y^{m}-\frac{1}{\left(m-1\right)!}D_{x}^{n+1}D_{y}^{m-1}\Psi\left(0,0\right)x^{n-\alpha}y^{m}\end{pmatrix}\right]$$

Rearranging the index sum, one can rewrite the equation as

$$D_{x'}^{\alpha}\Psi'\left(x',y'\right)\approx D_{x}^{\alpha}\Psi\left(x,y\right)+\theta\sum_{n,m}^{\infty}$$
$$\left[\begin{pmatrix}\frac{\left(n-\alpha\right)}{\Gamma\left(n-\alpha+1\right)m!}D_{x}^{n}D_{y}^{m}\Psi\left(0,0\right)x^{n-\alpha-1}y^{m+1}-\frac{1}{\Gamma\left(n-\alpha+1\right)m!}D_{x}^{n}D_{y}^{m+1}\Psi\left(0,0\right)x^{n-\alpha+1}y^{m}\\+\frac{n+1}{\Gamma\left(n-\alpha+2\right)m!}D_{x}^{n}D_{y}^{m+1}\Psi\left(0,0\right)x^{n-\alpha+1}y^{m}-\frac{1}{\Gamma\left(n-\alpha+1\right)m!}D_{x}^{n+1}D_{y}^{m}\Psi\left(0,0\right)x^{n-\alpha}y^{m+1}\end{pmatrix}\right]$$

Following the manipulations in (Malkawi & Rousan, 2004) one can simplify the above term and reach the final equation

$$D_{x'}^{\alpha}\Psi'\left(x',y'\right)\approx D_{x}^{\alpha}\Psi\left(x,y\right)+\theta\left(\alpha D_{x}^{\alpha-1}D_{y}^{1}\Psi\left(x,y\right)+\frac{x^{-\alpha-1}y}{\Gamma\left(-\alpha\right)}\Psi\left(0,y\right)\right).$$

In a similar way, one can derive the transformation of the other component $D_{y'}^{\alpha}\Psi'\left(x',y'\right)$

$$D_{y'}^{\alpha}\Psi'\left(x',y'\right)\approx D_{y}^{\alpha}\Psi\left(x,y\right)-\theta\left(\alpha D_{y}^{\alpha-1}D_{x}^{1}\Psi\left(x,y\right)+\frac{y^{-\alpha-1}x}{\Gamma\left(-\alpha\right)}\Psi\left(x,0\right)\right).$$

The above results agree completely with the previous general result shown in (4.11) and (4.12).

CONCLUSION

In section 4, a derivation of the transformation of the Riemann-Liouville and Caputo partial fractional derivatives under spatial rotation is performed. In addition, the transformation of the Riemann-Liouville and Caputo partial fractional derivatives of a scalar function are carried out. One observes no fundamental difference between the transformations of the Riemann-Liouville and Caputo partial fractional derivatives. The fact that the Caputo fractional derivative of a constant vanishes does not lead to a significant difference in its transformation. The difference one observes is that in the Riemann-Liouville transformation, there is a term that depends on the value of the function at the boundary. While, for the Caputo transformation, a similar term depends on the normal derivative of the function at the boundary. Also, one observes that the Riemann-Liouville and Caputo partial fractional derivatives of a scalar function do not transform as the components of a gradient vector, as in the case of the integer-order, $\alpha=1$. Thus, the Riemann-Liouville and Caputo partial fractional derivatives of a scalar function are found to have a new form of transformation, different from the scalar, vector, and tensor quantities. They represent new entities that deserve to be studied thoroughly in future work.

Next, few examples of scalar functions are considered, such as the constant scalar function $\Psi(x,y)=1$. According to (4.11) and (4.12), the Riemann-Liouville transformations of the fractional derivatives are, for infinitesimal $\theta \ll 1$,

$$D_{x'}^{\alpha}\left(1\right) \approx D_{x}^{\alpha}\left(1\right) + \theta\left(\frac{x^{-\alpha-1}y}{\Gamma\left(-\alpha\right)}\right),$$

$$D_{y'}^{\alpha}\left(1\right) \approx D_{y}^{\alpha}\left(1\right) - \theta\left(\frac{y^{-\alpha-1}x}{\Gamma\left(-\alpha\right)}\right).$$

One can easily check that the result is as expected, for example

$$D_{x'}^{\alpha}\left(1\right) = \frac{x'^{-\alpha}}{\Gamma\left(-\alpha\right)} \approx \frac{\left(x+\theta y\right)^{-\alpha}}{\Gamma\left(-\alpha\right)} \approx D_{x}^{\alpha}\left(1\right) + \theta\left(\frac{x^{-\alpha-1}y}{\Gamma\left(-\alpha\right)}\right),$$

as required. It is easy to show that the following quantities are invariant,

$$x^{\alpha}D_{x}^{\alpha}\left(1\right), y^{\alpha}D_{y}^{\alpha}\left(1\right), y^{-\alpha}D_{x}^{\alpha}\left(1\right) - x^{-\alpha}D_{y}^{\alpha}\left(1\right), x^{2+\alpha}D_{x}^{\alpha}\left(1\right) + y^{2+\alpha}D_{y}^{\alpha}\left(1\right).$$

One can use $(dx)^{\alpha}$ and $(dy)^{\alpha}$ instead of x^{α} and y^{α} in the above invariant forms. The Caputo partial fractional derivatives of a constant scalar field vanish and the transformations in (4.15) and (4.16) are satisfied.

Consider the scalar function $\Psi(x,y)=x^2+y^2$. The Riemann-Liouville and Caputo x partial fractional derivatives are, respectively

$$D_{x}^{\alpha}\Psi\left(x,y\right) = \frac{2}{\Gamma\left(3-\alpha\right)}x^{2-\alpha} + \frac{x^{-\alpha}y^{2}}{\Gamma\left(1-\alpha\right)},$$

$$ {}^{C}D_{x}^{\alpha}\Psi\left(x,y\right)=\frac{2}{\Gamma\left(3-\alpha\right)}x^{2-\alpha}. $$

Again it is easy to check that the transformations of the partial fractional derivatives, for infinitesimal angle θ≪1, agree with (4.11), (4.15), for both the Riemann-Liouville and Caputo fractional derivatives. A similar transformation holds for D_{y}^{α} and ${}^{C}D_{y}^{\alpha}$ with *x* and *y* interchanged and θ→ -θ.

One can easily check that the quantities

$$ x^{\alpha}D_{x}^{\alpha}\Psi\left(x,y\right)+y^{\alpha}D_{y}^{\alpha}\Psi\left(x,y\right),\ x^{\alpha}y^{\alpha}D_{x}^{\alpha}D_{y}^{\alpha}\Psi\left(x,y\right),x^{\alpha}\ {}^{C}D_{x}^{\alpha}\Psi\left(x,y\right)+y^{\alpha}\ {}^{C}D_{y}^{\alpha}\Psi\left(x,y\right). \tag{6.1} $$

are invariant under spatial rotation. The above terms may not exhaust all possible invariant terms. Again, one can use $(dx)^{\alpha}$ and $(dy)^{\alpha}$ instead of x^{α} and y^{α} in the above invariant forms.

For the function $\Psi(x,y)=(x^2+y^2)^2$, one can check all the transformations are correct. However, the quantities in (6.1) are not invariant separately. Nevertheless, it is possible to find a linear combination of the first two quantities to form an invariant quantity, namely,

$$ x^{\alpha}D_{x}^{\alpha}\Psi\left(x,y\right)+y^{\alpha}D_{y}^{\alpha}\Psi\left(x,y\right)+A\left(x^{\alpha}y^{\alpha}D_{x}^{\alpha}D_{y}^{\alpha}\Psi\left(x,y\right)\right), $$

where *A* is some constant to make the combination invariant. Other forms of general scalar fields need to be studied individually and invariant terms could be generated.

One of the applications to this work would be in writing interaction terms that are required to be invariant under space rotation. Thus, one can preserve Lorentz invariance in constructing interaction terms using fractional derivatives. For example, consider a particle of mass *m* and speed υ moving in a simple harmonic potential field of $\Psi(x,y)=A(x^2+y^2)$, where *A* is some constant. The classical Lagrangian representing this interaction is simply $L=\frac{m}{2}v^{2}-\Psi\left(x,y\right)$. Now one can consider adding few more terms to this Lagrangian representing new interactions, such as

$$ L=\frac{m}{2}v^{2}-\Psi\left(x,y\right)+B_{1}\left(x^{\alpha}D_{x}^{\alpha}\Psi\left(x,y\right)+y^{\alpha}D_{y}^{\alpha}\Psi\left(x,y\right)\right)+B_{2}x^{\alpha}y^{\alpha}D_{x}^{\alpha}D_{y}^{\alpha}\Psi\left(x,y\right), $$

where B_1 and B_2 are constants and 0<α<1. Here we treat time as a standard quantity (not fractionalized). The new terms preserve Lorentz transformation and thus are invariant under spatial rotation. The new terms represent new interactions with the scalar potential field $\Psi(x,y)$ that need to be explored and there physical meaning needs interpretation. For the scalar function, $\Psi(x,y)=(x^2+y^2)^2$ one requires the constants $B_1=B_2$ to preserve invariance in the Lagrangian interaction.

The results derived in this work could also be used to investigate the transformation properties of fractional differential equations and their possible solutions. In addition, by following similar steps to section 4, one can also investigate the transformation properties of the fractional derivatives of vector fields under space rotation in a similar manner to the scalar field case. However, this part has not been investigated so far and left for future work.

REFERENCES

Adda, F. B. (1997). Geometric Interpretation of the Fractional Derivative. *Journal of Fractional Calculus, 11*, 21–52.

Adda, F. B. (2001). The Differentiability in the Fractional Calculus. *Nonlinear Analysis*, 47.

Agrawal, O. P. (2002). Formulation of Euler-Lagrange Equations for Fractional Variational Problems. *Journal of Mathematical Analysis and Applications, 272*(1), 272. doi:10.1016/S0022-247X(02)00180-4

Arfken, G. B., Weber, H. J., & Harris, F. H. (2012). *Mathematical Methods for Physicists*. Elsevier Academic Press.

Baker, A. (2003). *Matrix Groups: An Introduction to Lie Group Theory*. Springer.

Buckwar, E., & Luchko, Y. (1998). Invariance of a Partial Differential Equation of Fractional Order under the Lie Group of Scaling Transformations. *Journal of Mathematical Analysis and Applications, 227*(1), 81–97. doi:10.1006/jmaa.1998.6078

Calcagni, G. (2012). Geometry of Fractional Spaces, *Adv. Theoretical and Mathematical Physics, 16*(2), 549–644. doi:10.4310/ATMP.2012.v16.n2.a5

Carpinteri, A., & Mainardi, F. (1997). *Fractals and Fractional Calculus in Continuum Mechanics*. Wien: Springer. doi:10.1007/978-3-7091-2664-6

Cottrill-Shepherd, K., & Naber, M. (2001). Fractional Differential Forms. *Mathematics and Physics, 42*(5), 2203–2212. doi:10.1063/1.1364688

Engheta, N. (1998). Fractional Curl Operator in Electromagnetics. *Microwave and Optical Technology Letters, 17*(2), 86–91. doi:10.1002/(SICI)1098-2760(19980205)17:2<86::AID-MOP4>3.0.CO;2-E

Gazizov, R. K., Kasatkin, A. A., & Lukashchuk, S. Y. (2009). Symmetry Properties of Fractional Diffusion Equations. *Physica Scripta, T136*, 014016. doi:10.1088/0031-8949/2009/T136/014016

Gazizov, R. K., Kasatkin, A. A., & Lukashchuk, S. Y. (2011). *Group-Invariant Solutions of Fractional Differential Equations*. Nonlinear Science and Complexity.

Gazizov, R. K., Kasatkin, A. A., & Lukashchuk, S. Y. (2012). Fractional Differential Equations: Change of Variables and Nonlocal Symmetries. *Ufa Mathematical Journal., 4*(4), 54–67.

Golmankhaneh, A. K., Ali, M. Y., & Baleanu, D. (2012). On the Fractional Hamilton and Lagrange Mechanics. *International Journal of Theoretical Physics, 51*(9), 51. doi:10.100710773-012-1169-8

Hanyga, A. (2002). Multi-Dimensional Solutions of Space-Time-Fractional Diffusion Equations. *Proceedings of the Royal Society of London, 458*(2018), 429–450. doi:10.1098/rspa.2001.0893

Helgason, S. (2001). *Differential Geometry, Lie Groups, and Symmetric Spaces*. American Mathematical Society. doi:10.1090/gsm/034

Herrmann, R. (2011). *Fractional Calculus. An Introduction for Physicists*. Singapore: World Scientific. doi:10.1142/8072

Herzallah, M. A. E., Muslih, S. I., Baleanu, D., & Eqab, M. R. (2011). Hamilton-Jacobi and Fractional Like Action with Time Scaling. *Nonlinear Dynamics*, *66*(4), 549–555. doi:10.100711071-010-9933-x

Hilfer, R. (2000). *Applications of Fractional Calculus in Physics*. Singapore: World Scientific. doi:10.1142/3779

Jeffreys, H. (1931). *Cartesian Tensors*. Cambridge University Press.

Klimek, M. (2002). Lagrangian and Hamiltonian Fractional Sequential Mechanics. *Czechoslovak Journal of Physics*, 52.

Machado, J. A. T. (2003). A Probabilistic Interpretation of the Fractional Order Differentiation. *Fractional Calculus & Applied Analysis*, *8*, 73–80.

Malkawi E. (2015). Spatial Rotation of the Fractional Derivative in Two-Dimensional Space. *Advances in Mathematical Physics*. doi:10.1155/2015/719173

Malkawi, E. (2016). Spatial Rotation in Cartesian Coordinates. *The Pioneer Journal of Mathematics and Mathematical Sciences.*, *16*(2), 123–135.

Malkawi, E. (2018). A Study of the Caputo Fractional Differential Operator under Two-Dimensional Space Rotation. *The Pioneer Journal of Mathematics and Mathematical Sciences*, *22*(1), 1–9.

Malkawi, E., & Rousan, A. A. (2004). Transformation of Fractional Derivatives under Space Rotation. *International Journal of Applied Mathematics*, *16*(2), 175–185.

Meerschaert, M. M., Mortensen, J., & Wheatcraft, S. W. (2006). Fractional Vector Calculus for Fractional Advection-Dispersion. *Physica A*, *367*, 181–190. doi:10.1016/j.physa.2005.11.015

Miller, K. S., & Ross, B. (1993). *An Introduction to the Fractional Calculus and Fractional Differential Equations*. John Wiley.

Monsrefi-Torbati, M., & Hammond, J. K. (1998). Physical and Geometrical Interpretation of Fractional Operators. *Journal of the Franklin Institute*, *335*(6), 1077–1086. doi:10.1016/S0016-0032(97)00048-3

Muslih, S. I., & Baleanu, D. (2005). Hamiltonian Formulation of Systems with Linear Velocities within Riemann-Liouville Fractional Derivatives. *Journal of Mathematical Analysis and Applications*, *304*(2), 599–606. doi:10.1016/j.jmaa.2004.09.043

Naqvi, Q. A., & Abbas, M. (2004). Complex and Higher Order Fractional Curl Operator in Electromagnetics. *Optics Communications*, *241*(4-6), 349–355. doi:10.1016/j.optcom.2004.07.028

Nigmatullin, R. R. (1992). A Fractional Integral and its Physical Interpretation. *Theoretical and Mathematical Physics*, *90*(3), 242–251. doi:10.1007/BF01036529

Nishimoto, K. (1989). *Fractional Calculus: Integrations and Differentiations of Arbitrary Order*. New Haven: University of New Haven Press.

Oldham, K. B., & Spanier, J. (1974). *The Fractional Calculus; Theory and Applications of Differentiation and Integration to Arbitrary Order*. New York: Academic Press.

Podlubny, I. (1999). *Fractional Differential Equations*. San Diego: Academic Press.

Podlubny, I. (2002). Geometric and Physical Interpretation of Fractional Integration and Fractional Differentiation. *Fractional Calculus & Applied Analysis*, *5*(4).

Riewe, F. (1996). Nonconservative Lagrangian and Hamiltonian Mechanics. *Physical Review. E*, *53*(2), 1890–1899. doi:10.1103/PhysRevE.53.1890 PMID:9964451

Rutman, R. S. (1995). On Physical Interpretations of Fractional Integration and Differentiation. *Theoretical and Mathematical Physics*, *105*(3), 1509–1519. doi:10.1007/BF02070871

Samko, S. G., Kilbas, A. A., & Marichev, O. I. (1993). *Fractional Integrals and Derivatives: Theory and Applications*. New York: Gordon and Breach.

Tarasov, V. E. (2005). Fractional Generalization of Gradient Systems. *Letters in Mathematical Physics*, *73*(1), 49–58. doi:10.100711005-005-8444-z

Tarasov, V. E. (2013). Review of Some Promising Fractional Physical Models. *International Journal of Modern Physics B*, *27*(09), 1330005. doi:10.1142/S0217979213300053

West, B., Bologna, M., & Grigolini, P. (2003). *Physics of Fractal Operators*. New York: Springer. doi:10.1007/978-0-387-21746-8

Wigner, E. P. (1959). *Group Theory and its Application to the Quantum Mechanics of Atomic Spectra*. New York: Academic Press.

Yong, C., Zhen-Ya, Y., & Hong-Qing, Z. (2003). Applications of Fractional Exterior Differential in Three-Dimensional Space. *Applied Mathematics and Mechanics*, *24*(3), 256–260. doi:10.1007/BF02438263

Zaslavsky, G. M. (2005). *Hamiltonian Chaos and Fractional Dynamics*. Oxford: Oxford University Press.

Chapter 10
On the Approximate Solution of Caputo–Riesz–Feller Fractional Diffusion Equation

Samir Shamseldeen
https://orcid.org/0000-0001-7228-9120
Faculty of Engineering, Mansoura University, Egypt

Ahmed Elsaid
https://orcid.org/0000-0002-8248-3119
Faculty of Engineering, Mansoura University, Egypt

Seham Madkour
Faculty of Engineering, Mansoura University, Egypt

ABSTRACT

In this work, a space-time fractional diffusion equation with spatial Riesz-Feller fractional derivative and Caputo fractional time derivative is introduced. The continuation of the solution of this fractional equation to the solution of the corresponding integer order equation is proved. Also, a very useful Riesz-Feller fractional derivative is proved; the property is essential in applying iterative methods specially for complex exponential and/or real trigonometric functions. The analytic series solution of the problem is obtained via the optimal homotopy analysis method (OHAM). Numerical simulations are presented to validate the method and to highlight the effect of changing the fractional derivative parameters on the behavior of the obtained solutions. The results in this work are originally extracted from the author's work.

INTRODUCTION

Fractional derivatives have been used for building and invistigating numerous models and systems in various branches of applied sciences. For example, in describing physical models (Al-khedhairi, Matouk, & Khan, 2019); (El-Sayed A. M., 1996); (Shamseldeen, 2018), in biological systems (Elsadany &

DOI: 10.4018/978-1-7998-3122-8.ch010

Matouk, 2015) (Al-Khedhairi, Askar, Matouk, Elsadany, & Ghazel, 2018) (Matouk & Elsadany, 2016), in economic and finance (Khana & Kumar, 2016) (Hegazi, Ahmed, & Matouk, 2013), in engineering systems (El-Sayed, Nour, Elsaid, Matouk, & Elsonbaty, 2016) and in some interdisciplinary fields of science (Bagley & Calico, 1991) (Kader, Latif, & Nour, 2017) (Abdel Latif, El-Shazly, Baleanu, Elsaid, & Nour, 2018) (Shamseldeen, Elsaid, & Madkour, 2019a) (Shamseldeen, Elsaid, & Madkour, 2019b) (Taghavian & Tavazoei, 2019) (Taghavian & Tavazoei, 2018).

Anomalous diffusion is one process that has been successfully modeled by fractional derivatives. This type of diffusion is characterized by the nonlinear dependence of the mean square displacement $x(t)$ of a diffusing particle over time t: $x^2(t)\propto k_\alpha t^\alpha$ and it can be interpreted as the Lévy stable densities. On the other hand, in the case of classical diffusion, linear dependence $x^2(t)\propto kt$ occurs and it follows Gaussian statistics and Fick's second law for running processes at time t (Metzler & Klafter, 2000). Anomalous diffusion is described by fractional partial differential equations (FPDEs) by replacing classical derivatives with derivatives of fractional order. A type of anomalous diffusion that received great attention is the fractional diffusion equation with spatial Riesz and Riesz-Feller fractional derivatives (Elsaid, 2011)-(Gorenflo, Mainardi, Moretti, Pagnini, & Paradisi, 2002). These fractional derivative operators are used to model phenomena as anomalous diffusion (Metzler & Klafter, 2000), discrete random walk models (Gorenflo, Mainardi, Moretti, Pagnini, & Paradisi, 2002)- (Gorenflo & Vivoli, 2003), and continuous random walk models (Gorenflo & Mainardi, 2003) in the form of FPDEs.

General solutions are obtained for some classes of linear FPDEs, using integral transforms. For example, in (Agrawal, 2002) a general solution is given for a time fractional diffusion-wave equation defined in a bounded space domain. The fractional time derivative is described in the Caputo sense. The finite sine Laplace transform technique is used to obtain the solution. In (Jiang, Liu, Turner, & Burrage, 2012) analytic techniques are introduced to solve certain types of multi-term time–space Caputo–Riesz fractional advection–diffusion equations on a finite domain using the equivalent relationship between the Laplacian operator and the Riesz fractional derivative. In (Mainardi, Pagnini, & Luchko, 2007) a fundamental solution is investigated for the space-time fractional diffusion-wave equation, which is obtained from the standard diffusion equation by replacing the second-order space derivative with a Riesz-Feller derivative and the first-order time derivative with a Caputo derivative. In (Haubold, Mathai, & Saxena, 2011) a solution of a unified reaction–diffusion equation with the Riemann–Liouville fractional derivative as the time-derivative and Riesz–Feller derivative as the space-derivative in terms of H-function. In (Garg, Sharma, & Manohar, 2013) a closed form solution in terms of Mittag-Leffler function is obtained for a space-time fractional telegraph equation with Riesz–Feller derivative as the space-derivative. In (Huang & Liu, 2005) a solution of a space-time fractional advection-dispersion equation is derived in terms of the Green functions.

The proposed above methods work in case of linear FPDEs. However, for the nonlinear case the numerical methods and the semi-analytic approaches have the advantage. Several numerical techniques are introduced in the literature, and we state some of the known methods here. In (Yang, Liu, & Turner, 2010) the matrix transform method with the method of lines are used to obtain the numerical solution of a fractional partial differential equation with Riesz space fractional derivatives on a finite domain. An explicit finite-difference approximation is used to solve the variable-order nonlinear space fractional diffusion equation (Lin, Liu, Anh, & Turner, 2009) . Crank–Nicolson method for the fractional diffusion equation with the Riesz fractional derivative in (Celik & Duman, 2012) while the improved matrix transform method with Pade approximation are used in (Rahman, Mahmood, & Younis, 2014). Yet, semi-analytic solutions for this type of problems include only the series solution by the variational iteration method approach (see (Elsaid, 2010) and (Ervin & Roop, 2007)) and by the homotopy analysis method (HAM), see for example (Elsaid, 2011) and (Zhang, Wei, Tang, & He, 2013). In (Ray & Sahoo, 2015) the homotpoy analysis method with the Fourier transform is used to obtain and analytical approximate solutions of Riesz fractional diffusion equation and Riesz fractional advection–dispersion equation.

As seen in the literature review, only few articles dealt with applying iterative techniques to these types of FPDEs due to the difficulty of repeated application of Riesz or Riesz-Feller fractional derivatives to the solution components. The authors recently have discussed the genraized diffusion and wave fractional problems using the semianalytic homotopy analysis method and the results obtained in (Elsaid, Shamseldeen, & Madkour, 2016b) are introduced in this chapter.

We consider the space-time fractional diffusion equation of the form

$$^{C}D_{x}^{\beta}u\left(x,t\right)=k\left(u\right)R_{x}^{\alpha,\theta}u\left(x,t\right)+g\left(u\right),-\infty<x\left\langle\infty,t\right\rangle0,u\left(x,0\right)=f\left(x\right),\tag{1}$$

where $^{C}D_{t}^{\beta}$ denotes the Caputo fractional derivative (in time) of order β, and $R_{x}^{\alpha,\theta}$ denotes the Riesz-Feller fractional derivative (in space) of order α and skewness θ. Parameters α, θ, and β are real numbers restricted to

$$0<\alpha\leq 2,\alpha\neq 1,\left|\theta\right|\leq\min\left(\alpha,2-\alpha\right),0<\beta\leq 1,\tag{2}$$

and the two functions *k* and *g* are continuous functions in *u*. We aim to establish the continuation of the solution of equation (1) to the exact solution of the corresponding equation in Riesz fractional derivative as the skewness parameter approaches zero. This objective is carried out theoretically then via approximate series solution obtained iteratively by applying the OHAM.

This chapter is organized as follows. In section two, basic definitions of fractional derivative operators involved are presented. Proof of solution continuation is presented in section three. The OHAM is illustrated in section four. The results of numerical experiments are presented in section five. Section six contains the conclusion of this work.

FRACTIONAL DERIVATIVES AND INTEGRALS

Definition 1. *A real function f(x), x>0, is said to be in the space* C_μ, $\mu\in\mathbb{R}$, *if there exists a real number p>μ, such that* $f(x)=x^pf_1(x)$, *where* $f_1(x)\in C(0,\infty)$, *and it is said to be in the space* C_μ^m *if* $f^m\in C_\mu$, $m\in\mathbb{N}$.

Definition 2. *The Riemann-Liouville fractional integral operator of order α≥0 of a function* $f(x)\in C_\mu$, $\mu\geq -1$ *is defined as*

$$\begin{cases}J^{\alpha}f\left(x\right)=\dfrac{1}{\Gamma\left(\alpha\right)}\displaystyle\int_0^x(x-\tau)^{\alpha-1}f\left(\tau\right)d\tau,\alpha>0,x>0,\\ J^{0}f\left(x\right)=f\left(x\right).\end{cases}\tag{2}$$

The operator J^α satisfies the following properties. For $f \in C_\mu$, $\mu \geq -1$, $\alpha, \beta \geq 0$ and $\gamma > -1$:

$$J^\alpha J^\beta f(x) = J^{\alpha+\beta} f(x),$$

$$J^\alpha J^\beta f(x) = J^\beta J^\alpha f(x),$$

$$J^\alpha x^\gamma = \frac{\Gamma(\gamma+1)}{\Gamma(\gamma+\alpha+1)} x^{\alpha+\gamma}.$$

Definition 3. *The fractional derivative in Caputo sense of* $f(x) \in C_{-1}^m$, $m \in \mathbb{N}$, *x>0 is defined as*

$$^C D_x^\beta f(x) = \begin{cases} J^{m-\beta} \dfrac{d^m}{dx^m} f(x), & m-1 < \beta < m, \\ \dfrac{d^m}{dx^m} f(x), & \beta = m. \end{cases} \tag{3}$$

Some basic properties of Caputo fractional derivative are listed in the following lemma

Lemma 1. *If* $m-1 < \alpha \leq m$, $m \in \mathbb{N}$ and $f \in C_\mu^m$, $\mu \geq -1$, then:

$$D^\alpha(c) = 0, (c \text{ is a constant}) \tag{4}$$

$$D^\alpha x^\gamma = \begin{cases} \dfrac{\Gamma(\gamma+1)}{\Gamma(\gamma-\alpha+1)} x^{\alpha+\gamma}, & \gamma > \alpha - 1, \\ 0, & \gamma \leq \alpha - 1, \end{cases}$$

$$D^\alpha J^\alpha f(x) = f(x), \tag{5}$$

and

$$J^\alpha D^\alpha f(x) = f(x) - \sum_{k=0}^{m-1} f^{(k)}(0) \frac{x^k}{k!}, x > 0. \tag{6}$$

Definition 4. *The Riesz-Feller partial fractional derivative* $R_x^{\alpha,\theta}$ *is defined as* (Gorenflo, Mainardi, Moretti, Pagnini, & Paradisi, 2002)

$$R_x^{\alpha,\theta} u(x) = -\left[c_+ D_+^\alpha u(x) + c_- D_-^\alpha u(x)\right], 0 < \alpha < 2, \alpha \neq 1 \tag{7}$$

where $D_{\pm}^{\alpha}u(x)$ are the Weyl fractional derivatives defined as

$$D_{\pm}^{\alpha}u(x)=\begin{cases}\pm\dfrac{d}{dx}W_{\pm}^{1-\alpha}u(x),0<\alpha<1\\ \dfrac{d^2}{dx^2}W_{\pm}^{2-\alpha}u(x),1<\alpha<2\end{cases}, \tag{8}$$

and $W_{\pm}^{\beta}$ denote the Weyl fractional integrals of order β>0, given by

$$\begin{aligned}W_{+}^{\beta}u(x)&=\frac{1}{\Gamma(\beta)}\int_{-\infty}^{x}(x-z)^{\beta-1}u(z)\,dz,\\ W_{-}^{\beta}u(x)&=\frac{1}{\Gamma(\beta)}\int_{x}^{\infty}(z-x)^{\beta-1}u(z)\,dz.\end{aligned} \tag{9}$$

The coefficients c_+ and c_- have the following forms

$$c_{\pm}=\frac{\sin\left((\alpha\mp\theta)\pi/2\right)}{\sin(\alpha\pi)}, \tag{10}$$

where |θ|≤min(α,2–α). When α=0 the Weyl fractional derivative degenerates into the identity operator

$$D_{\pm}^{0}u(x)=Iu(x)=u(x). \tag{11}$$

For continuity we get

$$D_{\pm}^{1}u(x)=\pm\frac{d}{dx}u(x),D_{\pm}^{2}u(x)=\frac{d^2}{dx^2}u(x). \tag{12}$$

Evidently, in case α=2 and θ=0 we define

$$R_{x}^{\alpha,0}u(x)=\frac{d^2}{dx^2}u(x). \tag{13}$$

For the case α=1 and θ=0 we have

$$R_{x}^{1,0}u(x)=\frac{d}{dx}Hu(x)=\frac{d}{dx}\frac{1}{\pi}\int_{-\infty}^{\infty}\frac{u(z)}{z-x}dz, \tag{14}$$

where H is the Hilbert transform and the integral is understood in the Cauchy principal value sense. For the case $\alpha=1$ and $\theta=\pm 1$ we obtain the first derivative

$$R_x^{1,\pm 1}u(x) = \pm \frac{d}{dx}u(x). \tag{15}$$

The special case $\theta=0$ is known as Riesz fractional derivative.

OPTIMAL HOMOTOPY ANALYSIS METHOD (OHAM)

In this section, we illustrate the basic steps of the classical homotopy analysis method (HAM). Consider the following nonlinear equation of the form

$$N[u(x,t)]=0, \tag{16}$$

where N is a nonlinear operator, $u(x,t)$ is the unknown function and x and t denote spatial and temporal independent variables, respectively. Liao (Liao S., 2003) generalized the traditional homotopy method to construct the so-called zero-order deformation equation given by

$$(1-p)\mathcal{L}\left[\phi(x,t;p) - u_0(x,t)\right] = p\hbar H(x,t) N\left[\phi(x,t;p)\right], \tag{17}$$

where $p\in[0,1]$ is an embedding parameter, $\hbar$ is a nonzero auxiliary parameter, $H(x,t)$ is an auxiliary function, $\mathcal{L}$ is an auxiliary linear operator, $u_0(x,t)$ is an initial guess of $u(x,t)$ and $\phi(x,t;p)$ is an unknown function. Obviously, when $p=0$ and $p=1$ we have $\phi(x,t;0) = u_0(x,t)$, $\phi(x,t;1) = u(x,t)$, respectively. Thus, as p increases from 0 to 1, the solution $\phi(x,t;p)$ varies from the initial guess $u_0(x,t)$ to the required solution $u(x,t)$. By expanding $\phi(x,t;p)$ in Taylor series with respect to p, we have

$$\phi(x,t;p) = u_0(x,t) + \sum_{m=1}^{\infty} u_m(x,t)p^m, \tag{18}$$

where

$$u_m(x,t) = \frac{1}{m!}\frac{\partial^m \phi(x,t;p)}{\partial p^m}\Big|_{p=0}. \tag{19}$$

If the auxiliary linear operator $\mathcal{L}$, the initial guess and the auxiliary parameter $\hbar$ and the auxiliary function are so properly chosen, then, as proved by Liao (Liao S., 2003), series (18) converges at $p=1$ and one has

$$u(x,t) = u_0(x,t) + \sum_{m=1}^{\infty} u_m(x,t), \tag{20}$$

which must be one of solutions of the original nonlinear equation. Using definition (19), the governing equation of the HAM can be deduced from the zero-order deformation equation (17) as follows. Define the vector

$$\overrightarrow{u_n} = \{u_0(x,t), u_1(x,t), u_2(x,t), \ldots, u_n(x,t)\} \tag{21}$$

From equation (17), the so-called *m*th-order deformation equation is given by

$$\mathcal{L}\left[u_m(x,t) - \chi_m u_{m-1}(x,t)\right] = \hbar H(x,t) \Re_m\left[\vec{u}_{m-1}(x,t)\right], \tag{22}$$

where

$$\Re_m\left[\vec{u}_{m-1}\right] = \frac{1}{(m-1)!} \frac{\partial^{m-1} N\left[\phi(x,t;p)\right]}{\partial p^{m-1}} \Big|_{p=0}, \tag{23}$$

and

$$\chi_m = \begin{cases} 0, m \leq 1, \\ 1, m > 1. \end{cases} \tag{24}$$

Applying the inverse operator $\mathcal{L}^{-1}$ to both sides of (22), $u_m(x,t)$ can be calculated by symbolic computations software. The HAM has been successfully applied to solve various classes of equations and applied problems (Liao S., 2005), (Bouremel, 2007), (Abbasbandy & Zakaria, 2008), (Cang, Tan, Xu, & Liao, 2009), (Ellahi, Raza, & Vafai, 2012)

In the classical HAM, choosing the value of parameter $\hbar$ depends on inspecting the graph of the quantity of interest; the solution or one of its derivatives. Yet, when $H(x,t)$ is fixed, it is obvious that $u_m(x,t)$ contains only one control parameter $\hbar$. Thus, by constructing a formula for the residual error, the OHAM solution is obtained by choosing the value for parameter $\hbar$ that minimizes the error. Here, the averaged residual error defined for ordinary differential equations in (Liao S., 2010) is generalized to the case of two variable partial differential equations in the following form

$$E_m(\hbar) = \frac{1}{MK} \sum_{i=0}^{M} \sum_{j=0}^{K} \left[N\left(\sum_{n=0}^{m} u_n\left(\frac{i}{M}, \frac{j}{K} \right) \right) \right]^2, \tag{25}$$

which is a nonlinear algebraic equation of one unknown; the convergence-control parameter $\hbar$. Thus the optimal value of $\hbar$ is determined by the minimum of the averaged residual error E_m to ensure the fast convergence of the homotopy series [see (Elsaid, Shamseldeen, & Madkour, 2016a), (Elsaid, Shamseldeen, & Madkour, 2016b) (Elsaid, Shamseldeen, & Madkour, 2016c) and (Shamseldeen, Elsaid, & Madkour, 2019a)]

CONTINUATION OF THE SOLUTION

In this section, we define Riesz-Feller fractional derivative on unbounded domain in Caputo sense. This definition is considered as a generalization of the definition given in (El-Sayed & Gaber, 2006) for Riesz fractional derivative on bounded domains.

Definition 5. *The Riesz-Feller fractional derivative in Caputo sense is defined by*

$$ {}^{C}R_{x}^{\alpha,\theta}f\left(x\right)=-[(c_{+})^{C}D_{+}^{\alpha}+\left(c_{-}\right)^{C}D_{-}^{\alpha}\big]f\left(x\right),0<\alpha<2,\alpha\neq1, \tag{26}$$

where ${}^{C}D_{+}^{\alpha}$ and ${}^{C}D_{-}^{\alpha}$ are Weyl fractional derivatives defined in Caputo sense for α∈(0,2), α≠1 as

$$ {}^{C}D_{\pm}^{\alpha}f\left(x\right)=\begin{cases}\pm W_{\pm}^{1-\alpha}\dfrac{d}{dx}f\left(x\right), & 0<\alpha<1,\\ W_{\pm}^{2-\alpha}\dfrac{d^{2}}{dx^{2}}f\left(x\right), & 1<\alpha<2,\end{cases} \tag{27}$$

and defined for α=0, 1, and 2 as

Lemma 2. Let *f*(*x*) belong to the class of "good functions", a class of all functions which are infinitely differentiable everywhere, (Miller & Ross, 1993). Then for α∈(0,2), α≠1,

$$ {}^{C}R_{x}^{\alpha,\theta}f\left(x\right)=R_{x}^{\alpha,\theta}f\left(x\right). \tag{28}$$

Proof. Consider the case α∈(0,1), Riesz-Feller fractional derivative can be written in the form

$$ R_{x}^{\alpha,\theta}f\left(x\right)=\frac{-d}{dx}\left[\int_{0}^{\infty}\frac{c_{+}f\left(x-\lambda\right)-c_{-}f\left(x+\lambda\right)}{\lambda^{\alpha}}d\lambda\right], \tag{29}$$

which can be written as

$$R_x^{\alpha,\theta} f(x) = \int_0^{\infty} \frac{c_- f'(x+\lambda) - c_+ f'(x-\lambda)}{\lambda^\alpha} d\lambda. \tag{30}$$

For ${}^C R_x^{\alpha,\theta}$, we have

$${}^C R_x^{\alpha,\theta} f(x) = c_+ \int_{-\infty}^{x} \frac{f'(z)}{(x-z)^\alpha} dz - c_- \int_x^{\infty} \frac{f'(y)}{(y-x)^\alpha} dy, \tag{31}$$

which by substitution yields

$${}^C R_x^{\alpha,\theta} f(x) = \int_0^{\infty} \frac{c_- f'(x+\lambda) - c_+ f'(x-\lambda)}{\lambda^\alpha} d\lambda, \tag{32}$$

which is the same as (30). A similar argument holds for the case $\alpha \in (1,2)$.

As the equivalence between the Riemann and Caputo definitions of Riesz-Feller fractional derivative is deduced, we discuss the continuation of the solution obtained.

Theorem 1. If $f(x)$ is a function in $L^1(-\infty,\infty)$, then the exact solution $u_\beta^{\alpha,\theta}$ of the space-time fractional diffusion *equation (1)* is given by

$$u_\beta^{\alpha,\theta}(x,t) = \frac{1}{2\pi} \int_{-\infty}^{\infty} \int_{-\infty}^{\infty} E_\beta\left(-\omega^\alpha t^\beta e^{-i\theta\pi/2}\right) f(v) e^{i\omega(x-v)} d\omega dv, \tag{33}$$

where $E_\beta(z)$ is the one parameter Mittag-Leffler function defined as (Podlubny, 1998)

$$E_\beta(z) = \sum_{k=0}^{\infty} \frac{z^k}{\Gamma(1+k\beta)}, \beta > 0 \tag{34}$$

Proof. The joint Fourier-Laplace tranform method is used to find the solution and this can be found in (Shamseldeen, Elsaid, & Madkour, 2019a)

Theorem 2. *Let* $\alpha \in (1,2)$, $\beta \in (0,1)$, $|\theta| \leq \min(\alpha, 2-\alpha)$, $f(x)$ *be a function in* $L^1(-\infty,\infty)$,*and* $u_\beta^{\alpha,\theta}$ *be the solution (33) of the space-time fractional problem (1), then*

$$\lim_{\theta \to 0} u_\beta^{\alpha,\theta}(x,t) = u_\beta^{\alpha}(x,t), \tag{35}$$

where $u_\beta^{\alpha}(x,t)$ is the exact solution of Riesz fractional diffusion equation with the same limits for parameters α and β.

Proof. Consider the set of functions

$$\varphi_n(\omega) = \frac{1}{2\pi} E_\beta(-\omega^\alpha t^\beta e^{-i\pi/2n})_{-\infty}^{\infty} f(v) e^{i\omega(x-v)} dv,$$

$$\omega \in (-\infty, \infty), n = 1, 2, \ldots \tag{36}$$

Then

$$\left|\varphi_n(\omega)\right| \le \frac{1}{2\pi} \left|E_\beta\left(-\omega^\alpha t^\beta e^{-i\pi/2n}\right)\right|_{-\infty}^{\infty} \left|f(v)\right| dv,$$

and as $f(x) \in L^1(-\infty,\infty)$, then there exists $M>0$ such that

$$\left|\varphi_n(\omega)\right| \le \frac{M}{2\pi} \left|E_\beta\left(-\omega^\alpha t^\beta e^{-i\pi/2n}\right)\right|. \tag{37}$$

From (Zhang & Liu, 2007)Theorem (1.6), there exists $K_1>0$ such that

$$\left|E_\beta(-z)\right| \le \frac{K_1}{1+|z|}. \tag{38}$$

Then

$$\left|\varphi_n(\omega)\right| \le \frac{1}{2\pi} \frac{MK_1}{1+\left|\omega^\alpha t^\beta\right|}, \omega \in (-\infty, \infty), n = 1, 2, \ldots \tag{39}$$

By Lebesgue dominated convergence theorem and setting $\theta = \frac{1}{n}$, we have

$$\lim_{\theta \to 0} u_\beta^{\alpha,\theta}(x,t) = \lim_{n\to\infty -\infty}^{\infty} \varphi_n(\omega) d\omega$$

$$= _{-\infty}^{\infty} \lim_{n\to\infty} \varphi_n(\omega) d\omega.$$

$$= \frac{1}{2\pi} \int_{-\infty}^{\infty}\int_{-\infty}^{\infty} E_\beta\left(-\omega^\alpha t^\beta\right) f(v) e^{i\omega(x-v)} d\omega dv$$

$$= u_{\beta}^{\alpha}\left(x,t\right), \tag{40}$$

which is the exact solution of the Riesz fractional diffusion equation 35.

Similar arguments hold to prove the continuation of the solution when both parameters α and β tend to their integer limits.

NUMERICAL SIMULATION

In this section, we consider linear and nonlinear problems to illustrate the efficiency of the OHAM to this type of problems and to illustrate the continuation of the solution we proved in section 4. Yet, to apply the recursive technique of OHAM to problem (1), a repeated evaluation of Riesz-Feller fractional derivative to solution components is needed. This is accomplished by using a property of Riesz-Feller fractional derivative that is proved in the following lemma.

Lemma 3. *Let* $\alpha\in(0,2)$, $\alpha\neq1$, $|\theta|\leq\min(\alpha,2-\alpha)$, *and* $f(x)$ *be a function in* $L^1(-\infty,\infty)$. *Then*

$$R_x^{\alpha,\theta}\left(e^{i\omega x}\right) = -\omega^{\alpha}e^{i\left(\omega x-\theta\frac{\pi}{2}\right)}$$

Proof. Case $\alpha\in(0,1)$. From the definition of Riesz-Feller fractional derivative and by substitution, we can write

$$R_x^{\alpha,\theta}\left(e^{i\omega x}\right) = -\frac{d}{dx}\left[\int_0^{\infty}\frac{c_+e^{i\omega\left(x-\tau\right)}-c_-e^{i\omega\left(x+\tau\right)}}{\Gamma\left(1-\alpha\right)\tau^{\alpha}}d\tau\right], \tag{41}$$

which reduces to

$$R_x^{\alpha,\theta}\left(e^{i\omega x}\right) = \frac{-i\omega}{\Gamma\left(1-\alpha\right)}e^{i\omega x}\left(c_+\frac{\Gamma\left(1-\alpha\right)}{\left(i\omega\right)^{1-\alpha}}-c_-\frac{\Gamma\left(1-\alpha\right)}{\left(-i\omega\right)^{1-\alpha}}\right), \tag{42}$$

$$R_x^{\alpha,\theta}\left(e^{i\omega x}\right) = -\omega^{\alpha}e^{i\omega x}(c_+i^{\alpha}-c_-\left(-i\right)^{\alpha}) \tag{43}$$

Substituting for c_+ and c_- we have

$$R_x^{\alpha,\theta}\left(e^{i\omega x}\right) = -\omega^{\alpha}e^{i\omega x}\frac{\sin\left(\left(\alpha-\theta\right)\pi/2\right)e^{i\pi\alpha/2}+\sin\left(\left(\alpha+\theta\right)\pi/2\right)e^{-i\pi\alpha/2}}{\sin\left(\alpha\pi\right)}, \tag{44}$$

Utilizing Euler form and trigonometric identities, we have

$$R_x^{\alpha,\theta}\left(e^{i\omega x}\right) = -\omega^{\alpha} e^{i\left(\omega x - \theta\frac{\pi}{2}\right)}.$$

Case α∈(1,2)

$$R_x^{\alpha,\theta}\left(e^{i\omega x}\right) = -\frac{d^2}{dx^2}\int_0^{\infty}\frac{c_+ e^{i\omega\left(x-\tau\right)} - c_- e^{i\omega\left(x+\tau\right)}}{\Gamma\left(2-\alpha\right)\tau^{\alpha-1}}d\tau, \tag{45}$$

which reduces to

$$R_x^{\alpha,\theta}\left(e^{i\omega x}\right) = \frac{\omega^2}{\Gamma\left(2-\alpha\right)} e^{i\omega x}\left(c_+\frac{\Gamma\left(2-\alpha\right)}{\left(i\omega\right)^{2-\alpha}} - c_-\frac{\Gamma\left(2-\alpha\right)}{\left(-i\omega\right)^{2-\alpha}}\right). \tag{46}$$

Carrying out the same procedure, the lemma follows.

Example 1. *Consider problem (1) with* $k(u)=1$, $g(u)=au$ *and* $f(x)=\epsilon\sin(\pi x/L)$, *i.e.,*

$${}^{C}D_t^{\beta}u\left(x,t\right) = R_x^{\alpha,\theta}u\left(x,t\right) + au\left(x,t\right), -\infty < x\left\langle\infty, t\right\rangle 0, u\left(x,0\right) = \epsilon\sin\left(\pi x / L\right), \tag{47}$$

where α,ϵ, *and L are real constants.*

By means of OHAM, consider the linear operator

$$\mathcal{L}\left[\phi\left(x,t;p\right)\right] = {}^{C}D_t^{\beta}\phi\left(x,t;p\right), \tag{48}$$

and the operator

$$N\left[\phi\left(x,t;p\right)\right] = {}^{C}D_t^{\beta}\phi - R_x^{\alpha,\theta}\phi - a\ \phi. \tag{49}$$

Then the *m*th-order deformation equation, with $H(x,t)=1$, for this problem can be written as

$$\mathcal{L}\left[u_m\left(x,t\right) - \chi_m u_{m-1}\left(x,t\right)\right] = \hbar\Re_m\left[u_{m-1}\left(x,t\right)\right], \tag{50}$$

where

$$\Re_m\left[u_{m-1}\right] = {}^{C}D_t^{\beta}u_{m-1} - R_x^{\alpha,\theta}u_{m-1} - a\ u_{m-1}. \tag{51}$$

Now, apply the inverse operator $\mathcal{L}^{-1} = J_t^{\mu}$ to both sides of (50) to obtain

$$u_m(x,t) = \chi_m u_{m-1}(x,t) + \left(u_m(x,0) - \chi_m u_{m-1}(x,0)\right) + \hbar J_t^{\beta}\left[\Re_m\left[u_{m-1}(x,t)\right]\right], \tag{52}$$

where

$$u_0(x,t) = u(x,0) = \epsilon \sin\left(\frac{\pi x}{L}\right), \text{and } u_m(x,0) = 0, m > 1. \tag{53}$$

When we refer to equations (51) and (52) with the conditions (53), we can get

$$u_1(x,t) = \frac{\hbar t^{\beta}\epsilon}{\Gamma[1+\beta]}\left(-a\sin\left(\frac{\pi x}{L}\right) + \left(\frac{\pi}{L}\right)^{\alpha}\sin\left(\frac{\pi x}{L} - \frac{\pi\theta}{2}\right)\right),$$

$$u_2 = \frac{\hbar(1+\hbar)t^{\beta}\int}{\Gamma(1+2\beta)}\left(-a\sin\left(\frac{\pi x}{L}\right) + \left(\frac{\pi}{L}\right)^{\alpha}\sin\left(\frac{\pi x}{L} - \frac{\pi\theta}{2}\right)\right) + \frac{\hbar^2 t^{2\beta}\int}{\Gamma(1+2\beta)}$$

$$\left(\left(a^2\sin\left(\frac{\pi x}{L}\right) + \left(\frac{\pi}{L}\right)^{\alpha}\left(\left(\frac{\pi}{L}\right)^{\alpha}\sin\left(\frac{\pi x}{L} - \pi\theta\right) - 2a\sin\left(\frac{\pi x}{L} - \frac{\pi\theta}{2}\right)\right)\right)\right).$$

The series solution is given by $u=u_0+u_1+u_2+u_3+\ldots$. Figure 1 shows the effect of the skewness parameter θ on the behavior of the solution . The plot represents the sum of the first five terms in the OHAM series when a=1.0, ϵ=1.0, L=3.0, t=0.2, 0<x<π and the fractional parameters β=1.0, α=1.5, and θ=0, 0.5, and -0.5.

Figure 1. The behavior of the OHAM solution of (47) at t=0.2, 0<x<π, and the fractional parameter α=1.5 and β=1, the skewness θ=0, 0.25, -0.25

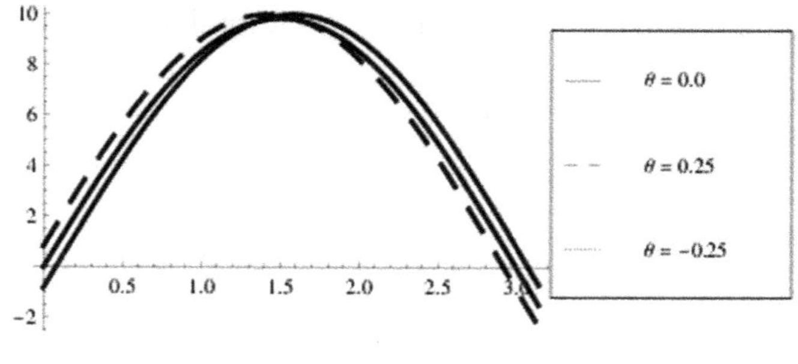

Figure 2. The behavior of the OHAM solution of (47) at t=0.2, 0<x<π, and the fractional parameter θ=1, β=1, and α=1.8, 1.95, compared with the exact solution

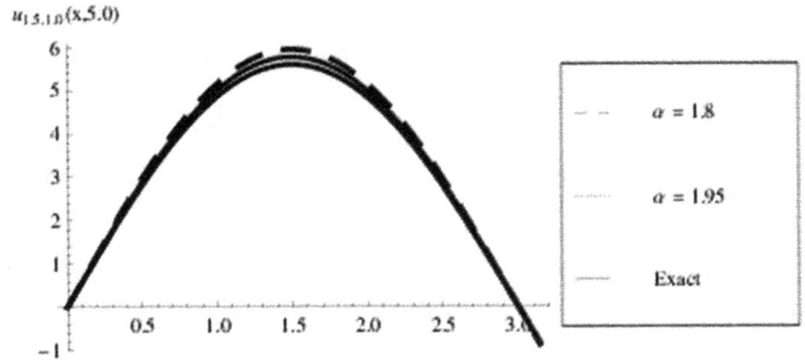

Figure 3. The behavior of the OHAM solution of (47) at t=0.2, 0<x<π, and the fractional parameter θ=0, α=2.0, and β=0.5, 0.7, 0.9, compared with the exact solution

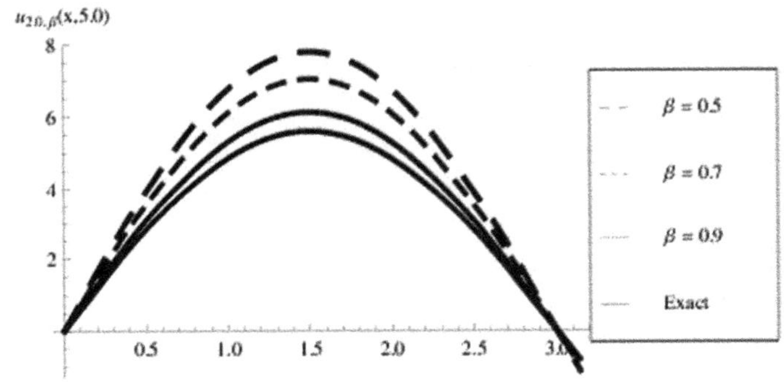

Table 1. The estimated optimal convergence parameter ħ and the corresponding residual error E_m at different fractional parameters used in Example 1

Fractional Parameters **α, θ, β**			**Optimal ħ**	**Residual Error (E_4)**
α=1.5, β=1.0	θ=	0.00	-0.990381137	4.6032E-10
		0.25	-1.001711905	2.5426E-5
		-0.25	-1.0023206958	2.54262E-6
β=1.0, θ=0	α=	1.80	-0.9339901124	2.2375E-7
		1.90	-0.9303835167	3.4254E-7
		1.95	-0.9285827510	4.2025E-7
α=2.0, θ=0	β=	0.5	-0.9099205063	9.7340E-9
		0.7	-0.9151677201	7.4716E-8
		0.9	-0.9225579219	3.0301E-8

The solution (the first five terms in the OHAM series) behavior as the parameters θ, α and β change at a fixed time t=5.0and in the interval $0\leq x\leq\pi$ shown in Figures 1, 2 and 3. The amplitude of the solution decreases α and/or β increases . As α reaches 2 and/or β increases up to unity, the series solution approximately coincides with the exact solution of the corresponding integer order equation $u = e^{-\left(\pi^2/L^2-a\right)t}\sin\left(\pi x / L\right)$, represented by the solid line in the graph, refer to Figures 1 and 2, and this numerical result supports the continuation of the solution. The series displayed in plots is the partial sum of the first five terms. The optimal convergence parameter $\hbar$ in each case is obtained by minimizing the residual error E_m displayed in (25), in the interval $0\leq x\leq\pi$ and $0\leq t\leq 6.0$, and the results presented in table 1.

Example 2. *Consider the nonlinear fractional problem (1) with k(u)=u, g(u)=4u²−u* and $f(x)$=sin(2x), i.e.,

$$ {}^{C}D_t^{\beta}u\left(x,t\right) = u\left(x,t\right)R_x^{\alpha,\theta}u\left(x,t\right) + 4u^2 - u, -\infty < x\left\langle\infty,t\right\rangle 0, u\left(x,0\right) = \sin\left(2x\right), \tag{54} $$

Here, let us chose the auxiliary linear operator as

$$ \mathcal{L}\left[\phi\right] = \frac{\partial}{\partial t}\left(\phi\right), \tag{55} $$

and the nonlinear operator N as

$$ N\left[\phi\right] = {}^{C}D_t^{\beta}\phi - \phi R_x^{\alpha,\theta}\left(\phi\right) - 4\phi^2 + \phi. \tag{56} $$

Then, mth-order deformation equation for this problem is given by

$$ \mathcal{L}\left[u_m\left(x,t\right) - \chi_m u_{m-1}\left(x,t\right)\right] = \hbar H\left(x,t\right)\Re_m\left[u_{m-1}\left(x,t\right)\right], \tag{57} $$

where $\Re_m\left[u_{m-1}\left(x,t\right)\right]$ is given by

$$ \Re_m\left[u_{m-1}\left(x,t\right)\right] = {}^{C}D_t^{\beta}\left(u_{m-1}\right) - \sum_{k=0}^{m-1}u_k R_x^{\alpha,\theta}\left(u_{m-1-k}\right) - 4\sum_{k=0}^{m-1}u_k u_{m-1-k} + u_{m-1}, \tag{58} $$

We choose $H(x,t)$=1 and u_0=sin(2x). Following same procedure in the previous example to obtain the series solution terms u_0=sin(2x),

$$ u_1 = \frac{\hbar t^{\beta}}{2\Gamma\left(1+\beta\right)}\left(-4 + 4\cos\left(4x\right) + 2^{\alpha}\cos\left(\frac{\pi\theta}{2}\right) - 2^{\alpha}\cos\left(4x - \frac{\pi\theta}{2}\right) + 2\sin\left(2x\right)\right), $$

and to find the next terms in the solution series, use the following recursive formula which is valid for $m>1$,

$$u_m = \left(1+\hbar\right)u_{m-1} + \hbar J_t^{\beta}\left[-\sum_{k=0}^{m-1} u_k R_x^{\alpha,\,\theta}\left(u_{m-1-k}\right) - 4\sum_{k=0}^{m-1} u_k u_{m-1-k} + u_{m-1}\right], \tag{59}$$

then, the solution is thus obtained as $u=u_0+u_1+u_2+u_3+\ldots$

Figures 4, 5, and 6 show the solution (the first five terms in the OHAM series) behavior as the parameters θ, α and β change at a fixed time t=0.2 and in the interval $0\leq x\leq \pi/2$. As α and/or β increases, the amplitude of the solution decreases . As α tends to 2 and/or β goes up to unity, the series solution approximately coincides with the exact solution of the corresponding integer order equation $u=e^{-t}\sin(2x)$, represented by the solid line in the graph, and this confirms the continuation of the solution numerically. The series displayed in plots is the partial sum of the first five terms; n=5 (summing u_0 to u_4). The optimal convergence parameter $\hbar$ in each case is obtained by minimizing the residual error E_m displayed in (25) in the interval $0\leq x\leq 2$ and $0\leq t\leq 0.4$, and the obtained results shown in Table 2.

Figure 4. The behavior of the OHAM solution of (54) at t=0.2, 0<x<π/2, and the fractional parameter α=1.5 and β=1, the skewness θ=0, 0.5, -0.5

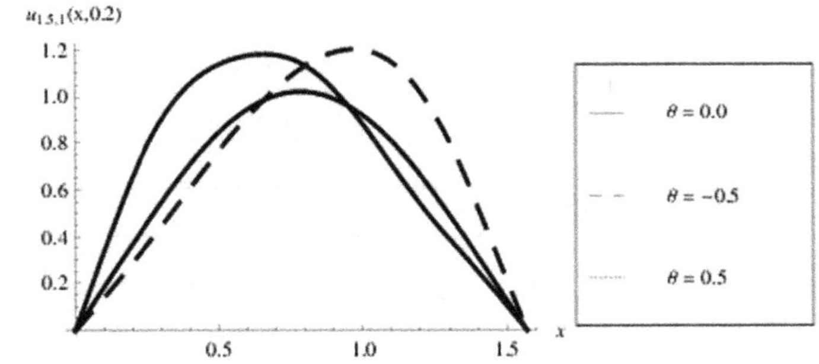

Figure 5. The behavior of the OHAM solution of (54) at t=0.2, 0<x<π/2, and the fractional parameter θ=0, β=2, and α=1.9, 1.95, 1.99, compared with the exact solution

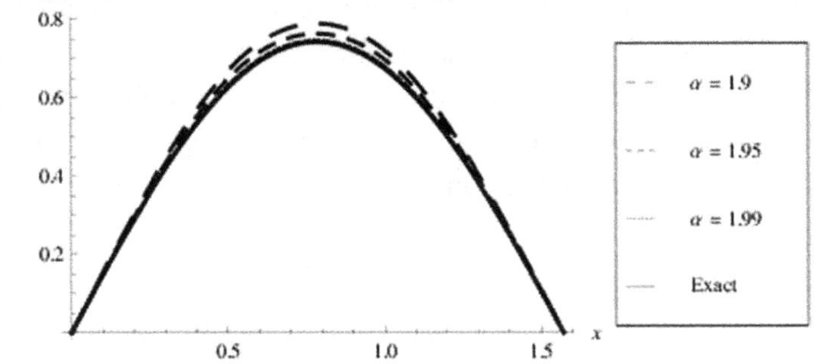

Figure 6. The behavior of the OHAM solution of (54) at t=0.2, 0<x<π/2, and the fractional parameter θ=0, α=2.0, and β=0.5, 0.7, 0.95, compared with the exact solution

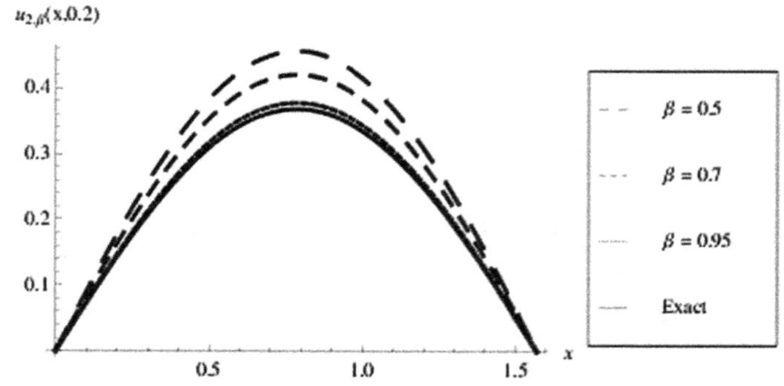

Table 2.

Fractional Parameters α, θ, β			Optimal ħ	Residual Error (E_4)
α=1.5, β=1.0	θ=	0.0	-0.831733694	2.3498E-8
		0.5	-0.954364814	1.0016E-5
		-0.5	-0.637748947	7.4733E-6
β=1.0, θ=0	α=	1.92	-0.589337116	2.3160E-6
		1.95	-0.589337116	2.1595E-6
		1.99	-0.639152091	7.6926E-6
α=2.0, θ=0	β=	0.50	-0.858637427	2.9308E-9
		0.70	-0.929790675	7.6259E-12
		0.95	-0.973269800	2.7+29E-12

CONCLUSION

In this chapter, a definition of the fractional-order Riesz-Feller derivative in the Caputo sense is proposed and its equivalence with the classical definition is proved. Then, the continuation of the solution of the fractional order anomalous diffusion equation with Riesz-Feller spatial derivative to the corresponding Reisz problem is proved. The iterative series solution for the fractional equation is obtained using the OHAM by the aid of the Riesz-Feller fractional operator property applied on trignometric functions. The advantage of using this technique is the ability to estimate an approximation to the residual error. The results obtained illustrate graphically the continuation of the solution we proved theoretically.

REFERENCES

Abbasbandy, S., & Zakaria, F. S. (2008). Soliton solutions for the fifth-order KdV equation with the homotopy analysis method. *Nonlinear Dynamics*, *51*(1-2), 83–87. doi:10.100711071-006-9193-y

Abdel Latif, M. S., El-Shazly, E., Baleanu, D., Elsaid, A., & Nour, H. M. (2018). Some new soliton-like and doubly periodic-like solutions of Fisher equation with time-dependent coefficients. *Modern Physics Letters B*, *32*(33), 1850413. doi:10.1142/S0217984918504134

Agrawal, O. P. (2002). Solution for a fractional diffusion-wave equation defined in a bounded domain. *Nonlinear Dynamics*, *29*(1/4), 145–155. doi:10.1023/A:1016539022492

Al-Khedhairi, A., Askar, S. S., Matouk, A. E., Elsadany, A., & Ghazel, M. (2018). Dynamics, Chaos Control, and Synchronization in a Fractional-Order Samardzija-Greller Population System with Order Lying in (0, 2). *Complexity*, *2018*, 2018. doi:10.1155/2018/6719341

Al-khedhairi, A., Matouk, A. E., & Khan, I. (2019). Chaotic dynamics and chaos control for the fractional-order geomagnetic field model. *Chaos, Solitons \& Fractals, 128*, 390-401.

Bagley, R. L., & Calico, R. A. (1991). Fractional order state equations for the control of viscoelasticallydamped structures. *Journal of Guidance, Control, and Dynamics*, *14*(2), 304–311. doi:10.2514/3.20641

Bouremel, Y. (2007). Explicit series solution for the Glauert-jet problem by means of the homotopy analysis method. *Communications in Nonlinear Science and Numerical Simulation*, *12*(5), 714–724. doi:10.1016/j.cnsns.2005.07.001

Cang, J., Tan, Y., Xu, H., & Liao, S. J. (2009). Series solutions of non-linear Riccati differential equations with fractional order. *Chaos, Solitons & Fractals, 40*, 1-9.

Celik, C., & Duman, M. (2012). Crank--Nicolson method for the fractional diffusion equation with the Riesz fractional derivative. *Journal of Computational Physics*, *231*(4), 1743–1750. doi:10.1016/j.jcp.2011.11.008

El-Sayed, A. M. (1996). Fractional-order diffusion-wave equation. *International Journal of Theoretical Physics*, *35*(2), 311–322. doi:10.1007/BF02083817

El-Sayed, A. M., & Gaber, M. (2006). On the finite Caputo and finite Riesz derivatives. *Electronic Journal of Theoretical Physics*, *3*, 81–95.

El-Sayed, A. M., Nour, H. M., Elsaid, A., Matouk, A. E., & Elsonbaty, A. (2016). Dynamical behaviors, circuit realization, chaos control, and synchronization of a new fractional order hyperchaotic system. *Applied Mathematical Modelling*, *40*(5-6), 3516–3534. doi:10.1016/j.apm.2015.10.010

Ellahi, R., Raza, M., & Vafai, K. (2012). Series solutions of non-Newtonian nanofluids with Reynolds' model and Vogel's model by means of the homotopy analysis method. *Mathematical and Computer Modelling*, *55*(7-8), 1876–1891. doi:10.1016/j.mcm.2011.11.043

Elsadany, A. A., & Matouk, A. E. (2015). Dynamical behaviors of fractional-order Lotka--Volterra predator--prey model and its discretization. *Journal of Applied Mathematics and Computing*, *49*(1-2), 269–283. doi:10.100712190-014-0838-6

Elsaid, A. (2010). The variational iteration method for solving Riesz fractional partial differential equations. *Computers & Mathematics with Applications, 60*, 1940-1947.

Elsaid, A. (2011). Homotopy analysis method for solving a class of fractional partial differential equations. *Communications in Nonlinear Science and Numerical Simulation*, *16*(9), 3655–3664. doi:10.1016/j.cnsns.2010.12.040

Elsaid, A., Shamseldeen, S., & Madkour, S. (2016a). Iterative Solution of Fractional Diffusion Equation Modelling Anomalous Diffusion. *Applications and Applied Mathematics*, *11*, 815–827.

Elsaid, A., Shamseldeen, S., & Madkour, S. (2016b). Semianalytic Solution of Space-Time Fractional Diffusion Equation. *International Journal of Differential Equations*, *2016*, 2016. doi:10.1155/2016/2371837

Elsaid, A., Shamseldeen, S., & Madkour, S. (2016c). Solution of Fractional Wave Equation by the Optimal Homotopy Analysis Method. *European Journal of Pure and Applied Mathematics*, 586-601.

Ervin, V. J., & Roop, J. P. (2007). Variational solution of fractional advection dispersion equations on bounded domains. *Numerical Methods for Partial Differential Equations*, *23*(2), 256–281. doi:10.1002/num.20169

Garg, M., Sharma, A., & Manohar, P. (2013). Solution of generalized space-time fractional telegraph equation with composite and Riesz-Feller fractional derivatives. *International Journal of Pure and Applied Mathematics*, *83*(5), 685–691. doi:10.12732/ijpam.v83i5.12

Gorenflo, R., & Mainardi, F. (2003). Fractional diffusion processes: probability distributions and continuous time random walk. In *Processes with Long-Range Correlations* (pp. 148–166). Springer. doi:10.1007/3-540-44832-2_8

Gorenflo, R., Mainardi, F., Moretti, D., Pagnini, G., & Paradisi, P. (2002). Discrete random walk models for space--time fractional diffusion. *Chemical Physics*, *284*(1-2), 521–541. doi:10.1016/S0301-0104(02)00714-0

Gorenflo, R., & Vivoli, A. (2003). Fully discrete random walks for space--time fractional diffusion equations. *Signal Processing*, *83*(11), 2411–2420. doi:10.1016/S0165-1684(03)00193-2

Haubold, H. J., Mathai, A. M., & Saxena, R. K. (2011). Further solutions of fractional reaction--diffusion equations in terms of the H-function. *Journal of Computational and Applied Mathematics*, *235*(5), 1311–1316. doi:10.1016/j.cam.2010.08.016

Hegazi, A. S., Ahmed, E.-S. M., & Matouk, A. E. (2013). On Dynamical Behaviors and Chaos Control of the Fractional-Order Financial System. In Chaos and Complexity Theory for Management: Nonlinear Dynamics (pp. 34-49). IGI Global. doi:10.4018/978-1-4666-2509-9.ch002

Huang, F., & Liu, F. (2005). The fundamental solution of the space-time fractional advection-dispersion equation. *Journal of Applied Mathematics and Computing*, *18*(1-2), 339–350. doi:10.1007/BF02936577

Jiang, H., Liu, F., Turner, I., & Burrage, K. (2012). Analytical solutions for the multi-term time--space Caputo--Riesz fractional advection--diffusion equations on a finite domain. *Journal of Mathematical Analysis and Applications*, *389*(2), 1117–1127. doi:10.1016/j.jmaa.2011.12.055

Kader, A. H., Latif, M. S., & Nour, H. M. (2017). Some New Exact Solutions of the Modified KdV Equation Using Lie Point Symmetry Method. *International Journal of Applied and Computational Mathematics*, *3*(S1), 1163–1171. doi:10.100740819-017-0408-3

Khana, A., & Kumar, S. (2016). Anti-synchronization of chaotic financial systems by using fuzzy logic constant controller. *Indian Journal of Industrial and Applied Mathematics, 7*(2), 136–147. doi:10.5958/1945-919X.2016.00013.X

Liao, S. (2003). *Beyond perturbation: introduction to the homotopy analysis method.* CRC press. doi:10.1201/9780203491164

Liao, S. (2005). Comparison between the homotopy analysis method and homotopy perturbation method. *Applied Mathematics and Computation, 169*(2), 1186–1194. doi:10.1016/j.amc.2004.10.058

Liao, S. (2010). An optimal homotopy-analysis approach for strongly nonlinear differential equations. *Communications in Nonlinear Science and Numerical Simulation, 15*(8), 2003–2016. doi:10.1016/j.cnsns.2009.09.002

Lin, R., Liu, F., Anh, V., & Turner, I. (2009). Stability and convergence of a new explicit finite-difference approximation for the variable-order nonlinear fractional diffusion equation. *Applied Mathematics and Computation, 212*(2), 435–445. doi:10.1016/j.amc.2009.02.047

Luchko, Y., & Gorenflo, R. (1998). *The initial value problem for some fractional differential equations with the Caputo derivatives.* Academic Press.

Mainardi, F., Pagnini, G., & Luchko, Y. (2007). The fundamental solution of the space-time fractional diffusion equation. *Fractional Calculus & Applied Analysis, 4*, 153–192.

Matouk, A. E., & Elsadany, A. A. (2016). Dynamical analysis, stabilization and discretization of a chaotic fractional-order GLV model. *Nonlinear Dynamics, 85*(3), 1597–1612. doi:10.100711071-016-2781-6

Metzler, R., & Klafter, J. (2000). The random walk's guide to anomalous diffusion: A fractional dynamics approach. *Physics Reports, 339*(1), 1–77. doi:10.1016/S0370-1573(00)00070-3

Miller, K. S., & Ross, B. (1993). *An introduction to the fractional calculus and fractional differential equations.* Academic Press.

Podlubny, I. (1998). *Fractional differential equations: an introduction to fractional derivatives, fractional differential equations, to methods of their solution and some of their applications* (Vol. 198). Academic Press.

Rahman, M., Mahmood, A., & Younis, M. (2014). Improved and more feasible numerical methods for Riesz space fractional partial differential equations. *Applied Mathematics and Computation, 237*, 264–273. doi:10.1016/j.amc.2014.03.103

Ray, S. S., & Sahoo, S. (2015). Analytical approximate solutions of Riesz fractional diffusion equation and Riesz fractional advection--dispersion equation involving nonlocal space fractional derivatives. *Mathematical Methods in the Applied Sciences, 38*(13), 2840–2849. doi:10.1002/mma.3267

Shamseldeen, S. (2018). Approximate solution of space and time fractional higher order phase field equation. *Physica A, 494*, 308–316. doi:10.1016/j.physa.2017.12.056

Shamseldeen, S., Elsaid, A., & Madkour, S. (2019a). Caputo-Riesz-Feller fractional wave equation: Analytic and approximate solutions and their continuation. *Journal of Applied Mathematics and Computing*, *59*(1-2), 423–444. doi:10.100712190-018-1186-8

Shamseldeen, S., Elsaid, A., & Madkour, S. (2019b). On The Solution of Nonlinear Generalized Caputo-Riesz Fractional EFK and KS Equations. *Eurasian Journal of Mathematical and Computer Applications*, *7*(2), 62–78. doi:10.32523/2306-6172-2019-7-2-62-78

Taghavian, H., & Tavazoei, M. S. (2018). Exact solution of linear fractional distributed order systems with exponential order weight functions. In *Mathematical Techniques of Fractional Order Systems* (pp. 103–132). Elsevier. doi:10.1016/B978-0-12-813592-1.00004-0

Taghavian, H., & Tavazoei, M. S. (2019). Algebraic Conditions for Stability Analysis of Linear Time-Invariant Distributed Order Dynamic Systems: A Lagrange Inversion Theorem Approach. *Asian Journal of Control*, *21*(2), 879–890. doi:10.1002/asjc.1780

Yang, Q., Liu, F., & Turner, I. (2010). Numerical methods for fractional partial differential equations with Riesz space fractional derivatives. *Applied Mathematical Modelling*, *34*(1), 200–218. doi:10.1016/j.apm.2009.04.006

Zhang, H., & Liu, F. (2007). The fundamental solutions of the space, space-time Riesz fractional partial differential equations with periodic conditions. *Numerical Mathematics-English Series*, *16*, 181–192.

Zhang, X., Wei, L., Tang, B., & He, Y. (2013). Homotopy analysis method for space-time fractional differential equations. *International Journal of Numerical Methods for Heat & Fluid Flow*, *23*, 1063-1075.

Section 4
New Trends in Applied Mathematics

Chapter 11
Generalized Fractional Ostrowski's Type Inequalities Involving Riemann-Liouville Fractional Integration

Ather Qayyum
Institute of Southern Punjab Multan, Pakistan

Muhammad Shoaib
Higher Colleges of Technology, UAE

ABSTRACT

Fractional calculus has applications in many practical problems such as electromagnetic waves, visco-elastic systems, quantum evolution of complex systems, diffusion waves, physics, engineering, finance, social sciences, economics, mathematical biology, and chaos theory. Theories of inequality are growing rapidly; the fractional version of Ostrowski's type inequality is one of them. In this chapter, the author firstly presents some generalized Montgomery identities for Riemann-Liouville fractional integrals, which are designed by using a new and special type of Peano kernels. Secondly, inequalities via convex function are also discussed. In addition, some general fractional representation formulae are studied for a function in terms of the fractional Riemann-Liouville integrals of different orders and its ordinary derivatives. Moreover, by utilizing Montgomery identities, some new fractional versions of Ostrowski's type integral inequalities are established. Some well-known results are deduced as special cases from the results are developed here.

INTRODUCTION

Fractional calculus was introduced by J. Liouville in the beginning of the 19th century. However, it wasn't put into practical use until the last quarter of the 19th century when it was first used by Heaviside in the analysis of electrical transmission lines. Since then, the interest in fractional calculus has been growing

DOI: 10.4018/978-1-7998-3122-8.ch011

continually owing to its applications in different branches of sciences. Fractional calculus has important applications in science and technology, e.g. in biological systems (Ahmed & Elgazzar, 2007; Matouk et al, 2015; Elsadany & Matouk, 2015; Matouk, 2009; A El-Sayed et al, 1996, Matouk & Elsadany, 2016; Selvam et al, 2017; Selvam et al, 2018a-b; Ameen & Novati, 2017; Al-Khedhairi et al, 2018), economic and financial systems (Laskin, 2000; Chen, 2008; Chen et al, 2011; Zhen et al, 2011; Hegazi et al, 2013b; Khan & Kumar, 2016), social models (Ahmad & El-Khazali, 2007; Ali, 2019), physical models (Heaviside, 1971; El-Sayed, 1996; Kusnezov et al, 1999; Hilfer, 2000; Khan & Kumar, 2018; Saqib et al, 2018a-d, Khan & Kumar, 2019a; Al-Khedhairi et al, 2019a; Al-Khedhairi et al, 2019b) and engineering systems (Sun et al, 1984; Petras, 2009; Matouk, 2009a; Matouk, 2009b; Matouk 2010; Matouk, 2011a; Matouk, 2011b; Hegazi & Matouk, 2011; Hegazi et al, 2011; Hegazi & Matouk, 2013; Hegazi et al, 2013a; Matouk & Elsadany, 2014; Banu & Nasir, 2015; Nasir & Singh, 2015; Matouk, 2016; El-Sayed et al, 2016a; El-Sayed et al, 2016b; Khan & Kumar, 2019b). Furthermore, fractional derivatives have also useful applications in some interdisciplinary fields of science (Ben Adda F, 1997; Caputo, 1967 ; Ichise et al, 1971; Mainardi, 2010; Samet et al, 2018; Kusnezov et al, 1999; Dragomir, 2005; Mitrinović et al, 1994; Anastassiou et al, 2009; Podlubny, 1999 & Song et al, 2010; Laskin, 2000; Heaviside, 1971; Hilfer 2000; Bagley et al, 1991; Qayyum et al, 2015; Qayyum et al, 2019; Taghavian & Tavazoei, 2017; Taghavian & Tavazoei, 2018; Taghavian & Tavazoei, 2019).

Anastassiou et al, 2009 presented the fractional version of Ostrowski's type inequality. Sarikaya et al, 2014 gave a generalization of fractional version of Ostrowski inequality. Some authors (Qayyum et al, 2015 & Qayyum et al, 2019) and (Sarikaya et al, 2012 & Ahmed et al, 2007) provided novel inequalities with the help of Riemann-Liouville Fractional integration (RLFI). Further, some inequalities involving Liouville Caputo Fractional Derivatives are obtained with applications given by Samet et al, 2018.

In this chapter, we will use the Riemann Liouville Fractional Integral (RLFI) to design some new fractional version of Ostrowski's type integral inequalities. A number of special cases of the classical Ostrowski's inequalities are derived by using the results developed in this literature.

PRELIMINARIES

We will give some definitions which are useful in understanding our derivations and results.

Definition 1. The Riemann Liouville Fractional Integral (RLFI)

The Riemann Liouville Fractional Integral (RLFI) operator of order $\sigma \geq 0$ is defined by Anastassiou et al, 2009 as,

$$J_a^{\sigma} f(x) = \frac{1}{\Gamma(\sigma)} \int_a^x (x-s)^{\sigma-1} f(s)\, ds \tag{1}$$

Where $J_a^0 f(x) = f(x)$.

Definition 2. Ostrowski inequality

Consider f a real valued continuous and differentiable mapping on $I^{\circ} \in I \subseteq R$ and let $a,b \in I^{o}$ with $a<b$. If $f'_{\infty} = \sup_{t\in[a,b]} \left|f''(s)\right| < \infty$, then

$$\left|f(x) - \frac{1}{b-a}\int_a^b f(s)\,ds\right| \le \left[\frac{1}{4} + \frac{\left(x - \dfrac{a+b}{2}\right)^2}{(b-a)^2}\right](b-a)f'_{\infty}, \tag{2}$$

This inequality is known as Ostrowski inequality (Ostrowski, 1938). Generalizations of this inequality are available in (Mitrinović et al, 1994).

Definition 3. Montgomery's identity

Consider the real valued function f defined on $[a,b]$. The function f is continuous and differentiable on (a,b). Let f' is also integrable on $[a,b]$, then

$$f(x) = \frac{1}{b-a}\int_a^b f(s)\,ds + \frac{1}{b-a}\int_a^b P_1(x,s) f'(s)\,ds, \tag{3}$$

where $P_1(x,s)$ is defined by

$$P_1(x,s) = \begin{cases} s-a & if\; a \le s \le x \le b \\ s-b & if\; a \le x < s \le b. \end{cases} \tag{3.1}$$

This identity is known as Montgomery's identity.

Definition 4. Extension of the kernel $P_1(x,s)$.

Define $\Omega_1(x,s)$ as

$$\Omega_1(x,s) = \begin{cases} \dfrac{1}{b-a}\left[s - \left(a + \varrho\dfrac{b-a}{2}\right)\right] & if\; a \le s \le x \\ \dfrac{1}{b-a}\left[s - \left(b - \varrho\dfrac{b-a}{2}\right)\right] & if\; x < s \le b. \end{cases} \tag{4}$$

The kernel $\Omega_1(x,s)$ is an extension of $P_1(x,s)$ which will be adapted to define the fractional version of $P_1(x,s)$.

Definition 5. The Fractional version of $\Omega_1(x,s)$.

The fractional version of $\Omega_1(x,s)$ is defined as

$$P_2(x,s)=\begin{cases}\dfrac{s-a}{b-a}(b-x)^{1-\sigma}\,\Gamma(\sigma) & if\; a\le s\le x\le b\\ \dfrac{s-b}{b-a}(b-x)^{1-\sigma}\,\Gamma(\sigma) & if\; a\le x< s\le b.\end{cases} \tag{4.1}$$

Definition 6. Extension of fractional peano type kernel

Consider $\Omega_2(x,.)$: $[a,b]\to\mathbb{R}$,

$$\begin{cases}\dfrac{1}{b-a}(b-x)^{1-\sigma}\,\Gamma(\sigma)\left(s-\left(a+\varrho\dfrac{b-a}{2}\right)\right) & if\; a\le s\le x\\ \dfrac{1}{b-a}(b-x)^{1-\sigma}\,\Gamma(\sigma)\left(s-\left(b-\varrho\dfrac{b-a}{2}\right)\right) & if\; x< s\le b\end{cases} \tag{5}$$

for all

$$x\in\left[a+\varrho\frac{b-a}{2},b-\varrho\frac{b-a}{2}\right]$$

and $\varrho\in[0,1]$. The kernel $\Omega_2(x,s)$ is an extension of fractional peano type kernel.

Definition 7. Convex function

A function $f(y)$ is convex if for any two points y_1 and y_2 in $[a,b]$ and any $\alpha\in(0,1)$,

$$f\left(\alpha y_1+(1-\alpha)y_2\right)\le\alpha f(y_1)+(1-\alpha)f(y_2).$$

Lemma 1. Let $\Omega_2(x,.)$: $[b,c]\to\mathbb{R}$, be the fractional peano type kernel, then

$$\forall x\in\left[b+\varrho\frac{c-b}{2},b-\varrho\frac{c-b}{2}\right]$$

and $\varrho\in[0,1]$, we have

$$\left(1-\varrho\right)f\left(x\right)=\frac{\Gamma\left(\sigma\right)}{c-b}\left(c-x\right)^{1-\sigma}J_{a}^{\sigma}\left(f\left(c\right)\right)-\frac{\varrho}{2}\left(c-x\right)^{1-\sigma}\left(c-b\right)^{\sigma-2}f\left(b\right)$$
$$-J_{b}^{\sigma-1}\left(\Omega_{2}\left(x,c\right)f\left(c\right)\right)+J_{b}^{\sigma}\left(\Omega_{2}\left(x,c\right)f'\left(c\right)\right) \tag{6}$$

where σ≥1.

Proof. From (5)

$$\Gamma\left(\sigma\right)J_{b}^{\sigma}\left(\Omega_{2}\left(x,c\right)f'\left(c\right)\right)=\int_{b}^{c}\left(c-s\right)^{\sigma-1}\Omega_{2}\left(x,s\right)f'\left(s\right)ds$$
$$=\frac{1}{c-b}\int_{b}^{x}\left(c-s\right)^{\sigma-1}\left(s-\left(b+\varrho\frac{c-b}{2}\right)\right)f'\left(s\right)ds \tag{7}$$
$$+\frac{1}{c-b}\int_{x}^{c}\left(c-s\right)^{\sigma-1}\left(s-\left(c-\varrho\frac{c-b}{2}\right)\right)f'\left(s\right)ds.$$

The desired identity can easily be obtained from (7).

A FRACTIONAL VERSION OF OSTROWSKI TYPE INEQUALITY-I

By using the identity proved in Lemma 1, we state and prove the following inequality:

Theorem 1. Let *f* real valued function defined on [*b*,*c*], is differentiable on (*b*,*c*) such that $f'\left(b\right)\in L_{1}\left(b,c\right)$. If $\left|f'\left(x\right)\right|$ is bounded on the interval [*b*,*c*], then the following inequality holds:

$$\left|\left(1-\varrho\right)f\left(x\right)+\frac{\varrho}{2}\left(c-x\right)^{1-\sigma}\left(c-b\right)^{\sigma-2}f\left(b\right)-\frac{\Gamma\left(\sigma\right)}{c-b}\left(c-x\right)^{1-\sigma}J_{b}^{\sigma}\left(f\left(c\right)\right)+J_{b}^{\sigma-1}\left(\Omega_{2}\left(x,c\right)f\left(c\right)\right)\right|$$
$$\leq\frac{M}{\sigma\left(\sigma+1\right)}\left\{\frac{2\left(c-x\right)\left(\sigma+1\right)}{c-b}\left(\frac{b+c}{2}-x\right)-\frac{2}{c-b}\left(c-x\right)^{2}+\left(c-b\right)^{\sigma}\left(c-x\right)^{1-\sigma}-\frac{\varrho}{2}\left(\sigma+1\right)\left(c-b\right)^{\sigma}\left(c-x\right)^{1-\sigma}\right\}, \tag{8}$$

where σ≥1.

Proof. By using Lemma 1, we get

$$\left|(1-\varrho)f(x)+\frac{\varrho}{2}(c-x)^{1-\sigma}(c-b)^{\sigma-2}f(b)-\frac{\Gamma(\sigma)}{c-b}(c-x)^{1-\sigma}J_b^{\sigma}(f(c))+J_b^{\sigma-1}(\Omega_2(x,c)f(c))\right|$$
$$\leq\frac{1}{\Gamma(\sigma)}\left|\int_b^c(c-s)^{\sigma-1}\Omega_2(x,s)f'(s)ds\right|\leq\frac{M}{\Gamma(\sigma)}\int_b^c(c-s)^{\sigma-1}\left|\Omega_2(x,s)\right|ds \tag{9}$$
$$=M(c-x)^{1-\sigma}\left\{\int_b^x\frac{1}{c-b}(c-s)^{\sigma-1}\left(s-\left(b+\varrho\frac{c-b}{2}\right)\right)ds+\int_x^c\frac{1}{c-b}(c-s)^{\sigma-1}\left(\left(c-\varrho\frac{c-b}{2}\right)-s\right)ds\right\}.$$

After simple calculation, we get (8).

Remark 1. Consider ϱ=0 in Theorem 1, it reduces to the result obtained by Anastassiou et.al 2009.

FRACTIONAL VERSION OF OSTROWSKI TYPE INEQUALITY VIA CONVEXITY

Theorem 2. Let f be a real valued convex function on $[b,c]$. Then for any $x\in(b,c)$, , then

$$\frac{1}{\sigma(\sigma+1)}\left\{-\frac{1}{c-b}(\sigma+1)\left[x-\left(b+\varrho\frac{c-b}{2}\right)\right](b-x)-\frac{\varrho(\sigma+1)}{2}\times(c-b)^{\sigma}(c-x)^{1-\sigma}-\frac{1}{c-b}(c-x)^{2}\right.$$
$$\left.+(c-b)^{\sigma}(c-x)^{1-\sigma}]f_{-}'(x)-\left[\frac{1}{c-b}(\sigma+1)\left(\left(c-\varrho\frac{c-b}{2}\right)-x\right)(c-x)+\frac{1}{c-b}(c-x)^{2}\right]f_{+}'(x)\right\}$$
$$\leq\frac{\Gamma(\sigma)}{c-b}(c-x)^{1-\sigma}J_b^{\sigma}(f(c))-J_b^{\sigma-1}(\Omega_2(x,c)f(c))-\frac{\varrho}{2}(c-x)^{1-\sigma}(c-b)^{\sigma-2}f(b)-(1-\varrho)f(x). \tag{10}$$

Proof. From Lemma 2, we have

$$(1-\varrho)f(x)-\frac{\Gamma(\sigma)}{c-b}(c-x)^{1-\sigma}J_b^{\sigma}(f(c))+\frac{\varrho}{2}(c-x)^{1-\sigma}(c-b)^{\sigma-2}f(b)+J_b^{\sigma-1}(\Omega_2(x,c)f(c))$$
$$=\frac{1}{c-b}(c-x)^{1-\sigma}\left\{\int_b^x(c-s)^{\sigma-1}\left[s-\left(b+\varrho\frac{c-b}{2}\right)\right]f'(s)ds-\int_x^c(c-s)^{\sigma-1}\left[\left(c-\varrho\frac{c-b}{2}\right)-s\right]f'(s)ds\right\}. \tag{11}$$

The convexity of the function f implies that for any $x\in(b,c)$ we have

$$f'(s)\leq f_{-}'(x),s\in[b,x], \tag{12}$$

$$f'(s)\geq f_{+}'(x),s\in[x,c]. \tag{13}$$

By using the above inequalities, we get the required result given in (10).

Corollary 1. Let σ=1, ϱ=0 in (10), we get Dragomir's result given in Dragomir, 2005).

Now, we present another new result.

Theorem 3. Let *f* real valued function defined on [*b*,*c*]. Then

$$\frac{\Gamma(\sigma)}{c-b}(c-x)^{1-\sigma} J_b^{\sigma}(f(c)) - J_b^{\sigma-1}(\Omega_2(x,b)f(c)) - \frac{\varrho}{2}(c-x)^{1-\sigma}(c-b)^{\sigma-2} f(b) - (1-\varrho)f(x)$$
$$\leq \frac{1}{\sigma(\sigma+1)}\left\{\frac{1}{c-b}(\sigma+1)\left(\left(c-\varrho\frac{c-b}{2}\right)-x\right)(c-x) - \frac{1}{c-b}(c-x)^2) f_-'(c) - (-((c-b)^{\sigma}(c-x)^{1-\sigma}\right. \tag{14}$$
$$\left. -\frac{1}{c-b}(c-x)^2 - \frac{1}{c-b}(\sigma+1)\left(x-\left(a+\varrho\frac{c-b}{2}\right)\right)(b-x) - \frac{\varrho(\sigma+1)}{2}(c-b)^{\sigma}(c-x)^{1-\sigma})) f_+'(b)\right\}.$$

for any $x \in (b,c)$.

Proof. Suppose $f_+'(b) < \infty$ and $f_-'(c) < \infty$. The convexity of the function *f* implies that for any $x \in (b,c)$ we have

$$f'(s) \geq f_+'(b), s \in [b,x], \tag{15}$$

$$f'(s) \leq f_-'(c), s \in [x,c]. \tag{16}$$

Multiplying (15) with

$$(c-s)^{\sigma-1}\left(s-\left(b+\varrho\frac{c-b}{2}\right)\right) \geq 0$$

integrate on [*b*,*x*] we get

$$\int_b^x (c-s)^{\alpha-1}\left(s-\left(b+\varrho\frac{c-b}{2}\right)\right) f'(s)\,ds$$
$$\geq \int_b^x (c-s)^{\alpha-1}\left(s-\left(b+\varrho\frac{c-b}{2}\right)\right) f_+'(b)\,ds = \frac{1}{\alpha(\alpha+1)}\left\{(c-b)^{\alpha+1} - (c-x)^{\alpha+1}\right. \tag{17}$$
$$\left. -(\alpha+1)\left(x-\left(b+\varrho\frac{c-b}{2}\right)\right)(c-x)^{\alpha} - \frac{\varrho(\alpha+1)}{2}(c-b)^{\alpha+1}\right\} f_+'(b).$$

and by multiplying (16) with

$$\left(c-s\right)^{\sigma-1}\left(\left(c-\varrho\frac{c-b}{2}\right)-s\right),$$

integrate on $[x,c]$, we get

$$\begin{aligned}&\int_x^c\left(c-s\right)^{\alpha-1}\left(\left(c-\varrho\frac{c-b}{2}\right)-s\right)f'\left(s\right)ds\leq\int_x^c\left(c-s\right)^{\alpha-1}\left(s-\left(c-\varrho\frac{c-b}{2}\right)\right)f_-'\left(b\right)\,ds\\&=\frac{1}{\alpha\left(\alpha+1\right)}\left(\left(\alpha+1\right)\left(\left(c-\varrho\frac{c-b}{2}\right)-x\right)\left(c-x\right)^{\alpha}-\left(c-x\right)^{\alpha+1}\right)f_-'\left(c\right).\end{aligned}\tag{18}$$

By subtracting (17) and (18) and using (11), we get the desired result (14).

DERIVATION OF SOME NEW IDENTITIES

In this section, we will develop some identities which involve Riemann-Liouville fractional integral (RLFI) for functions which are n times differentiable.

The Case σ≥1

Theorem 4. Let f real valued absolutely continuous mapping and $\omega(x,.)$: $[a,b]\to\mathbb{R}$, be the fractional Peano type kernel which is given by

$$\omega\left(x,s\right)=\begin{cases}\dfrac{1}{b-a}\left(b-x\right)^{1-\sigma}\Gamma\left(\sigma\right)\left(s-\left(a+\varrho\dfrac{b-a}{2}\right)\right) & if\ a\leq s\leq x\\ \dfrac{1}{b-a}\left(b-x\right)^{1-\sigma}\Gamma\left(\sigma\right)\left(s-\left(b-\varrho\dfrac{b-a}{2}\right)\right) & if\ x<s\leq b\end{cases}\tag{19}$$

for all

$$x\in\left[a+\varrho\frac{b-a}{2},b-\varrho\frac{b-a}{2}\right]$$

and $\varrho\in[0,1]$, then for $\sigma\geq1$.

$$\begin{aligned}J_a^{\sigma}\left(\omega\left(x,b\right)f'\left(b\right)\right)&=\left(1-\varrho\right)f\left(x\right)+\frac{\varrho}{2}\left(b-x\right)^{1-\sigma}\left(b-a\right)^{\sigma-1}f\left(a\right)\\&\quad-\frac{\left(b-x\right)^{1-\sigma}}{b-a}\Gamma\left(\sigma\right)J_a^{\sigma}\left(f\left(b\right)\right)+J_a^{\sigma-1}\left(\omega\left(x,b\right)f\left(b\right)\right).\end{aligned}\tag{20}$$

Proof. Using (19), we find that

$$\begin{aligned} J_a^{\sigma}\left(\omega\left(x,b\right)f'\left(b\right)\right) &= \frac{1}{\Gamma\left(\sigma\right)}\int_a^b \left(b-s\right)^{\sigma-1}\omega\left(x,s\right)f'\left(s\right)ds \\ &= \frac{\left(b-x\right)^{1-\sigma}}{b-a}\int_a^x \left(b-s\right)^{\sigma-1}\left(s-\left(a+\varrho\frac{b-a}{2}\right)\right)f'\left(s\right)ds \\ &+ \frac{\left(b-x\right)^{1-\sigma}}{b-a}\int_x^b \left(b-s\right)^{\sigma-1}\left(s-\left(b-\varrho\frac{b-a}{2}\right)\right)f'\left(s\right)ds. \end{aligned} \tag{21}$$

$\int ( )\ (( ))$

After simplification, we get the desired identity given in (20).

The Case σ>2

Theorem 5. Let f be a real valued twice differentiable function with f'' integrable on $[a,b]$, then for

$$x \in \left[a+\varrho\frac{b-a}{2}, b-\varrho\frac{b-a}{2}\right],$$

we get

$$\begin{aligned} J_a^{\sigma}\left(\omega\left(x,b\right)f''\left(b\right)\right) &= \left(1-\varrho\right)\left(\sigma-1\right)\left(b-x\right)^{-1}f\left(x\right) \\ &-2\frac{\left(b-x\right)^{1-\sigma}}{b-a}\Gamma\left(\sigma\right)J_a^{\sigma-1}\left(f\left(b\right)\right)+J_a^{\sigma-2}\left(\omega\left(x,b\right)f\left(b\right)\right)+\frac{\varrho}{2} \\ &\times\left[\left(b-x\right)^{1-\sigma}\left(b-a\right)^{\sigma-1}f'\left(a\right)+\left(\sigma-1\right)\left(b-x\right)^{1-\sigma}\left(b-a\right)^{\sigma-2}f\left(a\right)\right] \\ &+\left(b-x\right)^{1-\sigma}\left(b-a\right)^{\sigma-2}f\left(a\right)-\left(1-\varrho\right)f'\left(x\right). \end{aligned} \tag{22}$$

Proof. If σ>2, and $f'': [a,b] \to \mathbb{R}$ integrable on $[a,b]$, then we have

$$\begin{aligned}
&\underset{a}{J^{\sigma}}\left(\omega\left(x,b\right)f''\left(b\right)\right)=\frac{1}{\Gamma\left(\sigma\right)}\int_{a}^{b}\left(b-s\right)^{\sigma-1}\omega\left(x,s\right)f''\left(s\right)ds\\
&=\frac{\left(b-x\right)^{1-\alpha}}{b-a}\left[\left(b-x\right)^{\alpha-1}\left(x-\left(a+\varrho\frac{b-a}{2}\right)\right)f'\left(x\right)+\frac{\varrho\left(b-a\right)^{\alpha}f'\left(a\right)}{2}\right.\\
&+\left(\alpha-1\right)\left(x-\left(a+\varrho\frac{b-a}{2}\right)\right)\left(b-x\right)^{\alpha-2}f\left(x\right)\\
&+\frac{\varrho}{2}\left(\alpha-1\right)\left(b-a\right)^{\alpha-1}f\left(a\right)-\left(b-x\right)^{\alpha-1}\times\left(x-\left(b-\varrho\frac{b-a}{2}\right)\right)f'\left(x\right)-\int_{a}^{b}\left(b-s\right)^{\alpha-1}f'\left(s\right)ds\\
&-\left(\alpha-1\right)\left(b-x\right)^{\alpha-2}\left(x-\left(b-\varrho\frac{b-a}{2}\right)\right)f\left(x\right)-\left(\alpha-1\right)\int_{a}^{b}\left(b-s\right)^{\alpha-2}f\left(s\right)ds+\left(\alpha-1\right)\left(\alpha-2\right)\\
&\left.\times\left\{\int_{a}^{x}\left(s-\left(a+\varrho\frac{b-a}{2}\right)\right)\left(b-s\right)^{\alpha-3}f\left(s\right)ds\int_{x}^{b}\left(b-s\right)^{\alpha-3}\left(s-\left(b-\varrho\frac{b-a}{2}\right)\right)f\left(s\right)ds\right\}\right].
\end{aligned}$$

Applying elementary analysis operations, we obtain the inequality (22).

The Case σ>3

Theorem 6. Let f be a three times differentiable real valued function with

$$x\in\left[a+\varrho\frac{b-a}{2},b-\varrho\frac{b-a}{2}\right].$$

The third derivative of f is integrable on $[a,b]$, then

$$\begin{aligned}
&\underset{a}{J^{\sigma}}\left(\omega\left(x,b\right)f'''\left(b\right)\right)=\left(1-\varrho\right)\left[f''\left(x\right)+\left(\sigma-1\right)\left(b-x\right)^{-1}f'\left(x\right)\right.\\
&\left.+\left(\sigma-1\right)\left(\sigma-2\right)\left(b-x\right)^{-2}f\left(x\right)\right]+\frac{\varrho}{2}\left(b-x\right)^{1-\sigma}\left[\left(b-a\right)^{\sigma-1}f''\left(a\right)+\left(\sigma-1\right)\left(b-a\right)^{\sigma-2}f'\left(a\right)\right.\\
&\left.+\left(\sigma-1\right)\left(\sigma-2\right)\left(b-a\right)^{\sigma-3}f\left(a\right)\right]+\left(b-x\right)^{1-\sigma}\left(b-a\right)^{\sigma-2}f'\left(a\right)\\
&+2\left(\sigma-1\right)\left(b-x\right)^{1-\sigma}\left(b-a\right)^{\sigma-3}f\left(a\right)-3\frac{\left(b-x\right)^{1-\sigma}}{b-a}\Gamma\left(\sigma\right)\underset{a}{J^{\sigma-2}}\left(f\left(b\right)\right)+\underset{a}{J^{\sigma-3}}\left(\omega\left(x,b\right)f\left(b\right)\right).
\end{aligned}\tag{23}$$

Proof. For σ>3, we have

$$\Gamma\left(\sigma\right)\underset{a}{J^{\sigma}}\left(\omega\left(x,b\right)f'''\left(b\right)\right)=\frac{1}{\Gamma\left(\sigma\right)}\int_{a}^{b}\left(b-s\right)^{\sigma-1}\omega\left(x,s\right)f'''\left(s\right)ds.\tag{23.1}$$

If we use the equality (1) and also apply similar methods in the case σ>2 we obtain the equality (23). Hence, the proof is completed.

Similarly, we obtain the following identity for the *n*th derivative.

The Case σ>n

Theorem 7. Let f be a real valued function which is n- times differentiable with $f^{(n)}$ also integrable on $[a,b]$ then

$$\begin{aligned} \underset{a}{J^{\sigma}}\left(\omega(x,b)f^{(n)}(b)\right) &= (1-\varrho)\sum_{k=0}^{n-1}\frac{\Gamma(\sigma)(b-x)^{-(n-k-1)}}{\Gamma\left(\sigma-(n-k-1)\right)}f^{(k)}(x) \\ &+\frac{\varrho}{2}(b-x)^{1-\sigma}\sum_{k=0}^{n-1}\frac{\Gamma(\sigma)(b-a)^{\sigma-n+k}}{\Gamma\left(\sigma-(n-k-1)\right)}f^{(k)}(a) \\ &+(b-x)^{1-\sigma}\Gamma(\sigma)\sum_{k=0}^{n-1}\frac{k(b-a)^{\sigma-k-1}}{\Gamma(\sigma-k+1)}f^{(n-k-1)}(a) \\ &-\frac{(b-x)^{1-\sigma}}{b-a}\Gamma(\sigma)n\underset{a}{J^{\sigma-n+1}}f(b)+\underset{a}{J^{\sigma-n}}\left(\omega(x,b)f(b)\right), \end{aligned} \tag{24}$$

where σ>*n*∈*N*, and

$$x\in\left[a+\varrho\frac{b-a}{2},b-\varrho\frac{b-a}{2}\right]$$

$$(\ )\sum^{(\ )}$$

Proof. By integrating by parts, we find that

$$\begin{aligned} &\underset{a}{J^{\sigma}}\left(\omega(x,b)f^{(n)}(b)\right)=\frac{1}{\Gamma(\sigma)}\int_a^b(b-s)^{\sigma-1}\omega(x,s)f^{(v)}(s)\,ds \\ &=\frac{(b-x)^{1-\sigma}}{b-a}\left[\int_a^x(b-s)^{\sigma-1}\left(s-\left(a+\varrho\frac{b-a}{2}\right)\right)f^{(n)}(s)\,ds+\int_x^b(b-s)^{\sigma-1}\left(s-\left(b-\varrho\frac{b-a}{2}\right)\right)f^{(n)}(s)\,ds\right] \\ &=(1-\varrho)f^{(n-1)}(x)+\frac{\varrho}{2}(b-x)^{1-\sigma}(b-a)^{\sigma-1}f^{(n-1)}(a)-\frac{(b-x)^{1-\sigma}}{b-a}\Gamma(\sigma)\underset{a}{J^{\sigma}}f^{(n-1)}(b)+\underset{a}{J^{\sigma-1}}\left(\omega(x,b)f^{(n-1)}(b)\right). \end{aligned}$$

We observe that

$$J_a^{\sigma-1}\left(\omega(x,b)f^{(n-1)}(b)\right)=(\sigma-1)(1-\varrho)(b-x)^{-1}f^{(n-2)}(x)$$
$$+\frac{\varrho}{2}(\sigma-1)(b-x)^{1-\sigma}(b-a)^{\sigma-2}f^{(n-2)}(a)-\frac{(b-x)^{1-\sigma}}{b-a}\Gamma(\sigma)J_a^{\sigma-1}f^{(n-2)}(b)+J_a^{\sigma-2}\left(\omega(x,b)f^{(n-2)}(b)\right),$$

and

$$\begin{aligned}J_a^{\sigma-2}\left(\omega(x,b)f^{(n-2)}(b)\right)&=(\sigma-1)(\sigma-2)(1-\varrho)(b-x)^{-2}f^{(n-3)}(x)\\&+\frac{\varrho}{2}(\sigma-1)(\sigma-2)(b-x)^{1-\sigma}(b-a)^{\sigma-3}f^{(n-3)}(a)\\&-\frac{(b-x)^{1-\sigma}}{b-a}\Gamma(\sigma)J_a^{\sigma-2}f^{(n-3)}(b)+J_a^{\sigma-3}\left(\omega(x,b)f^{(n-3)}(b)\right).\end{aligned}$$

$$\begin{aligned}J_a^{\sigma-(n-1)}\left(\omega(x,b)f'(b)\right)&=(\sigma-1)(\sigma-2)\cdots(\sigma-(n-1))(1-\varrho)(b-x)^{-(n-1)}f(x)\\&+\frac{\varrho}{2}(\sigma-1)(\sigma-2)\cdots(\sigma-(n-1))(b-x)^{1-\sigma}(b-a)^{\sigma-n}f(a)\\&-\frac{(b-x)^{1-\sigma}}{b-a}\Gamma(\sigma)J_a^{\sigma-(n-1)}f(b)+J_a^{\sigma-n}\left(\omega(x,b)f(b)\right).\end{aligned}$$

This gives us the identity

$$\begin{aligned}J_a^{\sigma}\left(\omega(x,b)f^{(n)}(b)\right)&=(1-\varrho)\sum_{k=0}^{n-1}\frac{\Gamma(\sigma)}{\Gamma(\sigma-(n-k-1))}(b-x)^{-(n-k-1)}f^{(k)}(x)\\&+\frac{\varrho}{2}(b-x)^{1-\sigma}\sum_{k=0}^{n-1}\frac{\Gamma(\sigma)(b-a)^{\sigma-n+k}}{\Gamma(\sigma-(n-k-1))}f^{(k)}(a)\\&-\frac{(b-x)^{1-\sigma}}{b-a}\Gamma(\sigma)\sum_{k=0}^{n-1}J_a^{\sigma-k}f^{(n-k-1)}(b)+J_a^{\sigma-n}\left(\omega(x,b)f(b)\right).\end{aligned}\tag{25}$$

Also, we have

$$\sum_{k=0}^{n-1}J_a^{\sigma-k}f^{(n-k-1)}(b)=-\sum_{k=1}^{n-1}\frac{k(b-a)^{\sigma-k}}{\Gamma(\sigma-k+1)}f^{(n-k-1)}(a)+nJ_a^{\sigma-n+1}f(b).\tag{26}$$

If we substitute (25) in (26), we obtain the desired result given in (24).
Note: By using different values of ϱ in Theorem 4, 5, 6, 7, 8 and 9, we get the variety of results.

AN OSTROWSKI TYPE FRACTIONAL INEQUALITY-II

We now use (20) developed in section 3, to obtain Ostrowski type fractional inequality involving Riemann Liouville Fractional Integral (RLFI).

Theorem 8. Let f be a real valued differentiable function on (b,c) such that $f'(c)\in L_1(b,c)$. If $|f'(x)|$ is bounded by M, for every $b\leq x\leq c$ then the following inequality holds:

$$\Bigg|(1-\varrho)(c-x)^0 f(x)-\frac{(c-x)^{1-\sigma}}{c-b}\Gamma(\sigma)J_b^{\sigma}(f(c))$$
$$+J_b^{\sigma-1}(\omega(x,c)f(c))+\frac{\varrho}{2}(c-x)^{1-\sigma}(c-b)^{\sigma-1}f(b)\Bigg|\leq\frac{M(c-x)^{1-\sigma}}{c-b}\Bigg\{-\frac{(c-b)^{\sigma+1}}{\sigma+1}\left(\left(1-\frac{\varrho}{2}\right)^{\sigma+1}-1\right)$$
$$+\frac{(c-b)^{\sigma+1}}{\sigma}\left(\left(1-\frac{\varrho}{2}\right)^{\sigma+1}-\left(1-\frac{\varrho}{2}\right)\right)+\frac{1}{\sigma+1}\left[(c-x)^{\sigma+1}-(c-b)^{\sigma+1}\left(1-\frac{\varrho}{2}\right)^{\sigma+1}\right]$$
$$-\frac{1}{\sigma}(c-b)\left[\left(1-\frac{\varrho}{2}\right)(c-x)^{\sigma}-(c-b)^{\sigma}\left(1-\frac{\varrho}{2}\right)^{\sigma+1}\right]-\frac{1}{\sigma+1}\left(\left(\frac{c-b}{2}\right)^{\sigma+1}-(c-x)^{\sigma+1}\right)-\frac{1}{\sigma+1}\left(\frac{c-b}{2}\right)^{\sigma+1}$$
$$\times\left(\varrho^{\sigma+1}-1\right)+\left(\varrho\frac{c-b}{2}\right)\left(\left(\frac{c-b}{2}\right)^{\sigma}-(c-x)^{\sigma}\right)+\frac{\varrho}{\sigma}\left(\frac{c-b}{2}\right)^{\sigma+1}\left(\varrho^{\sigma}-1\right)+\frac{1}{\sigma(\sigma+1)}\left(\varrho\frac{c-b}{2}\right)^{\sigma+1}\Bigg\} \tag{27}$$

or

$$\left|(1-\varrho)(c-x)^0 f(x)-\frac{(c-x)^{1-\sigma}}{c-b}\Gamma(\sigma)J_b^{\sigma}(f(c))+J_b^{\sigma-1}(\omega(x,c)f(c))+\frac{\varrho}{2}(c-x)^{1-\sigma}(c-b)^{\sigma-1}f(b)\right|$$
$$\leq\frac{M(c-x)^{1-\sigma}}{c-b}\Bigg\{-\frac{1}{\sigma+1}\left[\left(1-\frac{\varrho}{2}\right)^{\sigma+1}-1\right](c-b)^{\sigma+1}+\frac{1}{\sigma}\left[\left(1-\frac{\varrho}{2}\right)^{\sigma+1}-\left(1-\frac{\varrho}{2}\right)\right](c-b)^{\sigma+1}$$
$$+\frac{1}{\sigma+1}(c-b)^{\sigma+1}\left[\left(\frac{1}{2}\right)^{\sigma+1}-\left(1-\frac{\varrho}{2}\right)^{\sigma+1}\right]-\frac{(c-b)^{\sigma+1}}{\sigma}\left[\left(1-\frac{\varrho}{2}\right)\left(\frac{1}{2}\right)^{\sigma}-\left(1-\frac{\varrho}{2}\right)^{\sigma+1}\right]$$
$$+\frac{1}{\sigma+1}\left[(c-x)^{\sigma+1}-\left(\frac{c-b}{2}\right)^{\sigma+1}\right]-\frac{c-b}{\sigma}\left(1-\frac{\varrho}{2}\right)\times\left[(c-x)^{\sigma}-\left(\frac{c-b}{2}\right)^{\sigma}\right]$$
$$-\frac{1}{\sigma+1}\left[\left(\varrho\frac{c-b}{2}\right)^{\sigma+1}-(c-x)^{\sigma+1}\right]+\left(\varrho\frac{c-b}{2}\right)\times\left[\left(\varrho\frac{c-b}{2}\right)^{\sigma}-(c-x)^{\sigma}\right]+\frac{1}{\sigma(\sigma+1)}\left(\varrho\frac{c-b}{2}\right)^{\sigma+1}\Bigg\}. \tag{28}$$

Proof. From Theorem 4, we have

$$\left|(1-\varrho)(c-x)^{0} f(x)-\frac{(c-x)^{1-\sigma}}{c-b}\Gamma(\sigma) J_{b}^{\sigma}(f(c))\right.$$
$$\left.+J_{b}^{\sigma-1}(\omega(x,c) f(c))+\frac{\varrho}{2}(c-x)^{1-\sigma}(c-b)^{\sigma-1} f(b)\right|=\left|J_{b}^{\sigma}(\omega(x,c) f'(c))\right|.$$

From (1) and (21), we get

$$\frac{1}{\Gamma(\sigma)}\left|\int_{b}^{c}(c-s)^{\sigma-1}\omega(x,s) f'(s)\,ds\right| \leq \frac{1}{\Gamma(\sigma)}\int_{b}^{c}(c-s)^{\sigma-1}\left|\omega(x,s)\right|\left|f'(s)\right| ds$$
$$\leq \frac{M}{\Gamma(\sigma)}\int_{b}^{c}(c-s)^{\sigma-1}\left|\omega(x,s)\right| ds \leq \frac{M(c-x)^{1-\sigma}}{c-b}\left[\int_{b}^{x}\left|(c-s)^{\sigma-1}\left(s-\left(b+\varrho\frac{c-b}{2}\right)\right)\right| ds\right. \qquad (29$$
$$\left.+\int_{x}^{b}\left|(c-s)^{\sigma-1}\left(\left(c-\varrho\frac{c-b}{2}\right)-s\right)\right| ds\right].$$

We have two cases:

- For $x \in \left[b+\varrho\frac{c-b}{2}, \frac{b+c}{2}\right]$ and ϱ∈[0,1]

$$\int_{b}^{x}\left|(c-s)^{\sigma-1}\left(s-\left(b+\varrho\frac{c-b}{2}\right)\right)\right| ds+\int_{x}^{c}\left|(c-s)^{\sigma-1}\left(\left(c-\varrho\frac{c-b}{2}\right)-s\right)\right| ds$$
$$=\left[\int_{b}^{b+\varrho\frac{c-b}{2}}(c-s)^{\sigma-1}\left(b+\varrho\frac{c-b}{2}-s\right) ds+\int_{b+\varrho\frac{b-c}{2}}^{x}(c-s)^{\sigma-1}\right.$$
$$\times\left(t-\left(b+\varrho\frac{c-b}{2}\right)\right) ds+\int_{x}^{\frac{b+c}{2}}(c-s)^{\sigma-1}\left(c-\varrho\frac{c-b}{2}-s\right) ds \qquad (30)$$
$$\left.+\int_{\frac{b+c}{2}}^{c-\varrho\frac{c-b}{2}}(c-s)^{\sigma-1}\left(c-\varrho\frac{c-b}{2}-s\right) ds+\int_{c-\varrho\frac{c-b}{2}}^{c}(c-s)^{\sigma-1}\left(s-\left(c-\varrho\frac{b-b}{2}\right)\right) ds\right].$$

Now, using (29) and (30), we get (27).
For ϱ=0, we get the Anastassiou's result given in [11].

- For $x \in \left[\frac{b+c}{2}, c-\varrho\frac{c-b}{2}\right]$ and ϱ∈[0,1]

$$\int_b^x \left|\left(c-s\right)^{\sigma-1}\left(s-\left(b+\varrho\,\frac{c-b}{2}\right)\right)\right| ds + \int_x^c \left|\left(c-s\right)^{\sigma-1}\left(\left(c-\varrho\,\frac{c-b}{2}\right)-s\right)\right| ds$$

$$= \left[\int_b^{b+\varrho\frac{c-b}{2}} \left(c-s\right)^{\sigma-1}\left(b+\varrho\,\frac{c-b}{2}-s\right) ds + \int_{b+\varrho\frac{c-b}{2}}^{\frac{b+c}{2}} \left(c-s\right)^{\sigma-1}\left(s-\left(b+\varrho\,\frac{c-b}{2}\right)\right) ds + \int_{\frac{b+c}{2}}^{x} \left(c-s\right)^{\sigma-1}\right.$$

$$\left.\times\left(s-\left(b+\varrho\,\frac{c-b}{2}\right)\right) ds + \int_x^{c-\varrho\frac{c-b}{2}} \left(c-s\right)^{\sigma-1}\left(b-\varrho\,\frac{c-b}{2}-s\right) ds + \int_{c-\varrho\frac{c-b}{2}}^{c} \left(c-s\right)^{\sigma-1}\left(s-\left(c-\varrho\,\frac{c-b}{2}\right)\right) ds\right].$$

Hence, we get (28)

For ϱ=0, again we get Anastassiou's result [11].

AN OSTROWSKI TYPE FRACTIONAL INEQUALITY-III

In this section, we derive Ostrowski type fractional inequality for twice differentiable mappings by using the identity given (22).

Theorem 9. Let the real valued function *f* be twice differentiable on (b,c) such that $f''(c) \in L_1(b,c)$. If $\left|f''(x)\right| \le f''_{\infty}$ for every $x \in [b,c]$ and $\sigma \ge 1$, then

$$\Bigg|\left(1-\varrho\right)\left(\sigma-1\right)\left(c-x\right)^{-1}f\left(x\right)-2\frac{\left(c-x\right)^{1-\sigma}}{c-b}\Gamma\left(\sigma\right)J_{b}^{\sigma-1}\left(f\left(c\right)\right)$$
$$+J_{b}^{\sigma-2}\left(\omega\left(x,c\right)f\left(c\right)\right)+\frac{\varrho}{2}\left\{\left(c-x\right)^{1-\sigma}\left(c-b\right)^{\sigma-1}f'\left(b\right)+\left(\sigma-1\right)\left(c-x\right)^{1-\sigma}\left(c-b\right)^{\sigma-2}f\left(b\right)\right\}$$
$$+\left(c-x\right)^{1-\sigma}\left(c-b\right)^{\sigma-2}f\left(b\right)+\left(1-\varrho\right)f'\left(x\right)\Bigg|\leq f_{\infty}''\frac{\left(c-x\right)^{1-\sigma}}{b-b}\left\{-\frac{\left(c-b\right)^{\sigma+1}}{\sigma+1}\left(\left(1-\frac{\varrho}{2}\right)^{\sigma+1}-1\right)\right.$$
$$+\frac{\left(c-b\right)^{\sigma+1}}{\sigma}\left(\left(1-\frac{\varrho}{2}\right)^{\sigma+1}-\left(1-\frac{\varrho}{2}\right)\right)+\frac{1}{\sigma+1}\left(\left(c-x\right)^{\sigma+1}-\left(c-b\right)^{\sigma+1}\left(1-\frac{\varrho}{2}\right)^{\sigma+1}\right) \tag{31}$$
$$-\frac{1}{\sigma}\left(\left(1-\frac{\varrho}{2}\right)\left(c-b\right)\left(c-x\right)^{\sigma}-\left(c-b\right)^{\sigma+1}\left(1-\frac{\varrho}{2}\right)^{\sigma+1}\right)-\frac{1}{\sigma+1}\left(\left(\frac{c-b}{2}\right)^{\sigma+1}-\left(c-x\right)^{\sigma+1}\right)$$
$$+\frac{1}{\sigma}\left(\varrho\frac{c-b}{2}\right)\times\left(\left(\frac{c-b}{2}\right)^{\sigma}-\left(c-x\right)^{\sigma}\right)-\frac{1}{\sigma+1}\left(\left(\varrho\frac{c-b}{2}\right)^{\sigma+1}-\left(\frac{c-b}{2}\right)^{\sigma+1}\right)$$
$$\left.+\frac{1}{\sigma}\left(\varrho\frac{c-b}{2}\right)\left(\left(\varrho\frac{c-b}{2}\right)^{\sigma}-\left(\frac{c-b}{2}\right)^{\sigma}\right)-\frac{1}{\sigma+1}\left(\varrho\frac{c-b}{2}\right)^{\sigma+1}+\frac{1}{\sigma}\left(\varrho\frac{c-b}{2}\right)^{\sigma+1}\right\}$$

or

$$\Bigg|\left(1-\varrho\right)\left(\sigma-1\right)\left(c-x\right)^{-1}f\left(x\right)-2\frac{\left(c-x\right)^{1-\sigma}}{c-b}\Gamma\left(\sigma\right)J_{b}^{\sigma-1}\left(f\left(c\right)\right)$$
$$+J_{b}^{\sigma-2}\left(\omega\left(x,c\right)f\left(c\right)\right)+\frac{\varrho}{2}\left[\left(c-x\right)^{1-\sigma}\left(c-b\right)^{\sigma-1}f'\left(b\right)+\left(\sigma-1\right)\left(c-x\right)^{1-\sigma}\left(c-b\right)^{\sigma-2}f\left(b\right)\right]$$
$$+\left(c-x\right)^{1-\sigma}\left(c-b\right)^{\sigma-2}f\left(b\right)+\left(1-\varrho\right)f'\left(x\right)\Bigg|\leq f_{\infty}''\frac{\left(c-x\right)^{1-\sigma}}{c-b}\left\{-\frac{1}{\sigma+1}\left(\left(1-\frac{\varrho}{2}\right)^{\sigma+1}-1\right)\left(c-b\right)^{\sigma+1}\right.$$
$$+\frac{1}{\sigma}\left(\left(1-\frac{\varrho}{2}\right)^{\sigma+1}-\left(1-\frac{\varrho}{2}\right)\right)\left(c-b\right)^{\sigma+1}+\frac{1}{\sigma+1}\left(c-b\right)^{\sigma+1}\left(\left(\frac{1}{2}\right)^{\sigma+1}-\left(1-\frac{\varrho}{2}\right)^{\sigma+1}\right) \tag{32}$$
$$-\frac{1}{\sigma}\left(c-b\right)^{\sigma+1}\left(\left(1-\frac{\varrho}{2}\right)\left(\frac{1}{2}\right)^{\sigma}-\left(1-\frac{\varrho}{2}\right)^{\sigma+1}\right)+\frac{1}{\sigma+1}\left(\left(c-x\right)^{\sigma+1}-\left(\frac{c-b}{2}\right)^{\sigma+1}\right)-\frac{1}{\sigma}\left(1-\frac{\varrho}{2}\right)$$
$$\times\left(c-b\right)\left(\left(c-x\right)^{\sigma}-\left(\frac{c-b}{2}\right)^{\sigma}\right)-\frac{1}{\sigma+1}\times\left(\left(\varrho\frac{c-b}{2}\right)^{\sigma+1}-\left(c-x\right)^{\sigma+1}\right)$$
$$\left.+\frac{1}{\sigma}\left(\varrho\frac{c-b}{2}\right)\left(\left(\varrho\frac{c-b}{2}\right)^{\sigma}-\left(c-x\right)^{\sigma}\right)-\frac{1}{\sigma+1}\left(\varrho\frac{c-b}{2}\right)^{\sigma+1}+\frac{1}{\sigma}\left(\varrho\frac{c-b}{2}\right)\left(\varrho\frac{c-b}{2}\right)^{\sigma}\right\}$$

holds.

Proof. From (22), we get

$$\Bigg|(1-\varrho)(\sigma-1)(c-x)^{-1}f(x)-2\frac{(c-x)^{1-\sigma}}{c-b}\Gamma(\sigma)J_b^{\sigma-1}(f(c))$$
$$+J_b^{\sigma-2}(\omega(x,c)f(c))+\frac{\varrho}{2}\left\{(c-x)^{1-\sigma}(c-b)^{\sigma-1}f'(b)+(\sigma-1)(c-x)^{1-\sigma}(c-b)^{\sigma-2}f(b)\right\}$$
$$+(c-x)^{1-\sigma}(c-b)^{\sigma-2}f(b)+(1-\varrho)f'(x)\Bigg|=\left|J_b^{\sigma}(\omega(x,c)f''(c))\right|.$$

Now from (1) and (27), we obtain

$$\frac{1}{``(\sigma)}\left|\int_b^c(c-s)^{\sigma-1}\omega(x,s)f''(s)\,ds\right|$$
$$\leq\frac{1}{\Gamma(\sigma)}\int_b^c(c-s)^{\sigma-1}|\omega(x,s)||f''(t)|\,ds\leq\frac{f''_\infty}{\Gamma(\sigma)}\int_b^c(c-s)^{\sigma-1}|\omega(x,s)|\,ds$$
$$\leq f''_\infty\frac{(c-x)^{1-\sigma}}{c-b}\left\{\int_b^x\left|(c-s)^{\sigma-1}\left(s-\left(b+\varrho\frac{c-b}{2}\right)\right)\right|ds+\int_x^c\left|(c-s)^{\sigma-1}\left(\left(b-\varrho\frac{c-b}{2}\right)-s\right)\right|ds\right\}.$$

After simplification above integrations, we get (31) and (32).

REFERENCES

Ahmad, W. M., & El-Khazali, R. (2007). Fractional-order Dynamical Models of Love. *Chaos, Solitons, and Fractals*, *33*(4), 1367–1375. doi:10.1016/j.chaos.2006.01.098

Ahmad, W. M., & El-Khazali, R. (2007). Fractional-order dynamical models of love. *Chaos, Solitons, and Fractals*, *33*(4), 1367–1375. doi:10.1016/j.chaos.2006.01.098

Ahmed, E., & Elgazzar, A. S. (2007). On Fractional Order Differential Equations Model for Nonlocal Epidemics. *Physica A*, *379*(2), 607–614. doi:10.1016/j.physa.2007.01.010

Al-Khedhairi, A., Askar, S. S., Matouk, A. E., Elsadany, A., & Ghazel, M. (2018). Dynamics, chaos control, and synchronization in a fractional-order Samardzija-Greller population system with order lying in (0,2). *Complexity*.

Al-khedhairi, A., Matouk, A. E., & Askar, S. S. (2019a). Computations of synchronization conditions in some fractional-order chaotic and hyperchaotic systems. *Pramana Journal of Physics, 92*, 72.

Al-khedhairi, A., Matouk, A. E., & Khan, I. (2019b). Chaotic dynamics and chaos control for the fractional-order geomagnetic field model. *Chaos, Solitons, and Fractals*, *128*, 1–12. doi:10.1016/j.chaos.2019.07.019

Ali, H. M. (2019). *New approximate solutions to fractional smoking model using the generalized Mittag-Leffler function method*. doi:10.18576/pfda/010101

Ameen, I., & Novati, P. (2017). The solution of fractional order epidemic model by implicit Adams methods. *Applied Mathematical Modelling*, *43*, 78–84. doi:10.1016/j.apm.2016.10.054

Anastassiou, G., Hooshmandas, M. R., Ghasemi, A., & Moftakharzadeh, F. (2009). Montgomery identities for fractional integrals and related fractional inequalities. *Journal of Inequalities in Pure and Applied Mathematics*, *10*(4), 97.

Bagley, R. L., & Calico, R. A. (1991). Fractional order State Equations for the Control of Viscoelastically Damped Structures. *Journal of Guidance, Control, and Dynamics*, *14*(2), 304–311. doi:10.2514/3.20641

Ben Adda, F. (1997). Geometric interpretation of the fractional derivative. *J. Fract. Calc*, *11*, 21–52.

Caputo, M. (1967). Linear models of dissipation whose Q is almost frequency independent-II. *Geophysical Journal of the Royal Astronomical Society*, *13*(5), 529–539. doi:10.1111/j.1365-246X.1967.tb02303.x

Chen, L., Chai, Y., & Wu, R. (2011). Control and Synchronization of Fractional-Order Financial System Based on Linear Control. *Discrete Dynamics in Nature and Society*, *2011*, 1–21. doi:10.1155/2011/958393

Chen, W. C. (2008). Nonlinear dynamics and chaos in a fractional-order financial system. *Chaos, Solitons, and Fractals*, *36*(5), 1305–1314. doi:10.1016/j.chaos.2006.07.051

Dragomir, S. S. (2005). *An Ostrowski type inequality for convex functions*. Univerzitet u Beogradu Publikacije Elektrotehni ckog Fakulteta. Serija Matematika, 16.

El-Sayed, A. M. (1996). Fractional-order Diffusion-wave Equation. *International Journal of Theoretical Physics*, *35*(2), 311–322. doi:10.1007/BF02083817

Elsadany, A. A., & Matouk, A. E. (2015). Dynamical behaviors of fractional-order Lotka– Volterra predator–prey model and its discretization. *Journal of Applied Mathematics and Computing*, *49*(1-2), 269–283. doi:10.100712190-014-0838-6

Heaviside, O. (1971). Electromagnetic theory. New York: Chelsea.

Hegazi, A. S., Ahmed, E., & Matouk, A. E. (2011). The effect of fractional order on synchronization of two fractional order chaotic and hyperchaotic systems. *J. Fract. Calc. Appl.*, *1*(3), 1–15.

Hegazi, A. S., Ahmed, E., & Matouk, A. E. (2013a). On chaos control and synchronization of the commensurate fractional order Liu system. *Communications in Nonlinear Science and Numerical Simulation*, *18*(5), 1193–1202. doi:10.1016/j.cnsns.2012.09.026

Hegazi, A. S., Ahmed, E., & Matouk, A. E. (2013b). On Dynamical Behaviors and Chaos Control of the Fractional-order Financial System. In S. Banerjee (Ed.), *Chaos and Complexity Theory for Management: Nonlinear Dynamics* (pp. 34–49). IGI Global. doi:10.4018/978-1-4666-2509-9.ch002

Hegazi, A. S., & Matouk, A. E. (2011). Dynamical behaviors and synchronization in the fractional order hyperchaotic Chen system. *Applied Mathematics Letters, 24*(11), 1938–1944. doi:10.1016/j.aml.2011.05.025

Hegazi, A. S., & Matouk, A. E. (2013). Chaos control of the modified Autonomous Van der Pol-Duffing circuits via active control. In S. Banerjee & L. Rondoni (Eds.), *Applications of Chaos and Nonlinear Dynamics in Science and Engineering* (Vol. 3, pp. 185–202). Berlin: Springer. doi:10.1007/978-3-642-34017-8_7

Hilfer, R. (Ed.). (2000). *Applications of fractional calculus in physics*. Upper Saddle River, NJ: World Scientific. doi:10.1142/3779

Ichise, M., Nagayanagi, Y., & Kojima, T. (1971). An analog simulation of noninteger order transfer functions for analysis of electrode process. *J. Electroanal. Chem., 33*(2), 253–265. doi:10.1016/S0022-0728(71)80115-8

Khan, A & Kumar, S. (2016). Anti-synchronization of chaotic finance systems using fuzzy logic constant controller. *Indian Journal of Industrial and Applied Mathematics*. Doi:10.5958/1945-919X.2016.00013X

Khan, A., & Kumar, S. (2018). Study of chaos in satellite systems. *Pramana Journal of Physics, 90*(1), 13. doi:10.100712043-017-1502-0

Khan, A & Kumar, S. (2019a). T-S fuzzy modeling and predictive control and synchronization of chaotic satellite systems. *International Journal of Modeling and Simulation, 39*(3).

Kusnezov, D., Bulgac, A., & Dang, G. D. (1999). Quantum Levy Processes and Fractional Kinetics. *Physical Review Letters, 82*(6), 1136–1139. doi:10.1103/PhysRevLett.82.1136

Laskin, N. (2000). Fractional Market Dynamics. *Physica A, 287*(3-4), 482–492. doi:10.1016/S0378-4371(00)00387-3

Mainardi, F. (2010). *Fractional Calculus and Waves in Linear Viscoelasticity: An Introduction to Mathematical Models*. London: Imperial College Press. doi:10.1142/p614

Matouk, A. E. (2009). Stability Conditions, Hyperchaos and control in a novel fractional order hyperchaotic system. *Physics Letters. [Part A], 373*(25), 2166–2173. doi:10.1016/j.physleta.2009.04.032

Matouk, A. E. (2009a). Chaos synchronization between two different fractional systems of Lorenz family. *Mathematical Problems in Engineering, 2009*, 1–11. doi:10.1155/2009/572724

Matouk, A. E. (2009b). Stability conditions, hyper chaos and control in a novel fractional order hyper chaotic system. *Physics Letters. [Part A], 373*(25), 2166–2173. doi:10.1016/j.physleta.2009.04.032

Matouk, A. E. (2010). Dynamical behaviors, linear feedback control and synchronization of the fractional order Liu system. *Journal of Nonlinear Systems and Applications, 1*(3), 135–140.

Matouk, A. E. (2011a). Chaos, feedback control and synchronization of a fractional-order modified Autonomous Van der Pol-Duffing circuit. *Communications in Nonlinear Science and Numerical Simulation, 16*(2), 975–986. doi:10.1016/j.cnsns.2010.04.027

Matouk, A. E. (2011b). Synchronization in integer and fractional order chaotic systems. In S. Banerjee (Ed.), *Chaos Synchronization and Cryptography for Secure Communications: Applications for Encryption* (pp. 127–151). IGI Global. doi:10.4018/978-1-61520-737-4.ch007

Matouk, A. E., & Elsadany, A. A. (2016). Dynamical analysis, stabilization and discretization of a chaotic fractional-order GLV model. *Nonlinear Dynamics, 85*(3), 1597–1612. doi:10.100711071-016-2781-6

Matouk, A. E., Elsadany, A. A., Ahmed, E., & Agiza, H. N. (2015). Dynamical behavior of fractional-order Hastings–Powell food chain model and its discretization. *Communications in Nonlinear Science and Numerical Simulation, 27*(1-3), 153–167. doi:10.1016/j.cnsns.2015.03.004

Mitrinović, D. S., Pecarić, J. E., & Fink, A. M. (1994). *Inequalities for Functions and Their Integrals and Derivatives*. Dordrecht: Kluwer Academic Publishers.

Ostrowski, A. M. (1938). Uber die Absolutabweichung einer di erentienbaren Funktionen von ihren Integralimittelwert. *Commentarii Mathematici Helvetici, 10*(1), 226–227. doi:10.1007/BF01214290

Petras, I. (2009). Chaos in the fractional-order Volta's system: Modeling and simulation. *Nonlinear Dynamics, 57*(1-2), 157–170. doi:10.100711071-008-9429-0

Podlubny, I. (1999). *Fractional differential equations: an introduction to fractional derivatives, fractional differential equations, to methods of their solution and some of their applications*. New York: Academic Press.

Qayyum, A., Faye. I.& Shoaib. M. (2015). On new generalized inequalities via Riemann-Liouville Fractional integration. *Journal of Fractional Calculus and Applications, 6*(1), 91-100.

Qayyum, A., Faye. I., Shoaib. M. & Erden, S. (2019). Generalized fractional Ostrowski type inequalities for higher order derivatives. *Communication in Mathematical Modeling and Application, 4*(2).

Samet, B., & Aydi, H. (2018). On Some Inequalities Involving Liouville–Caputo Fractional Derivatives and Applications to Special Means of Real Numbers. *Mathematics, 6*(10), 193–202. doi:10.3390/math6100193

Saqib, M., Ali, F., Khan, I., Sheikh, N. A., & Khan, A. (2018a). Entropy Generation in Different Types of Fractionalized Nanofluids. *Arabian Journal for Science and Engineering*, 1–10.

Saqib, M., Ali, F., Khan, I., Sheikh, N. A., & Shafie, S. B. (2018b). Convection in ethylene glycol-based molybdenum disulfide nanofluid. *Journal of Thermal Analysis and Calorimetry*, 1–10.

Saqib, M., Khan, I., & Shafie, S. (2018c). Application of Atangana–Baleanu fractional derivative to MHD channel flow of CMC-based-CNT's nanofluid through a porous medium. *Chaos, Solitons, and Fractals, 116*, 79–85. doi:10.1016/j.chaos.2018.09.007

Saqib, M., Khan, I., & Shafie, S. (2018d). Natural convection channel flow of CMC-based CNTs nanofluid. *The European Physical Journal Plus, 133*(12), 549. doi:10.1140/epjp/i2018-12340-3

Sarikaya, M. Z., & Ogunmez, H. (2012). On new inequalities via Riemann-Liouville Fractional integration. Abstract and Applied Analysis, Article ID 428983. doi:10.1155/2012/428983

Sarikaya, M. Z., & Yaldiz, H. (2013). On weighted montogomery identities for Riemann-Luville fractional integral. *Konuralp Journal of Mathematics*, *1*(1), 48–53.

Selvam, G. M., Janagaraj, A., & S, B. J. (2017). Stability in a Discrete Fractional Order Prey Predator System with Functional Response. *International Journal of Advanced Research in Computer Science and Software Engineering*, *7*(5), 630–633. doi:10.23956/ijarcsse/SV7I5/0336

Selvam, G. M., & Janagaraj, A., R. (2018a). Numerical Analysis of a Fractional Order Discrete Prey – Predator System with Functional Response. *International Journal of Engineering & Technology*, *7*(4.10), 681–684.

Selvam, G. M., & Janagaraj, A. R. (2018b). Allee effect and Holling type - II response in a discrete fractional order prey - predator model. IOP Conference Series. *Journal of Physics*, *1139*, 1–7.

Song, L., Xu, S., & Yang, J. (2010). Dynamical Models of Happiness with Fractional Order. *Communications in Nonlinear Science and Numerical Simulation*, *15*(3), 616–628. doi:10.1016/j.cnsns.2009.04.029

Sun, H. H., Abdelwahab, A. A., & Onaral, B. (1984). Linear approximation of transfer function with a pole of fractional order. *IEEE Transactions on Automatic Control*, *29*(5), 441–444. doi:10.1109/TAC.1984.1103551

Taghavian, H., & Tavazoei, M. S. (2017). Robust Stability Analysis of Distributed-Order Linear Time-Invariant Systems With Uncertain Order Weight Functions and Uncertain Dynamic Matrices. *Journal of Dynamic Systems, Measurement, and Control*, *139*(12), 121010. doi:10.1115/1.4037268

Taghavian, H., & Tavazoei, M. S. (2018). Exact solution of linear fractional distributed order systems with exponential order weight functions. In *Mathematical Techniques of Fractional Order Systems* (pp. 103–132). Elsevier Publishing. doi:10.1016/B978-0-12-813592-1.00004-0

Taghavian, H., & Tavazoei, M. S. (2019). Algebraic Conditions for Stability Analysis of Linear Time-Invariant Distributed Order Dynamic Systems: A Lagrange Inversion Theorem Approach. *Asian Journal of Control*, *21*(2), 879–890. doi:10.1002/asjc.1780

Zhen, W., Xia, H., & Guodong, S. (2011). Analysis of nonlinear dynamics and chaos in a fractional order financial system with time delay. *Computers & Mathematics with Applications (Oxford, England)*, *62*(3), 1531–1539. doi:10.1016/j.camwa.2011.04.057

Chapter 12
The Invariant Subspace Method for Solving Fractional Partial Differential Equations

Mohamed Soror Abdel Latif
https://orcid.org/0000-0002-9013-1017
Mansoura University, Egypt

Abass Hassan Abdel Kader
Mansoura University, Egypt

ABSTRACT

In this chapter, the authors discuss the effectiveness of the invariant subspace method (ISM) for solving fractional partial differential equations. For this purpose, they have chosen a nonlinear time fractional partial differential equation (PDE) with variable coefficients to be investigated through this method. One-, two-, and three-dimensional invariant subspace classifications have been performed for this equation. Some new exact solutions have been obtained using the ISM. Also, the authors give a comparison between this method and the homogeneous balance principle (HBP).

INTRODUCTION

Nonlinear PDEs plays a very important rule in modeling many phenomena in science and engineering. So, many methods have been proposed for getting exact solutions of PDEs. Such as, Lie symmetry method (Olver, 2000; Abdel Kader et al, 2017; El-Bialy et al, 2017; Abdel Kader et al, 2018; Abdel Latif et al, 2018; Abdel Latif et al, 2018a), the extended tanh function method (Fan, 2000; Abdel Latif, 2011), the Hirota bilinear method (Hirota, 2004), the HBP (Rui, 2017) and the ISM (Galaktionov & Svirshchevskii, 2007; Ma & Liu, 2012).

Fractional-order modeling in the form of fractional PDES is better than integer-order modeling of biological (Ahmed & Elgazzar, 2007; Elsadany & Matouk, 2015; Matouk et al, 2015; Matouk & Elsadany, 2016; Al-Khedhairi et al, 2018), economic (Laskin, 2000), financial (Chen, 2008; Chen et al, 2011; Zhen

DOI: 10.4018/978-1-7998-3122-8.ch012

et al, 2011; Hegazi et al, 2013b), psychological (Ahmad & El-Khazali, 2007), physical (Heaviside, 1971; El-Sayed, 1996; Kusnezov et al, 1999; Hilfer, 2000; Al-Khedhairi et al, 2019a; Al-Khedhairi et al, 2019b) and engineering systems (Sun et al, 1984; Petras, 2009; Matouk, 2009a; Matouk, 2009b; Matouk 2010; Matouk, 2011a; Matouk, 2011b; Hegazi & Matouk, 2011; Hegazi et al, 2011; Hegazi & Matouk, 2013; Hegazi et al, 2013a; Matouk & Elsadany, 2014; Matouk, 2016; El-Sayed et al, 2016a; El-Sayed et al, 2016b) where memory effects are very important. Furthermore, fractional derivatives have also useful applications in some interdisciplinary fields of science (Elsaid et al, 2016; Elsaid et al, 2016a; Elsaid et al, 2016b; Ichise et al, 1971; Bagley & Torvik, 1983; Bagley & Calico, 1991; Ben Adda F, 1997; Magin, 2006; Tavazoei et al, 2009; Tavazoei, 2010; Mainardi, 2010; Latif et al, 2013).

The ISM can be also used for getting exact solutions of fractional PDEs. Recently, it is extensively used for getting exact solutions of fractional PDEs (Sahadevan & Parakash, 2017; Jiang et al, 2018). In this chapter, we investigate the effectiveness of the ISM and make a comparison between it and the HBP.

INVARIANT SUBSPACE METHOD (SAHADEVAN & PARAKASH, 2017)

The ISM can be used for solving the following fractional PDE with order n

$$ {}_0D_t^{\alpha} u = F\left[u\right] = H\left(x, u, u_x, u_{xx}, \ldots\right). \tag{1} $$

The ISM can be summarized in the following steps:

Step 1: assume the solution of Equation (1) in the form

$$ u\left(x,t\right) = \sum_{i=1}^{j} f_i\left(t\right)\varphi_i\left(x\right), \tag{2} $$

where j depends upon the dimension of invariant subspace ($j \leq 2n+1$).

Step 2: determine the functions $\varphi_i(x)$ as follows:

Solve the determining equation

$$ \left(\frac{d^j}{dx^j} + c_{j-1}\left(x\right)\frac{d^{j-1}}{dx^{j-1}} + \ldots + c_1\left(x\right)\frac{d}{dx} + c_0\left(x\right)\right)F\left[y\left(x\right)\right] = 0, \tag{3} $$

to determine the coefficients $c_0\left(x\right), \ldots, c_{j-1}\left(x\right)$.

Solve the ordinary differential equation

$$\left(\frac{d^j}{dx^j}+c_{j-1}\left(x\right)\frac{d^{j-1}}{dx^{j-1}}+\ldots+c_1\left(x\right)\frac{d}{dx}+c_0\left(x\right)\right)y=0, \tag{4}$$

to obtain the solution

$$y=\sum_{i=1}^{j}C_i\varphi_i\left(x\right), \tag{5}$$

where, C_i, i=1,…,k are arbitrary constants. Hence, we obtain the functions $\varphi_i(x)$.

Step 3: substitute Equation (2) into Equation (1), to get

$${}_0D_t^{\alpha}\left(\sum_{i=1}^{j}f_i\left(t\right)\varphi_i\left(x\right)\right)=F\left[\sum_{i=1}^{j}f_i\left(t\right)\varphi_i\left(x\right)\right]=\sum_{i=1}^{j}\omega_i\left(t\right)\varphi_i\left(x\right),$$

or, equivalently

$$\sum_{i=1}^{j}{}_0D_t^{\alpha}\left(f_i\left(t\right)\right)\varphi_i\left(x\right)=\sum_{i=1}^{j}\omega_i\left(t\right)\varphi_i\left(x\right),$$

which implies that

$${}_0D_t^{\alpha}\left(f_i\left(t\right)\right)=\omega_i\left(t\right), i=1,\ldots,j. \tag{6}$$

Step 4: solving the system of fractional ordinary differential equations (ODEs) (6), we obtain the solution of Equation (1).

In the following sections, we examine the following example using the ISM:

Example 1: Consider a nonlinear time fractional PDE with diffusion term as follows (Rui, 2017)

$${}_0D_t^{\alpha}u=f'\left(u\right)u_x^2+f\left(u\right)u_{xx}+\delta u^2 \tag{7}$$

where u=$u(t,x)$, the $f(u)$ is diffusion term and δ is a constant. In Equation (7) (In most cases) the fractional derivative ${}_0D_t^{\alpha}u$ is defined in the Riemann-Liouville sense. The fractional Riemann-Liouville derivative is given by

$${}_0D_t^\alpha f(t) = \frac{1}{\Gamma(1-\alpha)}\left(\frac{d}{dt}\right)\int_0^t (t-\tau)^{-\alpha} f(\tau)\,d\tau, 0<\alpha<1.$$

In only one case (given in Table 2), ${}_0D_t^\alpha u$ will be in the Caputo sense which is defined by

$${}_0D_t^\alpha f(t) = \frac{1}{\Gamma(1-\alpha)}\int_0^t (t-\tau)^{-\alpha} f'(\tau)\,d\tau, 0<\alpha<1.$$

One-Dimensional Invariant Subspace Classification of Equation (7)

Here, Equation (4) is given by

$$y' + c_0(x)y = 0. \tag{8}$$

Also,

$$F[y] = f'(y)(y')^2 + f(y)y'' + \delta y^2. \tag{9}$$

Substituting Equation (9) and Equation (8) into Equation (3) (with j=1), we get

$$f(y)c_0'' + c_0\left(y\left(\delta - 3c_0' f'(y)\right) - 2f(y)c_0'\right) + yc_0^3\left(yf''(y) + 2f'(y)\right) = 0, \tag{10}$$

Solutions of Equation (10) are given in the following cases:

Case 1: $c_0(x) = \sqrt{\frac{2\delta}{q_1}}\tan\left(\sqrt{\frac{\delta}{2q_1}}x + r_1\right), f(y) = \frac{1}{4}q_1 y,$

Case 2: $c_0(x) = \sqrt{\frac{\delta}{2q_1}}\tan\left(\sqrt{\frac{2\delta}{q_1}}x + r_2\right), f(y) = q_1 y,$

Case 3: $c_0(x) = s_1, f(y) = \frac{-\delta}{2s_1^2}y + r_4 + \frac{r_5}{y},$

Case 4: $c_0(x) = \frac{c_2}{x + c_1}, f(y) = c_3 y^{-1-\frac{1}{c_2}} + c_4 y^{-2/c_2}, \delta = 0,$

where r_i, c_i, q_i, s_i, i=1,2,…,5, are constants.

For each case, solving Equation (4), we get the one dimensional invariant subspaces $\{\varphi_1\}$ of Equation (7) which are given in Table 1.

Table 1. One dimensional invariant subspaces for Equation (7)

Case	φ1
1	$\cos^2\left(\sqrt{\frac{\delta}{2q_1}}x + r_1\right)$
2	$\sqrt{\cos\left(\sqrt{\frac{2\delta}{q_1}}x + r_2\right)}$
3	$e^{-s_1 x}$
4	$\left(c_1 + x\right)^{-c_2}$

The solution of Equation (7) can be expressed in the form

$$u(x,t) = L(t)\varphi_1. \tag{11}$$

Substituting Equation (11) into Equation (7), we obtain the determining equation

$${}_0D_t^{\alpha} L(t)\varphi_1 = f'(u)L(t)^2 \varphi_{1x}^2 + f(u)L(t)\varphi_{1_{xx}} + \delta L(t)^2 \varphi_1^2 \tag{12}$$

For each case, if we substitute $f(u)$ and φ_1 into Equation (12), we get some determining equations. These determining equations and their solutions are shown in Table 2.

Table 2. The determining equations (12) and their solutions for each case

Case	Determining Equations	Solution
1	${}_0D_t^{\alpha} L(t) = \frac{3\delta}{4} L(t)^2$	$L(t) = \frac{4\Gamma(1-\alpha)}{3\delta\Gamma(1-2\alpha)} t^{-\alpha}$
2	${}_0D_t^{\alpha} L(t) = 0$	For Caputo differential operator, $L(t)=C$
3	${}_0D_t^{\alpha} L(t) = r_4 s_1^2 L(t)$	$L(t) = r_6 t^{\alpha-1} E_{\alpha,\alpha}\left(r_4 s_1^2 t^{\alpha}\right)$
4	${}_0D_t^{\alpha} L(t) = \left(-1 + c_2\right) c_2 c_4 L(t)^{1-\frac{2}{c_2}}$	$L(t) = t^{\frac{\alpha c_2}{2}} \left(\left(c_2 - 1\right) c_2 c_4 \frac{\Gamma\left(\frac{c_2\alpha}{2} + 1 - \alpha\right)}{\Gamma\left(\frac{c_2\alpha}{2} + 1\right)}\right)^{c_2/2}$

Hence, we obtain the following solutions of Equation (7)

$$u_1\left(x,t\right)=\frac{4\Gamma\left(1-\alpha\right)}{3\delta\Gamma\left(1-2\alpha\right)}t^{-\alpha}\cos^2\left(\sqrt{\frac{\delta}{2q_1}}x+r_1\right),$$

$$u_2\left(x,t\right)=C\sqrt{\cos\left(\sqrt{\frac{2\delta}{q_1}}x+r_2\right)},$$

$$u_3\left(x,t\right)=r_6t^{\alpha-1}E_{\alpha,\alpha}\left(r_4s_1^2t^{\alpha}\right)e^{-s_1x},$$

$$u_4\left(x,t\right)=t^{\frac{\alpha c_2}{2}}\left(\left(c_2-1\right)c_2c_4\frac{\Gamma\left(\frac{c_2\alpha}{2}-\alpha+1\right)}{\Gamma\left(\frac{c_2\alpha}{2}+1\right)}\right)^{c2/2}\left(c_1+x\right)^{-c2}.$$

Two-Dimensional Invariant Subspace Classification of Equation (7)

In this case, Equation (4) takes the form

$$y''+c_1\left(x\right)y'+c_0\left(x\right)y=0. \tag{13}$$

Substituting Equation (9) and Equation (13) into Equation (3) (with j=2), we obtain

$$\begin{aligned}&-yf\left(y\right)c_0''+y'\left(-4yc_0'f'\left(y\right)-2f\left(y\right)c_0'-7yc_0c_1f'\left(y\right)-f\left(y\right)c_1''+2c_1f\left(y\right)c_1'\right)\\&+\left(y'\right)^2\left(6yc_0f''\left(y\right)-3c_0f'\left(y\right)-4c_1'f'\left(y\right)+4c_1^2f'\left(y\right)+2\delta\right)-2yc_0f\left(y\right)c_1'\\&+3y^2c_0^2f'\left(y\right)+3\delta y^2c_0-5c_1f''\left(y\right)\left(y'\right)^3+f^{(3)}\left(y\right)\left(y'\right)^4=0.\end{aligned} \tag{14}$$

Equating the coefficients of $y',\left(y'\right)^2,\ldots$ with zero, we get the following overdetermined system of ODEs

$$f^{(3)}\left(y\right)=0,$$

$$c_1f''\left(y\right)=0,$$

$$6y c_0 f''(y) - 3c_0 f'(y) - 4c_1' f'(y) + 4c_1^2 f'(y) + 2\delta = 0,$$
$$-f(y)(2c_0' + c_1'' - 2c_1 c_1') - yf'(y)(4c_0' + 7c_0 c_1) = 0,$$

$$3y^2 c_0 (c_0 f'(y) + \delta) - yf(y)(c_0'' + 2c_0 c_1') = 0. \tag{15}$$

The system (15) can be solved as given in the following cases

Case 1: $c_0(x) = c_1(x) = \delta = 0, f(y) = b_2 y^2 + b_3 y + b_4,$
Case 2: $c_0(x) = c_1(x) = \delta = 0, f(y) = b_2 y^2 + b_3 y + b_4,$
Case 3: $c_0(x) = 0, c_1(x) = -c_8, f(y) = \dfrac{-\delta}{2c_8^2} y + c_9,$

where c_i, i=5,…,9, are constants. For each case, solving Equation (13), we get the two dimensional invariant subspaces $\{\varphi_1, \varphi_2\}$ of Equation (7) which are shown in Table 3.

Table 3. Two dimensional invariant subspaces for Equation (7)

Case	φ_1	φ_2
1	1	x
2	1	$\sin(c_5 x + c_6)$
3	1	$e^{c_8 x}$

The solution of Equation (7) can be expressed in the form

$$u(x,t) = L(t)\varphi_1 + G(t)\varphi_2. \tag{16}$$

Substituting Equation (16) into Equation (7), we get the determining equation

$$\begin{aligned} {}_0D_t^\alpha L(t)\varphi_1 + {}_0D_t^\alpha G(t)\varphi_2 &= f'(u)\left(L(t)\varphi_{1_x} + G(t)\varphi_{2_x}\right)^2 \\ &+ f(u)\left(L(t)\varphi_{1_{xx}} + G(t)\varphi_{2_{xx}}\right) + \delta\left(L(t)\varphi_1 + G(t)\varphi_2\right)^2. \end{aligned} \tag{17}$$

For each case, if we substitute $f(u)$ and φ_1, φ_2 into Equation (17), we get some determining equations. These determining equations and their solutions are shown in Table 4.

Hence, we obtain the solutions of Equation (3) as follows:

Table 4. The determining equations (17) and their solutions for each case

Case	Determining Equations	Solution
1	$_0D_t^\alpha G(t)=2b_2G(t)^3$ $_0D_t^\alpha L(t)=b_3G(t)^2+2b_2L(t)G(t)^2$	$G(t)=\frac{\pm1}{\sqrt{2b_2}}\sqrt{\frac{\Gamma\left(1-\frac{\alpha}{2}\right)}{\Gamma\left(1-\frac{3\alpha}{2}\right)}}t^{\frac{-\alpha}{2}}$ $L(t)=\frac{b_3\Gamma(1-\alpha)\Gamma\left(1-\frac{\alpha}{2}\right)}{2b_2\left(\Gamma\left(1-\frac{3\alpha}{2}\right)-\Gamma(1-\alpha)\Gamma\left(1-\frac{\alpha}{2}\right)\right)}$ For b_3=0, $L(t)=c_{11}t^{-\frac{\alpha}{2}}$
2	$_0D_t^\alpha L(t)=\delta L(t)^2+\frac{\delta}{2}G(t)^2$ $_0D_t^\alpha G(t)=\left(-c_5^2c_7+\frac{3}{2}\delta L(t)\right)G(t)$	For c_7=0, $L(t)=\frac{2\Gamma(1-\alpha)}{3\delta\Gamma(1-2\alpha)}t^{-\alpha}$ $G(t)=\pm\frac{2\Gamma(1-\alpha)}{3\delta\Gamma(1-2\alpha)}t^{-\alpha}$
3	$_0D_t^\alpha L(t)=\delta L(t)^2$ $_0D_t^\alpha G(t)=c_8^2c_9G(t)+\frac{3}{2}\delta L(t)G(t)$	*For* c_9=0 $L(t)=\frac{\Gamma(1-\alpha)}{\delta\Gamma(1-2\alpha)}t^{-\alpha}$ $G(t)=c_{10}t^{\gamma_1}$ *With constraint:* $\frac{3\Gamma(1-\alpha)}{\Gamma(1-2\alpha)}-\frac{2\Gamma(\gamma_1+1)}{\Gamma(\gamma_1-\alpha+1)}=0.$

$$u_6(x,t)=c_{11}t^{-\frac{\alpha}{2}}+\frac{\pm1}{\sqrt{2b_2}}\sqrt{\frac{\Gamma\left(1-\frac{\alpha}{2}\right)}{\Gamma\left(1-\frac{3\alpha}{2}\right)}}t^{\frac{-1}{2}\alpha}x\ ,$$

where c_{11} is a constant.

$$u_7(x,t)=\frac{2\Gamma(1-\alpha)}{3\delta\Gamma(1-2\alpha)}t^{-\alpha}\pm\frac{2\Gamma(1-\alpha)}{3\delta\Gamma(1-2\alpha)}t^{-\alpha}\sin(c_5x+c_6).$$

$$u_8(x,t)=\frac{\Gamma(1-\alpha)}{\delta\Gamma(1-2\alpha)}t^{-\alpha}+c_{10}t^{\gamma_1}e^{c_8x},$$

where γ_1 is the roots of

$$\frac{3\Gamma(1-\alpha)}{\Gamma(1-2\alpha)}-\frac{2\Gamma(\gamma_1+1)}{\Gamma(\gamma_1-\alpha+1)}=0.$$

Three-Dimensional Invariant Subspace Classification of Equation (7)

In this case, Equation (4) is given by

$$y''' + c_2(x)y'' + c_1(x)y' + c_0(x)y = 0. \tag{18}$$

Substituting Equation (9) and Equation (18) into Equation (3) (with j=3), we get

$$\begin{aligned}
&-yf(y)c_0'' + y'(-5yc_0'f'(y) - 2f(y)c_0' + yc_0c_2f'(y) - f(y)c_1'' + 2c_1f(y)c_2' \\
&-12c_1f'(y)y'' - 5c_2'f'(y)y'' + c_2^2f'(y)y'' + 15f''(y)(y'')^2 + 6\delta y'') \\
&+(y')^2(-10yc_0f''(y) - 4c_0f'(y) - 5c_1'f'(y) + c_1c_2f'(y) - 4c_2f''(y)y'' + 2\delta c_2) \\
&+2yc_0f(y)c_2' - 10yc_0f'(y)y'' - \delta y^2c_0 - 2f(y)c_1'y'' + (y')^3\left(10f^{(3)}(y)y'' - 9c_1f''(y)\right) \\
&-f(y)c_2''y'' + 2c_2f(y)c_2'y'' + c_2f^{(3)}(y)(y')^4 - 7c_2f'(y)(y'')^2 + f^{(4)}(y)(y')^5 = 0.
\end{aligned} \tag{19}$$

Equating the coefficients of $y', (y')^2, \ldots, y'', (y'')^2$, with zero, we get overdetermined system of ODEs which has the following solutions

Case 1: $c_0(x) = c_2(x) = 0, c_1(x) = \dfrac{\delta}{2b_1}, f(y) = b_1y + b_2$

Case 2: $c_0(x) = c_1(x) = c_2(x) = \delta = 0, f(y) = c_{11}y,$

Case 3: $c_0(x) = c_2(x) = 0, c_1(x) = -c_{12}, \delta = -2c_{12}c_{13}, f(y) = c_{13}y,$

where c_{12}, c_{13} are constants. For each case, solving Equation (18), we get the three dimension invariant subspaces $\{\varphi_1, \varphi_2, \varphi_3\}$ of Equation (7) which are shown in Table 5.

Table 5. Three dimensional invariant subspaces for Equation (7)

Case	φ_1	φ_2	φ_3
1	1	$\cos\left(\sqrt{\dfrac{\delta}{2b_1}}x\right)$	$\sin\left(\sqrt{\dfrac{\delta}{2b_1}}x\right)$
2	1	x	x^2
3	1	$e^{\sqrt{c_{12}}\,x}$	$e^{-\sqrt{c_{12}}\,x}$

The solution of Equation (7) can be expressed in the form

$$u(x,t) = L(t)\varphi_1 + G(t)\varphi_2 + V(t)\varphi_3. \tag{20}$$

Substituting Equation (20) into Equation (7), we obtain the determining equation

$$\begin{aligned}&{}_0D_t^\alpha L(t)\varphi_1 + {}_0D_t^\alpha G(t)\varphi_2 + {}_0D_t^\alpha V(t)\varphi_3 = f'(u)\left(L(t)\varphi_{1_x} + G(t)\varphi_{2_x} + V(t)\varphi_{3_x}\right)^2 \\ &+f(u)\left(L(t)\varphi_{1_{xx}} + G(t)\varphi_{2_{xx}} + V(t)\varphi_{3_{xx}}\right) + \delta\left(L(t)\varphi_1 + G(t)\varphi_2 + V(t)\varphi_3\right)^2.\end{aligned} \tag{21}$$

For each case, if we substitute $f(u)$ and φ_1, φ_2, φ_3 into Equation (21), we obtain determining equations. These determining equations and their solutions are shown in Table 6.

Hence, we obtain the solutions of Equation (3) as follows:

$$u_9(x,t) = \frac{2\Gamma(1-\alpha)}{3\delta\Gamma(1-2\alpha)}t^{-\alpha} + b_3 t^{-\alpha}\cos\left(\sqrt{\frac{\delta}{2b_1}}x\right) \pm \frac{1}{3\delta}\sqrt{4\frac{\Gamma(1-\alpha)^2}{\Gamma(1-2\alpha)^2} - 9\delta^2 b_3^2\, t^{-\alpha}}\,\sin\left(\sqrt{\frac{\delta}{2b_1}}x\right),$$

Table 6. The determining equations (21) and their solutions for each case

Case	Determining Equations	Solution
1	${}_0D_t^\alpha L(t) = \delta L(t)^2 + \frac{\delta}{2}G(t)^2 + \frac{\delta}{2}V(t)^2$ ${}_0D_t^\alpha G(t) = \delta\left(-\frac{b_2}{2b_1} + \frac{3}{2}L(t)\right)G(t)$ ${}_0D_t^\alpha V(t) = \delta\left(-\frac{b_2}{2b_1} + \frac{3}{2}L(t)\right)V(t)$	For $b_2=0$, $L(t) = \frac{2\Gamma(1-\alpha)}{3\delta\Gamma(1-2\alpha)}t^{-\alpha}$ $G(t)=b_3t^{-\alpha}$ $V(t) = \pm\frac{1}{3\delta}\sqrt{4\frac{\Gamma(1-\alpha)^2}{\Gamma(1-2\alpha)^2} - 9\delta^2 b_3^2\, t^{-\alpha}}$
2	${}_0D_t^\alpha L(t) = c_{11}G(t)^2 + 2c_{11}V(t)L(t)$ ${}_0D_t^\alpha G(t) = 6c_{11}G(t)V(t)$ ${}_0D_t^\alpha V(t) = 6c_{11}V(t)^2$	$L(t) = \frac{3c_{11}b_3^2\Gamma(1-2\alpha)}{2\Gamma(1-\alpha)}t^{-\alpha}$ $V(t) = \frac{\Gamma(1-\alpha)}{6c_{11}\Gamma(1-2\alpha)}t^{-\alpha}$
3	${}_0D_t^\alpha L(t) = -2c_{12}c_{13}\left(2G(t)V(t) + L(t)^2\right)$ ${}_0D_t^\alpha G(t) = -3c_{12}c_{13}G(t)L(t)$ ${}_0D_t^\alpha V(t) = -3c_{12}c_{13}V(t)L(t)$	$L(t) = \frac{-\Gamma(1-\alpha)}{3c_{12}c_{13}\Gamma(1-2\alpha)}t^{-\alpha}$ $G(t) = \frac{\Gamma(1-\alpha)^2}{36b_4c_{12}^2c_{13}^2\Gamma(1-2\alpha)^2}t^{-\alpha}$ $V(t)=b_4t^{\alpha}$

$$u_{10}\left(x,t\right)=\frac{3c_{11}b_3^2\Gamma\left(1-2\alpha\right)}{2\Gamma\left(1-\alpha\right)}t^{-\alpha}+b_3t^{-\alpha}x+\frac{\Gamma\left(1-\alpha\right)}{6c_{11}\Gamma\left(1-2\alpha\right)}t^{-\alpha}x^2,$$

$$u_{11}\left(x,t\right)=\frac{-\Gamma\left(1-\alpha\right)}{3c_{12}c_{13}\Gamma\left(1-2\alpha\right)}t^{-\alpha}+\frac{\Gamma\left(1-\alpha\right)^2}{36b_4c_{12}^2c_{13}^2\Gamma\left(1-2\alpha\right)^2}t^{-\alpha}e^{\sqrt{c_{12}}\,x}+b_4t^{-\alpha}e^{-\sqrt{c_{12}}\,x},$$

where b_3,b_4 are constants.

The Homogeneous Balanced Method (Rui, 2017)

In this section, we introduce the HBP for solving fractional partial differential equations. The homogeneous balanced method can be summarized in the following steps

Step 1: the solution of Equation (1) can be expressed as

$$u=\sum_{i=0}^{n}a_ib\left(t\right)\left(c\left(x\right)\right)^i, \tag{22}$$

where a_i, $i=0,1,\ldots,n$ are some constants to be determined later, $b(t)$ can be chosen from the following three functions:

- The power function t^{d_i}, where d_i, $i=0,1,\ldots,n$ are some constants to be determined from the computations in the following steps.
- The Mittag-Leffler function with one parameter $E_\alpha(\delta t^\alpha)$.
- The product of the power function and the Mittag-Leffler function with two parameter (i.e. $t^{\beta-1}E_{\alpha,\beta}\left(\delta t^\alpha\right)$).

The function $c(x)$ can be chosen from the transcendental functions.

Step 2: the value of n can be computed from the balance between the highest power of the function $c(x)$ (after choosing it) in the two sides of Equation (1).

Step 3: substitute Equation (22) into Equation (1) and compare both sides of Equation (1), to get the values of a_i, $i=0,1,\ldots,n$ and d_i, $i=0,1,\ldots,n$.

To make a comparison between the ISM and the HBP we take the same example discussed in the preceding sections. Here we cannot make the classification done using the ISM for the function $f(u)$. We can take only some examples with predefined values of $f(u)$.

Example 2 (Rui, 2017): Consider the following special form of Equation (7) (when $f(u)=\beta u$)

$$ {}_0D_t^\alpha u = \beta u_x^2 + \beta u u_{xx} + \delta u^2. \tag{23} $$

After investigating Equation (23) using the HBP, the following solutions of Equation (23) are obtained [12]:

Case 1: when $\delta=2\beta$

$$ u = \frac{\Gamma(1-\alpha)}{3\beta\Gamma(1-2\alpha)} t^{-\alpha}(1 \pm \sin x), \alpha \neq \frac{1}{2}, $$

$$ u = \frac{\Gamma(1-\alpha)}{3\beta\Gamma(1-2\alpha)} t^{-\alpha}(1 \pm \cos x), \alpha \neq \frac{1}{2}. $$

Case 2: when $\delta= -2\beta$

$$ u = -\frac{\Gamma(1-\alpha)}{2\beta\Gamma(1-2\alpha)} t^{-\alpha} + a_1 t^{d_1} e^x, \alpha \neq \frac{1}{2}, $$

where the value of d_1 can be evaluated from the relation $\frac{\Gamma(d_1+1)}{\Gamma(d_1+1-\alpha)} = \frac{3\Gamma(1-\alpha)}{2\Gamma(1-2\alpha)}$.

Case 3: when $\delta=0$

$$ u = \frac{3\beta a_1^2\Gamma(1-2\alpha)}{2\Gamma(1-\alpha)} t^{-\alpha} + a_1 t^{-\alpha} x + \frac{\Gamma(1-\alpha)}{6\beta\Gamma(1-2\alpha)}, \alpha \neq \frac{1}{2}. $$

CONCLUSION

In this chapter, we have investigated the effectiveness of the ISM for getting exact solutions of nonlinear fractional PDEs. It can be effectively used in making a classification for fractional PDEs with variable coefficients as done in Example 1. Also, it is systematic and does need any guesses. On the other hand, in the HBP, we need some experience in determining the value of n and the function $c(x)$.

REFERENCES

Abdel Kader, A. H., Abdel Latif, M. S., El Bialy, F., & Elsaid, A. (2018). Symmetry analysis and some new exact solutions of some nonlinear KdV-like equations. *Asian-European Journal of Mathematics, 11*(03), 1850040. doi:10.1142/S1793557118500407

Abdel Kader, A. H., Abdel Latif, M. S., & Nour, H. M. (2017). Some New Exact Solutions of the Modified KdV Equation Using Lie Point Symmetry Method. *International Journal of Applied and Computational Mathematics, 3*(1), 1163–1171. doi:10.100740819-017-0408-3

Abdel Latif, M. S. (2011). Some exact solutions of KdV equation with variable coefficients. *Communications in Nonlinear Science and Numerical Simulation, 16*(4), 1783–1786. doi:10.1016/j.cnsns.2010.07.023

Abdel Latif, M. S., El-Shazly, E., Baleanu, D., Elsaid, A., & Nour, H. M. (2018). Some new soliton-like and doubly periodic-like solutions of Fisher equation with time-dependent coefficients. *Modern Physics Letters B, 32*(33), 1850413. doi:10.1142/S0217984918504134

Abdel Latif, M. S., Shazly, E. E., Elsaid, A., & Nour, H. M. (2018a). Group classification and some new periodic-like and soliton-like solutions of the generalised Fisher equation with time-variable coefficients. *International Journal of Dynamical Systems and Differential Equations, 8*(4), 313–325. doi:10.1504/IJDSDE.2018.096218

Ahmad, W. M., & El-Khazali, R. (2007). Fractional-order dynamical models of love. *Chaos, Solitons, and Fractals, 33*(4), 1367–1375. doi:10.1016/j.chaos.2006.01.098

Ahmed, E., & Elgazzar, A. S. (2007). On fractional order differential equations model for nonlocal epidemics. *Physica A, 379*(2), 607–614. doi:10.1016/j.physa.2007.01.010

Al-Khedhairi, A., Askar, S. S., Matouk, A. E., Elsadany, A., & Ghazel, M. (2018). Dynamics, chaos control, and synchronization in a fractional-order Samardzija-Greller population system with order lying in (0,2). *Complexity.*

Al-khedhairi, A., Matouk, A. E., & Askar, S. S. (2019a). Computations of synchronization conditions in some fractional-order chaotic and hyperchaotic systems. *Pramana Journal of Physics, 92*, 72.

Al-khedhairi, A., Matouk, A. E., & Khan, I. (2019b). Chaotic dynamics and chaos control for the fractional-order geomagnetic field model. *Chaos, Solitons, and Fractals, 128*, 1–12. doi:10.1016/j.chaos.2019.07.019

Bagley, R. L., & Calico, R. A. (1991). Fractional order state equations for the control of viscoelastically damped structures. *Journal of Guidance, Control, and Dynamics, 1991*(14), 304–311. doi:10.2514/3.20641

Bagley, R. L., & Torvik, P. J. (1983). A theoretical basis for the application of fractional calculus to viscoelasticity. *Journal of Rheology (New York, N.Y.), 27*(3), 201–210. doi:10.1122/1.549724

Ben Adda, F. (1997). Geometric interpretation of the fractional derivative. *J. Fract. Calc, 11*, 21–52.

Chen, L., Chai, Y., & Wu, R. (2011). Control and Synchronization of Fractional-Order Financial System Based on Linear Control. *Discrete Dynamics in Nature and Society, 2011*, 1–21. doi:10.1155/2011/958393

Chen, W. C. (2008). Nonlinear dynamics and chaos in a fractional-order financial system. *Chaos, Solitons, and Fractals*, *36*(5), 1305–1314. doi:10.1016/j.chaos.2006.07.051

El-Bialy, F., Abdel Latif, M. S., & Elsaid, A. (2017). Symmetry analysis and some new exact solutions of the (2+ 1)-dimensional Burgers equations. *Acta Physica Polonica B*, *48*(11), 2031–2043. doi:10.5506/APhysPolB.48.2031

El-Sayed, A. M. A. (1996). Fractional-order diffusion-wave equation. *International Journal of Theoretical Physics*, *35*(2), 311–322. doi:10.1007/BF02083817

El-Sayed, A. M. A., Elsonbaty, A., Elsadany, A. A., & Matouk, A. E. (2016b). Dynamical analysis and circuit simulation of a new fractional-order hyperchaotic system and its discretization. International Journal of Bifurcation and Chaos, 26. doi:10.1142/S0218127416502229

El-Sayed, A. M. A., Nour, H. M., Elsaid, A., Matouk, A. E., & Elsonbaty, A. (2016a). Dynamical Behaviors, Circuit Realization, Chaos Control and Synchronization of a New Fractional Order Hyperchaotic System. *Applied Mathematical Modelling*, *40*(5-6), 3516–3534. doi:10.1016/j.apm.2015.10.010

Elsadany, A. A., & Matouk, A. E. (2015). Dynamical behaviors of fractional-order Lotka– Volterra predator–prey model and its discretization. *Journal of Applied Mathematics and Computing*, *49*(1-2), 269–283. doi:10.100712190-014-0838-6

Elsaid, A., Abdel Latif, M. S., & Maneea, M. (2016). Similarity solutions of fractional order heat equations with variable coefficients. *Miskolc Mathematical Notes*, *17*(1), 245–254. doi:10.18514/MMN.2016.1610

Elsaid, A., Latif, A., & Maneea, M. (2016b). Similarity solutions for multiterm time-fractional diffusion equation. *Advances in Mathematical Physics*, *2016*, 2016. doi:10.1155/2016/7304659

Elsaid, A., Latif, M. S. A., & Maneea, M. (2016a). Similarity Solutions for Solving Riesz Fractional Partial Differential Equations. *Progress in Fractional Differentiation and Applications*, *2*(4), 293-298.

Fan, E. (2000). Extended tanh-function method and its applications to nonlinear equations. *Physics Letters. [Part A]*, *277*(4-5), 212–218. doi:10.1016/S0375-9601(00)00725-8

Galaktionov, V., & Svirshchevskii, S. (2007). *Exact Solutions and Invariant Subspaces of Nonlinear Partial Differential Equations in Mechanics and Physics*. London: Chapman & Hall/CRC.

Heaviside, O. (1971). Electromagnetic theory. New York: Chelsea.

Hegazi, A. S., Ahmed, E., & Matouk, A. E. (2011). The effect of fractional order on synchronization of two fractional order chaotic and hyperchaotic systems. *J. Fract. Calc. Appl.*, *1*(3), 1–15.

Hegazi, A. S., Ahmed, E., & Matouk, A. E. (2013a). On chaos control and synchronization of the commensurate fractional order Liu system. *Communications in Nonlinear Science and Numerical Simulation*, *18*(5), 1193–1202. doi:10.1016/j.cnsns.2012.09.026

Hegazi, A. S., Ahmed, E., & Matouk, A. E. (2013b). On Dynamical Behaviors and Chaos Control of the Fractional-order Financial System. In S. Banerjee (Ed.), *Chaos and Complexity Theory for Management: Nonlinear Dynamics* (pp. 34–49). IGI Global. doi:10.4018/978-1-4666-2509-9.ch002

Hegazi, A. S., & Matouk, A. E. (2011). Dynamical behaviors and synchronization in the fractional order hyperchaotic Chen system. *Applied Mathematics Letters, 24*(11), 1938–1944. doi:10.1016/j.aml.2011.05.025

Hegazi, A. S., & Matouk, A. E. (2013). Chaos control of the modified Autonomous Van der Pol-Duffing circuits via active control. In S. Banerjee & L. Rondoni (Eds.), *Applications of Chaos and Nonlinear Dynamics in Science and Engineering* (Vol. 3, pp. 185–202). Berlin: Springer. doi:10.1007/978-3-642-34017-8_7

Hilfer, R. (Ed.). (2000). *Applications of fractional calculus in physics*. Upper Saddle River, NJ: World Scientific. doi:10.1142/3779

Hirota, R. (2004). *The direct method in soliton theory* (Vol. 155). Cambridge University Press. doi:10.1017/CBO9780511543043

Ichise, M., Nagayanagi, Y., & Kojima, T. (1971). An analog simulation of noninteger order transfer functions for analysis of electrode process. *J. Electroanal. Chem., 33*(2), 253–265. doi:10.1016/S0022-0728(71)80115-8

Jiang, J., Feng, Y., & Li, S. (2018). Exact solutions to the fractional differential equations with mixed partial derivatives. *Axioms, 7*(1), 10. doi:10.3390/axioms7010010

Kusnezov, D., Bulgac, A., & Dang, G. D. (1999). Quantum levy processes and fractional kinetics. *Physical Review Letters, 82*(6), 1136–1139. doi:10.1103/PhysRevLett.82.1136

Laskin, N. (2000). Fractional market dynamics. *Physica A, 287*(3-4), 482–492. doi:10.1016/S0378-4371(00)00387-3

Latif, M. A., Dragomir, S. S., & Matouk, A. E. (2013). New inequalities of Ostrowski type for Co-ordinated s-Convex functions via fractional integrals. *Journal of Fractional Calculus and Applications, 4*(1), 22-36.

Ma, W. X., & Liu, Y. (2012). Invariant subspaces and exact solutions of a class of dispersive evolution equations. *Communications in Nonlinear Science and Numerical Simulation, 17*(10), 3795–3801. doi:10.1016/j.cnsns.2012.02.024

Magin, R. L. (2006). *Fractional Calculus in Bioengineering*. Begell House.

Mainardi, F. (2010). *Fractional Calculus and Waves in Linear Viscoelasticity: An Introduction to Mathematical Models*. London: Imperial College Press. doi:10.1142/p614

Matouk, A. E. (2009a). Chaos synchronization between two different fractional systems of Lorenz family. *Mathematical Problems in Engineering, 2009*, 1–11. doi:10.1155/2009/572724

Matouk, A. E. (2009b). Stability conditions, hyperchaos and control in a novel fractional order hyperchaotic system. *Physics Letters. [Part A], 373*(25), 2166–2173. doi:10.1016/j.physleta.2009.04.032

Matouk, A. E. (2010). Dynamical behaviors, linear feedback control and synchronization of the fractional order Liu system. *Journal of Nonlinear Systems and Applications, 1*(3), 135–140.

Matouk, A. E. (2011a). Chaos, feedback control and synchronization of a fractional-order modified Autonomous Van der Pol-Duffing circuit. *Communications in Nonlinear Science and Numerical Simulation, 16*(2), 975–986. doi:10.1016/j.cnsns.2010.04.027

Matouk, A. E. (2011b). Synchronization in integer and fractional order chaotic systems. In S. Banerjee (Ed.), *Chaos Synchronization and Cryptography for Secure Communications: Applications for Encryption* (pp. 127–151). IGI Global. doi:10.4018/978-1-61520-737-4.ch007

Matouk, A. E. (2016). Chaos synchronization of a fractional-order modified Van der Pol Duffing system via new linear control, backstepping control and Takagi-Sugeno fuzzy approaches. *Complexity, 21*(S1), 116–124. doi:10.1002/cplx.21719

Matouk, A. E., & Elsadany, A. A. (2014). Achieving synchronization between the fractional-order hyperchaotic Novel and Chen systems via a new nonlinear control technique. *Applied Mathematics Letters, 29*, 30–35. doi:10.1016/j.aml.2013.10.010

Matouk, A. E., & Elsadany, A. A. (2016). Dynamical analysis, stabilization and discretization of a chaotic fractional-order GLV model. *Nonlinear Dynamics, 85*(3), 1597–1612. doi:10.100711071-016-2781-6

Matouk, A. E., Elsadany, A. A., Ahmed, E., & Agiza, H. N. (2015). Dynamical behavior of fractional-order Hastings–Powell food chain model and its discretization. *Communications in Nonlinear Science and Numerical Simulation, 27*(1-3), 153–167. doi:10.1016/j.cnsns.2015.03.004

Olver, P. (2000). *Applications of Lie Groups to Differential Equations*. New York: Springer.

Petras, I. (2009). Chaos in the fractional-order Volta's system: Modeling and simulation. *Nonlinear Dynamics, 57*(1-2), 157–170. doi:10.100711071-008-9429-0

Rui, W. (2017). Applications of homogenous balanced principle on investigating exact solutions to a series of time fractional nonlinear PDEs. *Communications in Nonlinear Science and Numerical Simulation, 47*, 253–266. doi:10.1016/j.cnsns.2016.11.018

Sahadevan, R., & Prakash, P. (2017). On Lie symmetry analysis and invariant subspace methods of coupled time fractional partial differential equations. *Chaos, Solitons, and Fractals, 104*, 107–120. doi:10.1016/j.chaos.2017.07.019

Sun, H. H., Abdelwahab, A. A., & Onaral, B. (1984). Linear approximation of transfer function with a pole of fractional order. *IEEE Transactions on Automatic Control, 29*(5), 441–444. doi:10.1109/TAC.1984.1103551

Tavazoei, M. S. (2010). A note on fractional-order derivatives of periodic functions. *Automatica, 46*(5), 945–948. doi:10.1016/j.automatica.2010.02.023

Tavazoei, M. S., Haeri, M., & Attari, M. (2009). A proof for nonexistence of periodic solutions in time invariant fractional order systems. *Automatica, 45*(8), 1886–1890. doi:10.1016/j.automatica.2009.04.001

Zhen, W., Xia, H., & Guodong, S. (2011). Analysis of nonlinear dynamics and chaos in a fractional order financial system with time delay. *Computers & Mathematics with Applications (Oxford, England), 62*(3), 1531–1539. doi:10.1016/j.camwa.2011.04.057

Chapter 13
Apparent Measure, Integral of Real Order, and Relative Dimension

Faycal Adda
Community College of Qatar, Qatar

ABSTRACT

In this chapter, the author introduces a concept of apparent measure in R^n and associates the concept of relative dimension (of real order) that depends on the geometry of the object to measure and on the distance that separates it from an observer. At the end, the author discusses the relative dimension of a Cantor set. This measure enables us to provide a geometric interpretation of the Riemann-Liouville's integral of order $\alpha\epsilon$ *⊢]0,1], and based on this interpretation, the author introduces a modification on the Riemann-Liouville's integral to make it symmetrical and then introduces a new fractional derivative that exploits at the same time the right and the left fractional derivatives.*

INTRODUCTION

The theoretical and practical interest of fractional operator appears naturally in several fields of sciences, economic and engineering, including fluid flow, rheology, modeling, signal processing, diffusion problem, chaos control problem (Oldham & Spanier, 1974; Srivastava & Saigo, 1987; Samko et al, 1987 ; Le Méhauté, 1991; Podlubny, 1999; Matouk, 2009a; Matouk, 2009b; Matouk 2010; Matouk, 2011a; Matouk, 2011b; Hegazi & Matouk, 2011; Hegazi et al, 2011; Hegazi & Matouk, 2013; Hegazi et al, 2013a; Hegazi et al, 2013b; Matouk & Elsadany, 2014; Matouk et al, 2015; Matouk, 2016; Matouk & Elsadany, 2016; El-Sayed et al, 2016a; El-Sayed et al, 2016b; Al-Khedhairi et al, 2018; Al-Khedhairi et al, 2019a; Al-Khedhairi et al, 2019b) in spite of the lack of a geometrical interpretation of this operator (Ben Adda, 1997). In this chapter, we give a geometrical interpretation of the Riemann-Liouville's integral (Oldham & Spanier, 1974; Samko et al, 1987; Riemann, 1990; Dugowson, 1997) of order $\alpha \in]0,1]$ as a surface, which can be evaluated and compared to the classical surface of integral function. We introduce

DOI: 10.4018/978-1-7998-3122-8.ch013

a concept of "apparent" measure, with respect to a point x_0, that changes value according to the distance of the object to be measured from x_0. We introduce and prove some properties of this measure. The concept of relative dimension follows from this apparent measure. For example, we discuss the "small irregular" Cantor set having Lebesgue measure zero and Häusdorff measure one. By the aid of the "apparent" measure we interpret the Riemann-Liouville's integral and we modify it.

1. MEASURE OF REAL ORDER

1.1. Measure in R

Let $I_{\mathbb{R}}$ be the set of all bounded open intervals of $\mathbb{R}$, of the form $I=]a,b[$, where $a\leq b$. In the case where $a=b$ we have $\varnothing \in I_{\mathbb{R}}$. We define the length

$L : I_{\mathbb{R}} \rightarrow \overline{\mathbb{R}}_+$ by $L(]a,b[)=b-a$.

Definition 1. For any $x_0\in\mathbb{R}$ we call the length of order $\alpha \in \mathbb{R}_+^*$, the map $\mu_{x_0}^{\alpha} : I_{\mathbb{R}} \rightarrow \overline{\mathbb{R}}_+$, defined by

$$\mu_{x_0}^{\alpha}\left(\left]a,b\right[\right) = \begin{cases} \left\| x_0 - a\right|^{\alpha} - \left|x_0 - b\right|^{\alpha}\right| if\ x_0 \notin \left]a,b\right[\\ \left|x_0 - a\right|^{\alpha} - \left|x_0 - b\right|^{\alpha}\ if\ x_0 \in \left]a,b\right[\end{cases} \tag{1}$$

Remarks.

1. If $x_0 \notin \left]a,b\right[$, the length

$$L\left(\left]a,b\right[\right) = b - a = b - x_0 + x_0 - a = \left\| x_0 - a\right| - \left|x_0 - b\right\| = \mu_{x_0}^{1}\left(\left]a,b\right[\right)$$

and if $x_0\in]a,b[$, we have

$$L\left(\left]a,b\right[\right) = \left(x_0 - a\right) + \left(b - x_0\right) = \mu_{x_0}^{1}\left(\left]a,b\right[\right).$$

In one hand this length L is independent of the point x_0, it is intrinsic. On the other hand the length of real order $\mu_{x_0}^{\alpha}$ for $\alpha\in \mathbb{R}_+^*$ depends on the distance that separates x_0 and the boundaries of the interval $]a,b[$. For example, for $\alpha\in]0,1[$, the farther the point x_0 is from the interval $]a,b[$, the smaller the length of order α of this interval appears (thanks to the property of the function x^{α} for $\alpha\in]0,1[$). Because of this dependence of the position of the point x_0, it can be called "apparent" length. It mimics the impression an observer gets for the measure of an object from his position located in the position x_0: the object appears smaller to the observer when it gets further from him. The extension of this concept will be done later on.

2. Obviously we have $\mu_{x_0}^{\alpha}(\varnothing)=0$, and the function can be considered as a density measure indeed:
   a. If $x_0 \geq b$, we denote

$$\mu_{x_0}^{\alpha}(]a,b[)=(x_0-a)^{\alpha}-(x_0-b)^{\alpha}=\int_a^b \alpha(x_0-t)^{\alpha-1}dt=\int_a^b d\mu_{x_0}^{\alpha}(t) \quad (2)$$

with

$$d\mu_{x_0}^{\alpha}(t)=\alpha|x_0-t|^{\alpha-1}dt \quad (3)$$

   b. If $x_0 \leq a$, we denote

$$\mu_{x_0}^{\alpha}(]a,b[)=(b-x_0)^{\alpha}-(a-x_0)^{\alpha}=\int_a^b \alpha(t-x_0)^{\alpha-1}dt=\int_a^b d\mu_{x_0}^{\alpha}(t) \quad (4)$$

with

$$d\mu_{x_0}^{\alpha}(t)=\alpha|x_0-t|^{\alpha-1}dt \quad (5)$$

   c. If $x_0 \in [a,b]$, we have

$$\begin{aligned}\mu_{x_0}^{\alpha}(]a,b[)&=(x_0-a)^{\alpha}+(b-x_0)^{\alpha}\\&=\int_a^{x_0}\alpha(x_0-t)^{\alpha-1}dt+\int_{x_0}^b \alpha(t-x_0)^{\alpha-1}dt=\int_a^b d\mu_{x_0}^{\alpha}(t)\end{aligned} \quad (6)$$

where $\mu_{x_0}^{\alpha}$ is a measure of the density $f(t)=\alpha|t-x_0|^{\alpha-1}$, it can be extended to any Borel set ("apparent" length of a Borel set).

1.2. Properties of $\mu_{x_0}^{\alpha}$

Let $]a,b[$be a non-empty interval of $\mathbb{R}$, based on the characteristics of the graph of the function $g(x_0)=\mu_{x_0}^{\alpha}(]a,b[)$, it is not difficult to figure out that the map

$$x_0 \to g(x_0)=\mu_{x_0}^{\alpha}(]a,b[)$$

is increasing in $\left]-\infty, \frac{a+b}{2}\right[$ and decreasing in $\left]\frac{a+b}{2}, +\infty\right[$ for α∈]0,1[and conversely for α∈]1,+∞[, whereas it is constant for α=1. The measure of order α of the interval]*a*,*b*[is maximal when the point x_0 is at the center of the interval]*a*,*b*[and α∈]0,1[, and it becomes minimal at this point for α∈]1,+∞[.

In the following some simple properties of this measure can be presented without proof as follow.

Properties 1. Let $\left]a,b\right[\subset \mathbb{R}$, with *a*<*b*, x_0∈ℝ, we have

1. (Scaling) For all x_0∈ℝ, and all $\lambda \in \mathbb{R}_+^*$, $\alpha \in \mathbb{R}_+^*$,

$$\mu_{\lambda x_0}^{\alpha}\left(\left]\lambda a, \lambda b\right[\right) = \lambda^{\alpha} \mu_{x_0}^{\alpha}\left(\left]a,b\right[\right) \tag{7}$$

2. (Translation invariance) For all *c*∈ℝ, $\alpha \in \mathbb{R}_+^*$, we have

$$\mu_{x_0+c}^{\alpha}\left(\left|a+c, b+c\right|\right) = \mu_{x_0}^{\alpha}\left(\left]a,b\right[\right) \tag{8}$$

3. For all $\alpha \in \mathbb{R}_+^*$, we have

$$\mu_{a}^{\alpha}\left(\left]a,b\right[\right) = \mu_{b}^{\alpha}\left(\left]a,b\right[\right) = \left(b-a\right)^{\alpha} \tag{9}$$

4. (Continuity) For all x_0∈ℝ, $\mu_{x_0}^{\alpha}\left(\left]a,b\right[\right)$ tends towards *L*(]*a*,*b*[) when α tends to 1.
5. If α∈]0,1[, we have

$$\mu_{a}^{\alpha}\left(\left]a,b\right[\right) = \mu_{b}^{\alpha}\left(\left]a,b\right[\right) \geq L\left(\left]a,b\right[\right);\ \text{if } b-a \leq 1 \tag{10}$$

$$\mu_{a}^{\alpha}\left(\left]a,b\right[\right) = \mu_{b}^{\alpha}\left(\left]a,b\right[\right) < L\left(\left]a,b\right[\right);\ \text{if } b-a > 1 \tag{11}$$

6. For all real numbers a', b', such that $a' > a$, $b' < b$ and $x_0 \in \mathbb{R} \setminus \left]a,b\right[$, we have

$$\mu_{x_0}^{\alpha}\left(\left]a,b\right[\right) > \mu_{x_0}^{\alpha}\left(\left]a',b'\right[\right) \tag{12}$$

Remark. Equation (7) is equivalent to for all

$$\rho>0,\ d\mu_{x}^{\alpha}\left(\rho t\right) = \rho^{\alpha} \mu_{x/\rho}^{\alpha}\left(t\right) \tag{13}$$

1.3. Measure of Real Order in R^n

We start this subsection with the definition of the measure of a simple set in $\mathbb{R}^n$. A brick P is defined in $\mathbb{R}^n$ by n inequalities of the form

$$a_i \leq x_i \leq b_i \left(i = 1, 2, \ldots, n\right) \tag{14}$$

where the a_i and b_i indicate given real numbers for all i=1,2,…,n.

Definition 2. Let P be a parallelepiped in $\mathbb{R}^n$, $\alpha \in \mathbb{R}_+^*$, and let $x_0 \in \mathbb{R}^n$. We call measure of order α of P with respect to x_0 the quantity

$$\mu_{x_0}^{\alpha}\left(P\right) = \prod_{i=1}^{n} \left| \left|x_{0,i} - a_i\right|^{\alpha} - \left|x_{0,i} - b_i\right|^{\alpha} \right| \text{if } x_0 \notin P \tag{15}$$

$$\mu_{x_0}^{\alpha}\left(P\right) = \prod_{i=1}^{n} \left(x_{0,i} - a_i\right)^{\alpha} + \left(b_i - x_{0,i}\right)^{\alpha} \text{ if } x_0 \in P \tag{16}$$

Remark. One can obtain an equivalent of Fubini's theorem for the product of measure of real order. We let the formulation to the reader. One can also build up in a traditional way the product measure on $\mathbb{R}^n$ that can lead to find out the generalization of the multiple integral of real order of Riemann Liouville introduced in (Dugowson, 1997, p. 459).

1.4. Measure of a Bounded Set of R^n

Let A be a subset of $\mathbb{R}^n$, and let 1_A be the characteristic function of A and $x_0 \in \mathbb{R}^n$, we can simply define the measure of a bounded subset of $\mathbb{R}^n$.

Definition 3. let $A \subset \mathbb{R}^n$, $\alpha \in \mathbb{R}_+^*$, we will say that A is α-measurable if its characteristic function is $\mu_{x_0}^{\alpha}$-integrable. The number

$$\mu_{x_0}^{\alpha}\left(A\right) = \int_{\mathbb{R}^n} 1_A\left(t\right) d\mu_{x_0}^{\alpha}\left(t\right) \tag{17}$$

is called the measure of order α of A with respect to x_0.

Remark. For n=2 this measure is called apparent area with respect to x_0, and for n=3 it is called apparent volume with respect to x_0. The various properties of this measure follow easily from the properties of the integral. One can also obtain a concept of apparent mass for a material system by following the same classical process as for the Lebesgue measure, and the same properties can be found. Recall finally that the measure of a set depends on the dimension of the space in which it is considered.

For example the measure of order α of a segment considered as a subset of ℝ is equal to its length of order α, but if it is considered as plunged in $\mathbb{R}^2$ then its measure of order α is equal to zero.

In the following we define a more precise measure of real order without the use of integral. First of all, we remind the definition of diameter of a set in a metric space.

Definition 4. Let (E,d) be a metric space. If $A \subset E$, $A \neq \varnothing$, the diameter of A, denoted *dian*(A), is the element of $[0,+\infty[$ defined by

$$diam\left(A\right) = \sup_{\left(x,y\right)\in A^2} d\left(x,y\right) \tag{18}$$

with $diam\left(\varnothing\right) = 0$

Definition 5. Let $x_0 \in \mathbb{R}^n$, $B \subset \mathbb{R}^n$ and let P_i be a brick of $\mathbb{R}^n$ for all $i \in N^*$, we define for δ>0

$$\mu_{x_0}^{\alpha,\delta}\left(B\right) = \inf\left\{\sum_{i=1}^{\infty}\mu_{x_0}^{\alpha}\left(P_i\right), B \subset \bigcup_{i=1}^{\infty} P_i, diam\left(P_i\right) \leq \delta\right\}$$

$$= inf\left\{\sum_{i=1}^{\infty}\prod_{j=1}^{n}\mu_{x_{0,j}}^{\alpha}\left(\left[a_{i,j}, b_{i,j}\right]\right), B \subset \bigcup_{i=1}^{\infty} P_i, diam\left(P_i\right) \leq \delta\right\} \tag{19}$$

with α∈]0,1]. The measure of order α of B with respect to x_0 is given by

$$\mu_{x_0}^{\alpha}\left(B\right) = \lim_{\delta\to 0}\mu_{x_0}^{\alpha,\delta}\left(B\right) = \sup_{\delta>0}\mu_{x_0}^{\alpha,\delta}\left(B\right) \tag{20}$$

Remarks.

1. The choice to let δ tend towards 0 enables us to have a concept of more precise measure where the selected covering follows the local geometry of the object.
2. The point x_0 represents the position from which the object is measured, this measure being therefore dependent on this position (Fig.1). The shape of the disc will have the same size at the same distance from the position x_0 from which the disc is measured .
3. The last definition coincides with definition 2 and 3 for α∈]0,1[

1.5. Properties of the Measure

The properties of the measure can be found as follow

Figure 1. The apparent measure of the disc for n=2. The farther the disc is from x_0 the smaller the surface of the disc appears (using the measure of order α with respect to x_0)

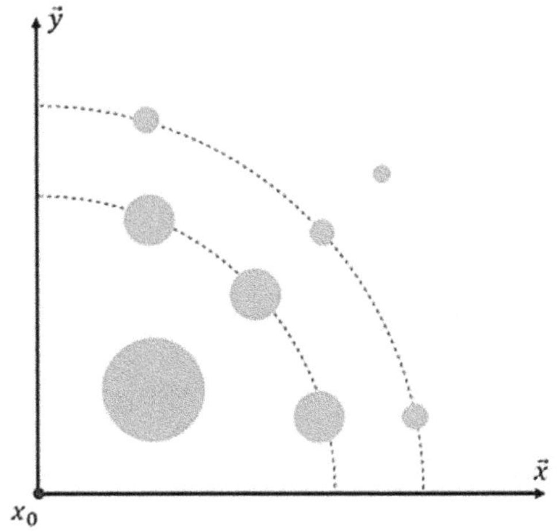

Theorem 1. Under the conditions of Definition 5 we have:

1. $\mu_{x_0}^{\alpha,\delta}$ is an outer measure
2. $\mu_{x_0}^{\alpha}$ is an outer measure
3. $\mu_{x_0}^{\alpha}$ is a Borel measure.

Proof.

1. The quantity $\mu_{x_0}^{\alpha,\delta}$ is an outer measure, indeed:
   a. We have $\mu_{x_0}^{\alpha,\delta} \in \overline{\mathbb{R}}_+$ for all $B \subset \mathbb{R}^n$
   b. The quantity

$$\mu_{x_0}^{\alpha,\delta}(\varnothing) = \inf\left\{\sum_{i=1}^{\infty}\mu_{x_0}^{\alpha}(P_i), \varnothing \subset \bigcup_{i=1}^{\infty}P_i, diam(P_i) \leq \delta\right\} = 0$$

Indeed, it is sufficient to take $P_i = \left\{x \in \mathbb{R}^n \;/\; a_i \leq x_i \leq a_i\right\}$.

   c. If $(A_k)_{k\geq 1}$ is a sequence of subsets of $\mathbb{R}^n$, lets prove that

$$\mu_{x_0}^{\alpha,\delta}\left(\bigcup_{k=1}^{\infty}A_k\right) \leq \sum_{k=1}^{\infty}\mu_{x_0}^{\alpha,\delta}(A_k)$$

If $\sum_{k=1}^{\infty}\mu_{x_0}^{\alpha,\delta}(A_k) = +\infty$, the inequality is obvious, if not we have $\mu_{x_0}^{\alpha,\delta}(A_k) < +\infty$ for all $k\geq 1$. There exists a sequence $(P_{i,k})_{i\geq 1}$ of bricks of $\mathbb{R}^n$ such that $A_k \subset \bigcup_{i=1}^{\infty}P_{i,k}$, and for all $N>0$ we obtain

$$\sum_{i=1}^{\infty}\mu_{x_0}^{\alpha}\left(P_{i,k}\right) < \mu_{x_0}^{\alpha,\delta}\left(A_k\right) + \frac{1}{N2^k}$$

Thus we have

$$\bigcup_{k=1}^{\infty} A_k \subset \bigcup_{i,k=1}^{\infty} P_{i,k}$$

a countable covering of bricks of $\mathbb{R}^n$. Then we have for all $N>0$

$$\mu_{x_0}^{\alpha,\delta}\left(\bigcup_{k=1}^{\infty} A_k\right) \leq \sum_{i=1}^{\infty}\sum_{k=1}^{\infty}\mu_{x_0}^{\alpha}\left(P_{i,k}\right) < \sum_{k=1}^{\infty}\mu_{x_0}^{\alpha,\delta}\left(A_k\right) + \frac{1}{N}$$

which gives the inequality.

2. $\mu_{x_0}^{\alpha}$ an outer measure, indeed, let $\left(A_k\right)_{k\geq 1} \subset \mathbb{R}^n$. Since

$$\mu_{x_0}^{\alpha,\delta}\left(\bigcup_{1}^{\infty} A_n\right) \leq \sum_{n=1}^{\infty}\mu_{x_0}^{\alpha,\delta}\left(A_n\right)$$

and since $\sup_{\delta>0}\mu_{x_0}^{\alpha,\delta}\left(A_k\right) = \mu_{x_0}^{\alpha}\left(A_k\right)$ we have then

$$\mu_{x_0}^{\alpha,\delta}\left(\bigcup_{1}^{\infty} A_n\right) \leq \sum_{n=1}^{\infty}\mu_{x_0}^{\alpha}\left(A_n\right)$$

When δ tends to 0, we obtain

$$\mu_{x_0}^{\alpha}\left(\bigcup_{1}^{\infty} A_n\right) \leq \sum_{n=1}^{\infty}\mu_{x_0}^{\alpha}\left(A_n\right)$$

3. $\mu_{x_0}^{\alpha}$ is a Borel measure. Indeed let A,B be two subsets of $\mathbb{R}^n$, such that $d(A,B)>0$, with

$$d\left(A,B\right) = \inf\left\{d\left(x,y\right), \forall x \in A, \forall y \in B\right\}$$

It is sufficient to consider $d(A,B)>4\delta$ for example to find an adequate covering that leads to the equality.

Properties 2. Let $\alpha\in]0,1]$, $x_0\in \mathbb{R}^n$, and let $\mu_{x_0}^{\alpha}$ be a measure on $\mathbb{R}^n$. We have

1. $\mu_{x_0}^{\alpha}\left(B\right)=0$ for all $B\subset\mathbb{R}^{n'}$, $n'<n$. (where $\mathbb{R}^n$ is isomorphic to $\mathbb{R}^{n'}\times\mathbb{R}^{n-n'}$ for any integer $n'<n$).
2. $\mu_{x_0}^{1}$ coincides with the Lebesgue's measure on $\mathbb{R}^n$.
3. For all λ>0, and all $B\subset\mathbb{R}^n$, we have $\mu_{\lambda x_0}^{\alpha}\left(\lambda B\right)=\lambda^n\mu_{x_0}^{\alpha}\left(B\right)$.

Proof.

1. We have

$$\mu_{x_0}^{\alpha,\delta}\left(B\right)=\inf\left\{\sum_{i=1}^{\infty}\prod_{j=1}^{n}\mu_{x_{0,j}}^{\alpha}\left(\left[a_{i,j},b_{i,j}\right]\right),B\subset\bigcup_{i=1}^{\infty}P_i,diam\left(P_i\right)\leq\delta\right\}$$

with α∈]0,1]. Since $n'<n$, P_i can be chosen in $\mathbb{R}^n$, the $n-n'$ remaining coordinated are null, which leads to the result.

This property means that the measure of a set depends on the dimension of the space in which is considered.

2. On $\mathbb{R}$ and for α=1 we have:

$$\mu_{x_0}^{\alpha}\left(P_i\right)=diam\left(P_i\right)$$

with $\mu_{x_0}^{1}\left(A\right)=H^1\left(A\right)$, and since the Häusdorff's measure, H^1 coincide with the Lebesgue's measure on $\mathbb{R}$, we deduce the result.

3. For all λ>0, we have

$$\mu_{\lambda x_0}^{\alpha,\delta}\left(\lambda B\right)=\inf\left\{\sum_{i=1}^{\infty}\prod_{j=1}^{n}\mu_{\lambda x_{0,j}}^{\alpha}\left(\left[\lambda a_{i,j},\lambda b_{i,j}\right]\right),B\subset\bigcup_{i=1}^{\infty}P_i,diam\left(P_i\right)\leq\delta\right\}$$

Since

$$\sum_{i=1}^{\infty}\prod_{j=1}^{n}\mu_{\lambda x_{0,j}}^{\alpha}\left(\left[\lambda a_{i,j},\lambda b_{i,j}\right]\right)=\lambda^n\sum_{i=1}^{\infty}\prod_{j=1}^{n}\mu_{x_{0,j}}^{\alpha}\left(\left[a_{i,j},b_{i,j}\right]\right)$$

we deduce that

$$\mu_{\lambda x_0}^{\alpha,\delta}\left(\lambda B\right)=\lambda^n\mu_{x_0}^{\alpha,\delta}\left(B\right)$$

Then when δ tends to 0, we have

$$\mu^{\alpha}_{\lambda x_0}\left(\lambda B\right)=\lambda^{n}\mu^{\alpha}_{x_0}\left(B\right)$$

Proposition 1. Let $\mathcal{P}_j$ be the projection of $\mathbb{R}^n$ according to the jth coordinates. Let $\alpha\in]0,1]$, $A\subset\mathbb{R}^n$ and let $x_0,x_1\in\mathbb{R}^n$,

If $\forall j\in\{1,2,3,\ldots,n\}$, $\inf\mathcal{P}_j\left(A\right)\geq x_{0,j}\geq x_{1,j}$ then $\mu^{\alpha}_{x_0}\left(A\right)\geq\mu^{\alpha}_{x_1}\left(A\right)$.

Proof. Let $i\in\mathbb{N}^*$, and P_i be a brick of $\mathbb{R}^n$ defined by the n inequalities $a_{i,j}\leq x\leq b_{i,j}$ for $j=1,\ldots,n$ with $a_{i,j}\leq b_{i,j}$ for all $j=1,\ldots,n$, such as $A\subset\bigcup_{i=1}^{\infty}P_i$ and $diam(P_i)<\delta$ for all $i\in\mathbb{N}^*$.

If $\forall j\in\{1,2,3,\ldots,n\}$ we have $\inf\mathcal{P}_j\left(A\right)\geq x_{0,j}>x_{1,j}$, we deduce that

$$\forall i,\inf\mathcal{P}_j\left(P_i\right)\geq x_{0,j}>x_{1,j},$$

then according to the properties of measure of real order in $\mathbb{R}$, we have

$$\mu^{\alpha}_{x_{0,j}}\left(\left[a_{i,j},b_{i,j}\right]\right)\geq\mu^{\alpha}_{x_{1,j}}\left(\left[a_{i,j},b_{i,j}\right]\right)$$

for all $(i,j)\in N^*\times\{1,\ldots,n\}$, therefore

$$\sum_{i=1}^{\infty}\prod_{j=1}^{n}\mu^{\alpha}_{x_{0,j}}\left(\left[a_{i,j},b_{i,j}\right]\right)\geq\sum_{i=1}^{\infty}\prod_{j=1}^{n}\mu^{\alpha}_{x_{1,j}}\left(\left[a_{i,j},b_{i,j}\right]\right)$$

that gives

$$\mu^{\alpha,\delta}_{x_0}\left(A\right)\geq\mu^{\alpha,\delta}_{x_1}\left(A\right)$$

which completes the proof.

Theorem 2. Let $f:B\subset\mathbb{R}^n\rightarrow\mathbb{R}^n$ be a function defined algebraically by

$$f(x)=\Lambda x+C \tag{21}$$

with C is a constant vector, Λ is a matrix

$$\begin{pmatrix} \lambda_1 & 0 & 0 \\ 0 & \ddots & 0 \\ 0 & 0 & \lambda_n \end{pmatrix}$$

with λ_i a positive real number for all $i=1,\ldots,n$. Let $x_0 \in B$, and let $A \subset B$. We have

$$\mu_{f(x_0)}^{\alpha}\left(f\left(A\right)\right) = \left(\prod_{j=1}^{n}\lambda_j^{\alpha}\right)\mu_{x_0}^{\alpha}\left(A\right) \tag{22}$$

Proof. Let $i \in \mathbb{N}^*$, and P_i be a brick of $\mathbb{R}^n$ defined by the n following inequalities $a_{i,j} < x_j < b_{i,j}$ for $j= 1,\ldots,n$ with $a_{i,j} \le b_{i,j}$ for all $j=1,\ldots,n$, such that $A \subset \bigcup_{i=1}^{\infty} P_i$ and $diam(P_i) < \delta$ for all $i \in \mathbb{N}^*$, $f(P_i)$ is a brick of R^n defined by the n inequalities

$\lambda_j a_{i,j} + c_j \le x_j \le \lambda_j b_{i,j} + c_j$ for $j=1,\ldots,n$

$$\mu_{f(x_0)}^{\alpha} f\left(P_i\right) = \prod_{j=1}^{n}\mu_{f(x_0)j}^{\alpha}\left(\left[\lambda_j a_{i,j} + c_j, \lambda_j b_{i,j} + c_j\right]\right) \tag{23}$$

$$= \prod_{j=1}^{n}\left|\left|f\left(x_0\right)_j - \left(\lambda_j b_{i,j} + c_j\right)\right|^{\alpha} - \left|f\left(x_0\right)_j - \left(\lambda_j a_{i,j} + c_j\right)\right|^{\alpha}\right| \tag{24}$$

$$= \prod_{j=1}^{n}\lambda_j^{\alpha}\left|\left|x_{0,j} - b_{i,j}\right|^{\alpha} - \left|x_{0,j} - a_{i,j}\right|^{\alpha}\right| \tag{25}$$

$$= \left(\prod_{j=1}^{n}\lambda_j^{\alpha}\right)\mu_{x_0}^{\alpha}\left(P_i\right) \tag{26}$$

which gives the result.

2. MEASURE AND DIMENSION OF REAL ORDER

The measure defined in the previous subsequence cannot define a concept of dimension of real order. In this purpose we build up the following measure:

Definition 6. Let $(P_i)_i \in \mathbb{N}^*$ be a sequence of parallelepipeds of $\mathbb{R}^n$ covering $B \subset \mathbb{R}^n$ and $x_0 \in \mathbb{R}^n$. Let $\beta \in [0,+\infty[$ and $\alpha \in]0,1]$. We define

$$H_{x_0}^{\beta,\alpha,\delta}(B) = \inf\left\{\sum_{i=1}^{\infty}\left(\sup_{1\le j\le n} \alpha_{x_{0,j}}^{\alpha}\left(\left[a_{i,j}, b_{i,j}\right]\right)\right)^{\beta}, B \subset \bigcup_{i=1}^{\infty} P_i, diam\left(P_i\right) \le \delta\right\} \tag{27}$$

the measure of order (α,β) of B compared to x_0 is given by

$$H_{x_0}^{\beta,\alpha}(B) = \lim_{\delta\to 0} H_{x_0}^{\beta,\alpha,\delta}(B) = \sup_{\delta>0} H_{x_0}^{\beta,\alpha,\delta}(B) \tag{28}$$

Theorem 3.

1. $H_{x_0}^{\beta,\alpha,\delta}$ is an outer measure
2. $H_{x_0}^{\beta,\alpha}$ is a measure
3. $H_{x_0}^{\beta,\alpha}$ coincide with the Häusdorff's measure of order β on $\mathbb{R}$ for $\alpha=1$.

Proof. For 1 and 2 the proof is similar to the proof of the theorem 1.

When $\alpha=1$, we have

$$\mu_{x_{0,j}}^{\alpha}\left(\left[a_{i,j}, b_{i,j}\right]\right) = Diam\left(\left[a_{i,j}, b_{i,j}\right]\right).$$

If we denote A^0 the interior of A, we have the following lemma

Lemma 1. Let A be a subset of $\mathbb{R}^n$, x_0 $\mathbb{R}^n \backslash A^0$, $\alpha \in]0,1]$ and let $0 \le \beta < \gamma < \infty$.

1. If $H_{x_0}^{\beta,\alpha}(A) < +\infty$ then $H_{x_0}^{\gamma,\alpha}(A) = 0$
2. If $H_{x_0}^{\gamma,\alpha}(A) > 0$ then $H_{x_0}^{\beta,\alpha}(A) = +\infty$

Proof.

1. If $H_{x_0}^{\beta,\alpha}(A) < +\infty$, then there exists a sequence $\left(P_n\right)_{n\ge 1}$ of bricks of $\mathbb{R}^n$ such that $A \subset \bigcup_{k=1}^{\infty} P_n$ with $diamP_n < \delta$, and we have for $k \in \mathbb{N}^*$

$$\sum_{i=1}^{\infty}\left(\sup_{1\le j\le n} \mu_{x_{0,j}}^{\alpha}\left(\left[a_{i,j}, b_{i,j}\right]\right)\right)^{\beta} \le H_{x_0}^{\beta,\alpha,\delta}(A) + \frac{1}{k} \le \sup_{\delta>0} H_{x_0}^{\beta,\alpha,\delta}(A) + \frac{1}{k}$$

According to the definition of measure we have

$$H_{x_0}^{\gamma,\alpha,\delta}\left(A\right) \leq \sum_{i=1}^{\infty}\left(\sup_{1\leq j\leq n} \mu_{x_{0,j}}^{\alpha}\left(\left[a_{i,j}, b_{i,j}\right]\right)\right)^{\gamma}$$
$$= \sum_{i=1}^{\infty}\left(\sup_{1\leq j\leq n} \mu_{x_{0,j}}^{\alpha}\left(\left[a_{i,j}, b_{i,j}\right]\right)\right)^{\beta}\left(\sup_{1\leq j\leq n} \mu_{x_{0,j}}^{\alpha}\left(\left[a_{i,j}, b_{i,j}\right]\right)\right)^{\gamma-\beta}$$

Since

$$\sup_{1\leq j\leq n} \mu_{x_{0,j}}^{\alpha}\left(\left[a_{i,j}, b_{i,j}\right]\right) \leq Diam\left(P_j\right), \forall_1 \leq j \leq n$$

then we obtain

$$H_{x_0}^{\gamma,\alpha,\delta}\left(A\right) \leq \delta^{\gamma-\beta}\sum_{i=1}^{\infty}\left(\sup_{1\leq j\leq n} \mu_{x_{0,j}}^{\alpha}\left(\left[a_{i,j}, b_{i,j}\right]\right)\right)^{\beta} \leq \delta^{\gamma-\beta}\left(H_{x_0}^{\beta,\alpha}\left(A\right) + \frac{1}{k}\right)$$

we deduce from it the result when δ tends to 0.

2. It is sufficient to remark that non 1 implies 2.

Thanks to this lemma we introduce the definition of relative dimension.

Definition 7. Let A be a subset of $\mathbb{R}^n$, $x_0 \in \mathbb{R}^n \setminus A^0$, $\alpha \in]0,1]$ and let $0 \leq \beta < \gamma < \infty$.

We define the relative dimension of a set A by

$$\dim\left(A\right) = \inf\left\{\gamma : H_{x_0}^{\gamma,\alpha}\left(A\right) = 0\right\} = \inf\left\{\beta : H_{x_0}^{\beta,\alpha}\left(A\right) < +\infty\right\} \quad (29)$$

$$= \sup\left\{\beta : H_{x_0}^{\beta,\alpha}\left(A\right) = +\infty\right\} = \sup\left\{\gamma : H_{x_0}^{\gamma,\alpha}\left(A\right) > 0\right\} \quad (30$$

Remark.

1. If $A \subset B \subset \mathbb{R}^n$, $x_0 \in \mathbb{R}^n$ and $x_0 \notin B$, we have $H_{x_0}^{\beta,\alpha}\left(A\right) \leq H_{x_0}^{\beta,\alpha}\left(B\right)$
2. If $d\left(x_{0,i}, P_i\left(A\right)\right) \neq 0$ for all i=1,2,…,n, then the comparison between the Definition 7 of the relative dimension and the Häusdorff dimension gives

$$\dim\left(A\right) \leq \dim_{\mathcal{H}}\left(A\right)$$

Until now, our "apparent" measure depends on the axis in which the object can be plunged (because of the use of parallelepipeds in Definition 6). To change this dependence, we propose the following definition.

Definition 8. Let $(O_i)_i \in \mathbb{N}^*$ be a sequence of open sets of $\mathbb{R}^n$ covering $B \subset \mathbb{R}^n$ and let $x_0 \in \mathbb{R}^n$.

Let $\beta \in [0,+\infty[$ and $\alpha \in]0,1]$ We define for $\delta > 0$

$$\mathcal{H}_{x_0}^{\beta,\alpha,\delta}(B) = \inf\left\{\sum_{i=1}^{\infty}\left(\left(D_{x_0}(O_i)\right)^{\alpha} - \left(d_{x_0}(O_i)\right)^{\alpha}\right)^{\beta}, B \subset \bigcup_{i=1}^{\infty} O_i, diam(O_i) \leq \delta\right\} \tag{31}$$

where $D_{x_0}(O_i) = \sup_{y \in O_i} d(x_0, y)$ and $d_{x_0}(O_i) = \inf_{y \in O_i} d(x_0, y)$.

The measure of order (α,β) of B with respect to x_0 is given by

$$\mathcal{H}_{x_0}^{\beta,\alpha}(B) = \sup_{\delta > 0} \mathcal{H}_{x_0}^{\beta,\alpha,\delta}(B) = \lim_{\delta \to 0} \mathcal{H}_{x_0}^{\beta,\alpha,\delta}(B) \tag{32}$$

As for the Häusdorff measure we have the following

Theorem 4.

1. $\mathcal{H}_{x_0}^{\beta,\alpha,\delta}$ is an outer measure
2. $\mathcal{H}_{x_0}^{\beta,\alpha}$ is a measure
3. $\mathcal{H}_{x_0}^{\beta,\alpha}$ coincides with the Häusdorff 's measure of order on $\mathbb{R}$ for $\alpha=1$.
4. $\mathcal{H}_{x_0}^{\beta,\alpha}(A) = 0$ if $A \subset \mathbb{R}^n$ and $\alpha\beta > n$.

Proof 4. It is not difficult to prove 1, 2 and 3.

Let m be an integer and let us consider the unit cube $Q \subset \mathbb{R}^n$. This cube can be decomposed into m^n cubes with side $\frac{1}{m}$. Therefore

$$\mathcal{H}_{x_0}^{\beta,\alpha,\delta}(Q) \leq \sum_{i=1}^{m^n}\left(\left(\frac{1}{m}\right)^{\alpha}\right)^{\beta} = m^n\left(\left(\frac{1}{m}\right)^{\alpha}\right)^{\beta} = m^{n-\alpha\beta} \tag{33}$$

and then we can conclude.

The Häusdorff dimension measures the metric size of any subset in a metric space. We can define an "apparent" Häusdorff measure if we put the "apparent" diameter given by

$$\mu_{x_0}^{\alpha}(O_i) = \left(D_{x0}(O_i)\right)^{\alpha} - \left(d_{x0}(O_i)\right)^{\alpha} \tag{34}$$

in place of $diam(O_i)$, and we have

Proposition 2. Let $\alpha\in]0,1]$, $A\subset\mathbb{R}^n$ and let $x_0,x_1\in\mathbb{R}^n$. Let $\left(O_i\right)_{i\in N^*}$ be a sequence of open sets of $\mathbb{R}^n$ covering A; If $\forall i\in\mathbb{N}^*$, $d(x_0,O_i)>d(x_1,O_i)$ then we have

$$\mathcal{H}_{x_0}^{\beta,\alpha}\left(A\right)\leq\mathcal{H}_{x_1}^{\beta,\alpha}\left(A\right) \tag{35}$$

where $d\left(x_i,O_i\right)=\inf_{y\in A} d\left(x_i,y\right)$

Proof. Let $\left(O_i\right)_{i\in\mathbb{N}^*}$ be a sequence of open sets of $\mathbb{R}^n$ covering $A\subset\mathbb{R}^n$ such that $diam(O_i)\leq\delta$, thanks to the properties of the apparent measure defined in the beginning of this chapter it can be said that if we have for all $i\in\mathbb{N}^*$, $d(x_0,O_i)>d(x_1,O_i)$ then

$$\left(\left(D_{x_0}\left(O_i\right)\right)^{\alpha}-\left(d_{x_0}\left(O_i\right)\right)^{\alpha}\right)^{\beta}<\left(\left(D_{x_1}\left(O_i\right)\right)^{\alpha}-\left(d_{x_1}\left(O_i\right)\right)^{\alpha}\right)^{\beta} \tag{36}$$

then

$$\sum_{i=1}^{\infty}\left(\left(D_{x_0}\left(O_i\right)\right)^{\alpha}-\left(d_{x_0}\left(O_i\right)\right)^{\alpha}\right)^{\beta}<\sum_{i=1}^{\infty}\left(\left(D_{x_1}\left(O_i\right)\right)^{\alpha}-\left(d_{x_1}\left(O_i\right)\right)^{\alpha}\right)^{\beta} \tag{37}$$

which gives the result.

2.1. Cantor Set in R

Let $0<\lambda<\frac{1}{2}$. We denote $I_{0,1}=[0,1]$, and let $I_{1,1}$ and $I_{1,2}$ be the intervals $[0,\lambda]$ and $[1-\lambda,1]$, respectively, obtained by taking off from the interval $[0,1]$ the interval $]\lambda,1-\lambda[$ with length $1-2\lambda$ and centered at the medium (Fig.2). We continue this process of selecting two subintervals of each interval $I_{1,1}$ and $I_{1,2}$, and so on. If we defined the intervals $I_{k-1,1},\ldots,I_{k-1,2^{k-1}}$, the intervals $I_{k,1},\ldots,I_{k,2^k}$ are obtained by deleting from the middle of each interval $I_{k-1,j}$ an interval of length

$$\left(1-2\lambda\right)diam\left(I_{k-1,j}\right)=\left(1-2\lambda\right)\lambda^{k-1}.$$

All the intervals $I_{k,j}$ thus obtained have length equal to λ^k. We define the limit set of this construction and we call it the Cantor set by

$$C\left(\lambda\right)=\bigcup_{k=0}^{\infty}\bigcup_{j=1}^{2^k}I_{k,j}$$

Then $C(\lambda)$ is an uncountable compact set without interior points and with zero Lebesgue measure, and Häusdorff 's measure equal to 1, meanwhile its Häusdorff 's dimension is known equal to

$$\dim_{\mathcal{H}} C\left(\lambda\right)=\frac{\log 2}{\log\dfrac{1}{\lambda}} \tag{38}$$

Figure 2. Cantor set

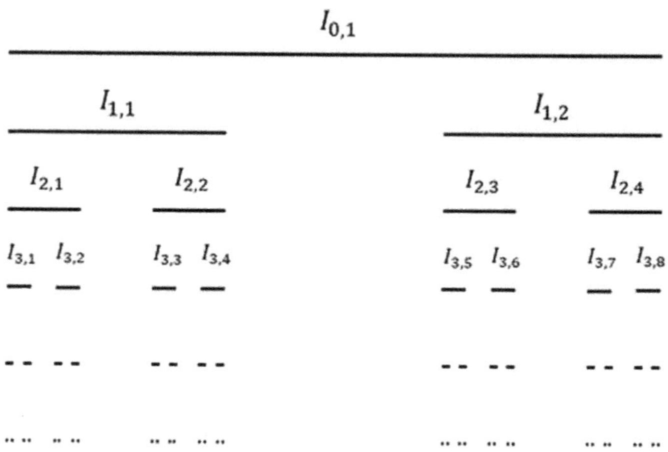

To determine the relative dimension of this set and to discuss its apparent measure, there are two cases:

1. Case where x_0=0, or x_0=1, or more generally if there exists $(k,j)\in\mathbb{N}\times\mathbb{N}^*$ such that x_0 belongs at the boundary of $I_{k,j}$.

It is known that for every $k\in\mathbb{N}^*$, we have $C\left(\lambda\right)\subset\bigcup_{j=1}^{2^k}I_{k,j}$, and then

$$H_{x_0}^{\beta,\alpha,\lambda^k}\left(C\left(\lambda\right)\right)\leq\sum_{j=1}^{2^k}\left(\mu_{x_0}^{\alpha}\left(I_{k,j}\right)\right)^{\beta}\leq\sum_{j=1}^{2^k}\left(\lambda^{k\alpha}\right)^{\beta}=\left(2\lambda^{\beta\alpha}\right)^{k} \tag{39}$$

In order for this upper bound to be useful, it should stay bounded as k tends to $+\infty$ and not equal to zero. The smallest value of β for which this happens is given by $2\lambda^{\beta\alpha}=1$, that is $\beta = \frac{\log 2}{\alpha \log\left(\frac{1}{\lambda}\right)}$ for $\alpha\in]0,1]$. Then for this choice we have

$$H_{x_0}^{\beta,\alpha}\left(C\left(\lambda\right)\right) \leq \lim_{k\to\infty}\sum_{j=1}^{2^k}\left(\lambda^{k,\alpha}\right)^{\beta} = 1 \tag{40}$$

Thus;

$$\dim H_{x_0}^{\beta,\alpha}\left(C\left(\lambda\right)\right) \leq \frac{\log 2}{\alpha \log\left(\frac{1}{\lambda}\right)} \tag{41}$$

As for the Häusdorff 's measure (Roger, 1970; Tricot, 1999) one can find a lower bounds of $H_{x_0}^{\beta,\alpha}\left(C\left(\lambda\right)\right)$. Indeed, let us consider an open recovering $\left(I_i\right)_{i\in\mathbb{N}}$ of $C(\lambda)$. Since $C(\lambda)$ is compact, a finite recovering $\left(I_j\right)_{j=1,\ldots,n}$ can be chosen. Since $C(\lambda)$ has no interior points, we can make I_j slightly larger if necessary, assuming that the end-points of each I_j are outside $C(\lambda)$. Then there is $\delta>0$ such that the distance from all these end-points to $C(\lambda)$ is at least δ.

For k enough large such that

$$diamI_j = \delta > \lambda^k = \mathcal{L}\left(I_{k,i}\right),$$

there exists j such that every interval $I_{k,i}$ is contained in I_j. Since for $\beta = \frac{\log 2}{\log\frac{1}{\lambda}}$ we can find a finite constant $c>1$ such that

$$c\left(\mu_{x_0}^{\alpha}\left(I_j\right)\right)^{\beta} \geq \sum_{I_{k,i}\subset I_j}\left(\mu_{x_0}^{\alpha}\left(I_{k,i}\right)\right)^{\beta} \tag{42}$$

hence

$$c\sum_j\left(\mu_{x_0}^{\alpha}\left(I_j\right)\right)^{\beta} \geq \sum_j\sum_{I_{k,i}\subset I_j}\left(\mu_{x_0}^{\alpha}\left(I_{k,i}\right)\right)^{\beta} \geq \sum_{i=1}^{2^k}\left(\mu_{x_0}^{\alpha}\left(I_{k,i}\right)\right)^{\beta} \tag{43}$$

and

$$\sum_{i=1}^{2^k}\left(\mu_{x_0}^{\alpha}\left(I_{k,i}\right)\right)^{\beta} \geq \sum_{i=1}^{2^k}\left(1-\left(1-\lambda^k\right)^{\alpha}\right)^{\beta} = 2^k\left(1-\left(1-\lambda^k\right)^{\alpha}\right)^{\beta} \simeq 2^k\alpha^{\beta}\lambda^{\beta k} \tag{44}$$

this lower bounds should remain bounded as k tends to infinity because $2\lambda^{\beta}=1$, for $\beta = \frac{\log 2}{\log\frac{1}{\lambda}}$, and we have for this value

$$\sum_{j}\left(\mu_{x_0}^{\alpha}\left(I_j\right)\right)^{\beta} \geq \frac{\alpha^{\beta}}{c} \tag{45}$$

which allows us to say that

$$H_{x_0}^{\beta,\alpha}\left(C\left(\lambda\right)\right) \geq \frac{\alpha^{\beta}}{c} \text{ for } \beta = \frac{\log 2}{\log\frac{1}{\lambda}} \tag{46}$$

which gives

$$\dim_{H_{x_0}^{\beta,\alpha}}\left(C\left(\lambda\right)\right) \geq \frac{\log 2}{\log\left(\frac{1}{\lambda}\right)} \tag{47}$$

Conclusion, if there is $(k,j)\in\mathbb{N}\times\mathbb{N}^*$ such that $x_0\in\partial I_{k,j}$, then

$$\frac{\log 2}{\log\left(\frac{1}{\lambda}\right)} \leq \dim_{H_{x_0}^{\beta,\alpha}}\left(C\left(\lambda\right)\right) \leq \frac{\log 2}{\alpha\log\left(\frac{1}{\lambda}\right)} \tag{48}$$

2. If $x_0\in[0,1]\backslash C(\lambda)$, a similar proof leads to the same conclusion.
   a. Case where $x_0 \notin \left[0,1\right]$, let us suppose $x_0<0$, the case $x_0>1$ will be the same. To find the upper bound, we have

$$H_{x_0}^{\beta,\alpha,\lambda^k}\left(C\left(\lambda\right)\right) \leq \sum_{j=1}^{2^k}\left(\mu_{x_0}^{\alpha}\left(I_{k,j}\right)\right)^{\beta} \leq \sum_{j=1}^{2^k}\left(\left(\lambda^k - x_0\right)^{\alpha} - \left(-x_0\right)^{\alpha}\right)^{\beta} = 2^k\left(\left(\lambda^k - x_0\right)^{\alpha} - \left(-x_0\right)^{\alpha}\right)^{\beta} \tag{49}$$

The last term;

$$2^k\left(\left(\lambda^k - x_0\right)^\alpha - \left(-x_0\right)^\alpha\right)^\beta$$

is equivalent to

$$\frac{\alpha^\beta}{\left(-x_0\right)^{(1-\alpha)\beta}}\left(2\lambda^\beta\right)^k.$$

The smallest value of β for which the upper bounds of $H_{x_0}^{\beta,\alpha}\left(C\left(\lambda\right)\right)$ should stay bounded as k tends to infinity is given by $2\lambda^\beta=1$, that is to say $\beta = \dfrac{\log 2}{\log\left(\dfrac{1}{\lambda}\right)}$.

For this choice we have

$$H_{x_0}^{\beta,\alpha}\left(C\left(\lambda\right)\right) \le \frac{\alpha^\beta}{\left(-x_0\right)^{(1-\alpha)\beta}} \tag{50}$$

therefore $\dim H_{x_0}^{\beta,\alpha}\left(C\left(\lambda\right)\right) \le \beta$.

The same covering chosen in 1. will give us a lower estimate of the relative dimension $\dim H_{x_0}^{\beta,\alpha}\left(C\left(\lambda\right)\right)$ by $\dfrac{\alpha^\beta}{c\left(-x_0\right)^{(1-\alpha)\beta}}$. For this over coverings, we just need to see that for $\beta = \dfrac{\log 2}{\log \dfrac{1}{\lambda}}$, we can find a finite constant $c>1$ such that

$$c\sum_j\left(\mu_{x_0}^\alpha\left(I_j\right)\right)^\beta \ge \sum_j \sum_{I_{k,i}\subset I_j}\left(\mu_{x_0}^\alpha\left(I_{k,i}\right)\right)^\beta \ge \sum_{i=1}^{2^k}\left(\mu_{x_0}^\alpha\left(I_{k,i}\right)\right)^\beta \tag{51}$$

With

$$\sum_{i=1}^{2^k}\left(\mu_{x_0}^\alpha\left(I_{k,i}\right)\right)^\beta \ge \sum_{i=1}^{2^k}\left(\left(1-x_0\right)^\alpha - \left(\left(1-x_0\right)-\left(\lambda^k\right)\right)^\alpha\right)^\beta \simeq \frac{\alpha^\beta\left(2\lambda^\beta\right)^k}{\left(1-x_0\right)^{(1-\alpha)\beta}} \tag{52}$$

which enables us to have

$$\dim_{H_{x_0}^{\beta,\alpha}}\left(C\left(\lambda\right)\right) \geq \frac{\log 2}{\log\left(\frac{1}{\lambda}\right)} \text{ and } H_{x_0}^{\beta,\alpha}\left(C\left(\lambda\right)\right) \geq \frac{\alpha^{\beta}}{c\left(1-x_0\right)^{(1-\alpha)\beta}} \tag{53}$$

for $\beta = \dfrac{\log 2}{\log\left(\frac{1}{\lambda}\right)}$.

For example:

If $x_0 = -1$, we find

$$\dim_{H_{x_0}^{\beta,\alpha}}\left(C\left(\lambda\right)\right) \geq \frac{\log 2}{\log\left(\frac{1}{\lambda}\right)} \text{ and } \frac{\alpha^{\beta}}{c2^{(1-\alpha)\beta}} \leq H_{-1}^{\beta,\alpha}\left(C\left(\lambda\right)\right) \leq \frac{\alpha^{\beta}}{c} \tag{54}$$

for $\beta = \dfrac{\log 2}{\log\left(\frac{1}{\lambda}\right)}$.

If $x_0 = -\dfrac{1}{2}$, we find

$$\dim_{H_{x_0}^{\beta,\alpha}}\left(C\left(\lambda\right)\right) \geq \frac{\log 2}{\log\left(\frac{1}{\lambda}\right)} \text{ and } \frac{\alpha^{\beta}}{c\left(\frac{3}{2}\right)^{(1-\alpha)\beta}} \leq H_{-\frac{1}{2}}^{\beta,\alpha}\left(C\left(\lambda\right)\right) \leq \frac{\alpha^{\beta}}{c\left(\frac{1}{2}\right)^{(1-\alpha)\beta}} \tag{55}$$

for the same β

Conclusion: If $x_0 \notin \left[0,1\right]$ the relative dimension of the Cantor set is constant and coincides with Häusdorff 's dimension of the Cantor, and if $x_0 \in \left[0,1\right] \cap \partial I_{k,j}$ for $(k,j)\in\mathbb{N}\times\mathbb{N}^*$, or if $x_0\in[0,1]\backslash C(\lambda)$, the relative dimension of the Cantor set verifies

$$\frac{\log 2}{\log\left(\frac{1}{\lambda}\right)} \leq \dim_{H_{x_0}^{\beta,\alpha}}\left(C\left(\lambda\right)\right) \leq \frac{\log 2}{\alpha\log\left(\frac{1}{\lambda}\right)}$$

which is a value greater than or equal to the Häusdorff 's dimension according to the value of $\alpha\in]0,1]$.

If $\alpha=1$ we find

$$\dim_{H_{x_0}^{\beta,1}}\left(C\left(\lambda\right)\right) = \dim_{\mathcal{H}^{\beta}}\left(C\left(\lambda\right)\right) \tag{56}$$

otherwise the relative dimension of the Cantor set is greater than the Häusdorff 's dimension.

Theorem 5. The relative dimension of the Cantor set is the value β that verifies:

1. If $x_0 \in [0,1] \setminus C(\lambda)$, or $x_0 \in [0,1] \cap \partial I_{k,j}$ for $(k,j) \in \mathbb{N} \times \mathbb{N}^*$, then

$$\frac{\log 2}{\log\left(\frac{1}{\lambda}\right)} \le \beta \le \frac{\log 2}{\alpha \log\left(\frac{1}{\lambda}\right)} \tag{57}$$

2. If $x_0 \notin [0,1]$, then

$$\beta = \frac{\log 2}{\log\left(\frac{1}{\lambda}\right)} \tag{58}$$

Remark.

1. For $x_0 = 0$ we have

$$\sum_{i=1}^{2^k} \mu_{x_0}^{\alpha}\left(I_{k,i}\right) = \lambda^{\alpha\beta k} + \left[\left(2\lambda^k + \left(1-2\lambda\right)\lambda^{k-1}\right)^{\alpha} - \left(\lambda^k + \left(1-2\lambda\right)\lambda^{k-1}\right)^{\alpha}\right]^{\beta} + \left[\left(5\lambda^k + 2\left(1-2\lambda\right)\lambda^{k-1}\right)^{\alpha} - \left(4\lambda^k + 2\left(1-2\lambda\right)\lambda^{k-1}\right)^{\alpha}\right]^{\beta} + \ldots \tag{59}$$

$$= \lambda^{\alpha\beta k}\left(1 + \left[\left(2 + \frac{\left(1-2\lambda\right)}{\lambda}\right)^{\alpha} - \left(1 + \frac{\left(1-2\lambda\right)}{\lambda}\right)^{\alpha}\right]^{\beta} + \left[\left(5 + \frac{2\left(1-2\lambda\right)}{\lambda}\right)^{\alpha} - \left(4 + \frac{2\left(1-2\lambda\right)}{\lambda}\right)^{\alpha}\right]^{\beta} + \ldots\right) \tag{60}$$

and then

$$\sum_{i=1}^{2^k} \mu_{x_0}^{\alpha}\left(I_{k,i}\right) = \lambda^{\alpha\beta k}\left(u_1 + u_2 + \ldots + u_{2^k}\right) \tag{61}$$

with (u_n) the sequence given by

$$u_n = \begin{cases} \left(\left(1+P_n\left(\frac{1}{\lambda}\right)\right)^{\alpha} - \left(P_n\left(\frac{1}{\lambda}\right)\right)^{\alpha}\right)^{\beta} & if\ n\ is\ odd\ number \\ \left(\left(P_n\left(\frac{1}{\lambda}\right)\right)^{\alpha} - \left(P_n\left(\frac{1}{\lambda}\right)-1\right)^{\alpha}\right)^{\beta} & if\ n\ is\ even\ number \end{cases} \tag{62}$$

where P_n is a polynomial, nevertheless we can see that for an even number of the form 2^k, we have $P_{2^k}\left(\frac{1}{\lambda}\right) = \frac{1}{\lambda^k}$ and this may lead to the following conclusion

a. If $\beta = \beta_2 = \dfrac{\log 2}{\alpha \log \frac{1}{\lambda}}$ then

$$\sum_{i=1}^{2^k} \left(\mu_{x_0}^{\alpha}\left(I_{k,i}\right)\right)^{\beta_2} = \lambda^{\alpha\beta_2 k}\left(u_1 + u_2 + \ldots + u_{2^k}\right) \to 0 \tag{63}$$

Indeed, by Cesaro's mean for the sequence (u_n) if it converges, we have

$$\frac{u_1 + u_2 + \ldots + u_{2^k}}{2^k} \approx u_{2^k} \to 0 \tag{64}$$

and we conclude $\dim_{H_{x_0}^{\beta_2,\alpha}}\left(C\left(\lambda\right)\right) \leq \beta_2$

b. If $\beta = \beta_1 = \dfrac{\log 2}{\log \frac{1}{\lambda}}$ then

$$\sum_{i=1}^{2^k} \left(\mu_{x_0}^{\alpha}\left(I_{k,i}\right)\right)^{\beta_1} = \lambda^{\alpha\beta_1 k}\left(u_1 + u_2 + \ldots + u_{2^k}\right) \to \alpha^{\beta_1} \tag{65}$$

indeed, we have

$$\frac{u_1 + u_2 + \ldots + u_{2^k}}{2^{k\alpha}} \approx u_{2^k} 2^{k(1-\alpha)} \tag{66}$$

where

$$u_{2^k} = \left(\left(\frac{1}{\lambda^k} \right)^{\alpha} - \left(\frac{1}{\lambda^k} - 1 \right)^{\alpha} \right)^{\beta_1},$$

then

$$\lim_{k \to \infty} u_{2^k} 2^{k(1-\alpha)} \approx \lim_{k \to \infty} 2^{k(1-\alpha)} \lambda^{k\beta_1(1-\alpha)} \alpha^{\beta_1}$$

and since $\lambda^{\beta_1} = \frac{1}{2}$ then we conclude

$$\lim_{k \to \infty} u_{2^k} 2^{k(1-\alpha)} = \alpha^{\beta_1}$$

we have then

$$\sum_{i=1}^{2^k} \left(\mu_{x_0}^{\alpha} \left(I_{k,i} \right) \right)^{\beta_1} \to \alpha^{\beta_1} > 0 \tag{67}$$

which means that for $x_0=0$ and $\alpha<1$ we have

$$\dim_{H_{x_0}^{\beta,\alpha}} \left(C\left(\lambda\right) \right) \geq \frac{\log 2}{\log\left(\frac{1}{\lambda} \right)}$$

2. To verify (51) for a special case, suppose that each interval I_j cover two intervals of the k-th generation intervals of the Cantor, and then by construction, we have

$$\mu_{x_0}^{\alpha} \left(I_j \right) \geq \mu_{x_0}^{\alpha} \left(I_{k,i} \right) \tag{68}$$

for all $I_{k,i} \subset I_j$, thus we get

$$\mu_{x_0}^{\alpha} \left(I_j \right) \geq \frac{1}{2} \sum_{I_{k,i} \subset I_j} \mu_{x_0}^{\alpha} \left(I_{k,i} \right) \tag{69}$$

The convexity of the function $f(x)=x^{\beta}$ for $\beta \in]0,1[$ gives

$$\left(\mu_{x_0}^{\alpha}\left(I_j\right)\right)^{\beta} \geq \left(\frac{1}{2}\sum_{I_{k,i}\subset I_j}\mu_{x_0}^{\alpha}\left(I_{k,i}\right)\right)^{\beta} \geq \frac{1}{2}\sum_{I_{k,i}\subset I_j}\left(\mu_{x_0}^{\alpha}\left(I_{k,i}\right)\right)^{\beta} \tag{70}$$

Then

$$4\left(\mu_{x_0}^{\alpha}\left(I_j\right)\right)^{\beta} \geq 2\sum_{I_{k,i}\subset I_j}\left(\mu_{x_0}^{\alpha}\left(I_{k,i}\right)\right)^{\beta} \geq \sum_{I_{k,i}\subset I_j}\left(\mu_{x_0}^{\alpha}\left(I_{k,i}\right)\right)^{\beta} \tag{71}$$

and we have a finite constant $c=4$.

3. INTEGRAL AND DERIVATIVE OF ORDER α∈]0,1]

In 1847 Riemann introduced the definition of integral of real order that is called today the Riemann-Liouville's integral of real order (Oldham & Spanier, 1974; Samko et al, 1987; Dugowson, 1997). This integral of real order is defined as follow:

Definition 9. Let f be a continuous function in $[a,b]$ with $-\infty<a<b<+\infty$, the integral of order $\alpha\in]0,+\infty[$ of the function f at the left of the point $x\in]a,b[$, if it exists, is defined by

$$I_a^{\alpha}f\left(x\right) = \frac{1}{\Gamma\left(\alpha\right)}\int_a^x\left(x-t\right)^{\alpha-1}f\left(t\right)dt. \tag{72}$$

and the integral of order α at the right of the point $x\in]a,b[$, if it exists, is given by

$$I_b^{\alpha}f\left(x\right) = \frac{1}{\Gamma\left(\alpha\right)}\int_x^b\left(t-x\right)^{\alpha-1}f\left(t\right)dt. \tag{73}$$

Remark. Using the concept of measure of real order that we introduced above for $\alpha\in]0,1]$, we can re-write the integral of real order of Riemann-Liouville and present it in another form. Indeed since the derivative with respect to t of

$$\frac{-\left(x-t\right)^{\alpha}}{\Gamma\left(\alpha+1\right)} \text{ is } \frac{\left(x-t\right)^{\alpha-1}}{\Gamma\left(\alpha\right)}$$

and the derivative respect to t of

$$\frac{\left(t-x\right)^{\alpha}}{\Gamma\left(\alpha+1\right)} \text{ is } \frac{\left(t-x\right)^{\alpha-1}}{\Gamma\left(\alpha\right)}$$

and since Γ(α+1)=αΓ(α) then one can write

$$ {}_{a}I_{x}^{\alpha}f\left(x\right)=\frac{1}{\Gamma\left(\alpha+1\right)}\int_{a}^{x}f\left(t\right)d\mu_{x}^{\alpha}\left(t\right)=\int_{a}^{x}f\left(t\right)d\eta_{x}^{\alpha}\left(t\right) \tag{74}$$

with

$$d\eta_{x}^{\alpha}\left(t\right)=\frac{1}{\Gamma\left(\alpha+1\right)}d\mu_{x}^{\alpha}\left(t\right),$$

and where

$$\mu_{x}^{\alpha}\left(t\right)=\frac{-1}{\Gamma\left(\alpha+1\right)}\left(x-t\right)^{\alpha},$$

$$ {}_{x}I_{b}^{\alpha}f\left(x\right)=\frac{1}{\Gamma\left(\alpha+1\right)}\int_{x}^{b}f\left(t\right)d\mu_{x}^{\alpha}\left(t\right)=\int_{x}^{b}f\left(t\right)d\eta_{x}^{\alpha}\left(t\right) \tag{75}$$

with

$$d\eta_{x}^{\alpha}\left(t\right)=\frac{1}{\Gamma\left(\alpha+1\right)}d\mu_{x}^{\alpha}\left(t\right),$$

and where

$$\mu_{x}^{\alpha}\left(t\right)=\frac{1}{\Gamma\left(\alpha+1\right)}\left(t-x\right)^{\alpha},$$

and we can say that the measure η_{x}^{α} is also a density measure (or a Radon measure) which satisfies the same properties that the "apparent" measure μ_{x}^{α}. If $d\eta_{x}^{\alpha}\left(t\right)=dF\left(t\right)$ then

$$F\left(t\right)=\frac{1}{\Gamma\left(\alpha+1\right)}\left(t-x\right)^{\alpha}$$

or

$$F\left(t\right) = \frac{-1}{\Gamma\left(\alpha+1\right)}\left(x-t\right)^{\alpha}$$

following the considered right or the left side.

4.1. Integral of Real Order α∈]0,1] of a Constant Function

Let $f(x)=C$ be a constant for all $x_0 \in [a,b]$. Let $x_0 \in]a,b[$, we have

$$I_a^{\alpha} f\left(x_0\right) = \int_a^{x_0} f\left(t\right) d\eta_{x_0}^{\alpha}\left(t\right) = C\int_a^{x_0} d\eta_{x_0}^{\alpha}\left(t\right) = C\eta_{x_0}^{\alpha}\left(\left]a, x_0\right[\right) = C\frac{\left(x_0 - a\right)^{\alpha}}{\Gamma\left(\alpha+1\right)} \quad (76)$$

This area is greater than the area $C(x_0-a)$ if $(x_0-a)\leq 1$ (see Fig. 3, b. with a.) and less than the area $C(x_0-a)$ if $(x_0-a)>1$ (see Fig. 3, c. with a.).

Figure 3. The graph a. represents the area distribution of the integral between a and x_0 of a constant function C. The graph b. represents the area distribution of the integral of real order between a and x_0 of a constant function C using the Riemann-Liouville's integral of real order for $x_0-a\leq 1$. The graph c. represents the area distribution of the integral of real order between a and x_0 of a constant function C using the Riemann-Liouville's integral of real order for $x_0-a>1$.

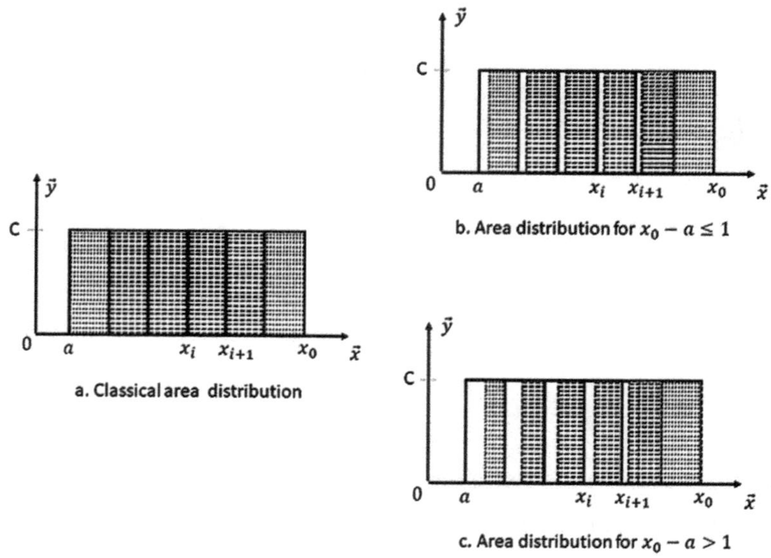

To understand the distribution of this area with respect to the apparent measure, let us consider a subdivision of interval $[a,x_0]$ into n sub intervals of equal size

$$a = x_1 < x_2 < \ldots < x_n = x_0 \tag{77}$$

we have then

$$I_a^\alpha f(x_0) = \int_a^{x_0} f(t)\, d\eta_{x_0}^\alpha(t) = \sum_{i=1}^{n} \int_{x_{i-1}}^{x_i} C d\eta_{x_0}^\alpha(t) = C\sum_{i=1}^{n} \eta_{x_0}^\alpha([x_{i-1}, x_i]) \tag{78}$$

4.2. Comments

The integral of real order is an antiderivative, and the area gives a representation of this antiderivative. Nevertheless, the area obtained depends on the value of α and of the interval length $[a,x_0]$, whether its length is greater or smaller than 1 (see Figure 3). In both cases there is always the following property

$$C\eta_{x_0}^\alpha([x_{n-1}, x_0]) > C\eta_{x_0}^\alpha([x_{n-2}, x_{n-1}]) > \ldots > C\eta_{x_0}^\alpha([x_{n-i}, x_{n-i+1}]) > \ldots > C\eta_{x_0}^\alpha([x_{i-1}, x_i]) \ldots > C\eta_{x_0}^\alpha([a, x_2]) \tag{79}$$

which means that the further x_0 is, the smaller the area of the rectangles obtained by the subdivision is.

4.3. Integral of Real Order of a Step Function

Let f be a step function in $[a,b]$, taking values ξ_k in $]x_k, x_{k+1}[$, and let $x_0 \in]a,b[$. We can then write f of the form

$$f(x) = \sum_{i=1}^{n} \xi_i 1_{E_i}(x) \tag{80}$$

where $E_i =]x_i, x_{i+1}[$ is a subdivision of interval $[a,b]$.

The integral of real order of the function f is given by:

$$I_a^\alpha f(x_0) = \int_a^{x_0} f(t)\, d\eta_{x_0}^\alpha(t) = \sum_{i=1}^{n} \int_{x_i}^{x_{i+1}} \xi_i\, d\eta_{x_0}^\alpha(t) = \sum_{i=1}^{n} \xi_i \eta_{x_0}^\alpha(]x_{i-1}, x_i[) \tag{81}$$

Therefore the expression "f admits an integral of real order on the left of the point x_0" means that the number

$$\sum_{i=1}^{n} \xi_i \eta_{x_0}^{\alpha}\left(\left]x_{i-1}, x_i\right[\right) \tag{82}$$

can be associated with the function *f* which represents a sum of area of rectangles with height ξ_i and base $\eta_{x_0}^{\alpha}\left(\left]x_{i-1}, x_i\right[\right)$, for $1 \leq i \leq n$ (see Fig. 4).

Figure 4. Area distribution for a step function using the Riemann-Liouville's integral of real order, the obtained area represents the sum of rectangles with height ξ_i the values of the step function in $\left]x_{i-1}, x_i\right[$ and base given by $\eta_{x_0}^{\alpha}\left(\left]x_{i-1}, x_i\right[\right)$, for $1 \leq i \leq n$

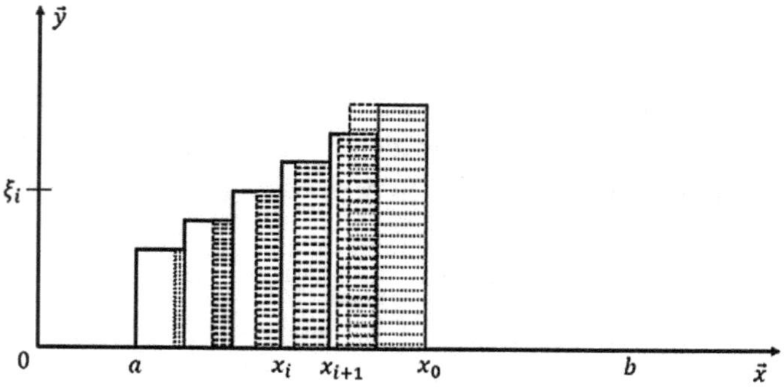

Remarks.

1. Intuitive considerations on the integral of real order:

The integral from *a* to x_0 of the product *f*(*t*) by the differential element $d\eta_{x_0^+}^{\alpha}$ represents a limit of the sum of n products

$$f\left(t_i\right) \times \eta_{x_0}^{\alpha}\left(\left[t_i, t_{i+1}\right]\right) \tag{83}$$

as n tends to the infinity, obtained by dividing the interval $[a,x_0]$ into *n* sub intervals,

$$t_1 = a, t_2 = a + \frac{x_0 - a}{n}, \ldots, t_i = a + \frac{\left(i-1\right)\left(x_0 - a\right)}{n}, \ldots, t_n = x_0 \tag{84}$$

Let *S* be the sum of *n* products

$$f\left(t_i\right) \times \eta_{x_0}^{\alpha}\left(\left[t_i, t_{i+1}\right]\right)$$

representing the area of *n* rectangles of base

$$\eta_{x_0}^{\alpha}\left(\left[t_i, t_{i+1}\right]\right) = \frac{1}{\Gamma\left(\alpha+1\right)}\left(\left(x_0 - t_i\right)^{\alpha} - \left(x_0 - t_{i+1}\right)^{\alpha}\right)$$
$$= \left(\frac{x_0 - a}{n}\right)^{\alpha}\left(\left(n - i + 1\right)^{\alpha} - \left(n - i\right)^{\alpha}\right)\frac{1}{\Gamma\left(\alpha+1\right)} \tag{85}$$

and height $f(t_i)$, we have then

$$S = \sum_{i=1}^{n} f\left(t_i\right)\left(\frac{x_0 - a}{n}\right)^{\alpha}\left(\left(n - i + 1\right)^{\alpha} - \left(n - i\right)^{\alpha}\right)\frac{1}{\Gamma\left(\alpha+1\right)} \tag{86}$$

As n tends to the infinity the base of the rectangle

$$\frac{1}{\Gamma\left(\alpha+1\right)}\left(\left(x_0 - t_i\right)^{\alpha} - \left(x_0 - t_{i+1}\right)^{\alpha}\right) \tag{87}$$

becomes the differential element

$$d\eta_{x_0}^{\alpha}\left(t\right) \tag{88}$$

with $\eta_{x_0}^{\alpha}\left(t\right) = \frac{\left(x_0 - t\right)^{\alpha}}{\Gamma\left(\alpha+1\right)}$ and the sum of n products $f\left(t_i\right) \times \eta_{x_0}^{\alpha}\left(\left[t_i, t_{i+1}\right]\right)$ becomes the integral of the "elementary products" of $f(t)$ by $d\eta_{x_0}^{\alpha}\left(t\right)$, where t vary from a to x_0.

2. Thanks to the measure $d\eta_{x_0}^{\alpha}$ in $\mathbb{R}$, we can rewrite several integrals of real order introduced in fractional calculus as follow in order to use the same interpretation of their integral of real order.

4.3.1. Weyl's Integral of Real Order (1917)

The Weyl integral of real order (Weyl, 1917), is given by

$$I_{-\infty}^{\alpha} f\left(x_0\right) = \frac{1}{\Gamma\left(\alpha\right)}\int_{-\infty}^{x_0}\left(x_0 - t\right)^{\alpha-1} f\left(t\right)dt = \int_{-\infty}^{x_0} f\left(t\right)d\eta_{x_0}^{\alpha}\left(t\right) \tag{89}$$

$$I_{+\infty}^{\alpha} f\left(x_0\right) = \frac{1}{\Gamma\left(\alpha\right)}\int_{x_0}^{+\infty}\left(t - x_0\right)^{\alpha-1} f\left(t\right)dt = \int_{x_0}^{+\infty} f\left(t\right)d\eta_{x_0}^{\alpha}\left(t\right) \tag{90}$$

The geometrical interpretation of Weyl integral of real order can be obtained using the same properties of the "apparent" measure μ_x^α. If the function $f(t)=c$ is a constant in the integral (89) then we have $I_{-\infty}^{\alpha} f(x_0) = c\ \ \eta_{x_0}^{\alpha}\left(\left]-\infty, x_0\right]\right)$ is not defined.

4.3.2. Erdelyi's Integral of Real Order (1964)

The Erdelyi's integral of real order (Eldélyi, 1964), is given by

$$I_0^\alpha f(x_0) = \frac{1}{\Gamma(\alpha)} \int_0^{x_0} (x_0 - t)^{\alpha-1} f(t)\, dt \tag{91}$$

$$I_{+\infty}^\alpha f(x_0) = \frac{1}{\Gamma(\alpha)} \int_{x_0}^{+\infty} (t - x_0)^{\alpha-1} f(t)\, dt \tag{92}$$

by using the concept of measurement of real order, we have

$$I_0^\alpha f(x_0) = \int_0^{x_0} f(t)\, d\eta_{x_0}^\alpha dt,\ \ I_{+\infty}^\alpha f(x_0) = \int_{x_0}^{+\infty} f(t)\, d\eta_{x_0}^\alpha dt \tag{93}$$

This notation leads to obtain a geometrical interpretation of this integral of real order similar to the Riemann-Liouville's integral of real order.

If the function $f(t)=c$ is a constant in the integral (93) gives $I_0^\alpha f(x_0) = c\ \ \eta_{x_0}^\alpha\left(\left[0, x_0\right]\right) < \infty$ meanwhile $I_{+\infty}^\alpha f(x_0) = c\ \ \eta_{x_0}^\alpha\left(\left[x_0, +\infty\right[\right)$ which is not defined.

4.3.3. Kober's Integral of Real Order (1940)

The Kober's integral of real order (Kober, 1940) is given by

$$I_0^{\beta,\alpha} f(x_0) = \frac{x_0^{-\beta-\alpha}}{\Gamma(\alpha)} \int_0^{x_0} t^\beta (x_0 - t)^{\alpha-1} f(t)\, dt \tag{94}$$

and

$$I_{+\infty}^{\beta,\alpha} f(x_0) = \frac{x_0^\beta}{\Gamma(\alpha)} \int_{x_0}^{+\infty} t^{-\beta-\alpha} (t - x_0)^{\alpha-1} f(t)\, dt \tag{95}$$

by using the concept of measurement of a real order, the geometrical interpretation of this integral is more complicated

$$I_{0}^{\beta,\alpha} f\left(x_{0}\right) = x_{0}^{-\beta-\alpha} \int_{0}^{x_{0}} t^{\beta} f\left(t\right) \eta_{x_{0}}^{\alpha}\, dt \qquad (96)$$

$$I_{+\infty}^{\beta,\alpha} f\left(x_{0}\right) = x_{0}^{\beta} \int_{x_{0}}^{+\infty} t^{-\beta-\alpha} f\left(t\right) \eta_{x_{0}}^{\alpha}\, dt \qquad (97)$$

but in general we can have a geometric interpretation of the integral of real order of the function $x^{\delta}f(x)$ for all $\delta \in \mathbb{R}$.

4.3.4. Caputo's Derivative of Real Order (1967)

The Caputo's derivative of real order (Caputo, 1967; Caputo, 1969) is defined by an integral of the form

$$f^{(\beta)}\left(x_{0}\right) = \frac{1}{\Gamma\left(n-\beta\right)} \int_{a}^{x_{0}} \frac{f^{(n)}\left(\tau\right)}{\left(x_{0}-\tau\right)^{(\beta+1-n)}}\, d\tau, \text{ where } n-1<\beta<n \text{ and } n \in \mathbb{N} - \{0\} \qquad (98)$$

To use the apparent measure let put $\alpha=\beta-n+1$, then we have $f^{(\beta)}\left(x_{0}\right) = \left(f^{(n)}\right)^{(\alpha)}\left(x_{0}\right)$ with $\alpha \in]0,1[$, this integral can be written with the notion of apparent measure as follow

$$f^{(\beta)}\left(x_{0}\right) = \int_{a}^{x_{0}} f^{(n)}\left(\tau\right) d\eta_{x_{0}}^{1-\alpha}\left(\tau\right) \text{for } 0<\alpha<1 \qquad (99)$$

which enables us to obtain a geometric interpretation of the Caputo's derivative of real order as being the area of real order of the derivative function $f^{(n)}(t)$.

4.4. Integral of Real Order: A Symmetric Definition

The geometric interpretation of the integral of real order of Riemann-Liouville showed that the area represented by this antiderivative of real order is more or less significant as the point x_0 is close or far from it. It follows that we have to modify this integral to exploit this significant information.

Definition 10. Let $f(x) \in L^{1}(a,b)$, $-\infty \leq a < b \leq +\infty$. We call integral of order $\alpha \in \mathbb{R}^{+}$ of f on the right of the point x (resp. on the left of the point x) the quantity

$$\mathcal{T}_{\tau^+}^{\alpha} f(x) = \frac{1}{\Gamma(\alpha)} \int_x^{x+\tau} \frac{f(t)}{(t-x)^{1-\alpha}} dt = \int_x^{x+\tau} f(t)\, d\eta_{x+}^{\alpha}(t) \qquad (100)$$

$$\mathcal{T}_{\tau^-}^{\alpha} f(x) = \frac{1}{\Gamma(\alpha)} \int_{x-\tau}^{x} \frac{f(t)}{(x-t)^{1-\alpha}} dt = \int_{x-\tau}^{x} f(t)\, d\eta_{x-}^{\alpha}(t). \qquad (101)$$

where $\tau>0$ is an arbitrary positive real and

$$d\eta_{x+}^{\alpha}(t) = \frac{1}{\Gamma(\alpha)} (t-x)^{\alpha-1}$$

and

$$d\eta_{x-}^{\alpha}(t) = \frac{1}{\Gamma(\alpha)} (x-t)^{\alpha-1}$$

Proposition 3. If n is a positive integer, and $f_n(t)=t^n$ we have then

$$\mathcal{T}_{\tau^-}^{\alpha} f_n(x) = \sum_{k=0}^{n} (x+\tau)^{n-k} \frac{n!}{(n-k)!} \mathcal{I}_{\tau^-}^{\alpha+k}(1) \qquad (102)$$

With

$$\mathcal{T}_{\tau^-}^{\alpha+k}(1) = \mathcal{T}_{\tau^+}^{\alpha+k}(1) = \frac{1}{\Gamma(\alpha+k)} \int_x^{x+\tau} \frac{1}{(t-x)^{1-(\alpha+k)}} dt = \frac{1}{\Gamma(\alpha+k)} \int_0^{\tau} \frac{1}{t^{1-(\alpha+k)}} dt = \frac{\tau^{\alpha+k}}{\Gamma(\alpha+k+1)}$$

Proof. By integration by part we obtain the following equality

$$\mathcal{T}_{\tau^+}^{\alpha} f_n(x) = (x+\tau)^n \mathcal{T}_{\tau^+}^{\alpha}(1) - n\mathcal{T}_{\tau^-}^{\alpha+1} f_{n-1}(x) \qquad (103)$$

and

$$\mathcal{T}_{\tau^-}^{\alpha} f_n(x) = (x-\tau)^n \mathcal{T}_{\tau^-}^{\alpha}(1) + n\mathcal{T}_{\tau^-}^{\alpha+1} f_{n-1}(x) \qquad (104)$$

The result is obtained by recurrence on n.

Remarks.

1. According to the value of *n* we find:
   a. For *n*=0, and α∈]0,1], we have

$$\mathcal{T}^{\alpha}_{\tau^+} f_0(x) = \mathcal{T}^{\alpha}_{\tau^-} f_0(x) = \frac{\tau^{\alpha}}{\Gamma(\alpha+1)} \tag{105}$$

   b. For *n*=1, we have

$$\mathcal{T}^{\alpha}_{\tau^+} f_1(x) = (x+\tau)\mathcal{T}^{\alpha}_{\tau^+}(1) - \mathcal{T}^{\alpha+1}_{\tau^+}(1) \tag{106}$$

and

$$\mathcal{T}^{\alpha}_{\tau^-} f_1(x) = (x-\tau)\mathcal{T}^{\alpha}_{\tau^-}(1) + \mathcal{T}^{\alpha+1}_{\tau^-}(1) \tag{107}$$

then

$$\mathcal{T}^{\alpha}_{\tau^+} f_1(x) \neq \mathcal{T}^{\alpha}_{\tau^-} f_1(x) \tag{108}$$

   c. For *n*=2, we have

$$\mathcal{T}^{\alpha}_{\tau^+} f_2(x) = (x+\tau)^2\, \mathcal{T}^{\alpha}_{\tau^+}(1) - 2(x+\tau)\mathcal{T}^{\alpha+1}_{\tau^+}(1) + 2\mathcal{T}^{\alpha+2}_{\tau^+}(1) \tag{109}$$

and

$$\mathcal{T}^{\alpha}_{\tau^-} f_2(x) = (x-\tau)^2\, \mathcal{T}^{\alpha}_{\tau^-}(1) - 2(x-\tau)\mathcal{T}^{\alpha+1}_{\tau^-}(1) + 2\mathcal{T}^{\alpha+2}_{\tau^-}(1) \tag{110}$$

then, we have

$$\mathcal{T}^{\alpha}_{\tau^+} f_2(x) \neq \mathcal{T}^{\alpha}_{\tau^-} f_2(x) \tag{111}$$

which allows us to say that ∀*n*>0, we have

$$\mathcal{T}^{\alpha}_{\tau^+} f_n(x) \neq \mathcal{T}^{\alpha}_{\tau^-} f_n(x) \tag{112}$$

2. Like for Riemann-Liouville's integral of real order (Kober, 1940; Eldélyi, 1964 ; Dugowson, 1997),, we have $\mathcal{T}^{\alpha}_{\tau^{\pm}} f(x)$ tends to *f*(*x*) when α tends to 0 for all function $f \in C^0$, we can then put the following equality by continuity

$$\mathcal{T}^{0}_{\tau^{\pm}} f(x) = f(x)$$

3. We can also consider the concept of a symmetric integral of real order given by

$$\mathcal{T}^{\alpha}_{\tau} f(x) = \frac{1}{2}\left(\mathcal{T}^{\alpha}_{\tau^{+}} f(x) + \mathcal{T}^{\alpha}_{\tau^{-}} f(x)\right) = \frac{1}{\Gamma(\alpha)} \int_{x-\tau}^{x+\tau} f(t) \left|x - t\right|^{\alpha-1} dt \tag{113}$$

4.5. Derivative of Real Order

The definition of the Riemann-Liouville's fractional derivative on the left and on the right of $x \in [a,b]$ is given by:

Let $p \geq 1$ be an integer given by $p-1 < \alpha \leq p$, we have

$$D^{\alpha}_{a} f(x) = \frac{d}{dx} I^{p-\alpha}_{a} f(x), \; D^{\alpha}_{b} f(x) = \frac{d}{dx} I^{p-\alpha}_{b} f(x) \tag{114}$$

Now one can introduce the fractional derivative associated with symmetric integral of real order.

Definition 11. Let $f(x) \in L^1(a,b)$, $-\infty \leq a < b \leq +\infty$. We define the fractional derivative of order $\alpha \in [0,1[$ of *f* on the right and on the left of the point *x* respectively, by the quantities

$$D^{\alpha}_{\tau^{+}} f(x) = \frac{d\mathcal{T}^{1-\alpha}_{\tau^{+}} f(x)}{dx} \text{ and } D^{\alpha}_{\tau^{-}} f(x) = \frac{d\mathcal{T}^{1-\alpha}_{\tau^{-}} f(x)}{dx} \tag{115}$$

and for $\alpha > 1$ by the quantities

$$D^{\alpha}_{\tau^{+}} f(x) = \frac{d^n\mathcal{T}^{n-\alpha}_{\tau^{+}} f(x)}{dx^n} \text{ and } D^{\alpha}_{\tau^{-}} f(x) = \frac{d^n\mathcal{T}^{n-\alpha}_{\tau^{-}} f(x)}{dx^n} \tag{116}$$

for $n-1 < \alpha \leq n$, and τ an arbitrary constant.

Remark. If α=0, we have

$$D^{0}_{\tau^{+}} f(x) = f(x+\tau) - f(x)$$

and

$$D^{0}_{\tau^-} f(x) = f(x) - f(x-\tau)$$

one can deduce the following

Corollary 1. If the function $f(x)$ is constant, then

$$D^{\alpha}_{\tau^+} f(x) = D^{\alpha}_{\tau^-} f(x) = 0 \tag{117}$$

We have a significant consequence of this new definition: the derivative of order $\alpha\in]0,1]$ of a constant function is equal to zero, being therefore consistent with traditional differential calculus.

Corollary 2. If n is a positive integer, then

$$D^{\alpha}_{\tau^+}(x^n) = \sum_{k=0}^{n-1} (-1)^k (x+\tau)^{n-k-1} \frac{n!}{(n-k-1)!} \mathcal{T}^{1-\alpha+k}_{\tau^+}(1) \tag{118}$$

and

$$D^{\alpha}_{\tau^-}(x^n) = \sum_{k=0}^{n-1} (x-\tau)^{n-k-1} \frac{n!}{(n-k-1)!} \mathcal{T}^{1-\alpha+k}_{\tau^-}(1) \tag{119}$$

Remark. A new definition of the derivation of a real order can be introduced, exploiting at the same time the fractional derivative of Riemann-Liouville modified on the left and on the right.

Definition 12. Let f be a function in $L^1(a,b)$, $-\infty \le a < b \le +\infty$. We call derivative of order $\alpha\in R^+$ of f the quantity

$$f^{(\alpha)}(x) = \frac{1}{2}\left(D^{\alpha}_{\tau^+} f(x) + D^{\alpha}_{\tau^-} f(x)\right) \tag{120}$$

This definition will be the subject of another work.

REFERENCES

Al-Khedhairi, A., Askar, S. S., Matouk, A. E., Elsadany, A., & Ghazel, M. (2018). Dynamics, chaos control, and synchronization in a fractional-order Samardzija-Greller population system with order lying in (0,2). *Complexity*.

Al-khedhairi, A., Matouk, A. E., & Askar, S. S. (2019a). Computations of synchronization conditions in some fractional-order chaotic and hyperchaotic systems. *Pramana Journal of Physics, 92*, 72.

Al-khedhairi, A., Matouk, A. E., & Khan, I. (2019b). Chaotic dynamics and chaos control for the fractional-order geomagnetic field model. *Chaos, Solitons, and Fractals, 128*, 390–401. doi:10.1016/j.chaos.2019.07.019

Ben Adda, F. (1997). Geometric interpretation of the fractional derivative. *J. Frac. Cal., 11*, 21–51.

Caputo, M. (1967). Linear models of dissipation whose Q is almost frequency independent-II. *Geophysical Journal of the Royal Astronomical Society, 13*(5), 529–539. doi:10.1111/j.1365-246X.1967.tb02303.x

Caputo, M. (1969). *Elasticité e dissipatizione*. Bologna: Zanichelli.

Dugowson, S. (1997). L'élaboration par Riemann d'une définition de la dérivation d'ordre non entier. *Revue d'Histoire de Mathématiques, Société mathématique de France, 3*, 49-97.

El-Sayed, A. M. A., Elsonbaty, A., Elsadany, A. A., & Matouk, A. E. (2016b). Dynamical analysis and circuit simulation of a new fractional-order hyperchaotic system and its discretization. International Journal of Bifurcation and Chaos, 26.

El-Sayed, A. M. A., Nour, H. M., Elsaid, A., Matouk, A. E., & Elsonbaty, A. (2016a). Dynamical Behaviors, Circuit Realization, Chaos Control and Synchronization of a New Fractional Order Hyperchaotic System. *Applied Mathematical Modelling, 40*(5-6), 3516–3534. doi:10.1016/j.apm.2015.10.010

Eldélyi, A. (1964). An integral equation involving Legendre function. *SIAM Journal on Applied Mathematics, 12*(1), 15–30. doi:10.1137/0112002

Hegazi, A. S., Ahmed, E., & Matouk, A. E. (2011). The effect of fractional order on synchronization of two fractional order chaotic and hyperchaotic systems. *J. Fract. Calc. Appl., 1*(3), 1–15.

Hegazi, A. S., Ahmed, E., & Matouk, A. E. (2013a). On chaos control and synchronization of the commensurate fractional order Liu system. *Communications in Nonlinear Science and Numerical Simulation, 18*(5), 1193–1202. doi:10.1016/j.cnsns.2012.09.026

Hegazi, A. S., Ahmed, E., & Matouk, A. E. (2013b). On Dynamical Behaviors and Chaos Control of the Fractional-order Financial System. In S. Banerjee (Ed.), *Chaos and Complexity Theory for Management: Nonlinear Dynamics* (pp. 34–49). IGI Global. doi:10.4018/978-1-4666-2509-9.ch002

Hegazi, A. S., & Matouk, A. E. (2011). Dynamical behaviors and synchronization in the fractional order hyperchaotic Chen system. *Applied Mathematics Letters, 24*(11), 1938–1944. doi:10.1016/j.aml.2011.05.025

Hegazi, A. S., & Matouk, A. E. (2013). Chaos control of the modified Autonomous Van der Pol-Duffing circuits via active control. In S. Banerjee & L. Rondoni (Eds.), *Applications of Chaos and Nonlinear Dynamics in Science and Engineering* (Vol. 3, pp. 185–202). Berlin: Springer. doi:10.1007/978-3-642-34017-8_7

Kober, H. (1940). On fractional integrals and derivatives. *Quart. J. Math. Oxford, 11*, 193-211.

Le Méhauté, A. (1991). Les Géométries fractales. *Traite des nouvelles technologies. Hermes.*

Matouk, A. E. (2009a). Chaos synchronization between two different fractional systems of Lorenz family. *Mathematical Problems in Engineering, 2009*, 1–11. doi:10.1155/2009/572724

Matouk, A. E. (2009b). Stability conditions, hyperchaos and control in a novel fractional order hyperchaotic system. *Physics Letters. [Part A], 373*(25), 2166–2173. doi:10.1016/j.physleta.2009.04.032

Matouk, A. E. (2010). Dynamical behaviors, linear feedback control and synchronization of the fractional order Liu system. *Journal of Nonlinear Systems and Applications, 1*(3), 135–140.

Matouk, A. E. (2011a). Chaos, feedback control and synchronization of a fractional-order modified Autonomous Van der Pol-Duffing circuit. *Communications in Nonlinear Science and Numerical Simulation, 16*(2), 975–986. doi:10.1016/j.cnsns.2010.04.027

Matouk, A. E. (2011b). Synchronization in integer and fractional order chaotic systems. In S. Banerjee (Ed.), *Chaos Synchronization and Cryptography for Secure Communications: Applications for Encryption* (pp. 127–151). IGI Global. doi:10.4018/978-1-61520-737-4.ch007

Matouk, A. E. (2016). Chaos synchronization of a fractional-order modified Van der Pol Duffing system via new linear control, backstepping control and Takagi-Sugeno fuzzy approaches. *Complexity, 21*(S1), 116–124. doi:10.1002/cplx.21719

Matouk, A. E., & Elsadany, A. A. (2014). Achieving synchronization between the fractional-order hyperchaotic Novel and Chen systems via a new nonlinear control technique. *Applied Mathematics Letters, 29*, 30–35. doi:10.1016/j.aml.2013.10.010

Matouk, A. E., & Elsadany, A. A. (2016). Dynamical analysis, stabilization and discretization of a chaotic fractional-order GLV model. *Nonlinear Dynamics, 85*(3), 1597–1612. doi:10.100711071-016-2781-6

Matouk, A. E., Elsadany, A. A., Ahmed, E., & Agiza, H. N. (2015). Dynamical behavior of fractional-order Hastings–Powell food chain model and its discretization. *Communications in Nonlinear Science and Numerical Simulation, 27*(1-3), 153–167. doi:10.1016/j.cnsns.2015.03.004

Oldham, K. B., & Spanier, J. (1974). *The Fractional calculus*. New York: Academic press.

Podlubny, I. (1999). *Fractional differential equations*. New York: Academic Press.

Riemann, B. (1990). Versuch einer allgemeinen auffassung der integration und differentiation (1847). Gesammelte mathematische Werke (H. Weber éd. avec le concours de R. Dedekind), Leipzig, 1876, p.353-366; 2 ième éd. 1892; rééd. Springer.

Roger. (1970). *Hausdorff measures*. Cambridge University Press.

Samko, S. G., Kilbas, A. A., & Marichev, O. I. (1987). *Fractional integral and derivatives, theory and application*. Minsk: Nauka I Teknika.

Srivastava, H. M., & Saigo, M. (1987). Multiplication of fractional calculus operators and boundary value problems involving the Euler-Darboux equation. *Journal of Mathematical Analysis and Applications*, *121*(2), 321–369. doi:10.1016/0022-247X(87)90251-4

Tricot, C. (1999). Courbe et dimension fractale (2nd ed.). Springer.

Weyl, H. (1917). Bermerkungen zum begriff des differentialquo tienten gebrochener ordnung, Vierteljschr. *Naturfosch. Gesellsch. Zurich*, *62*, 296–302.

Section 5
Applications of Fractional Calculus to Engineering

Chapter 14
Fractional-Order-Based Inferential Control:
A Case Study for the Level Control of Three Interacting Tank Processes

Abdul Wahid Nasir
https://orcid.org/0000-0002-0244-7290
CMR Institute of Technology, Bengaluru, India

Idamakanti Kasireddy
Vaagdevi College of Engineering, Warangal, India

Arun Kumar Singh
National Institute of Technology, Jamshedpur, India

ABSTRACT

This chapter presents the application of fractional differential operator in modelling and control of a three-tank interacting level process. In cases where the usage of sensors for the measurement of primary variable, which is the level of third tank in present case, is physically or economically not feasible, the measurement of secondary variable (i.e., second tank level) is used to determine the level of third tank for control purpose, known as inferential control scheme. The process is modeled and linearized around the operating points, resulting in third order plant, which is approximated to lower order integer and non-integer model. Both conventional integer order PI (IO-PI) & PID (IO-PID) and fractional order PI (FO-PI) & (FO-PID) controllers are implemented for this inferential control. Extensive simulation studies performed using MATLAB validate the supremacy of non-integer order model and controller over integer order model and controller. Genetic algorithm (GA) is being applied for both, firstly for reduced order model approximation and secondly for controller tuning.

DOI: 10.4018/978-1-7998-3122-8.ch014

INTRODUCTION

In Model Based Control as the term itself states, model of the plant is required for designing the control structure. Hence the performance of such model based control techniques depends considerably on the accuracy of plants being modelled. Accurate models will give improved controlled response whereas inaccurate models will degrade it. Some of the well-established control techniques falling in this category are Internal Model Control (*IMC*), Model Predictive Control (*MPC*) and Inferential Control etc.

The elementary concept of modelling lies in differential equations. Conventionally the order of differential operators incorporated in these differential equations used to be integer in nature. But in recent past a lot of studies has shown that if the order of these operators is generalized to non-integer (fractional), then capturing the dynamics of plants via models becomes more accurate and realistic (Elwakil, 2010; Efe, 2011; Kumar, 2013; Fahim et al, 2018). Research in the area of fractional order modelling and control has increased manifold in the last couple of decades due to the advent of various numerical techniques to deal with non-integer differential and integral operators (Oldham, 1974; Podlubny 1999). These non-integer order fundamental operators, where 'α' is the operator order and 'a' & 't' gives the limits of the operator is defined as follows:

$$
{}_aD_t^\alpha = \begin{cases} d^\alpha / dt^\alpha, & \alpha > 0 \\ 1, & \alpha = 0 \\ \int_a^t (d\tau)^{-\alpha}, & \alpha < 0 \end{cases}
$$

Different definitions of this fractional integro-differential is given by various experts namely Riemann-Liouville, Caputo, Grunwald-Lentnikov etc. (Monje et al, 2010). Hence control engineers are also making use of this concept in modelling and control (Monje et al, 2008; Wang et al, 2018; Shalaby, 2019), and are able to get satisfactory results. For carrying out all types of simulation work related to non-integer order plants, *FOMCON* toolbox (Tepljakov et al, 2011), has been used in this chapter.

A good number of cases have been reported in the literature where fractional order modelling has been exploited for IMC (Mamar & Rachid, 2013; Sondhi & Hote, 2014; Nasir & Singh, 2015) and MPC (Boudjehem & Boudjehem, 2012; Rhouma & Bouani, 2014; Joshi et al, 2014), but it has not been used to its potential for Inferential Control Scheme. Therefore, there is a need to investigate the usage of fractional calculus towards its application in Inferential Control. In this control scheme, the value of primary variable is inferred or estimated using the secondary variable, in situations where the measurement of primary variable is not available. Model of the plants plays a vital role in the estimation process. Accurate model of the plant will provide accurate estimation of primary controlled variable, which in turn will improve the control system performance. So, in present scenario non-integer modelling may be used to enhance controller performance owing to the accurate modelling of the plant. Generally, PI or PID type controller is used in Inferential Control, so additionally this scheme of control bears the potential to take advantage of fractional calculus by extending PI & PID to Fractional Order PI (FO-PI) and Fractional Order PID (FO-PID) respectively. Therefore, it is obvious that the utilization of Fractional Calculus (FC) will enhance the performance of Inferential Control, both from modelling as well as control point of view.

In light of the above discussion, a case study has been proposed in this chapter which deals with the level control of three interacting tank level process. Level control is one of the core control objective in various chemical, food and pharmaceutical industries. The control problem for such a level depends on how these tanks are coupled. Sometimes the tanks are so inter-connected that their level interacts, i.e. the level of one tank affects the dynamics of another tanks and vice-versa, a case of interacting process. In several such cases the level control problem of interacting tanks arises. So is the present case, where three tanks are inter-connected, with the objective of maintaining the level of 3rd tank at certain level. Sometimes, the physical conditions e.g. very high temperature, extreme chemical nature of process variable, etc. does not allow the installation at particular location. Due to these above mentioned reasons, if the sensor is absent to give the measurement of level of 3rd tank, then the sensor installed merely for monitoring the level of 2nd tank can be used to estimate the level of third tank for control purpose, known as Inferential Control Scheme. The level of 3rd tank can only be estimated from the level reading of 2nd tank, only if the plant models are available. In general FO modelling is exploited in two categories of system modelling, first one is input-output open loop data based modelling and the second one is the model order reduction of higher order systems to lower order modelling. In the present case, the later one is employed, to approximate the linearized 3rd order mathematical model of process to lower order integer and non-integer order models by minimizing error function using Genetic Algorithm (GA) to bring ease in control design and analysis. Both conventional integer order PI (IO-PI) & PID (IO-PID) and fractional order PI (FO-PI) & PID (FO-PID) controllers have been implemented for this Inferential Control. Extensive MATLAB simulations have been performed to have the comparative study among different control Inferential Control schemes arising due to the combinations of both integer & non-integer models and controllers.

INFERENTIAL CONTROL

In some control applications, measurement of process variable to be controlled is not available. If the process model is available and other supporting variable are available, then one can employ Inferential control scheme in such cases. The unmeasured controlled variable is known as primary variable and the supporting variable, with the help of which the measurement of primary variable is estimated is known as secondary variable. The implementation of such types of control schemes (Stephanopoulas, 1984), is effectively implemented in process industries and so is the case with present work. This control scheme can be best understood with the help of some block diagrams and equations given in the subsequent part of this section.

Refer to the block diagram of the process shown in Figure 1, with one unmeasured primary controlled variable y and one secondary measured output z. Manipulated variable is represented by m and d is the disturbance which affects both outputs. All the transfer functions G_{p1}, G_{p2}, G_{d1} and G_{d2} shown in the block diagram are assumed to be available. From Figure 1 following relations can be deduced.

$$y=G_{p1}m+G_{d1}d$$

$$z=G_{p2}m+G_{d2}d$$

From equation (3), '*d*' can be obtained as,

Figure 1. Block diagram of the plant

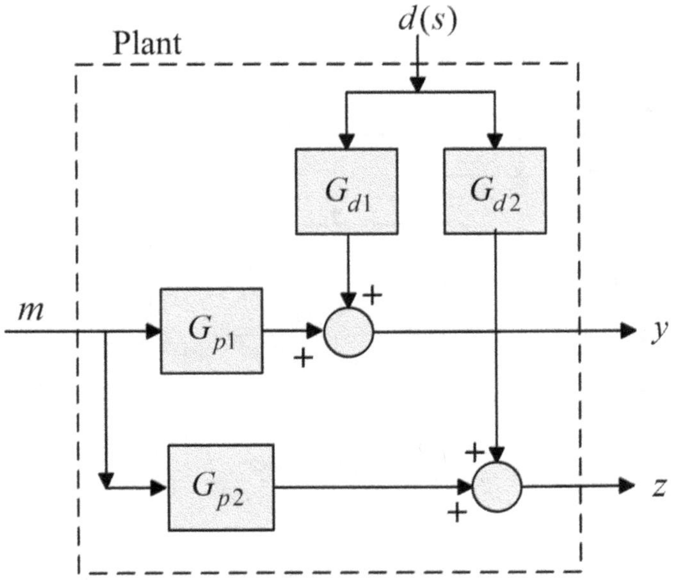

$$d = \frac{1}{G_{d2}} z - \frac{G_{p2}}{G_{d2}} m$$

Now substituting this value of '*d*' given by equation (4), in equation (2), results in following relationship,

$$y = \left[G_{p1} - \frac{G_{d1}}{G_{d2}} G_{p2} \right] m + \frac{G_{d1}}{G_{d2}} z$$

From the above equation (5), one can easily infer that y i.e. the unmeasured controlled output can be estimated in term of other measurable quantities like m & z, and can be realized using following block diagram as shown in Figure 2.

This estimated value of y is actually used to generate error signal for the controller after comparing it with desired set-point

GENETIC ALGORITHM

Genetic Algorithm is one of the most reliable and prominent optimization techniques among the various nature inspired evolutionary optimization methods that have flooded the research field (Brandstetter, 2014; Shabbir & Omenzetter, 2016; Jia & Yang, 2016; Dao et al, 2017). People from every section of science & technology are benefitting from this. The availability of fast and cheap computers has added feather to the cap regarding its implementation in the real world. The elementary principles of GA were first given by Holland (Holland, 1975), then later some literature appeared (Goldberg, 1989; Davis, 1991; Man et al, 1999), throwing some more lights on this noble idea. It is a stochastic nature based search

Figure 2. Inferential control scheme for the plant

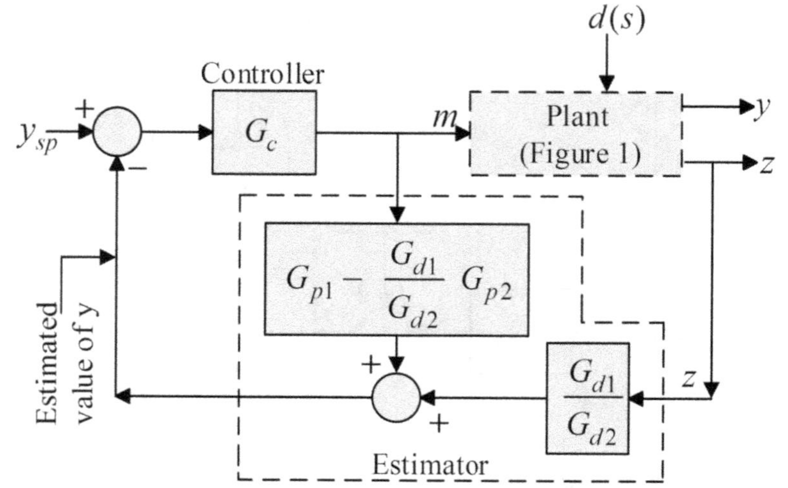

algorithm standing on the theory of natural evolution and selection which defends the fact that stronger individuals bear the greatest potential to fight against all odds in a competing environment and the weaker one gradually become extinct. Although there are some incidents where weaker individuals strive and are able to succeed, but these cases are limited to exceptions. Initially, a random set of parameters are assumed to be the possible solution of the problem and these random sets actually resemble the genes of chromosomes. One objective function is formulated which indicates the goodness of these chromosomes and this objective function is being minimized or maximized to obtain the optimized results.

Nowadays a general trend has developed to tune the controller parameters in control applications optimally by using some optimization technique, although various conventional methods of controller tuning are available because the former is able to produce comparatively improved results in majority of the cases. The implementation of GA in the current scheme of inferential control for a level process can be realized from following Figure 3, where 'C' is the controller used.

Table 1 gives the GA characteristics used to obtain the required results.

Table 1. GA Characteristics

Population Size	50
Fitness Scaling Function	Rank
Crossover Function	0.8
Crossover Fraction	Scattered
Migration Fraction	0.2
Ending Criterion	100 Iterations

Figure 3. Controller tuning of inferential control using GA

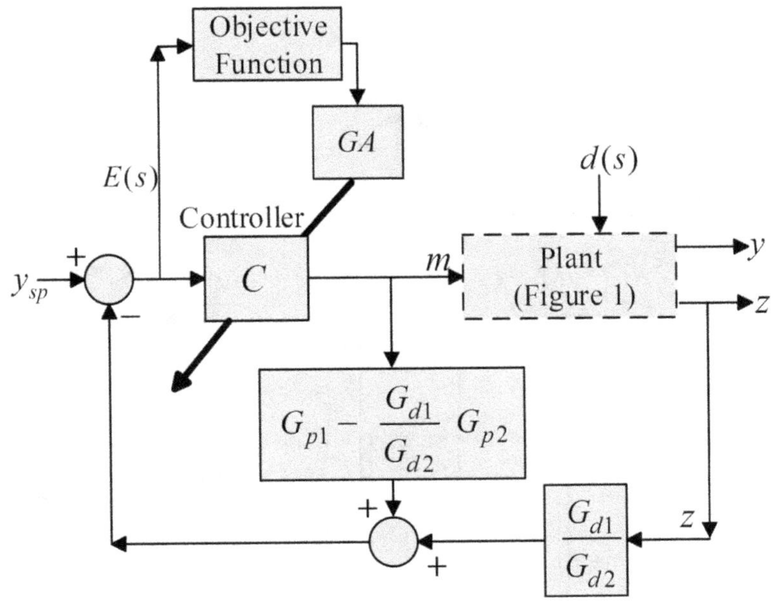

MATHEMATICAL MODELING OF THREE INTERACTING TANK PROCESS

Multi tank level control is considered to be a benchmark problem in studying the control strategy for interacting processes (Banu & Nasir; 2015). Consider the system shown in Figure 4, which represents a three tank interacting process. It consists of three identical cylindrical tanks Tank 1, Tank 2 and Tank 3, having same cross sectional area, A. Tank 1 and Tank 2 are inter-connected at the bottom through a manual control valve, having valve coefficient $\beta_{12,}$ using pipe of cross-sectional area α_1. Similarly Tank 2 and Tank 3 are inter-connected through another manual control valve, having valve coefficient $\beta_{23,}$ using pipe of cross-sectional area $\alpha_{2.}$ Also there is one drain outlet regulated through a manual control valve, having valve coefficient β3 using pipe of cross-sectional area $\alpha_{3.}$ The pump helps the liquid flow to the Tank1 and Tank 3 from sump through a control valve 1 and control valve 2 respectively. As the liquid enters the Tank 1 & Tank 2 the levels of all the three tanks rises in accordance to process dynamics. The control problem considered here is to maintain the liquid level of Tank 3, h_3 in desired range. For controlling the level of Tank 3, one should first have the level measurement of Tank 3. Since the Tank 3 level measurement, h_3 is not available, therefore it will be inferred from the level measurement of Tank 2, h_2 with the help of the technique as discussed in Section II. Hence in the present case h_3 is considered as unmeasurable controlled variable whose value is estimated using measurable secondary variable h_2 and is controlled by manipulating the liquid flow F_{in1} to the Tank 1. F_{in2}, is the constant flow to Tank 3, which is considered to be disturbance in the present case.

Assuming the fluid to be incompressible, the differential equations governing the dynamics of the plants based on conservation of mass are given below:

Figure 4. Three interacting Tank process

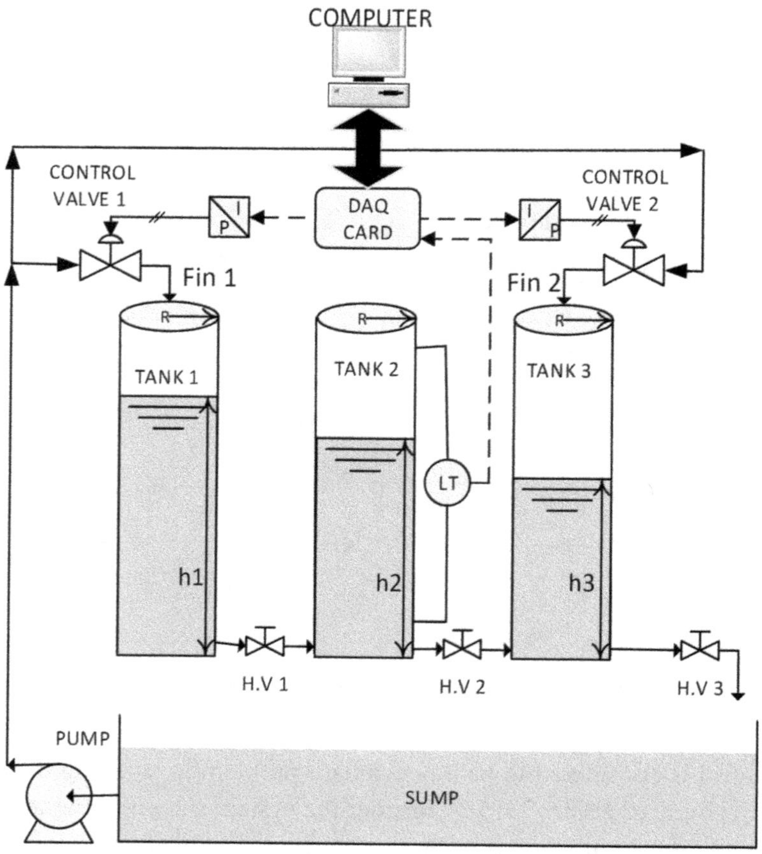

$$\frac{dh_1(t)}{dt} = -\frac{\beta_{12}\alpha_{12}}{A_1}\sqrt{2g\left(h_1(t) - h_2(t)\right)} + \frac{F_{in1}}{A_1}$$

$$\frac{dh_2(t)}{dt} = \frac{\beta_{12}\alpha_{12}}{A_2}\sqrt{2g\left(h_1(t) - h_2(t)\right)} - \frac{\beta_{23}\alpha_{23}}{A_2}\sqrt{2g\left(h_2(t) - h_3(t)\right)}$$

$$\frac{dh_3(t)}{dt} = \frac{\beta_{23}\alpha_{23}}{A_3}\sqrt{2g\left(h_2(t) - h_3(t)\right)} - \frac{\beta_3\alpha_3}{A_3}\sqrt{2gh_3(t)} + \frac{F_{in2}}{A_3}$$

where,

h_1, h_2, h_3 are the liquid levels of Tank 1, Tank 2 and Tank 3 respectively,
$A_1 = A_2 = A_3 = A$ = cross-sectional area of the cylindrical tank= 615cm^2,
$\alpha_1 = \alpha_2 = \alpha_3 = \alpha$ = cross-sectional area of the pipe connecting tanks = 5 cm^2,
β_{12}= valve ratio of the valve connecting Tank 1 and Tank2 = 0.9,

β_{23}= valve ratio of the valve connecting Tank 2 and Tank3 = 0.8,
β_3= valve ratio of the valve connecting Tank 2 and Tank3 = 0.3,

Since the plant under study is of non-linear in nature, therefore to implement the desired control scheme, it has to be linearized around the operating conditions. The operating conditions is as: $F_{in1}=300$ cm^2/sec; $F_{in2}=64$ cm^2/sec, $h_1=35.17$ cm, $h_2=32.93$ cm, and $h_3=30.03$ cm. For this purpose, Taylor's series is employed, and following state space model of the plant has been obtained as shown in equation (9) & equation (10).

$$\begin{bmatrix} \dot{h}_1 \\ \dot{h}_2 \\ \dot{h}_3 \end{bmatrix} = \begin{bmatrix} -0.108 & 0.108 & 0 \\ 0.108 & -0.193 & 0.085 \\ 0 & 0.085 & -0.095 \end{bmatrix} \begin{bmatrix} h_1 \\ h_2 \\ h_3 \end{bmatrix} + \begin{bmatrix} \frac{1}{615} & 0 & 0 \\ 0 & 0 & 0 \\ 0 & 0 & \frac{1}{615} \end{bmatrix} \begin{bmatrix} F_{in1} \\ F_{in2} \end{bmatrix}$$

$$Y = \begin{bmatrix} 0 & 1 & 1 \end{bmatrix} \begin{bmatrix} h_1 \\ h_2 \\ h_3 \end{bmatrix} + 0$$

Hence the transfer function relating the manipulated variable, F_{in1} & unmeasurable primary controlled variable, h_3 is given by equation (11), whereas the transfer function relating F_{in1} & measurable secondary variable, h_2 is given by equation (12).

$$G_{p1} = \frac{0.000015}{s^3 + 0.3954s^2 + 0.03041s + 0.000088}$$

$$G_{p2} = \frac{0.000176s + 0.000017}{s^3 + 0.3954s^2 + 0.03041s + 0.000088}$$

Similarly, the transfer function relating the disturbance, F_{in2} & unmeasurable primary controlled variable, h_3 is given by equation (13), whereas the transfer function relating F_{in2} & measurable secondary variable, h_2 is given by equation (14).

$$G_{d1} = \frac{0.001626s^2 + 0.000489s + 0.000015}{s^3 + 0.3954s^2 + 0.03041s + 0.000088}$$

$$G_{d2} = \frac{0.000138s + 0.000015}{s^3 + 0.3954s^2 + 0.03041s + 0.000088}$$

Reduced Order Model Approximation

Many papers have elaborated the concept of model order reduction and its application in control system (Feldmann & Freund, 1995; Chen, 1999; Lavania & Nagaria, 2016). The elementary idea of model order reduction is to represent the original system dynamics by approximating it with much smaller state-space dimension, so that the control system design and analysis becomes more convenient. This method has been widely accepted and is being widely used in various industrial applications. Therefore, the obtained third order transfer function for both process and disturbance of the plant is approximated as First Order Plus Dead Time (FOPDT) model using step error minimization technique. In this technique the same step input is given to both, the original third order model and the reduced model to generate the error as shown in following Figure 5.

Figure 5. Model order reduction using GA

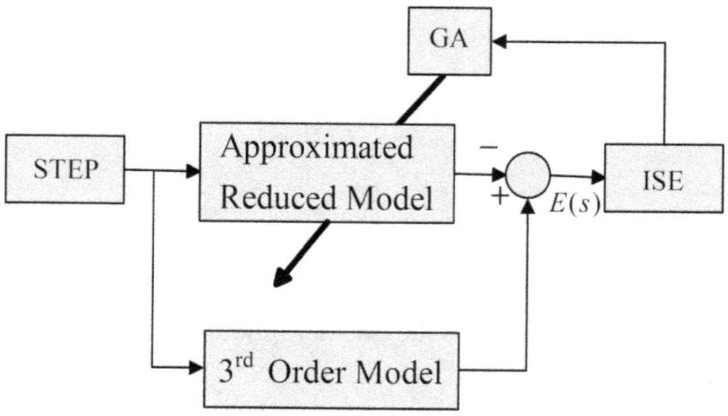

This error is used to construct objective function i.e. integral of squared error (ISE), in present case to be minimized using GA and the approximated model parameters for both reduced integer and non-integer order are obtained. The general form of reduced integer order, reduced non-integer lower order model and the objective function ISE is given by equation (15), equation (16), & equation (17) respectively.

$$G_{integer} = \frac{K}{\tau s + 1} e^{-Ls}$$

$$G_{non\text{-}integer} = \frac{K}{\tau s^{\alpha} + 1} e^{-Ls}$$

$$ISE = \int_{0}^{\infty} \{y_{sp}(t) - y(t)\}^2 dt = \int_{0}^{\infty} \{e(t)\}^2 dt$$

Table 2 gives approximated integer and non-integer order model transfer functions of original plants along with their corresponding value of ISE.

Table 2. Reduced Model Parameters with their ISE value

3rd Order T.F	Integer Order FOPDT Model	Non-Integer Order FOPDT Model
G_{p1}	$\frac{0.17}{328.29s+1}e^{-16.28s}$ ISE=0.00042	$\frac{0.17}{353.06s^{1.01}+1}e^{-12.63s}$ ISE=0.00020
G_{p2}	$\frac{0.192}{334.45s+1}e^{-0.71s}$ ISE=0.00274	$\frac{0.192}{368.64s^{1.016}+1}e^{-0.39s}$ ISE=0.00045
G_{d1}	$\frac{0.168}{307.45s+1}e^{-0.165s}$ ISE=0.00620	$\frac{0.17}{317.55s^{1.003}+1}e^{-0.008s}$ ISE=0.00431
G_{d2}	$\frac{0.17}{334.59s+1}e^{-2.25s}$ ISE=0.00038	$\frac{0.17}{334.59s^{1}+1}e^{-2.25s}$ ISE=0.00038

It can be noted that in case of disturbance transfer function model reduction of G_{d2}, both integer and non-integer FOPDT model is same having order as integer, because sometimes integer order model represent the special cases of fractional order model. Also to make the term proper, G_{d1}/G_{d2} in equation (5), a constraint is imposed on the order while deducing the reduced order model of the original 3rd order system. Figure 6, Figure 7, Figure 8 and Figure 9 represent the step responses for G_{p1}, G_{p2}, G_{d1} & G_{d2} third order transfer function along with their reduced integer and non-integer models respectively.

Figure 6. Open-Loop Step Response for G_{p1}

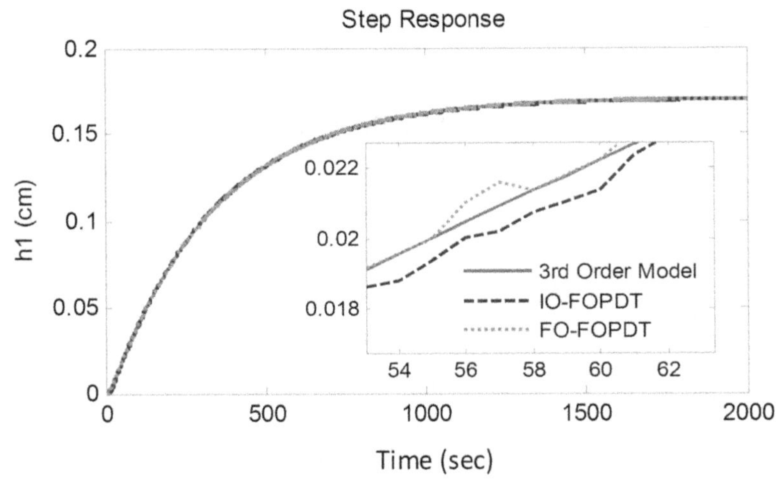

Figure 7. Open-Loop Step Response for G_{p2}

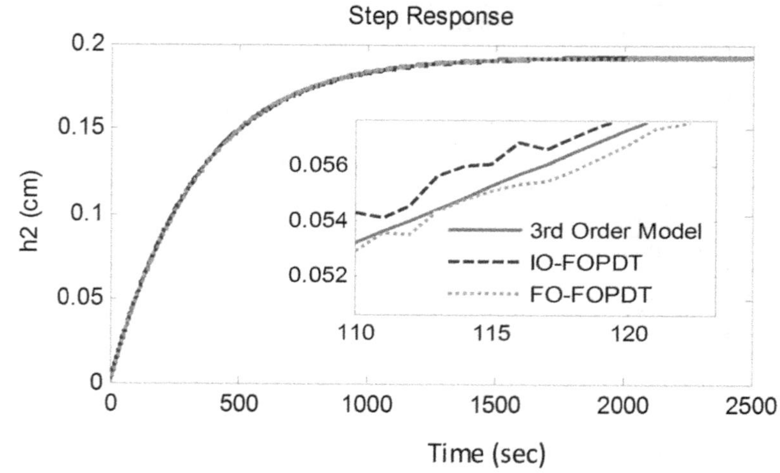

Figure 8. Open-Loop Step Response for G_{d1}

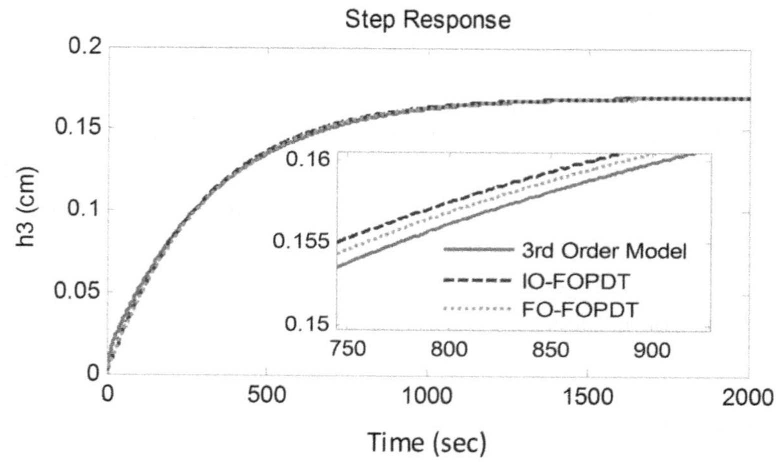

Figure 9. Open-Loop Step Response for G_{d2}

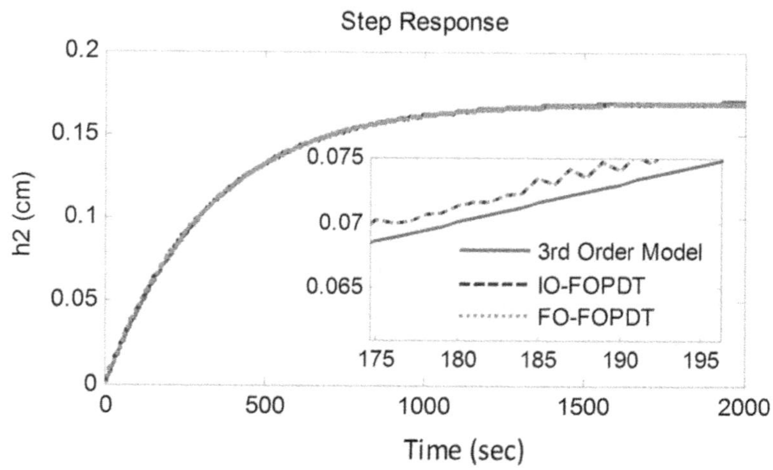

DIFFERENT SCHEMES OF INFERENTIAL CONTROL

Based on the nature of model of plant and controller in terms of their order, four combinations of inferential control loop can be considered (Chen et al, 2009), as shown below:

Scheme 1: Integer order model of plant with integer order controller.
Scheme 2: Integer order model of plant with non-integer order controller.
Scheme 3: Non-integer order model of plant with integer order controller.
Scheme 4: Non-integer order model of plant with non-integer order controller.

For implementing the Scheme 1 and Scheme 2 of the inferential control, the integer order models are used for the plants G_{p1}, G_{p2}, G_{d1} & G_{d2}, which are tabulated in 2nd column of Table 2. Similarly, for implementing the Scheme 3 and Scheme 4 of the inferential control, the non-integer order models are used for the plants G_{p1}, G_{p2}, G_{d1} & G_{d2}, which are tabulated in 3rd column of Table 2. Moving to control part, the conventional PI & PID controller are employed to meet the desired closed loop performance. The transfer function for integer order PI & PID controllers are given by equation (18) & (19) respectively, whereas for non-integer controllers they are given by equation (20) & (21) respectively.

$$C(s) = K_p + \frac{K_i}{s}$$

$$C(s) = K_p + \frac{K_i}{s^{\lambda}}$$

$$C(s) = K_p + \frac{K_i}{s} + K_d s$$

$$C(s) = K_p + \frac{K_i}{s^{\lambda}} + K_d s^{\mu}$$

Here also GA is used for obtaining the proportional gain (*Kp*), integral gain (*Ki*), derivative gain (*Kd*), as well as the order of integrator and differentiator i.e., λ & μ respectively, constituting the controller parameters. The controller parameters are optimized within the range of: $0 \le K_p \le 100$, $0 \le K_i \le 10$, $0 \le K_d \le 100$, $0 \le \lambda \le 2$ and $0 \le \mu \le 2$. As illustrated in Figure 3, the objective function for obtaining the optimal desired control performance is based on the error minimization criteria (Zhuang & Atherton, 1993; Dwyer, 2006; Mohammad, 2010). Prominent among them are as follows:

1. IAE: Integral of Absolute Error,

$$IAE = \int_0^{\infty} \left| e(t) \right| dt$$

2. ISE: Integral of Square Error,

$$ISE = \int_0^\infty e^2(t)\,dt$$

3. ITAE: Integral of Time multiplied Absolute Error,

$$ITAE = \int_0^\infty t\left|e(t)\right|dt$$

4. ITSE: Integral of Time multiplied Square Error,

$$ITSE = \int_0^\infty te^2(t)\,dt$$

The objective function which includes the time multiplication term and its higher term is able to suppress oscillations in the response at later stages of time, whereas the inclusion of higher power of error term helps to reduce large overshoot (Das et al, 2011).

RESULTS AND DISCUSSION

Optimal value of controller parameters was obtained in accordance with the performance index taken using GA. Table 3 and Table 4 show the controller parameters of integer and non-integer PI & PID respectively, obtained using different objective functions defined by equation (22) – equation (25), individually, along with their performance index values.

Figure 10, Figure 11, Figure 12 and Figure 13, represent the closed loop PI controlled performance responses when all the four schemes of inferential control being implemented taking ISE, IAE, ITSE and ITAE as objective function respectively subject to minimization. Similarly, Figure 14, Figure 15, Figure 16 and Figure 17, represent the closed loop PID controlled performance responses when all the four schemes of inferential control being implemented taking ISE, IAE, ITSE and ITAE as objective function respectively. All these servo performances has zoomed version of time response for a specific duration of time in the inset, so that the responses for each schemes can be identified easily.

Here Fractional calculus is exploited at two levels for improving the performance of Inferential control scheme. At first level, it is used in the model order reduction. The reduced non-integer model represents the plant dynamics more accurately than integer order model as observed from Table 2. Again at the 2nd level, the Fractional calculus is introduced in PI & PID controller design to further enhance the performance of the control system. Table 3 & Table 4 clearly indicates that the performance value is maximum for Scheme 1 and is gradually decreasing for Scheme 2, Scheme 3 with Scheme 4 being minimum, irrespective of the objective function taken i.e. ISE, IAE, ITSE or ITAE, which is well illustrated by Figure 10 – Figure 17. Since the integer order system represents the special case of non-integer order system, where the order of the system becomes integer number. Same is the case with reduced order modelling

Table 3. Integer & Non-integer PI Controller Parameters

Different Schemes	Controller Parameters	ISE
Controller Parameters Taking ISE as Performance Index		
Scheme 1	Kp=69.4, Ki=0.053	26640
Scheme 2	Kp=69.4, Ki=0.053, λ=1	26640
Scheme 3	Kp=88.9, Ki=0.083,	20812
Scheme 4	Kp=89.9, Ki=0.006, λ=1.706	20713
Controller Parameters Taking IAE as Performance Index		
Scheme 1	Kp=61.5, Ki=0.135	1121
Scheme 2	Kp=60.2, Ki=0.082, λ=1.084	1105
Scheme 3	Kp=80.5, Ki=0.174,	863
Scheme 4	Kp=79.7, Ki=0.159, λ=1.014	860
Controller Parameters Taking ITSE as Performance Index		
Scheme 1	Kp=65.7, Ki=0.1	387857
Scheme 2	Kp=67.3, Ki=0.013, λ=1.46	379094
Scheme 3	Kp=85.1, Ki=0.133,	230949
Scheme 4	Kp=86.8, Ki=0.022, λ=1.41	226818
Controller Parameters Taking ITAE as Performance Index		
Scheme 1	Kp=74.1, Ki=0.169	34052
Scheme 2	Kp=53, Ki=0.024, λ=1.31	24591
Scheme 3	Kp=82.3, Ki=0.185	16946
Scheme 4	Kp=72.6, Ki=0.142, λ=1.02	15332

Figure 10. Closed loop response with ISE optimized PI controller

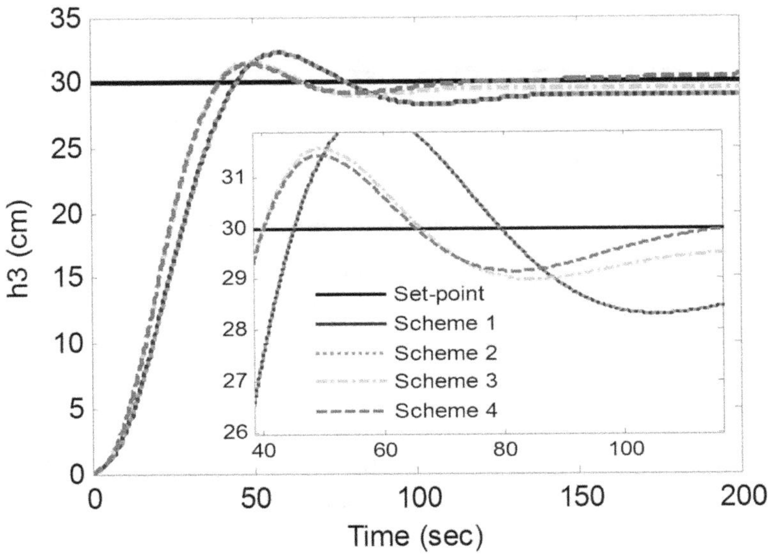

Table 4. Integer & Non-integer PID Controller Parameters

Different Schemes	Controller Parameters	ISE
Controller Parameters Taking ISE as Performance Index		
Scheme 1	K_p=71.7, K_i=0.074, K_d=100	25001
Scheme 2	Kp=57.7; Ki=0.229, λ=0.93, Kd=99.9, μ=0.532	24087
Scheme 3	Kp=91.3; Ki=0.098, Kd=99.9	19497
Scheme 4	Kp=79.8; Ki=0.045, λ=1.317, Kd=100, μ=0.561	18921
Controller Parameters Taking IAE as Performance Index		
Scheme 1	Kp=62.9; Ki=0.135, Kd=99.9	1017
Scheme 2	Kp=7.24; Ki=2.3, λ=0.58, Kd=99.9, μ=0.224	953
Scheme 3	Kp=77; Ki=0.162, Kd=93.4	788
Scheme 4	Kp=28.87; Ki=4.96, λ=0.474, Kd=99.99, μ=0.304	750
Controller Parameters Taking ITSE as Performance Index		
Scheme 1	Kp=68.9; Ki=0.119, Kd=99.9	327517
Scheme 2	Kp=47.9; Ki=0.396, λ=0.878, Kd=100, μ=0.428	292431
Scheme 3	Kp=89.9; Ki=0.162, Kd=93.8	195956
Scheme 4	Kp=69.9; Ki=0.99, λ=0.731, Kd=99.95, μ=0.498	175868
Controller Parameters Taking ITAE as Performance Index		
Scheme 1	Kp=66; Ki=0.139, Kd=99.9	21246
Scheme 2	Kp=38.2; Ki=1.3, λ=0.642, Kd=83.8, μ=0.415	16825
Scheme 3	Kp=76.9; Ki=0.179, Kd=85.7	14716
Scheme 4	Kp=22.03; Ki=4, λ=0.514, Kd=99.9, μ=0.26	11360

Figure 12. Closed loop response with ITSE optimized PI controller

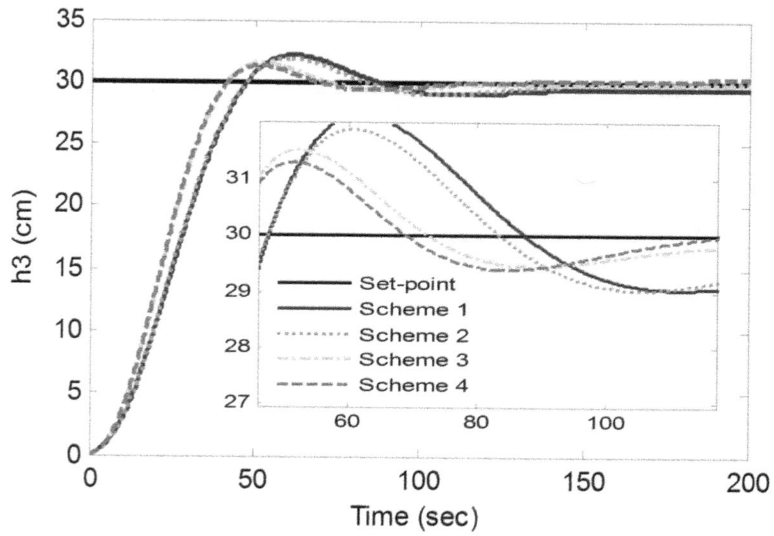

Figure 11. Closed loop response with IAE optimized PI controller

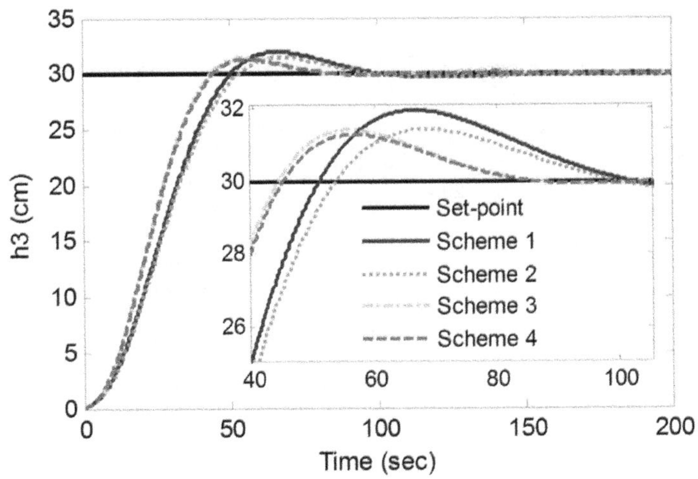

Figure 13. Closed loop response with ITAE optimized PI controller

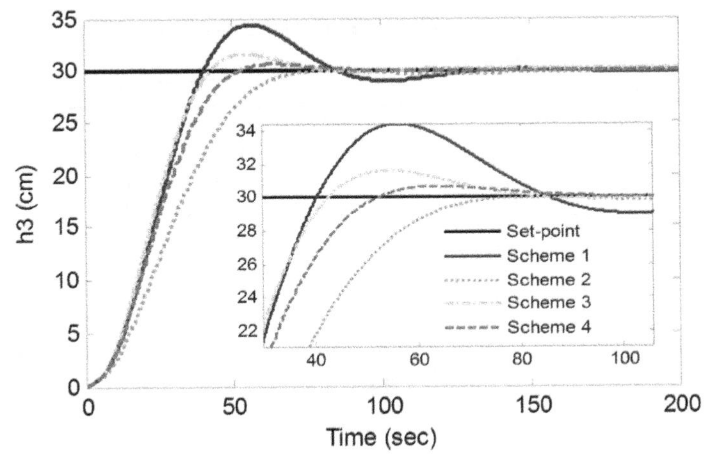

Figure 14. Closed loop response with ISE optimized PID controller

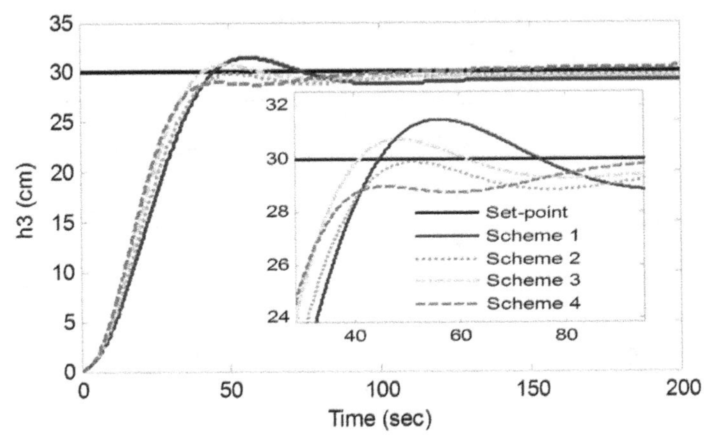

Figure 15. Closed loop response with IAE optimized PID controller

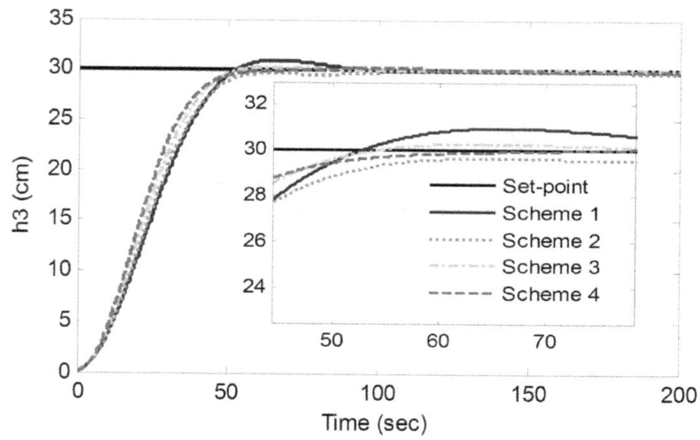

Figure 16. Closed loop response with ITSE optimized PID controller

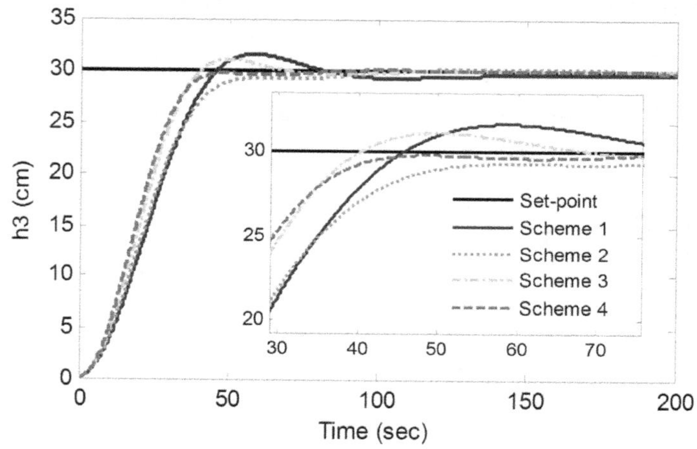

Figure 17. Closed loop response with ITAE optimized PID controller

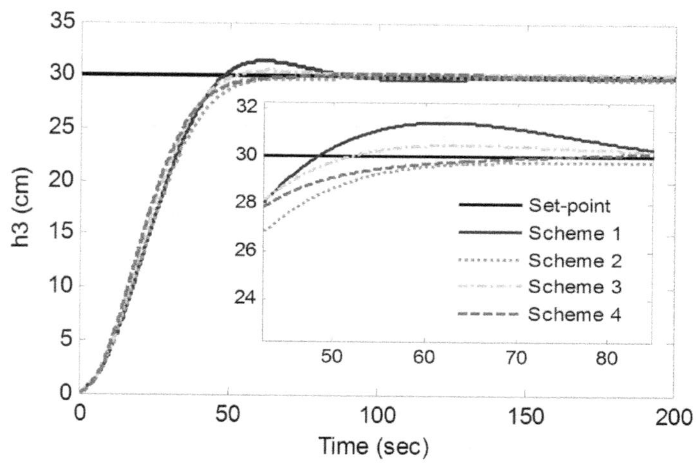

for Region G_{d2}, where reduced integer and non-integer models are same. Similar phenomenon occurs in PI controller designing by minimizing ISE, where the order of the integrator i.e. $\lambda=1$. After the overall analysis of results, it is observed that closed loop performance got improved via introduction of non-integer order in modelling as well as control, but the degree of improvement is comparatively higher when reduced non-integer models are employed for estimation as compared to non-integer controller. Also the ISE & ITSE optimized controllers, are not able to eliminate offset error, but IAE & ITAE optimized controllers are satisfactorily able to do so. Therefore, one can choose the performance index as per the desired closed loop performance.

CONCLUSION

In this paper, four different schemes of inferential control strategy arising from the combinations of different nature of plant and controller in term of their order (integer or non-integer), have been implemented for the level control of three interacting tank process. The controller parameter(s) for all these four schemes are obtained taking ISE, IAE, ITSE and ITAE individually as performance index. After going through the simulated results, it can be concluded that the inclusion of fractional calculus not only reduces the modelling error but also enhances the control performance for PI & PID controller The present scheme of modelling and control can be also carried out in the case where the real time plant data are available. Using these data plant can be directly modelled as lower order integer & non-integer models, and then the present scheme of inferential control can be implemented. This control technique can also be extended to other processes which include un-measurable primary controlled variable and secondary supporting measurable variable.

REFERENCES

Banu, U. S., & Nasir, A. W. (2015). Design, Analysis and Performance Evaluation of Fractional Order Proportional Integral for Three Interacting Tank Process in Frequency Domain considered as Third Order System. *IFAC-PapersOnLine*, *48*(30), 179–184. doi:10.1016/j.ifacol.2015.12.374

Boudjehem, D., & Boudjehem, B. B. (2012). A Fractional Model Predictive Control for Fractional Order Systems. In Fractional Dynamics and Control (pp. 59-71). New York: Springer.

Brandstetter, P., Dobrovsky, M., & Kuchar, M. (2014). Implementation of Genetic Algorithm in Control Structure of Induction Motor A.C. Drive. *Advances in Electrical and Computer Engineering*, *14*(4), 15–20. doi:10.4316/AECE.2014.04003

Chen, Y. (1999). *Model Order Reduction for Nonlinear System* (M. S. Thesis). Massachusetts Institute of Technology.

Chen, Y., Petras, I., & Xue, D. (2009). Fractional order control. In *American Control conference* (pp. 1397-1411). St. Louis, MO: Academic Press.

Dao, S. D., Abhary, K., & Marian, R. (2017). A bibliometric analysis of genetic algorithms throughout the history. *Computers & Industrial Engineering*, *110*, 395–403. doi:10.1016/j.cie.2017.06.009

Das, S., Saha, S., Das, S., & Gupta, A. (2011). On the selection of tuning methodology of FOPID controllers for the control of higher order processes. *ISA Transactions*, *50*(3), 376–388. doi:10.1016/j.isatra.2011.02.003 PMID:21420085

Davis, L. (1991). *Handbook of Genetic Algorithms*. New York: Van Nostrand Reinhold Company.

Dwyer, A. O. (2006). *Handbook of PI and PID controller tuning rules*. London, UK: Imperial College Press. doi:10.1142/p424

Efe, M. O. (2011). Fractional order systems in industrial automation – A Survey. *IEEE Transactions on Industrial Informatics*, *7*(4), 582–591. doi:10.1109/TII.2011.2166775

Elwakil, A. S. (2010). Fractional order circuits and systems magazine: An Emerging interdisciplinary research area. *IEEE Circuits and Systems*, *10*(4), 40–50.

Fahim, S. M., Ahmed, S., & Imtiaz, S. A. (2018). Fractional order model identification using the sinusoidal input. *ISA Transactions*, *83*, 35–41. doi:10.1016/j.isatra.2018.09.009 PMID:30243512

Feldmann, P., & Freund, R. W. (1995). Efficient Linear Circuit Analysis by Pade Approximation Via the Lanczos Process. *IEEE Transactions on Computer-Aided Design of Integrated Circuits and Systems*, *14*(5), 639–649. doi:10.1109/43.384428

Goldberg, D. E. (1989). *Genetic Algorithms in search Optimization and Machine Learning*. Boston, MA: Addison-Wesley.

Holland, J. H. (1975). *Adaption in Natural & Artificial Systems*. Cambridge, MA: MIT Press.

Jia, Y., & Yang, X. (2016). Optimization of control parameters based on genetic algorithms for spacecraft altitude tracking with input constraints. *Neurocomputing*, *177*, 334–341. doi:10.1016/j.neucom.2015.11.022

Joshi, M. M., Vyawahare, V. A., & Patil, M. D. (2014). Model predictive control for fractional-order system a modeling and approximation based analysis. In *2014 4th International Conference on Simulation and Modeling Methodologies Technologies and Applications* (pp. 361-372). 10.5220/0005038203610372

Lavania, S., & Nagaria, D. (2016). Evolutionary approach for model order reduction. *Perspectives on Science*, *8*, 361–363. doi:10.1016/j.pisc.2016.04.075

Mamar, B., & Rachid, M. (2013). Imc-pid-fractional-order-filter controllers design for integer order systems. *ISA Transactions*, *53*(5), 1620–1628. doi:10.1016/j.isatra.2014.05.007 PMID:24957276

Man, K. F., Tang, K. S., & Kwong, S. (1996). Genetic Algorithms: Concepts and Applications. *IEEE Transactions on Industrial Electronics*, *43*(5), 519–534. doi:10.1109/41.538609

Mohammad, S. T. (2010). Notes on integral performance indices in fractional-order control systems. *Journal of Process Control*, *20*(3), 85–291.

Monje, C., Vinagre, B. M., Feliu, V., & Chen, Y. Q. (2008). Tuning and auto-tuning of fractional order controllers for industry applications. *Control Engineering Practice*, *16*(7), 798–812. doi:10.1016/j.conengprac.2007.08.006

Monje, C. A., Chen, Y., Vinagre, B. M., Xue, D., & Feliu, V. (2010). *Fractional order systems and controls: Fundamental and applications*. London: Springer-Verlag.

Nasir, A. W., & Singh, A. K. (2015). IMC based fractional order controller for non-minimum phase system. In *2015 Annual IEEE India Conference* (pp. 1-6). New Delhi: IEEE. 10.1109/INDICON.2015.7443593

Oldham, K. B., & Spanier, J. (1974). *The Fractional Calculus*. New York: Academic Press.

Podlubny, I. (1999). Fractional-Order Systems and Controllers. *IEEE Transactions on Automatic Control, 44*(1), 208–214. doi:10.1109/9.739144

Rhouma, A., & Bouani, F. (2014). Robust model predictive control of uncertain fractional systems: A thermal application. *IET Control Theory & Applications, 8*(17), 1986–1994. doi:10.1049/iet-cta.2014.0703

Shabbir, F., & Omenzetter, P. (2016). Model updating using genetic algorithms with sequential niche technique. *Engineering Structures, 120*, 166–182. doi:10.1016/j.engstruct.2016.04.028

Shalaby, R., El-Hossainy, M., & Belal Abo-Zalam, B. (2019). Fractional order modeling and control for under-actuated inverted pendulum. *Communications in Nonlinear Science and Numerical Simulation, 74*, 97–121. doi:10.1016/j.cnsns.2019.02.023

Sondhi, S., & Hote, Y. V. (2014). Fractional IMC design for fractional order gas turbine model. In *2014 9th International Conference on Industrial and Information Systems* (pp. 1-5). 10.1109/ICIINFS.2014.7036585

Stephanopoulas, G. (1984). Chemical process control: An introduction to theory and practice. Pearson Education.

Tepljakov, A., Pettenkov, E., & Belikov, J. (2011). FOMCON: a MATLAB toolbox for fractional-order system identification and control. *International Journal of Microelectronics and Computer Science, 2*(2), 51-62.

Wang, X., Hoang, T. T. G., Pan, Z., & Chen, Y. (2018). Fractional - order modelling and control for two parallel PWM rectifiers. *IFAC-PapersOnLine, 51*(4), 54–59. doi:10.1016/j.ifacol.2018.06.026

Zhuang, M., & Atherton, D. P. (1993). Automatic tuning of optimum PID controllers. *IEE Proceedings. Control Theory and Applications, 140*(3), 216–224. doi:10.1049/ip-d.1993.0030

Chapter 15
On the Application of Fractional Derivatives to the Study of Memristor Dynamics

Rawid Banchuin
Siam University, Thailand

ABSTRACT

In this chapter, the authors report their work on the application of fractional derivative to the study of the memristor dynamic where the effects of the parasitic fractional elements of the memristor have been studied. The fractional differential equations of the memristor and the memristor-based circuits under the effects of the parasitic fractional elements have been formulated and solved both analytically and numerically. Such effects of the parasitic fractional elements have been studied via the simulations based on the obtained solutions where many interesting results have been proposed in the work. For example, it has been found that the parasitic fractional elements cause both charge and flux decay of the memristor and the impasse point breaking of the phase portraits between flux and charge of the memristor-based circuits similarly to the conventional parasitic elements. The effects of the order and the nonlinearity of the parasitic fractional elements have also been reported.

INTRODUCTION

Fractional calculus has important applications in science and technology, e.g. in biological systems (Ahmed & Elgazzar, 2007; Matouk et al, 2015; Elsadany & Matouk, 2015; Matouk et al, 2015; Matouk & Elsadany, 2016; Selvam et al, 2017; Selvam et al, 2018a-b; Ameen & Novati, 2017; Al-Khedhairi et al, 2018), economic and financial systems (Laskin, 2000; Chen, 2008; Chen et al, 2011; Zhen et al, 2011; Hegazi et al, 2013b; Khan & Kumar, 2016), social models (Ahmad & El-Khazali, 2007; Ali, 2019), physical models (Heaviside, 1971; El-Sayed, 1996; Kusnezov et al, 1999; Hilfer, 2000; Khan & Kumar, 2018; Saqib et al, 2018a-d, Khan & Kumar, 2019a; Al-Khedhairi et al, 2019a; Al-Khedhairi et al, 2019b) and engineering systems (Sun et al, 1984; Petras, 2009; Matouk, 2009a; Matouk, 2009b;

DOI: 10.4018/978-1-7998-3122-8.ch015

Matouk 2010; Matouk, 2011a; Matouk, 2011b; Hegazi & Matouk, 2011; Hegazi et al, 2011; Hegazi & Matouk, 2013; Hegazi et al, 2013a; Matouk & Elsadany, 2014; Banu & Nasir, 2015; Nasir & Singh, 2015; Matouk, 2016; El-Sayed et al, 2016a; El-Sayed et al, 2016b; Khan & Kumar, 2019b). Furthermore, fractional derivatives have useful applications in some interdisciplinary fields of science (Ichise et al, 1971; Bagley & Torvik, 1983; Bagley & Calico, 1991; Ben Adda F, 1997; Magin, 2006; Tavazoei et al, 2009; Tavazoei, 2010; Mainardi, 2010; Abdel Latif, 2011; El-Sayed & Salman, 2013; Latif et al, 2013; Abdel Kader et al, 2017; Abdel Kader et al, 2018; Abdel Latif et al, 2018; Ali & Ameen, 2019; Saqib et al, 2019a-b; Taghavian & Tavazoei, 2017; Taghavian & Tavazoei, 2018; Taghavian & Tavazoei, 2019). Fractional differential operators have also been applied in the studies on memristor (Machado, 2013; Yi-Fei & Yuan, 2013; Fouda & Radwan, 2013; Fouda & Radwan, 2015; Yu et al, 2015; Shi & Hu, 2017; Banchuin, 2018; Banchuin, 2019). Throughout this chapter our work on the application of fractional derivative to the study of memristor dynamic under the effects of parasitic fractional elements (Banchuin, 2019) will be reported by starting from some back ground on memristor.

BACKGROUND

Memristor is a nonlinear electrical circuit element which relates the flux (φ) and charge (q) through one of the following constitutive relations (Itoh & Chua, 2016)

$$\phi=f(q) \tag{1}$$

$$q=g(\phi) \tag{2}$$

where both $f()$ and $g()$ are differentiable scalar-valued functions. The memristor with its constitutive relation given by (1) and (2) can be referred to as the q-controlled memristor and the φ-controlled memristor. Without the parasitic elements, the ideal memristor can retains the values of φ and q at the moment that it is switched without any change as time goes by (Chua, 2012) due to the principle of the conservation of charge and flux) (Chua, 1969).

FLUX AND CHARGE DECAYING

In fact, the memristor with parasitic elements exhibits the decaying of both φ and q after being switched off (Itoh & Chua, 2016). Later, it has been shown that such decaying of both φ and q can also be occurred by the parasitic fractional elements (Banchuin, 2019). Consider the memristor with parasitic fractional capacitor depicted in Figure 1 where $W(\varphi)$ and C_α denote memductance of the memristor and pseudocapacitance of the parasitic fractional capacitor (Freeborn et al, 2013) respectively. Since it has been assumed that the memristor is φ-controlled and the memristor and its parasitic element share the same voltage but employ the current of equal amount with opposite direction as can be seen from Figure 1, we obtain the following fractional differential equation (FDE) (Banchuin, 2019). For more details on FDE, see (Podlubny, 1999).

$$D^{\alpha}[\phi] = -\frac{q}{C_{\alpha}} \tag{3}$$

which clearly shows that

$$q \propto - D^{\alpha}[\phi] \tag{4}$$

Noted that $D^{\alpha}[\ ]$ stands for the time fractional derivative operator.

Figure 1. The memristor with parasitic fractional capacitor
(Banchuin, 2019)

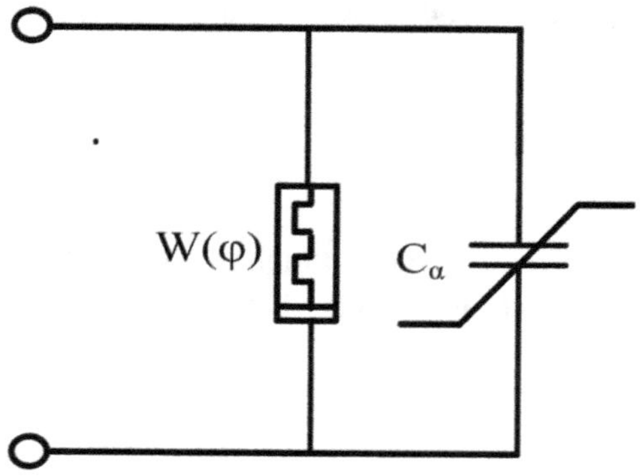

Since the time fractional derivative is also referred to as the rate of change similarly to conventional derivative, (4) shows that φ of the memristor affected by the parasitic fractional capacitor is decreased with respected to time when $q > 0$ and vice versa. Since it has been assumed that that $q = g(\varphi)$ is a strictly increasing function where $g(0) = 0$, it has been found that (Banchuin, 2019)

$$\lim_{t \to \infty}[\phi] = 0 \tag{5}$$

$$\lim_{t \to \infty}[q] = 0 \tag{6}$$

after applying the dynamic route theory (Chua, 1969).

It can be seen from (5) and (6) that the parasitic fractional capacitor causes the decaying of φ which in turn causes the decaying of q after the memristor is switched off (Banchuin, 2019).

Now, consider the effects of parasitic fractional inductor. The memristor with parasitic fractional inductor can be depicted in Figure 2 where $M(q)$ and L_{α} stand for memristance of the memristor and inductivity of the parasitic fractional inductor (Schäfer Krüger, 2006).

As we assumed that the memristor is q-controlled and the memristor and its parasitic element share the same current but employ the volatge of equal amount with opposite direction as can be seen from Figure 2, the following FDE can be obtained after the memristor is switched off (Banchuin, 2019)

Figure 2. The memristor with parasitic fractional inductor
(Banchuin, 2019)

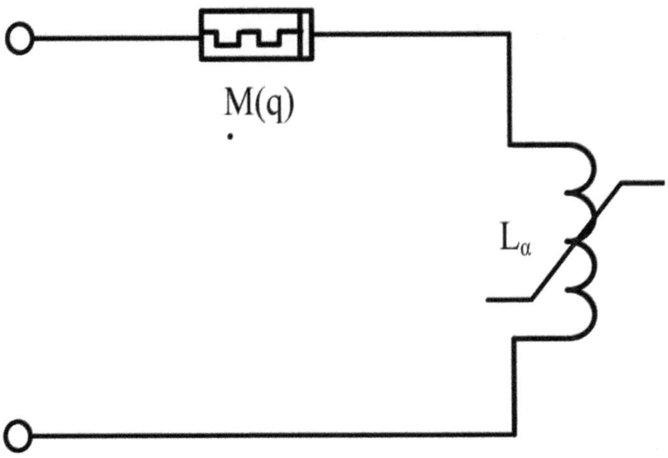

$$D^{\alpha}[q] = -\frac{\phi}{L_{\alpha}} \tag{7}$$

From (7), it can be seen that

$$\phi \propto - D^{\alpha}[q] \tag{8}$$

which shows that q is decreased with respected to time when $\varphi > 0$ and vice versa. By assuming that $\varphi = f(q)$ is a strictly increasing function with $f(0) = 0$ and applying the dynamic route theory, we have (Banchuin, 2019)

$$\lim_{t \to \infty}[q] = 0 \tag{9}$$

$$\lim_{t \to \infty}[\phi] = 0 \tag{10}$$

Therefore it can be seen that the parasitic fractional inductor induces the decaying of q which in turn causes the decaying of φ when the memristor is switched (Banchuin, 2019).

As a practical example, let $q= \phi 3\cdot$ As a result, the effect of the parasitic fractional capacitor can be analytically described by

$$D^{\alpha}[\phi] = \frac{\phi^{3}}{C_{\alpha}} \tag{11}$$

By using the Adomian decomposition [10], we have (Banchuin, 2019)

$$\phi = \phi_0 + \phi_1 + \phi_2 + \phi_3 \ldots \tag{12}$$

where φ_0 denotes the value of φ at $t = 0$ and

$$\phi_1 = -\frac{\phi_0^3 t^\alpha}{C_\alpha \Gamma(\alpha+1)}$$

$$\phi_2 = \frac{3\phi_0^5 t^{2\alpha}}{C_\alpha^2 \Gamma(2\alpha+1)} \tag{13}$$

$$\phi_3 = -\frac{\phi_0^7 t^{3\alpha}(3(\Gamma(\alpha+1))^2 + \Gamma(2\alpha+1))}{C_\alpha^3 \alpha (\Gamma(\alpha+1))^2 \Gamma(3\alpha)}$$

Noted that Γ() stands for the Gamma function (Mathai & Haubold, 2010).

For determining the numerical solution of (11), the conventional derivative based approximation of $D^\alpha[\phi]$ by the moments of a function (Pooseh et al, 2013) has been applied. As a result, we have (Banchuin, 2019)

$$\frac{d\phi}{dt} = \frac{-\frac{\phi^3}{C_\alpha} + \sum_{p=2}^{N}\left[\frac{\Gamma(p-1+\alpha)t^{1-p-\alpha}V_p}{\Gamma(2-\alpha)\Gamma(\alpha-1)(p-1)!}\right] - \frac{1+\sum_{p=2}^{N}\left[\frac{\Gamma(p-1+\alpha)}{\Gamma(\alpha)(p-1)!}\right]}{\Gamma(1-\alpha)} t^{-\alpha}\phi}{\frac{1+\sum_{p=1}^{N}\left[\frac{\Gamma(p-1+\alpha)}{\Gamma(\alpha-1)p!}\right]}{\Gamma(2-\alpha)} t^{1-\alpha}} \tag{14}$$

$$\frac{dV_2}{dt} = -\phi$$

$$\frac{dV_3}{dt} = -2t\phi$$

$$\frac{dV_N}{dt} = (1-N)t^{N-2}\phi$$

where $V_2(0) = 0$, $V_3(0) = 0, \ldots, V_N(0) = 0$.

In addition, we let $N = 7$ as it gives a satisfied approximation of the fractional derivative within a reasonable amount of computational effort (Pooseh et al, 2013). The numerically solved φ can be simulated against t for various values of α as depicted in Figure 3 where $\varphi_0 = 1$ and $C_\alpha = 0.01$ have been assumed.

Since $q= \phi 3\cdot$ q can be simulated as depicted in Figure 4. The decaying of both φ and can be clearly observed from these figures which confirms the above dynamic route theory based results. Moreover, we have also found that the decaying rates of φ and q which are faster than those associated with the conventional parasitic element, are inversely proportional to α (Banchuin, 2019). The mentioned faster decaying rates due to fractional parasitic element has been proved in (Banchuin, 2019).

Figure 3. φ v.s. t of the memristor with parasitic fractional capacitor/conventional parasitic capacitor (Banchuin, 2019)

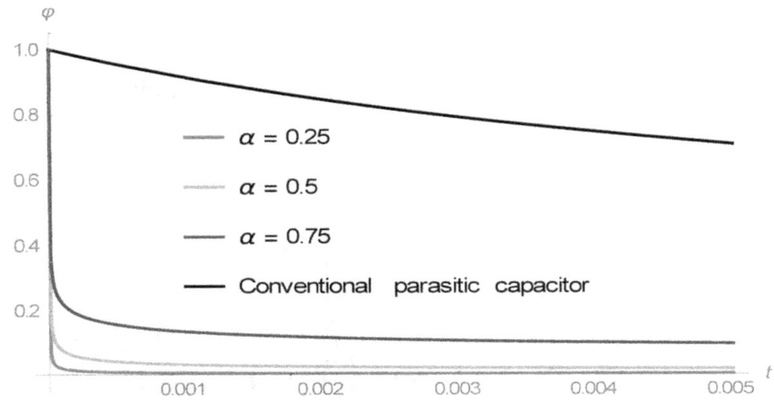

Figure 4. q v.s. t of the memristor with parasitic fractional capacitor/conventional parasitic capacitor (Banchuin, 2019)

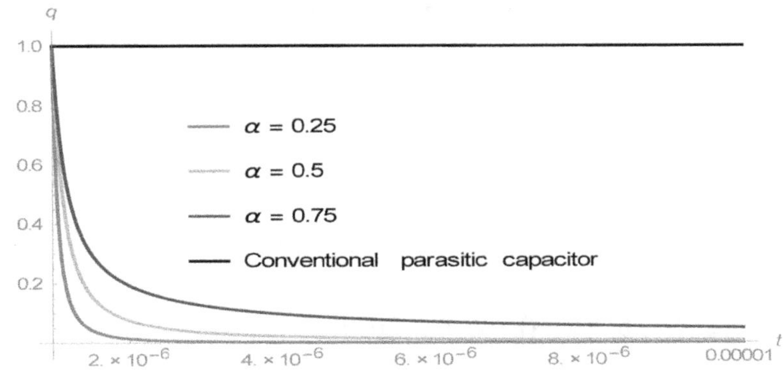

IMPASSE POINT BREAKING ON THE FLUX-CHARGE PLANE

The phase portraits between φ and q of the ideal φ-controlled memristor-inductor circuit exhibit discontinuous jumps at the impasse points. Recently, Banchuin proposed that such jumps at the impasse points can be broken by the effects of parasitic fractional element in (Banchuin, 2019). Consider the memristor-inductor circuit with parasitic fractional capacitor depicted in Figure 5. After performing the circuit analysis, we have (Banchuin, 2019)

$$\frac{dQ_L}{dt} = -\frac{\phi}{L} \tag{15}$$

$$D^{\alpha}[\phi] = \frac{Q_L - g(\phi)}{C_{\alpha}} \tag{16}$$

where Q_L= -q_L and q_L stands for the charge flowing through the inductor. Noted also that QL depends on q due to the Kirchhoff's charge law (KCHL) (Ithou & Chua, 2013).

Figure 5. The memristor-inductor circuit with parasitic fractional capacitor
(Banchuin, 2019)

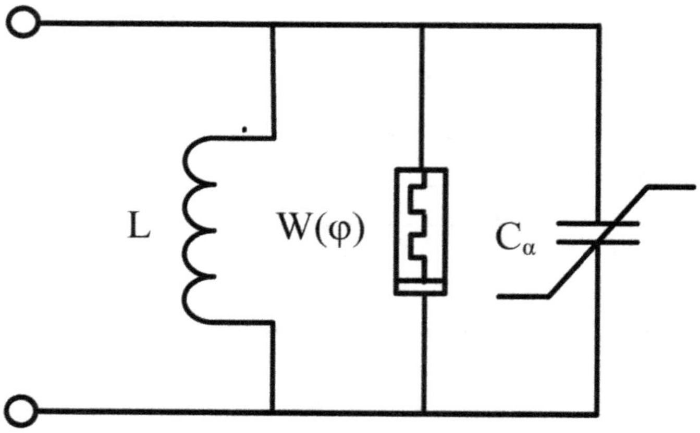

Since we let

$$g(\phi) = \frac{\phi^3}{3} - \phi \tag{17}$$

The following FDE can be obtained (Banchuin, 2019)

$$\frac{dQ_L}{dt} = -\frac{\phi}{L} \tag{18}$$

$$D^{\alpha}[\phi] = \frac{Q_L - (\phi^3 / 3) + \phi}{C_{\alpha}} \tag{19}$$

By keeping in mind that $dQ_L/dt \approx D^{0.999\ldots}[Q_L]$ and the fractional integration of order 0.999… can be approximated by the conventional one, the following solution can be obtained after applying the Adomian decomposition (Banchuin, 2019)

$$\phi = \phi_0 + \phi_1 + \phi_2 \dots \tag{20}$$

$$Q_L = Q_{L0} + Q_{L1} + Q_{L2} \dots \tag{21}$$

where $Q_{L0} = Q_L(0)$ and

$$\phi_1 = -\frac{t^\alpha (Q_{L0} + \phi_0 - (\phi_0^3 / 3))}{C_\alpha \Gamma(\alpha+1)}$$

$$\phi_2 = \frac{t^\alpha}{3C_\alpha^2}\left[\frac{(\phi_0^2 - 1)(\phi_0^3 - 3\phi_0 - 3Q_{L0})t^\alpha}{\Gamma(2\alpha+1)} - \frac{3C_\alpha \phi_0 t}{\Gamma(\alpha+2)}\right] \tag{22}$$

$$Q_{L1} = -\frac{\phi_0 t}{L}$$

$$Q_{L2} = \frac{t^{\alpha+1}(\phi_0^3 - 3\phi_0 - 3Q_{L0})}{3LC_\alpha \Gamma(\alpha+2)} \tag{23}$$

For determining the numerical solution, the conventional derivative based approximation of $D^\alpha[\phi]$ has been used thus we have (Banchuin, 2019)

$$\frac{dQ_L}{dt} = -\frac{\phi}{L}$$

$$\frac{d\phi}{dt} = \frac{\dfrac{Q_L}{C_\alpha} - \dfrac{\phi^3}{3C_\alpha} + \dfrac{\phi}{C_\alpha} + \displaystyle\sum_{p=2}^{N}\left[\frac{\Gamma(p-1+\alpha)t^{1-p-\alpha}V_p}{\Gamma(2-\alpha)\Gamma(\alpha-1)(p-1)!}\right] - \dfrac{1 + \displaystyle\sum_{p=2}^{N}\left[\frac{\Gamma(p-1+\alpha)}{\Gamma(\alpha)(p-1)!}\right]}{\Gamma(1-\alpha)}t^{-\alpha}}{\dfrac{1 + \displaystyle\sum_{p=1}^{N}\left[\frac{\Gamma(p-1+\alpha)}{\Gamma(\alpha-1)p!}\right]}{\Gamma(2-\alpha)}t^{1-\alpha}} \tag{24}$$

$$\frac{dV_2}{dt} = -\phi$$

$$\frac{dV_3}{dt} = -2t\phi$$

$$\frac{dV_N}{dt} = (1 - N)t^{N-2}\phi$$

By using with $N = 7$, φ and Q_L can be simulated against t based on the numerical solution of (24) as depicted in Figure 6 and 7 where $\phi 0 = 0.1$, $QL_0 = 0.1$, L = 1 and $C\alpha = 0.01$ have been assumed. These graphical results show that both φ and QL of both parasitic fractional capacitor and conventional parasitic capacitor affected circuits exhibit the oscillations. In particular, it has been stated in (Banchuin, 2019) that the parasitic fractional capacitor affected memristor-inductor circuit become a fractional Van der Pol oscillator (Misha et al, 2016; Barbosa et al, 2004; Shen et al, 2014). Moreover, the phase portraits between φ and QL simulated based on the numerical solution as depicted in Figure 8 shows that there exists no impasse point i.e. jump behavior. Therefore it can be seen that impasse points on the phase portraits can be broken by the effect of the parasitic fractional capacitor and the continuous oscillations of φ and QL can be obtained (Banchuin, 2019). It can also be seen that the memory and the linearity of the memristor-inductor-parasitic fractional capacitor circuit are respectively inversely and directly proportional to α.

Figure 6. φ v.s. t of the memristor-inductor circuit with parasitic fractional capacitor/conventional parasitic capacitor
(Banchuin, 2019)

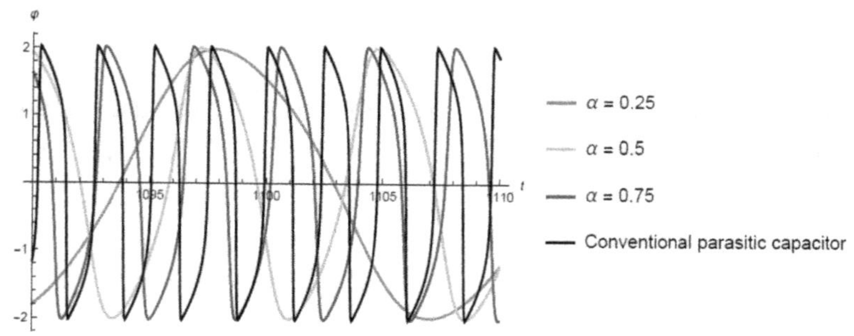

Figure 7. QL v.s. t of the memristor-inductor circuit with parasitic fractional capacitor/conventional parasitic capacitor
(Banchuin, 2019)

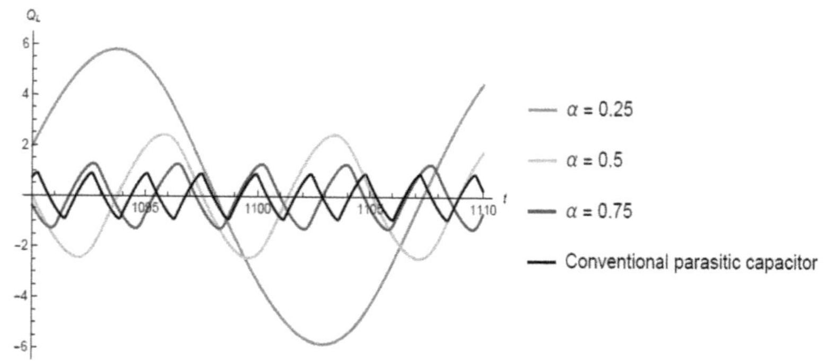

Figure 8. QL v.s. φ of the memristor-inductor circuit with parasitic fractional capacitor/conventional parasitic capacitor
(Banchuin, 2019)

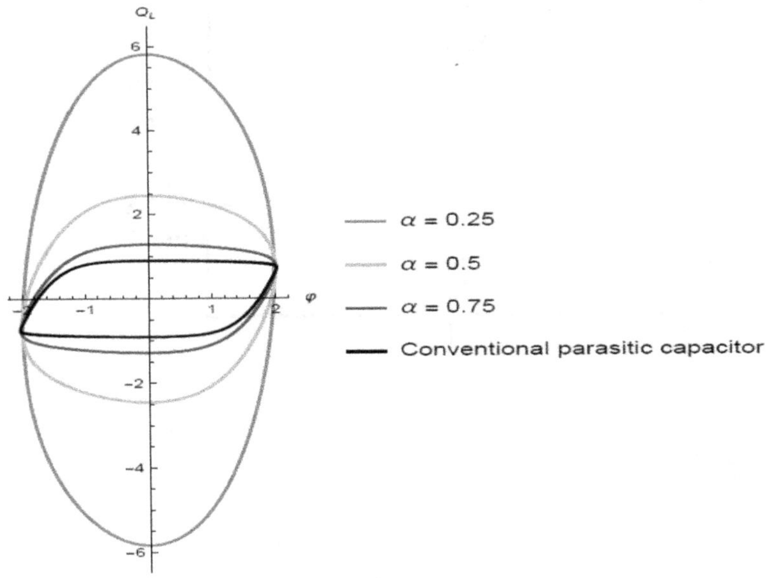

Now consider the memristor-capacitor circuit with the parasitic fractional inductor depicted in Figure 9. Since we assume that $f(q)=(q^3/3)-q$, we have (Banchuin, 2019)

$$\frac{d\Phi_C}{dt} = -\frac{q}{C} \tag{25}$$

$$D^{\alpha}[q] = \frac{\Phi_C - (q^3 / 3) + q}{L_{\alpha}} \tag{26}$$

where $\Phi_C = \phi C$ which denotes the flux of the capacitor, and ΦC depends on φ due to the Kirchhoff's flux law (KFLL) (Ithou & Chua, 2013).

By applying the Adomian decomposition, we have (Banchuin, 2019)

$$q = q_0 + q_1 + q_2 \cdots \tag{27}$$

$$\Phi_C = \Phi_{C0} + \Phi_{C1} + \Phi_{C2} \cdots \tag{28}$$

where $\Phi_{C0} = \Phi_C(0)$ and

Figure 9. The memristor-capacitor circuit with parasitic fractional inductor
(Banchuin, 2019)

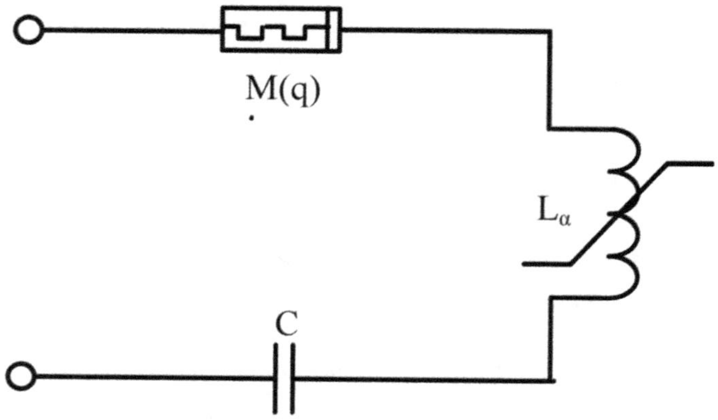

$$q_1 = -\frac{t^{\alpha}(\Phi_{C0} + q_0 - (q_0^3 / 3))}{L_{\alpha}\Gamma(\alpha+1)}$$

$$q_2 = \frac{t^{\alpha}}{3L_{\alpha}^2}\left[\frac{(q_0^2-1)(q_0^3 - 3q_0 - 3\Phi_{C0})t^{\alpha}}{\Gamma(2\alpha+1)} - \frac{3L_{\alpha}q_0 t}{\Gamma(\alpha+2)}\right] \tag{29}$$

$$\Phi_{C2} = \frac{t^{\alpha+1}(q_0^3 - 3q_0 - 3\Phi_{C0})}{3CL_{\alpha}\Gamma(\alpha+2)} \tag{30}$$

The resulting q, Φ_C and their phase portraits have been simulated as depicted in Figure 10-12 by numerically solving (25) and (26) simultaneously by using the conventional derivative based approximation of $D^{\alpha}[q]$ with $N = 7$. Noted also that $q(0) = 0.1$, $\Phi_C(0) = 0.1$, $C = 1$ and $L_{\alpha} = 0.01$ have been assumed (Banchuin, 2019). The continuous oscillations of both q and Φ_C can be observed. In addition, the impasse point breaking of the phase portraits between q and Φ_C can also be seen. This is because the parasitic fractional inductor affected circuit as such circuit become a fractional Van der Pol oscillator (Banchuin, 2019).

EFFECTS OF NONLINEARITY IN THE PARASITIC FRACTIONAL ELEMENT

In practice, there exists nonlinearity in the parasitic fractional element which yields the flux dependency in psuedocapacitance and charge dependency in inductivity. Thus the FDE of the memristor-inductor circuit affected by the parasitic fractional capacitor with nonlinearity can be given by (Banchuin, 2019)

$$\frac{dQ_L}{dt} = -\frac{\phi}{L} \tag{31}$$

Figure 10. q v.s. t of the memristor-capacitor circuit with parasitic fractional inductor/conventional parasitic inductor
(Banchuin, 2019)

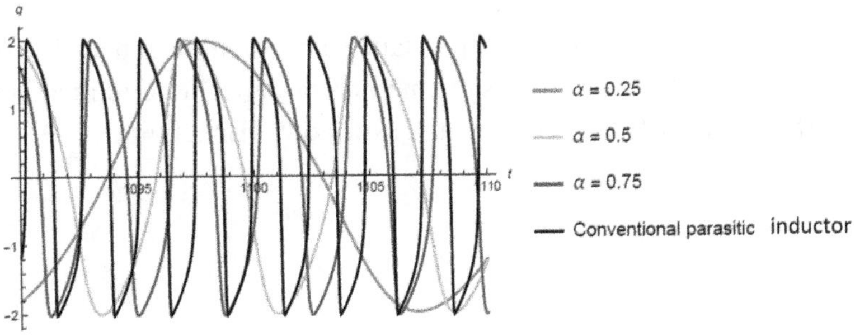

Figure 11. Φ_C v.s. t of the memristor-inductor circuit with parasitic fractional capacitor/conventional parasitic capacitor
(Banchuin, 2019)

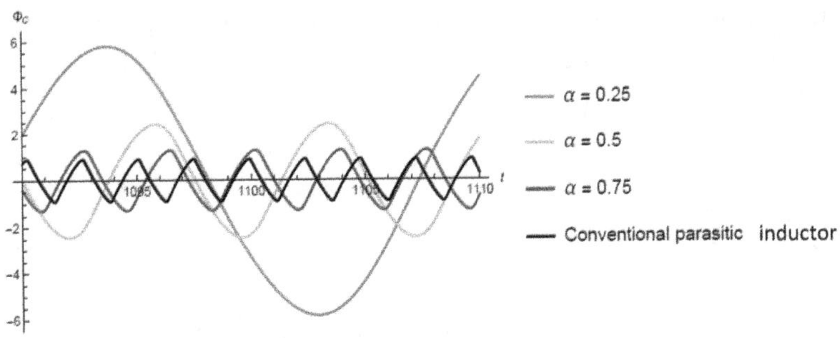

Figure 12. Φ_C v.s. q of the memristor-inductor circuit with parasitic fractional capacitor/conventional parasitic capacitor
(Banchuin, 2019)

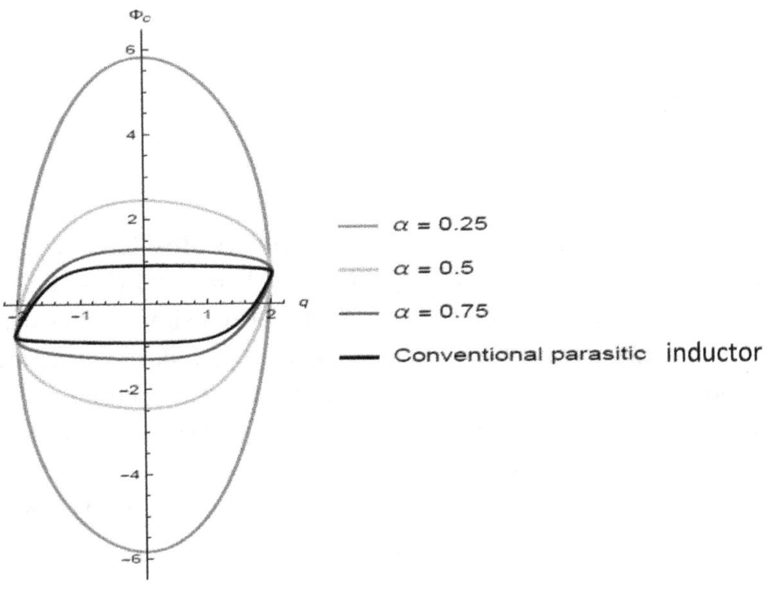

$$D^{\alpha}[\phi] = \frac{Q_L - g(\phi)}{C_{\alpha}(\phi)} \tag{32}$$

where $C_{\alpha}(\phi)$ stands for the φ dependent pseudocapacitance of the nonlinear parasitic fractional capacitor.

In addition, the system of FDEs that governs the memristor-capacitor circuit affected by the parasitic fractional inductor with nonlinearity can be obtained as follows (Banchuin, 2019)

$$\frac{d\Phi_C}{dt} = -\frac{q}{C} \tag{33}$$

$$D^{\alpha}[q] = \frac{\Phi_C - f(q)}{L_{\alpha}(q)} \tag{34}$$

where $L_{\alpha}(q)$ denotes the q dependent inductivity of the nonlinear parasitic fractional inductor.

By assuming that $C_{\alpha}(\phi)$=Cα(1+ ϕ2) where Cα means Cα(0) and using the Adomian decomposition, the analytical solution of (31) and (32) can also be given by (20) and (21) but with the following terms instead (Banchuin, 2019)

$$\phi_1 = -\frac{t^{\alpha}(Q_{L0} + \phi_0 - (\phi_0^3 / 3))}{C_{\alpha}(1+\phi_0^2)\Gamma(\alpha+1)}$$

$$\phi_2 = \frac{t^{\alpha}}{9C_{\alpha}^2(1+\phi_0^2)^3}\left[\frac{(\phi_0^4 + 6\phi_0^2 + 6\phi_0 Q_{L0} - 3Q_{L0})t^{\alpha}}{(\phi_0^3 - 3\phi_0 - 3Q_{L0})^{-1}\Gamma(2\alpha+1)} - \frac{9C_{\alpha}\phi_0 t(1+\phi_0^2)^2}{L\Gamma(\alpha+2)}\right] \tag{35}$$

$$Q_{L1} = -\frac{\phi_0 t}{L}$$

$$Q_{L2} = \frac{t^{\alpha+1}(\phi_0^3 - 3\phi_0 - 3Q_{L0})}{3LC_{\alpha}(1+\phi_0^2)\Gamma(\alpha+2)} \tag{36}$$

For determining the numerical solution, the conventional derivative based approximation of $D^{\alpha}[\phi]$ with N = 7 has been adopted where φ(0) = 0.1, QL(0) = 0.1, L = 1 and Cα = 0.01 have also been assumed. The resulting phase portraits between φ and QL have been simulated as depicted in Figure 13 where the impasse points breaking can also be seen. However, the wider loop areas and more roundish shapes of the phase portraits compared to those of the linear parasitic fractional capacitor based results displayed above can be observed. These differences in the sizes and shapes are caused by the nonlinearity of parasitic fractional capacitor (Banchuin, 2019).

Now, we let $L_{\alpha}(q)$=$L_{\alpha}(1+q^2)$ where L_{α}=$L_{\alpha}(0)$. As a result, the analytical solution of (33) and (34) can also be given by (27) and (28) but with (Banchuin, 2019)

Figure 13. Q_L v.s. φ of the memristor-inductor circuit with parasitic fractional capacitor/conventional parasitic capacitor with nonlinearity
(Banchuin, 2019)

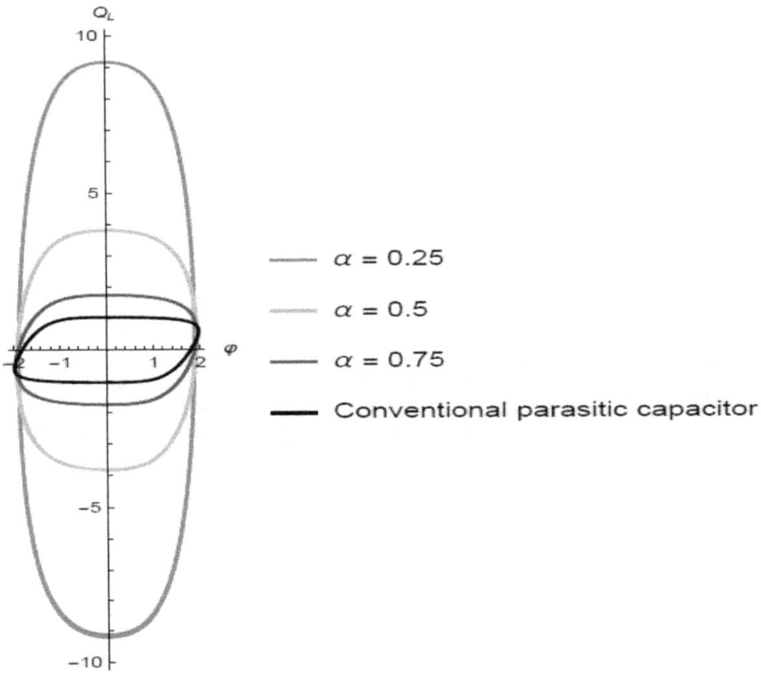

$$q_1 = -\frac{t^{\alpha}(\Phi_{C0} + q_0 - (q_0^3 / 3))}{L_{\alpha}(1 + q_0^2)\Gamma(\alpha + 1)}$$

$$q_2 = \frac{t^{\alpha}}{9L_{\alpha}^2(1 + q_0^2)^3}\left[\frac{(q_0^4 + 6q_0^2 + 6q_0\Phi_{C0} - 3\Phi_{C0})t^{\alpha}}{(q_0^3 - 3q_0 - 3\Phi_{C0})^{-1}\Gamma(2\alpha + 1)} - \frac{9L_{\alpha}q_0 t(1 + q_0^2)^2}{C\Gamma(\alpha + 2)}\right] \tag{37}$$

$$\Phi_{C1} = -\frac{q_0 t}{C}$$

$$\Phi_{C2} = \frac{t^{\alpha+1}(q_0^3 - 3q_0 - 3\Phi_{C0})}{3CL_{\alpha}(1 + q_0^2)\Gamma(\alpha + 2)} \tag{38}$$

By using the conventional derivative based approximation of $D^{\alpha}[q]$ the with $N = 7$ and assuming that $q(0) = 0.1$, $\Phi_C(0) = 0.1$, $C = 1$ and $L_{\alpha} = 0.01$, the phase portraits between q and Φ_C can be simulated as depicted in Figure 14 where the impasse points breaking can also be found. In addition, wider loop areas and more roundish shaped phase portraits compared to the above linear parasitic fractional inductor based counterparts can be observed. These differences are caused by the nonlinearity of parasitic fractional inductor (Banchuin, 2019).

Figure 14. Φ_C v.s. q of the memristor-inductor circuit with parasitic fractional capacitor/conventional parasitic capacitor with nonlinearity
(Banchuin, 2019)

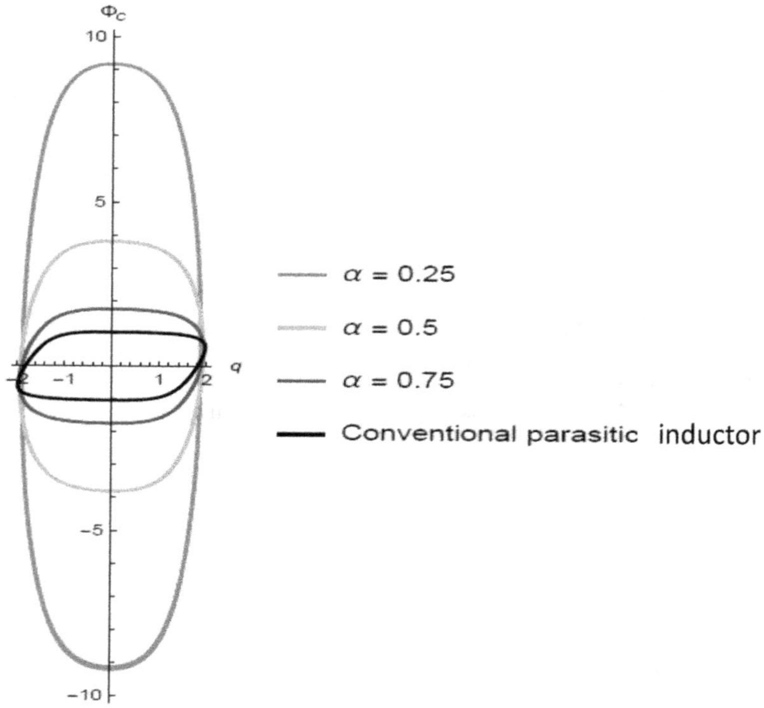

CONCLUSION

Fractional calculus has important applications in science and technology. Fractional differential operators have also been applied in the studies on memristor. Throughout this chapter our work on the application of fractional derivative to the study of memristor dynamic under the effects of parasitic fractional elements has been reported. In summary, such parasitic fractional elements cause flux and charge decaying along with impasse point breaking of the phase portraits of the memristor (Banchuin, 2019).

ACKNOWLEDGMENT

The author would like to acknowledge Prof. Ahmed E. Matouk for his valuable advice.

REFERENCES

Abdel Kader, A. H., Abdel Latif, M. S., El Bialy, F., & Elsaid, A. (2018). Symmetry analysis and some new exact solutions of some nonlinear KdV-like equations. *Asian-European Journal of Mathematics, 11*(03), 1850040. doi:10.1142/S1793557118500407

Abdel Kader, A. H., Abdel Latif, M. S., & Nour, H. M. (2017). Some New Exact Solutions of the Modified KdV Equation Using Lie Point Symmetry Method. *International Journal of Applied and Computational Mathematics, 3*(1), 1163–1171. doi:10.100740819-017-0408-3

Abdel Latif, M. S. (2011). Some exact solutions of KdV equation with variable coefficients. *Communications in Nonlinear Science and Numerical Simulation, 16*(4), 1783–1786. doi:10.1016/j.cnsns.2010.07.023

Abdel Latif, M. S., El-Shazly, E., Baleanu, D., Elsaid, A., & Nour, H. M. (2018). Some new soliton-like and doubly periodic-like solutions of Fisher equation with time-dependent coefficients. *Modern Physics Letters B, 32*(33), 1850413. doi:10.1142/S0217984918504134

Ahmad, W. M., & El-Khazali, R. (2007). Fractional-order dynamical models of love. *Chaos, Solitons, and Fractals, 33*(4), 1367–1375. doi:10.1016/j.chaos.2006.01.098

Ahmed, E., & Elgazzar, A. S. (2007). On fractional order differential equations model for nonlocal epidemics. *Physica A, 379*(2), 607–614. doi:10.1016/j.physa.2007.01.010

Al-Khedhairi, A., Askar, S. S., Matouk, A. E., Elsadany, A., & Ghazel, M. (2018). Dynamics, chaos control, and synchronization in a fractional-order Samardzija-Greller population system with order lying in (0,2). *Complexity*.

Al-khedhairi, A., Matouk, A. E., & Askar, S. S. (2019a). Computations of synchronization conditions in some fractional-order chaotic and hyperchaotic systems. *Pramana Journal of Physics, 92*, 72.

Al-khedhairi, A., Matouk, A. E., & Khan, I. (2019b). Chaotic dynamics and chaos control for the fractional-order geomagnetic field model. *Chaos, Solitons, and Fractals, 128*, 1–12. doi:10.1016/j.chaos.2019.07.019

Ali, H. M. (2019). New approximate solutions to fractional smoking model using the generalized Mittag-Leffler function method. *Progr. Fract. Differ. Appl., 5*(4), 1–9. doi:10.18576/pfda/010101

Ali, H. M., & Ameen, I. (2019). Mittag-Leffler approximation for the solution of nonlinear systems of fractional partial differential equations. *Sylwan, 163*(9), 17–32.

Ameen, I., & Novati, P. (2017). The solution of fractional order epidemic model by implicit Adams methods. *Applied Mathematical Modelling, 43*, 78–84. doi:10.1016/j.apm.2016.10.054

Bagley, R. L., & Calico, R. A. (1991). Fractional order state equations for the control of viscoelastically damped structures. *Journal of Guidance, Control, and Dynamics, 14*(2), 304–311. doi:10.2514/3.20641

Bagley, R. L., & Torvik, P. J. (1983). A theoretical basis for the application of fractional calculus to viscoelasticity. *Journal of Rheology (New York, N.Y.), 27*(3), 201–210. doi:10.1122/1.549724

Banchuin, R. (2018). On the memristances, parameters, and analysis of the fractional order memristor. *Active and Passive Electronic Components, 2018*, 1–14. doi:10.1155/2018/3408480

Banchuin, R. (2019). Effects of Parasitic Fractional Elements to the Dynamics of Memristor. *Journal of Electrical and Computer Engineering, 2019*, 1–13. doi:10.1155/2019/4190641

Banu, U. S., & Nasir, A. W. (2015). Design, Analysis and Performance Evaluation of Fractional Order Proportional Integral for Three Interacting Tank Process in Frequency Domain considered as Third Order System. *IFAC-PapersOnLine*, *48*(30), 179–184. doi:10.1016/j.ifacol.2015.12.374

Barbosa, R. S., Machado, J. T., Ferreira, I. M., & Tar, J. K. (2004, August). Dynamics of the fractional-order Van der Pol oscillator. In *Computational Cybernetics, 2004. ICCC 2004. Second IEEE International Conference on* (pp. 373-378). IEEE. 10.1109/ICCCYB.2004.1437752

Ben Adda, F. (1997). Geometric interpretation of the fractional derivative. *J. Fract. Calc*, *11*, 21–52.

Chen, L., Chai, Y., & Wu, R. (2011). Control and Synchronization of Fractional-Order Financial System Based on Linear Control. *Discrete Dynamics in Nature and Society*, *2011*, 1–21. doi:10.1155/2011/958393

Chen, W. C. (2008). Nonlinear dynamics and chaos in a fractional-order financial system. *Chaos, Solitons, and Fractals*, *36*(5), 1305–1314. doi:10.1016/j.chaos.2006.07.051

Chua, L. O. (1969). *Introduction to nonlinear network theory*. New York: McGraw-Hill.

Chua, L. O. (2012). The fourth element. *Proceedings of the IEEE*, *100*(6), 1920–1927. doi:10.1109/JPROC.2012.2190814

El-Sayed, A. M. A. (1996). Fractional-order diffusion-wave equation. *International Journal of Theoretical Physics*, *35*(2), 311–322. doi:10.1007/BF02083817

El-Sayed, A. M. A., Elsonbaty, A., Elsadany, A. A., & Matouk, A. E. (2016b). Dynamical analysis and circuit simulation of a new fractional-order hyperchaotic system and its discretization. International Journal of Bifurcation and Chaos, 26. doi:10.1142/S0218127416502229

El-Sayed, A. M. A., Nour, H. M., Elsaid, A., Matouk, A. E., & Elsonbaty, A. (2016a). Dynamical Behaviors, Circuit Realization, Chaos Control and Synchronization of a New Fractional Order Hyperchaotic System. *Applied Mathematical Modelling*, *40*(5-6), 3516–3534. doi:10.1016/j.apm.2015.10.010

El-Sayed, A. M. A., & Salman, S. M. (2013). On a discretization process of fractional order Riccati's differential equation. *J. Fract. Calc. Appl.*, *4*, 251–259.

Elsadany, A. A., & Matouk, A. E. (2015). Dynamical behaviors of fractional-order Lotka– Volterra predator–prey model and its discretization. *Journal of Applied Mathematics and Computing*, *49*(1-2), 269–283. doi:10.100712190-014-0838-6

Fouda, M. E., & Radwan, A. G. (2013). On the fractional-order memristor model. *Journal of Fractional Calculus and Applications*, *4*(1), 1-7.

Fouda, M. E., & Radwan, A. G. (2015). Fractional-order memristor response under dc and periodic signals. *Circuits, Systems, and Signal Processing*, *34*(3), 961–970. doi:10.100700034-014-9886-2

Freeborn, T. J., Maundy, B., & Elwakil, A. S. (2013). Measurement of supercapacitor fractional-order model parameters from voltage-excited step response. *IEEE Journal on Emerging and Selected Topics in Circuits and Systems*, *3*(3), 367–376. doi:10.1109/JETCAS.2013.2271433

George Maria Selvam, A. & Janagaraj, R. (2018a). Numerical Analysis of a Fractional Order Discrete Prey – Predator System with Functional Response. *International Journal of Engineering & Technology, 7*(4.10), 681–684.

George Maria Selvam, A., Janagaraj, R., & Britto Jacob, S. (2017). Stability in a Discrete Fractional Order Prey Predator System with Functional Response. *International Journal of Advanced Research in Computer Science and Software Engineering, 7*(5), 630–633. doi:10.23956/ijarcsse/SV7I5/0336

George Maria Selvam, A., Janagaraj, R., & Vignesh, D. (2018b). Allee effect and Holling type - II response in a discrete fractional order prey - predator model. *IOP Conference Series: Journal of Physics, 1139*, 1-7. 10.1088/1742-6596/1139/1/012003

Heaviside, O. (1971). Electromagnetic theory. New York: Chelsea.

Hegazi, A. S., Ahmed, E., & Matouk, A. E. (2011). The effect of fractional order on synchronization of two fractional order chaotic and hyperchaotic systems. *J. Fract. Calc. Appl., 1*(3), 1–15.

Hegazi, A. S., Ahmed, E., & Matouk, A. E. (2013a). On chaos control and synchronization of the commensurate fractional order Liu system. *Communications in Nonlinear Science and Numerical Simulation, 18*(5), 1193–1202. doi:10.1016/j.cnsns.2012.09.026

Hegazi, A. S., Ahmed, E., & Matouk, A. E. (2013b). On Dynamical Behaviors and Chaos Control of the Fractional-order Financial System. In S. Banerjee (Ed.), *Chaos and Complexity Theory for Management: Nonlinear Dynamics* (pp. 34–49). IGI Global. doi:10.4018/978-1-4666-2509-9.ch002

Hegazi, A. S., & Matouk, A. E. (2011). Dynamical behaviors and synchronization in the fractional order hyperchaotic Chen system. *Applied Mathematics Letters, 24*(11), 1938–1944. doi:10.1016/j.aml.2011.05.025

Hegazi, A. S., & Matouk, A. E. (2013). Chaos control of the modified Autonomous Van der Pol-Duffing circuits via active control. In S. Banerjee & L. Rondoni (Eds.), *Applications of Chaos and Nonlinear Dynamics in Science and Engineering* (Vol. 3, pp. 185–202). Berlin: Springer. doi:10.1007/978-3-642-34017-8_7

Hilfer, R. (Ed.). (2000). *Applications of fractional calculus in physics*. Upper Saddle River, NJ: World Scientific. doi:10.1142/3779

Ichise, M., Nagayanagi, Y., & Kojima, T. (1971). An analog simulation of noninteger order transfer functions for analysis of electrode process. *J. Electroanal. Chem., 33*(2), 253–265. doi:10.1016/S0022-0728(71)80115-8

Itoh, M., & Chua, L. O. (2013). Duality of memristor circuits. *International Journal of Bifurcation and Chaos in Applied Sciences and Engineering, 23*(1), 1330001. doi:10.1142/S0218127413300012

Itoh, M., & Chua, L. O. (2016). Parasitic effects on memristor dynamics. *International Journal of Bifurcation and Chaos in Applied Sciences and Engineering, 26*(06), 1630014. doi:10.1142/S0218127416300147

Khan, A & Kumar, S. (2016). Anti-synchronization of chaotic finance systems using fuzzy logic constant controller. *Indian Journal of Industrial and Applied Mathematics*. Doi:10.5958/1945-919X.2016.00013X

Khan, A., & Kumar, S. (2018). Study of chaos in satellite systems. *Pramana Journal of Physics, 90*(1), 13. doi:10.100712043-017-1502-0

Khan, A., & Kumar, S. (2019a). T-S fuzzy modeling and predictive control and synchronization of chaotic satellite systems. *International Journal of Modeling and Simulation, 39*(3).

Khan, A., & Kumar, S. (2019b). Measure of chaos and adaptive synchronization of chaotic satellite systems. *International Journal of Dynamics and Control, 7*(2), 536–546. doi:10.100740435-018-0481-4

Kusnezov, D., Bulgac, A., & Dang, G. D. (1999). Quantum levy processes and fractional kinetics. *Physical Review Letters, 82*(6), 1136–1139. doi:10.1103/PhysRevLett.82.1136

Laskin, N. (2000). Fractional market dynamics. *Physica A, 287*(3-4), 482–492. doi:10.1016/S0378-4371(00)00387-3

Latif, M. A., Dragomir, S. S., Matouk, A. E. (2013). New inequalities of Ostrowski type for Co-ordinated s-Convex functions via fractional integrals. *Journal of Fractional Calculus and Applications, 4*(1), 22-36.

Machado, J. T. (2013). Fractional generalization of memristor and higher order elements. *Communications in Nonlinear Science and Numerical Simulation, 18*(2), 264–275. doi:10.1016/j.cnsns.2012.07.014

Magin, R. L. (2006). *Fractional Calculus in Bioengineering*. Begell House.

Mainardi, F. (2010). *Fractional Calculus and Waves in Linear Viscoelasticity: An Introduction to Mathematical Models*. London: Imperial College Press. doi:10.1142/p614

Mathai, A. M., & Haubold, H. J. (2008). *Special functions for applied scientists*. New York: Springer. doi:10.1007/978-0-387-75894-7

Matouk, A. E. (2009a). Chaos synchronization between two different fractional systems of Lorenz family. *Mathematical Problems in Engineering, 2009*, 1–11. doi:10.1155/2009/572724

Matouk, A. E. (2009b). Stability conditions, hyperchaos and control in a novel fractional order hyperchaotic system. *Physics Letters. [Part A], 373*(25), 2166–2173. doi:10.1016/j.physleta.2009.04.032

Matouk, A. E. (2010). Dynamical behaviors, linear feedback control and synchronization of the fractional order Liu system. *Journal of Nonlinear Systems and Applications, 1*(3), 135–140.

Matouk, A. E. (2011a). Chaos, feedback control and synchronization of a fractional-order modified Autonomous Van der Pol-Duffing circuit. *Communications in Nonlinear Science and Numerical Simulation, 16*(2), 975–986. doi:10.1016/j.cnsns.2010.04.027

Matouk, A. E. (2011b). Synchronization in integer and fractional order chaotic systems. In S. Banerjee (Ed.), *Chaos Synchronization and Cryptography for Secure Communications: Applications for Encryption* (pp. 127–151). IGI Global. doi:10.4018/978-1-61520-737-4.ch007

Matouk, A. E. (2016). Chaos synchronization of a fractional-order modified Van der Pol Duffing system via new linear control, backstepping control and Takagi-Sugeno fuzzy approaches. *Complexity, 21*(S1), 116–124. doi:10.1002/cplx.21719

Matouk, A. E., & Elsadany, A. A. (2014). Achieving synchronization between the fractional-order hyperchaotic Novel and Chen systems via a new nonlinear control technique. *Applied Mathematics Letters, 29*, 30–35. doi:10.1016/j.aml.2013.10.010

Matouk, A. E., & Elsadany, A. A. (2016). Dynamical analysis, stabilization and discretization of a chaotic fractional-order GLV model. *Nonlinear Dynamics, 85*(3), 1597–1612. doi:10.100711071-016-2781-6

Matouk, A. E., Elsadany, A. A., Ahmed, E., & Agiza, H. N. (2015). Dynamical behavior of fractional-order Hastings–Powell food chain model and its discretization. *Communications in Nonlinear Science and Numerical Simulation, 27*(1-3), 153–167. doi:10.1016/j.cnsns.2015.03.004

Mishra, V., Das, S., Jafari, H., & Ong, S. H. (2016). Study of fractional order van der Pol equation. *Journal of King Saud University-Science, 28*(1), 55–60. doi:10.1016/j.jksus.2015.04.005

Nasir, A. W., & Singh, A. K. (2015). IMC based fractional order controller for non-minimum phase system. In *2015 Annual IEEE India Conference* (pp. 1-6). New Delhi: IEEE. 10.1109/INDICON.2015.7443593

Petras, I. (2009). Chaos in the fractional-order Volta's system: Modeling and simulation. *Nonlinear Dynamics, 57*(1-2), 157–170. doi:10.100711071-008-9429-0

Podlubny, I. (1999). *Fractional differential equations*. New York: Academic Press.

Pooseh, S., Almeida, R., & Torres, D. F. (2013). Numerical approximations of fractional derivatives with applications. *Asian Journal of Control, 15*(3), 698–712. doi:10.1002/asjc.617

Pu, Y. F., & Yuan, X. (2016). Fracmemristor: Fractional-order memristor. *IEEE Access: Practical Innovations, Open Solutions, 4*, 1872–1888. doi:10.1109/ACCESS.2016.2557818

Saqib, M., Ali, F., Khan, I., Sheikh, N. A., & Khan, A. (2018a). Entropy Generation in Different Types of Fractionalized Nanofluids. *Arabian Journal for Science and Engineering*, 1–10.

Saqib, M., Ali, F., Khan, I., Sheikh, N. A., & Shafie, S. B. (2018b). Convection in ethylene glycol-based molybdenum disulfide nanofluid. *Journal of Thermal Analysis and Calorimetry*, 1–10.

Saqib, M., Khan, I., & Shafie, S. (2018c). Application of Atangana–Baleanu fractional derivative to MHD channel flow of CMC-based-CNT's nanofluid through a porous medium. *Chaos, Solitons, and Fractals, 116*, 79–85. doi:10.1016/j.chaos.2018.09.007

Saqib, M., Khan, I., & Shafie, S. (2018d). Natural convection channel flow of CMC-based CNTs nanofluid. *The European Physical Journal Plus, 133*(12), 549. doi:10.1140/epjp/i2018-12340-3

Saqib, M., Khan, I., & Shafie, S. (2019a). Application of fractional differential equations to heat transfer in hybrid nanofluid: Modeling and solution via integral transforms. *Advances in Difference Equations, 2019*(52), 1–18. doi:10.118613662-019-1988-5

Saqib, M., Khan, I., & Shafie, S. (2019b). Generalized magnetic blood flow in a cylindrical tube with magnetite dusty particles. *Journal of Magnetism and Magnetic Materials, 484*, 490–496. doi:10.1016/j.jmmm.2019.03.032

Schäfer, I., & Krüger, K. (2006). Modelling of coils using fractional derivatives. *Journal of Magnetism and Magnetic Materials, 307*(1), 91–98. doi:10.1016/j.jmmm.2006.03.046

Shen, Y. J., Wei, P., & Yang, S. P. (2014). Primary resonance of fractional-order van der Pol oscillator. *Nonlinear Dynamics*, *77*(4), 1629–1642. doi:10.100711071-014-1405-2

Shi, M., & Hu, S. (2017). Pinched hysteresis loop characteristics of a fractional-order HP TiO_2 memristor. In *Intelligent Computing, Networked Control, and Their Engineering Applications* (pp. 705–713). Singapore: Springer. doi:10.1007/978-981-10-6373-2_70

Sun, H. H., Abdelwahab, A. A., & Onaral, B. (1984). Linear approximation of transfer function with a pole of fractional order. *IEEE Transactions on Automatic Control*, *29*(5), 441–444. doi:10.1109/TAC.1984.1103551

Taghavian, H., & Tavazoei, M. S. (2017). Robust Stability Analysis of Distributed-Order Linear Time-Invariant Systems With Uncertain Order Weight Functions and Uncertain Dynamic Matrices. *Journal of Dynamic Systems, Measurement, and Control*, *139*(12), 121010. doi:10.1115/1.4037268

Taghavian, H., & Tavazoei, M. S. (2018). Exact solution of linear fractional distributed order systems with exponential order weight functions. In *Mathematical Techniques of Fractional Order Systems* (pp. 103–132). Elsevier Publishing. doi:10.1016/B978-0-12-813592-1.00004-0

Taghavian, H., & Tavazoei, M. S. (2019). Algebraic Conditions for Stability Analysis of Linear Time-Invariant Distributed Order Dynamic Systems: A Lagrange Inversion Theorem Approach. *Asian Journal of Control*, *21*(2), 879–890. doi:10.1002/asjc.1780

Tavazoei, M. S. (2010). A note on fractional-order derivatives of periodic functions. *Automatica*, *46*(5), 945–948. doi:10.1016/j.automatica.2010.02.023

Tavazoei, M. S., Haeri, M., & Attari, M. (2009). A proof for nonexistence of periodic solutions in time invariant fractional order systems. *Automatica*, *45*(8), 1886–1890. doi:10.1016/j.automatica.2009.04.001

Yu, Y. J., Bao, B. C., Kang, H. Y., & Shi, M. (2015). Calculating area of fractional-order memristor pinched hysteresis loop. *Journal of Engineering (Stevenage, England)*, *2015*(11), 325–327. doi:10.1049/joe.2015.0154

Zhen, W., Xia, H., & Guodong, S. (2011). Analysis of nonlinear dynamics and chaos in a fractional order financial system with time delay. *Computers & Mathematics with Applications (Oxford, England)*, *62*(3), 1531–1539. doi:10.1016/j.camwa.2011.04.057

ADDITIONAL READING

Banchuin, R. (2017). Novel expressions for time domain responses of fractance device. *Cogent Engineering*, *4*(1), 1320823. doi:10.1080/23311916.2017.1320823

Chua, L. (1971). Memristor-the missing circuit element. *IEEE Transactions on Circuit Theory*, *18*(5), 507–519. doi:10.1109/TCT.1971.1083337

Garda, B., & Ogorzałek, M. J. (2016, September). Modelling the generic TiO2 memristor with the parasitic components. In *2016 International Conference on Signals and Electronic Systems (ICSES)* (pp. 173-176). IEEE. 10.1109/ICSES.2016.7593845

Itoh, M., & Chua, L. O. (2011). Memristor hamiltonian circuits. *International Journal of Bifurcation and Chaos in Applied Sciences and Engineering*, *21*(09), 2395–2425. doi:10.1142/S021812741103012X

Itoh, M., & Chua, L. O. (2014). Dynamics of memristor circuits. *International Journal of Bifurcation and Chaos in Applied Sciences and Engineering*, *24*(05), 1430015. doi:10.1142/S0218127414300158

Nakagawa, M., & Sorimachi, K. (1992). Basic characteristics of a fractance device. *IEICE Transactions on Fundamentals of Electronics, Communications and Computer Science*, *75*(12), 1814–1819.

Sah, M. P., Yang, C., Kim, H., Muthuswamy, B., Jevtic, J., & Chua, L. (2015). A generic model of memristors with parasitic components. *IEEE Transactions on Circuits and Systems. I, Regular Papers*, *62*(3), 891–898. doi:10.1109/TCSI.2014.2373674

Strukov, D. B., Snider, G. S., Stewart, D. R., & Williams, R. S. (2008). The missing memristor found. *Nature*, *453*(7191), 80–83. doi:10.1038/nature06932 PMID:18451858

KEY TERMS AND DEFINITIONS

Fractional Capacitor: A capacitor with arbitrary phase angle between 0 and π/2.

Fractional Inductor: An inductor with arbitrary phase angle between 0 and π/2.

Impasse Point: A discontinuous jump on the phase portrait.

Memristor: A nonlinear electrical circuit element which relates the flux and charge, theoretically postulated by L.O. Chua.

Parasitic Fractional Capacitor: An unintentionally occurred fractional capacitor intrinsic to a device.

Parasitic Fractional Inductor: An unintentionally occurred fractional inductor intrinsic to a device.

Phase Portrait: A graphical representation of comparative dynamic of variables.

Compilation of References

Abbasbandy, S., & Zakaria, F. S. (2008). Soliton solutions for the fifth-order KdV equation with the homotopy analysis method. *Nonlinear Dynamics*, *51*(1-2), 83–87. doi:10.100711071-006-9193-y

Abbas, S. (2010). Pseudo almost automorphic solutions of fractional order neutral differential equations. *Semigroup Forum*, *81*(3), 393–404. doi:10.100700233-010-9227-0

Abdel Kader, A. H., Abdel Latif, M. S., El Bialy, F., & Elsaid, A. (2018). Symmetry analysis and some new exact solutions of some nonlinear KdV-like equations. *Asian-European Journal of Mathematics*, *11*(03), 1850040. doi:10.1142/S1793557118500407

Abdel Kader, A. H., Abdel Latif, M. S., & Nour, H. M. (2017). Some New Exact Solutions of the Modified KdV Equation Using Lie Point Symmetry Method. *International Journal of Applied and Computational Mathematics*, *3*(1), 1163–1171. doi:10.100740819-017-0408-3

Abdel Latif, M. S. (2011). Some exact solutions of KdV equation with variable coefficients. *Communications in Nonlinear Science and Numerical Simulation*, *16*(4), 1783–1786. doi:10.1016/j.cnsns.2010.07.023

Abdel Latif, M. S., El-Shazly, E., Baleanu, D., Elsaid, A., & Nour, H. M. (2018). Some new soliton-like and doubly periodic-like solutions of Fisher equation with time-dependent coefficients. *Modern Physics Letters B*, *32*(33), 1850413. doi:10.1142/S0217984918504134

Abdel Latif, M. S., Shazly, E. E., Elsaid, A., & Nour, H. M. (2018a). Group classification and some new periodic-like and soliton-like solutions of the generalised Fisher equation with time-variable coefficients. *International Journal of Dynamical Systems and Differential Equations*, *8*(4), 313–325. doi:10.1504/IJDSDE.2018.096218

Abdullah, M., Butt, A. R., Raza, N., Alshomrani, A. S., & Alzahrani, A. K. (2018). Analysis of blood flow with nanoparticles induced by uniform magnetic field through a circular cylinder with fractional Caputo derivatives. *Journal of Magnetism and Magnetic Materials*, *446*, 28–36. doi:10.1016/j.jmmm.2017.08.074

Adda, F. B. (1997). Geometric Interpretation of the Fractional Derivative. *Journal of Fractional Calculus*, *11*, 21–52.

Adda, F. B. (2001). The Differentiability in the Fractional Calculus. *Nonlinear Analysis*, 47.

Agarwal, R. P., El-Sayed, A. M. A., & Salman, S. M. (2013). Fractional – Order Chua's system: Discretization, bifurcation and chaos. *Advances in Difference Equations*, *2013*(320).

Agarwal, R. P., Benchohra, M., & Hamani, S. (2009). Boundary value problems for fractional differential equations. *Georgian Mathematical Journal*, *16*(3), 401–411.

Agrawal, O. P. (2002). Formulation of Euler-Lagrange Equations for Fractional Variational Problems. *Journal of Mathematical Analysis and Applications*, *272*(1), 272. doi:10.1016/S0022-247X(02)00180-4

Agrawal, O. P. (2002). Solution for a fractional diffusion-wave equation defined in a bounded domain. *Nonlinear Dynamics, 29*(1/4), 145–155. doi:10.1023/A:1016539022492

Aguirregabiria, J. M., & Etxebarria, J. R. (1987). Fractal basin boundaries of a delay differential equation. *Physics Letters. [Part A], 122*(5), 241–244. doi:10.1016/0375-9601(87)90814-0

Ahmad, W. M., & El-Khazali, R. (2007). Fractional-order dynamical models of love. *Chaos, Solitons, and Fractals, 33*(4), 1367–1375. doi:10.1016/j.chaos.2006.01.098

Ahmed, C., El-Sayed, A. M. A., & El-Saka, H. A. A. (2006). On some RouthHurwitz conditions for fractional order differential equations and their applications in Lorenz, Rossler, Chua and Chen systems. *Physics Letters. [Part A], 358*(1), 1–4. doi:10.1016/j.physleta.2006.04.087

Ahmed, E., & Elgazzar, A. S. (2007). On fractional order differential equations model for nonlocal epidemics. *Physica A, 379*(2), 607–614. doi:10.1016/j.physa.2007.01.010

Ahmed, E., El-Sayed, A. M. A., El-Mesiry, E. M., & El-Saka, H. A. A. (2005). *Numerical solution for the fractional replicator equation. Internat. Modern Phys.* doi:10.1142/S0129183105007698

Ahmed, E., El-Sayed, A. M. A., & El-Saka, H. A. A. (2007). Equilibrium points, stability and numerical solutions of fractional-order predator-prey and rabies models. *Journal of Mathematical Analysis and Applications, 325*(1), 542–553. doi:10.1016/j.jmaa.2006.01.087

Ak Gumus, O., & Kose, H. (2012). On the stability of delay population dynamics related with Allee effects. *Math. Comput. Appl., 17*(1), 56–67.

Al Basir, F., & Roy, P. K. (2017). Stability analysis and optimal control of a fractional order model for HIV infection. *WSEAS Trans Math, 16*, 152-162. http://www.wseas.org/multimedia/journals/mathematics/2017/a385806-868.php

Ali, H. M. & Ameen, I. (2020). Save the pine forests of wilt disease using a fractional optimal control strategy, *Chaos, Solitons & Fractals, 132*, ID 109554. doi: 10.1016/j.chaos.2019.109554

Ali, H. M., Pereira, F. L., & Gama, S. M. (2016). A new approach to the Pontryagin maximum principle for nonlinear fractional optimal control problems. *Mathematical Methods in the Applied Sciences, 39*(13), 3640-3649. doi:10.1002/mma.3811

Ali, F., Saqib, M., Khan, I., & Sheikh, N. A. (2016). Application of Caputo-Fabrizio derivatives to MHD free convection flow of generalized Walters'-B fluid model. *The European Physical Journal Plus, 131*(10), 377. doi:10.1140/epjp/i2016-16377-x

Ali, F., Sheikh, N. A., Khan, I., & Saqib, M. (2017). Magnetic field effect on blood flow of Casson fluid in axisymmetric cylindrical tube: A fractional model. *Journal of Magnetism and Magnetic Materials, 423*, 327–336. doi:10.1016/j.jmmm.2016.09.125

Ali, H. M. (2019). New approximate solutions to fractional smoking model using the generalized Mittag-Leffler function method. *Progr. Fract. Differ. Appl., 5*(4), 1–9. doi:10.18576/pfda/010101

Ali, H. M., & Ameen, I. (2019). Mittag-Leffler approximation for the solution of nonlinear systems of fractional partial differential equations. *Sylwan, 163*(9), 17–32.

Al-khedhairi, A., Matouk, A. E., & Askar, S. S. (2019a). Computations of synchronization conditions in some fractional-order chaotic and hyperchaotic systems. *Pramana Journal of Physics, 92*, 72.

Al-khedhairi, A., Matouk, A. E., & Khan, I. (2019). Chaotic dynamics and chaos control for the fractional-order geomagnetic field model. *Chaos, Solitons \& Fractals, 128*, 390-401.

Al-Khedhairi, A., Askar, S. S., Matouk, A. E., Elsadany, A., & Ghazel, M. (2018). Dynamics, Chaos Control, and Synchronization in a Fractional-Order Samardzija-Greller Population System with Order Lying in (0, 2). *Complexity, 2018*, 2018. doi:10.1155/2018/6719341

Al-Khedhairi, A., Askar, S. S., Matouk, A. E., Elsadany, A., & Ghazel, M. (2018). Dynamics, chaos control, and synchronization in a fractional-order Samardzija-Greller population system with order lying in (0,2). *Complexity.*

Al-khedhairi, A., Matouk, A. E., & Khan, I. (2019b). Chaotic dynamics and chaos control for the fractional-order geomagnetic field model. *Chaos, Solitons, and Fractals, 128*, 390–401. doi:10.1016/j.chaos.2019.07.019

Allee, W. C. (1931). *Animal aggressions*. Chicago: University of Chicago press.

Ameen, I. (2017). *Fractional calculus: numerical methods and SIR models* (PhD thesis). University of Padova.

Ameen, I., Hidan, M., Mostefaoui, Z. & Ali, H. M. (2020). Fractional optimal control with fish consumption to prevent the risk of coronary heart disease. *Complexity*. . doi:10.1155/2020/9823753

Ameen, I., & Novati, P. (2017). The solution of fractional order epidemic model by implicit Adams methods. *Applied Mathematical Modelling, 43*, 78–84. doi:10.1016/j.apm.2016.10.054

an der Heiden, U. (1979). Delays in physiological systems. *Journal of Mathematical Biology, 8*(4), 345–364. doi:10.1007/BF00275831 PMID:541577

Anastassiou, G., Hooshmandas, M. R., Ghasemi, A., & Moftakharzadeh, F. (2009). Montgomery identities for fractional integrals and related fractional inequalities. *Journal of Inequalities in Pure and Applied Mathematics, 10*(4), 97.

Ankerhold, J., Grabert, H., & Ingold, G. L. (1995). Dissipative quantum systems with a potential barrier: General theory and the parabolic barrier. *Physical Review. E, 51*(5), 4267–4281. doi:10.1103/PhysRevE.51.4267 PMID:9963138

Arafa, A. A. M., Rida, S. Z., & Mohamed, H. (2014). An application of the homotopy analysis method to the transient behaviour of a biochemical reaction model. *Information Sciences Letters, 3*(1), 29. doi:10.12785/isl/030104

Arfken, G. B., Weber, H. J., & Harris, F. H. (2012). *Mathematical Methods for Physicists*. Elsevier Academic Press.

Atangana, A. (2018). Non validity of index law in fractional calculus: A fractional differential operator with Markovian and non-Markovian properties. *Physica A, 505*, 688–706. doi:10.1016/j.physa.2018.03.056

Atangana, A., & Alqahtani, R. T. (2016). Numerical approximation of the space-time Caputo-Fabrizio fractional derivative and application to groundwater pollution equation. *Adv. Difference Equations, 2016*(1), 156. doi:10.118613662-016-0871-x

Atangana, A., & Baleanu, D. (2017). Caputo-Fabrizio derivative applied to groundwater flow within confined aquifer. *Journal of Engineering Mechanics, 143*(5), D4016005. doi:10.1061/(ASCE)EM.1943-7889.0001091

Atangana, A., & Gómez-Aguilar, J. F. (2017). A new derivative with normal distribution kernel: Theory, methods and applications. *Physica A, 476*, 1–14. doi:10.1016/j.physa.2017.02.016

Atangana, A., & Gómez-Aguilar, J. F. (2018a). Fractional derivatives with no-index law property: Application to chaos and statistics. *Chaos, Solitons, and Fractals, 114*, 516–535. doi:10.1016/j.chaos.2018.07.033

Atangana, A., & Gómez-Aguilar, J. F. (2018b). Decolonisation of fractional calculus rules: Breaking commutativity and associativity to capture more natural phenomena. *The European Physical Journal Plus, 133*(4), 1–22. doi:10.1140/epjp/i2018-12021-3

Atangana, A., & Jain, S. (2018). A new numerical approximation of the fractal ordinary differential equation. *The European Physical Journal Plus*, *133*(2), 37. doi:10.1140/epjp/i2018-11895-1

Azhar, W. A., Vieru, D., & Fetecau, C. (2017). Free convection flow of some fractional nanofluids over a moving vertical plate with uniform heat flux and heat source. *Physics of Fluids*, *29*(8), 082001. doi:10.1063/1.4996034

Babakhani, A., & Daftardar-Gejji, V. (2002). On calculus of local fractional derivatives. *Journal of Mathematical Analysis and Applications*, *270*(1), 66–79. doi:10.1016/S0022-247X(02)00048-3

Bagley, R. L., & Calico, R. A. (1991). Fractional order state equations for the control of viscoelastically damped structures. *Journal of Guidance, Control, and Dynamics*, *14*(2), 304–311. doi:10.2514/3.20641

Bagley, R. L., & Torvik, P. J. (1983). A theoretical basis for the application of fractional calculus to viscoelasticity. *Journal of Rheology (New York, N.Y.)*, *27*(3), 201–210. doi:10.1122/1.549724

Baker, A. (2003). *Matrix Groups: An Introduction to Lie Group Theory*. Springer.

Balankin, A. S., Bory-Reyes, J., & Shapiro, M. (2016). Towards a physics on fractals: Differential vector calculus in three-dimensional continuum with fractal metric. *Physica A*, *444*, 345–359. doi:10.1016/j.physa.2015.10.035

Balankin, A. S., & Elizarraraz, B. E. (2012a). Map of fluid flow in fractal porous medium into fractal continuum flow. *Physical Review. E*, *85*(5), 056314. doi:10.1103/PhysRevE.85.056314 PMID:23004869

Balankin, A. S., & Elizarraraz, B. E. (2012b). Hydrodynamics of fractal continuum flow. *Physical Review. E*, *85*(2), 025302. doi:10.1103/PhysRevE.85.025302 PMID:22463270

Baleanu, D., Golmankhaneh, A. K., Nigmatullin, R., & Golmankhaneh, A. K. (2010). Fractional Newtonian mechanics. *Central European Journal of Physics*, *8*, 120–125.

Banchuin, R. (2018). On the memristances, parameters, and analysis of the fractional order memristor. *Active and Passive Electronic Components*, *2018*, 1–14. doi:10.1155/2018/3408480

Banchuin, R. (2019). Effects of Parasitic Fractional Elements to the Dynamics of Memristor. *Journal of Electrical and Computer Engineering*, *2019*, 1–13. doi:10.1155/2019/4190641

Banu, U. S., & Nasir, A. W. (2015). Design, Analysis and Performance Evaluation of Fractional Order Proportional Integral for Three Interacting Tank Process in Frequency Domain considered as Third Order System. *IFAC-PapersOnLine*, *48*(30), 179–184. doi:10.1016/j.ifacol.2015.12.374

Barbosa, R. S., Machado, J. T., Ferreira, I. M., & Tar, J. K. (2004, August). Dynamics of the fractional-order Van der Pol oscillator. In *Computational Cybernetics, 2004. ICCC 2004. Second IEEE International Conference on* (pp. 373-378). IEEE. 10.1109/ICCCYB.2004.1437752

Ben Adda, F. (1997). Geometric interpretation of the fractional derivative. *J. Frac. Cal.*, *11*, 21–51.

Ben Adda, F. (1997). Geometric interpretation of the fractional derivative. *J. Fract. Calc*, *11*, 21–52.

Bhatti, M. M., Zeeshan, A., & Ellahi, R. (2016a). Endoscope analysis on peristaltic blood flow of Sisko fluid with Titanium magneto-nanoparticles. *Computers in Biology and Medicine*, *78*, 29–41. doi:10.1016/j.compbiomed.2016.09.007 PMID:27643464

Bhatti, M. M., Zeeshan, A., & Ijaz, N. (2016b). Slip effects and endoscopy analysis on blood flow of particle-fluid suspension induced by peristaltic wave. *Journal of Molecular Liquids*, *218*, 240–245. doi:10.1016/j.molliq.2016.02.066

Blanquart, F. (2019). Evolutionary epidemiology models to predict the dynamics of antibiotic resistance. *Evolutionary Applications, 12*(3), 365–383. doi:10.1111/eva.12753 PMID:30828361

Bologna, M., Grigolini, P., & Riccardi, J. (1999). Lévy diffusion as an effect of sporadic randomness. *Physical Review. E, 60*(6), 6435–6442. doi:10.1103/PhysRevE.60.6435 PMID:11970559

Bonyah, E., Gomez-Aguilar, J. F., & Adu, A. (2018). Stability analysis and optimal control of a fractional human African trypanosomiasis model. *Chaos, Solitons & Fractals, 117*, 150-160.

Bonyah, E., Gómez-Aguilar, J. F., & Adu, A. (2018). Stability analysis and optimal control of a fractional human African trypanosomiasis model. *Chaos, Solitons, and Fractals, 117*, 150–160. doi:10.1016/j.chaos.2018.10.025

Boudjehem, D., & Boudjehem, B. B. (2012). A Fractional Model Predictive Control for Fractional Order Systems. In Fractional Dynamics and Control (pp. 59-71). New York: Springer.

Bouremel, Y. (2007). Explicit series solution for the Glauert-jet problem by means of the homotopy analysis method. *Communications in Nonlinear Science and Numerical Simulation, 12*(5), 714–724. doi:10.1016/j.cnsns.2005.07.001

Boyd, W. (1961). *Text-book of pathology; structure and function in diseases.* Academic Press.

Brandstetter, P., Dobrovsky, M., & Kuchar, M. (2014). Implementation of Genetic Algorithm in Control Structure of Induction Motor A.C. Drive. *Advances in Electrical and Computer Engineering, 14*(4), 15–20. doi:10.4316/AECE.2014.04003

Brodsky, S. J., Franke, V. A., Hiller, J. R., McCartor, G., Paston, S. A., & Prokhvatilov, E. V. (2004). A nonperturbative calculation of the electron's magnetic moment. *Nuclear Physics B, 703*(1-2), 333–362. doi:10.1016/j.nuclphysb.2004.10.027

Buckwar, E., & Luchko, Y. (1998). Invariance of a Partial Differential Equation of Fractional Order under the Lie Group of Scaling Transformations. *Journal of Mathematical Analysis and Applications, 227*(1), 81–97. doi:10.1006/jmaa.1998.6078

Budini, A. A., & Cáceres, M. O. (2004). Functional characterization of linear delay Langevin equations. *Physical Review. E, 70*(4), 046104. doi:10.1103/PhysRevE.70.046104 PMID:15600457

Cai, G., & Huang, J. (2007). A new finance chaotic attractor. *International Journal of Nonlinear Science, 3*(3), 213–220.

Calcagni, G. (2010). Fractal universe and quantum gravity. *Physical Review Letters, 104*(25), 251301. doi:10.1103/PhysRevLett.104.251301 PMID:20867360

Calcagni, G. (2012a). Geometry and field theory in multi-fractional spacetime. *Journal of High Energy Physics, 2012*(1), 65. doi:10.1007/JHEP01(2012)065

Calcagni, G. (2012b). Geometry of fractional spaces. *Advances in Theoretical and Mathematical Physics, 16*(2), 549–644. doi:10.4310/ATMP.2012.v16.n2.a5

Calcagni, G., Eichhorn, A., & Saueressig, F. (2013). Probing the quantum nature of spacetime by diffusion. *Physical Review. D, 87*(12), 124028. doi:10.1103/PhysRevD.87.124028

Calcagni, G., Magueijo, J., & Fernández, D. R. (2014). Varying electric charge in multiscale spacetimes. *Physical Review. D, 89*(2), 024021. doi:10.1103/PhysRevD.89.024021

Caldeira, A. O., & Leggett, A. J. (1981). Influence of dissipation on quantum tunneling in macroscopic systems. *Physical Review Letters, 46*(4), 211–214. doi:10.1103/PhysRevLett.46.211

Caldeira, A. O., & Leggett, A. J. (1983). Quantum tunnelling in a dissipative system. *Annals of Physics, 149*(2), 374–456. doi:10.1016/0003-4916(83)90202-6

Cang, J., Tan, Y., Xu, H., & Liao, S. J. (2009). Series solutions of non-linear Riccati differential equations with fractional order. *Chaos, Solitons & Fractals, 40*, 1-9.

Caputo, M. (1967). Linear models of dissipation whose Q is almost frequency independent-II. *Geophysical Journal of the Royal Astronomical Society, 13*(5), 529–539. doi:10.1111/j.1365-246X.1967.tb02303.x

Caputo, M. (1969). *Elasticité e dissipatizione*. Bologna: Zanichelli.

Caputo, M., & Fabrizio, M. (2015). A new definition of fractional derivative without singular kernel. *Prog. Fract. Differ. Appl., 1*, 73–85.

Carlton, J. M., Yowell, C. A., Sturrock, K. A., & Dame, J. B. (2001). Biomagnetic separation of contaminating host leukocytes from plasmodium-infected erythrocytes. *Experimental Parasitology, 97*(2), 111–114. doi:10.1006/expr.2001.4594 PMID:11281708

Carpinteri, A., & Mainardi, F. (1997). *Fractals and Fractional Calculus in Continuum Mechanics*. Wien: Springer. doi:10.1007/978-3-7091-2664-6

Carpinteri, A., & Sapora, A. (2010). Diffusion problems in fractal media defined on Cantor sets. *ZAMM-Journal of Applied Mathematics and Mechanics/Zeitschrift für Angewandte Mathematik und Mechanik. Applied Mathematics and Mechanics, 90*(3), 203–210.

Cattani, C. (2018). Sinc-fractional operator on Shannon wavelet space. *Frontiers in Physics, 6*, 118.

Celik, C., & Duman, M. (2012). Crank--Nicolson method for the fractional diffusion equation with the Riesz fractional derivative. *Journal of Computational Physics, 231*(4), 1743–1750. doi:10.1016/j.jcp.2011.11.008

Chen, Y. (1999). *Model Order Reduction for Nonlinear System* (M. S. Thesis). Massachusetts Institute of Technology.

Chen, Y., Petras, I., & Xue, D. (2009). Fractional order control. In *American Control conference* (pp. 1397-1411). St. Louis, MO: Academic Press.

Chen, D. Y., Liu, Y. X., Ma, X. Y., & Zhang, R. F. (2012). Control of a class of fractional-order chaotic systems via sliding mode. *Nonlinear Dynamics, 67*(1), 893901. doi:10.100711071-011-0002-x PMID:22757537

Chen, L., Chai, Y., & Wu, R. (2011). Control and Synchronization of Fractional-Order Financial System Based on Linear Control. *Discrete Dynamics in Nature and Society, 2011*, 1–21. doi:10.1155/2011/958393

Chen, W. (2006). Time–space fabric underlying anomalous diffusion. *Chaos, Solitons, and Fractals, 28*(4), 923–929. doi:10.1016/j.chaos.2005.08.199

Chen, W. (2008). Dynamics and control of a financial system with time-delayed feed backs. *Chaos, Solitons, and Fractals, 37*(4), 1198–1207. doi:10.1016/j.chaos.2006.10.016

Chen, W. C. (2008). Nonlinear dynamics and chaos in a fractional-order financial system. *Chaos, Solitons, and Fractals, 36*(5), 1305–1314. doi:10.1016/j.chaos.2006.07.051

Chen, W., Sun, H., Zhang, X., & Korošak, D. (2010b). Anomalous diffusion modeling by fractal and fractional derivatives. *Computers & Mathematics with Applications (Oxford, England), 59*(5), 1754–1758. doi:10.1016/j.camwa.2009.08.020

Chen, Y. Q., Vinagre, B. M., & Podlubny, I. (2003). A new discretization method for fractional-order differentiators via continued fraction expansion. *Proceedings of DETC, 3*, 761–769.

Chen, Y., Yan, Y., & Zhang, K. (2010a). On the local fractional derivative. *Journal of Mathematical Analysis and Applications, 362*(1), 17–33. doi:10.1016/j.jmaa.2009.08.014

Chua, L. O. (1969). *Introduction to nonlinear network theory*. New York: McGraw-Hill.

Chua, L. O. (2012). The fourth element. *Proceedings of the IEEE, 100*(6), 1920–1927. doi:10.1109/JPROC.2012.2190814

Coronel-Escamilla, A., Gómez-Aguilar, J. F., Baleanu, D., Córdova-Fraga, T., Escobar-Jiménez, R. F., Olivares-Peregrino, V. H., & Al Qurashi, M. M. (2017). Bateman-Feshbach Tikochinsky and Caldirola-Kanai oscillators with new fractional differentiation. Entropy, 19(2), 55.

Coronel-Escamilla, A., Gómez-Aguilar, J. F., López-López, M. G., Alvarado-Martínez, V. M., & Guerrero-Ramírez, G. V. (2016). Triple pendulum model involving fractional derivatives with different kernels. *Chaos, Solitons, and Fractals, 91*, 248–261. doi:10.1016/j.chaos.2016.06.007

Cottrill-Shepherd, K., & Naber, M. (2001). Fractional Differential Forms. *Mathematics and Physics, 42*(5), 2203–2212. doi:10.1063/1.1364688

Daftardar-Gejji, V., & Babakhani, A. (2004). Analysis of a system of fractional differential equations. *J. Math Appl Anal, 293*(2), 511–522. doi:10.1016/j.jmaa.2004.01.013

Dao, S. D., Abhary, K., & Marian, R. (2017). A bibliometric analysis of genetic algorithms throughout the history. *Computers & Industrial Engineering, 110*, 395–403. doi:10.1016/j.cie.2017.06.009

Darus, M., & Ibrahim, R. W. (2011). On the existence of univalent solutions for fractional integral equation of Volterra type in complex plane. *ROMAI J., 7*(1), 77–86.

Das, S. (2007). *Functional Fractional Calculus for System Identification and Controls*. Springer.

Das, S., & Gupta, P. A. (2011). mathematical model on fractional Lotka-Volterra equations. *Journal of Theoretical Biology, 277*(1), 16. doi:10.1016/j.jtbi.2011.01.034 PMID:21295596

Das, S., Saha, S., Das, S., & Gupta, A. (2011). On the selection of tuning methodology of FOPID controllers for the control of higher order processes. *ISA Transactions, 50*(3), 376–388. doi:10.1016/j.isatra.2011.02.003 PMID:21420085

David Logan, J. (2011). *A First Course in Differential Equations* (2nd ed.). New York: Springer. doi:10.1007/978-1-4419-7592-8

Davis, L. (1991). *Handbook of Genetic Algorithms*. New York: Van Nostrand Reinhold Company.

de Rafael, E. (2013). Update of the Electron and Muon g-Factors. *Nuclear Physics B - Proceedings Supplement, 234*, 193–198. doi:10.1016/j.nuclphysbps.2012.12.012

Debbouche, A., Nieto, J. J., & Torres, D. F. M. (2017). Optimal solutions to relaxation in multiple control problems of Sobolev type with nonlocal nonlinear fractional differential equations. *Journal of Optimization Theory and Applications, 174*(1), 7–31. doi:10.100710957-015-0743-7

Denysiuk, R., Silva, C. J., & Torres, D. F. M. (2015). Multiobjective approach to optimal control for a tuberculosis model. *Optimization Methods & Software, 30*(5), 893–910. doi:10.1080/10556788.2014.994704

Diekmann, O., van Gils, S. A., Verduyn Lunel, S. M., & Walther, H. O. (1955). *Delay Equations: Functional. Complex, and Nonlinear Analysis*. Berlin: Springer.

Diethelm & Ford. (n.d.). *The numerical solution of linear and non-linear fractional differential equations involving fractional derivatives several of several orders*. Numerical Analysis Report 379. Manchester Center for Numerical Computational Mathematics.

Diethelm, K. (2002). Predictor-corrector strategies for single- and multi-term fractional differential equations. In *Proceedings of the 5th HellenicEuropean Conference on Computer Mathematics and its Applications*. LEA Press.

Diethelm, K., & Freed, A. D. (1999). The FracPECE subroutine for the numerical solution of differential equations of fractional order. In S. Heinzel & T. Plesser (Eds.), Forschung und wissenschaftliches Rechnen. Gottingen: Gesellschaft für wissenschaftliche Datenverarbeitung (pp. 57–71). Academic Press.

Diethelm, K., & Freed, A. D. (1999). The FracPECE subroutine for the numerical solution of differential equations of fractional order. In S. Heinzel & T. Plesser (Eds.), Forschung und wissenschaftliches Rechnen. Gttingen: Gesellschaft für wissenschaftliche Datenverarbeitung (pp. 57–71). Academic Press.

Diethelm, K. (1997). An algorithm for the numerical solution of differential equations of fractional order. *Electronic Transactions on Numerical Analysis*, *5*, 1–6.

Diethelm, K., & Ford, N. J. (2002). Analysis of fractional differential equations. *Journal of Mathematical Analysis and Applications*, *265*(2), 229–248. doi:10.1006/jmaa.2000.7194

Diethelm, K., Ford, N. J., & Freed, A. D. (2002). A predictor-corrector approach for the numerical solution of fractional differential equations. *Nonlinear Dynamics*, *29*(1/4), 3–22. doi:10.1023/A:1016592219341

Diethelm, K., Ford, N. J., & Freed, A. D. (2004b). Detailed error analysis for a fractional Adams method. *Numerical Algorithms*, *36*(1), 31–52. doi:10.1023/B:NUMA.0000027736.85078.be

Diethelm, K., & Freed, A. (1999a). On the solution of nonlinear fractional order differential equations used in the modelling of viscoplasticity. In F. Keil, W. Mackens, H. Vo, & J. Werther (Eds.), *Scientific Computing in Chemical Engineering IIComputational Fluid Dynamics. Reaction Engineering, and Molecular Properties* (p. 217224). Heidelberg: Springer.

Diethelm, K., & Freed, A. (1999b). The FracPECE subroutine for the numerical solution of differential equations of fractional order. In S. Heinzel & T. Plesser (Eds.), *Forschung und wissenschaftliches Rechnen 1998* (pp. 57–71). Gottingen: Gesellschaft fur Wisseschaftliche Datenverarbeitung.

Dieuonn'e, X. (1960). *Foundations of modern analysis*. New York: Academic Press.

Ding, Y., Wang, Z., & Ye, H. (2011). Optimal control of a fractional-order HIV-immune system with memory. *IEEE Transactions on Control Systems Technology*, 20(3), 763-769. doi:10.1109/TCST.2011.2153203

Ding, J., & Yang, W. Y. H. (2009). A new modified hyperchaotic finance system and its control. *International Journal of Nonlinear Science*, *8*(1), 59–66.

Din, Q. (2017). Complexity and chaos control in a discrete-time prey-predator model. *Communications in Nonlinear Science and Numerical Simulation*, *49*, 113–134. doi:10.1016/j.cnsns.2017.01.025

Din, Q. (2018). Bifurcation analysis and chaos control in discrete-time glycolysis models. *Journal of Mathematical Chemistry*, *56*(3), 904–931. doi:10.100710910-017-0839-4

Din, Q., Elsadany, A. A., & Ibrahim, S. (2018). Bifurcation analysis and chaos control in a second order rational difference equations. *International Journal of Nonlinear Sciences and Numerical Simulation*, *19*(1), 53–68. doi:10.1515/ijnsns-2017-0077

Dragomir, S. S. (2005). *An Ostrowski type inequality for convex functions*. Univerzitet u Beogradu Publikacije Elektrotehni ckog Fakulteta. Serija Matematika, 16.

Driver, R. D. (1977). Ordinary and Delay Differential Equations. Springer-Verlag.

Duan, G.R. & Yu, H.H. (2013). *LMI in control systems analysis, design and applications.* CRC Press, Taylors and Francis Group.

Duarte, V., & Costa, J. S. (2002). Time domain implementations of non-integer order controllers. Proc. of Controlo 2002, 353-358.

Dugowson, S. (1997). L'élaboration par Riemann d'une définition de la dérivation d'ordre non entier. *Revue d'Histoire de Mathématiques, Société mathématique de France, 3*, 49-97.

Du, M., Wang, Z., & Hu, H. (2013). Measuring memory with the order of fractional derivative. *Scientific Reports, 3*(1), 1–3. doi:10.1038rep03431 PMID:24305503

Dwyer, A. O. (2006). *Handbook of PI and PID controller tuning rules.* London, UK: Imperial College Press. doi:10.1142/p424

Edelstein Keshet, L. (2005). *Mathematical Models in Biology.* New York: Society for Industrial and Applied Mathematics. doi:10.1137/1.9780898719147

Efe, M. O. (2011). Fractional order systems in industrial automation – A Survey. *IEEE Transactions on Industrial Informatics, 7*(4), 582–591. doi:10.1109/TII.2011.2166775

Eidelman, S., Giacomini, M., Ignatov, F. V., & Passera, M. (2007). The tau lepton anomalous magnetic moment. *Nuclear Physics B - Proceedings Supplements, 169*, 226-231. arXiv preprint hep-ph/0702026

Elabbasy, E. M., Elsadany, A. A., & Zhang, Y. (2014). Bifurcation analysis and chaos in a discrete reduced Lorenz system. *Applied Mathematics and Computation, 228*, 184–194. doi:10.1016/j.amc.2013.11.088

Elaidy, S. N. (2005). An introduction to difference equations (3rd ed.). Springer.

Elal, L., Sweilam, N. H., Nagy, A. M., & Almaghrebi, Y. S. (2016). Computational methods for the fractional optimal control HIV infection. *Journal of Fractional Calculus and Applications, 7*(2), 121–131.

Elaydi, S. N. (2008). *Discrete Chaos with Applications in Science and Engineering.* Chapman and Hall/CRC.

El-Bialy, F., Abdel Latif, M. S., & Elsaid, A. (2017). Symmetry analysis and some new exact solutions of the (2+ 1)-dimensional Burgers equations. *Acta Physica Polonica B, 48*(11), 2031–2043. doi:10.5506/APhysPolB.48.2031

Eldélyi, A. (1964). An integral equation involving Legendre function. *SIAM Journal on Applied Mathematics, 12*(1), 15–30. doi:10.1137/0112002

Ellahi, R., Rahman, S. U., Nadeem, S., & Vafai, K. (2015). The blood flow of Prandtl fluid through a tapered stenosed arteries in permeable walls with magnetic field. *Communications in Theoretical Physics, 63*(3), 353. doi:10.1088/0253-6102/63/3/353

Ellahi, R., Raza, M., & Vafai, K. (2012). Series solutions of non-Newtonian nanofluids with Reynolds' model and Vogel's model by means of the homotopy analysis method. *Mathematical and Computer Modelling, 55*(7-8), 1876–1891. doi:10.1016/j.mcm.2011.11.043

El-Raheem, Z. F., & Salman, S. M. (2014). On a discretization process of fractional-order logistic differential equation. *J. Egypt. Math. Soc., 22*(3), 407–412. doi:10.1016/j.joems.2013.09.001

Elsadany, A. A., & Matouk, A. E. (2015). Dynamical behaviors of fractional-order Lotka– Volterra predator–prey model and its discretization. *Journal of Applied Mathematics and Computing, 49*(1-2), 269–283. doi:10.100712190-014-0838-6

Elsaid, A. (2010). The variational iteration method for solving Riesz fractional partial differential equations. *Computers & Mathematics with Applications, 60*, 1940-1947.

Elsaid, A., Latif, M. S. A., & Maneea, M. (2016a). Similarity Solutions for Solving Riesz Fractional Partial Differential Equations. *Progress in Fractional Differentiation and Applications, 2*(4), 293-298.

Elsaid, A., Shamseldeen, S., & Madkour, S. (2016c). Solution of Fractional Wave Equation by the Optimal Homotopy Analysis Method. *European Journal of Pure and Applied Mathematics*, 586-601.

Elsaid, A. (2011). Homotopy analysis method for solving a class of fractional partial differential equations. *Communications in Nonlinear Science and Numerical Simulation, 16*(9), 3655–3664. doi:10.1016/j.cnsns.2010.12.040

Elsaid, A., Abdel Latif, M. S., & Maneea, M. (2016). Similarity solutions of fractional order heat equations with variable coefficients. *Miskolc Mathematical Notes, 17*(1), 245–254. doi:10.18514/MMN.2016.1610

Elsaid, A., Latif, A., & Maneea, M. (2016b). Similarity solutions for multiterm time-fractional diffusion equation. *Advances in Mathematical Physics, 2016*, 2016. doi:10.1155/2016/7304659

Elsaid, A., Shamseldeen, S., & Madkour, S. (2016a). Iterative Solution of Fractional Diffusion Equation Modelling Anomalous Diffusion. *Applications and Applied Mathematics, 11*, 815–827.

Elsaid, A., Shamseldeen, S., & Madkour, S. (2016b). Semianalytic Solution of Space-Time Fractional Diffusion Equation. *International Journal of Differential Equations, 2016*, 2016. doi:10.1155/2016/2371837

El-Sayed, A. M. A. (1996). Fractional differential difference equations. *J. Fract. Calc., 10*, 101–106.

El-Sayed, A. M. A., Agarwal, R. P., & Salman, S. M. (2013). Fractional-order Chua's system: discretization, bifurcation and chaos. *Journal of Advances in Difference Equations, 2013*, 320.

El-Sayed, A. M. A., El-Raheem, Z. F., & Salman, S. M. (2014). Discretization of forced Duffing system with fractional-order damping. *Journal of Advances in Difference Equations, 2014*(66).

El-Sayed, A. M. A., Elsonbaty, A., Elsadany, A. A., & Matouk, A. E. (2016b). Dynamical analysis and circuit simulation of a new fractional-order hyperchaotic system and its discretization. International Journal of Bifurcation and Chaos, 26.

El-Sayed, A. M. A., Elsonbaty, A., Elsadany, A. A., & Matouk, A. E. (2016b). Dynamical analysis and circuit simulation of a new fractional-order hyperchaotic system and its discretization. International Journal of Bifurcation and Chaos, 26. doi:10.1142/S0218127416502229

El-Sayed, A.M.A. (1988). Fractional order differential equation. *Kyungpook Math. J., 28*(2), 1822.

El-Sayed, A.M.A. (1998). Nonlinear functional-differential equations of arbitrary orders. *Nonlinear Anal., 33*(2).

El-Sayed, A.M.A., El-Mesiry, A.E.M., & El-Saka, H.A.A. (2007). On the fractional-order logistic equation. *Appl. Math. Lett., 20*(7), 817823.

El-Sayed, A.M.A., Gaafar, F.M. (2001). Fractional order differential equations with memory and fractional-order relaxation oscillation model. *Pure Math. Appl., 12*.

El-Sayed, A. M. A. (1992). On the fractional differential equations. *Applied Mathematics and Computation, 49*(2-3), 205213. doi:10.1016/0096-3003(92)90024-U

El-Sayed, A. M. A. (1996). Fractional-order diffusion-wave equation. *International Journal of Theoretical Physics, 35*(2), 311–322. doi:10.1007/BF02083817

El-Sayed, A. M. A. (1998). Nonlinear functional differential equations of arbitrary orders. Nonlinear Anal.: Theory. *Methods Appl.*, *33*, 181–186.

El-Sayed, A. M. A., Ahmed, E., & Herzallah, M. A. E. (2011). On the fractional-order differential Games with non-uniform interaction rate and Asymmetric games. *Journal of Fractional Calculus and Applications*, *1*(1), 1–9.

El-Sayed, A. M. A., El-Mesiry, E. M., & El-Saka, H. A. A. (2004). Numerical solution for multiterm fractional (arbitrary) orders differential equations. *Computational & Applied Mathematics*, *23*(1), 33–54. doi:10.1590/S0101-82052004000100002

El-Sayed, A. M. A., El-Mesiry, E. M., & El-Saka, H. A. A. (2007). On the fractional-order logistic equation. *J. Appl. Math. Letters*, *20*(7), 817–823. doi:10.1016/j.aml.2006.08.013

El-Sayed, A. M. A., Elsonbaty, A., Elsadany, A. A., & Matouk, A. E. (2016b). Dynamical analysis and circuit simulation of a new fractional-order hyperchaotic system and its discretization. *International Journal of Bifurcation and Chaos in Applied Sciences and Engineering*, 26.

El-Sayed, A. M. A., Gaafar, F. M., & Hashem, H. H. (2004). On the mayimal and minimal solutions of arbitrary orders nonlinear functional integral and differential equations. *Math. Sci. Res. J.*, *8*, 336–348.

El-Sayed, A. M. A., Nour, H. M., Elsaid, A., Matouk, A. E., & Elsonbaty, A. (2016a). Dynamical Behaviors, Circuit Realization, Chaos Control and Synchronization of a New Fractional Order Hyperchaotic System. *Applied Mathematical Modelling*, *40*(5-6), 3516–3534. doi:10.1016/j.apm.2015.10.010

El-Sayed, A. M. A., & Salman, S. M. (2013). On a discretization process of fractional order Riccati's differential equation. *J. Fract. Calc. Appl.*, *4*, 251–259.

El-Sayed, A. M. A., & Salman, S. M. (2013). On a discretization process of fractional-order Riccati differential equation. *Journal of Fractional Calculus and Applications*, *4*(2), 251–259.

El-Sayed, A. M. A., & Salman, S. M. (2013, July). The unified system between Lorenz and Chen systems: A discretization process. *Electronic Journal of Mathematical Analysis and Applications*, *1*(2), 318–325.

El-Sayed, A. M., & Gaber, M. (2006). On the finite Caputo and finite Riesz derivatives. *Electronic Journal of Theoretical Physics*, *3*, 81–95.

Elwakil, A. S. (2010). Fractional order circuits and systems magazine: An Emerging interdisciplinary research area. *IEEE Circuits and Systems*, *10*(4), 40–50.

Engheta, N. (1998). Fractional Curl Operator in Electromagnetics. *Microwave and Optical Technology Letters*, *17*(2), 86–91. doi:10.1002/(SICI)1098-2760(19980205)17:2<86::AID-MOP4>3.0.CO;2-E

Erjaee, G. H. (2009). On analytical justification of phase synchronization in different chaotic systems. *Chaos, Solitons, and Fractals*, *39*(3), 1195–1202. doi:10.1016/j.chaos.2007.06.006

Ervin, V. J., & Roop, J. P. (2007). Variational solution of fractional advection dispersion equations on bounded domains. *Numerical Methods for Partial Differential Equations*, *23*(2), 256–281. doi:10.1002/num.20169

Eyink, G. (1989). Quantum field-theory models on fractal spacetime. *Communications in Mathematical Physics*, *125*(4), 613–636. doi:10.1007/BF01228344

Fahim, S. M., Ahmed, S., & Imtiaz, S. A. (2018). Fractional order model identification using the sinusoidal input. *ISA Transactions*, *83*, 35–41. doi:10.1016/j.isatra.2018.09.009 PMID:30243512

Faieghi, M. R., Delavari, H., & Baleanu, D. (2013). A note on stability of sliding mode dynamics in suppression of fractional-order chaotic systems. *Computers & Mathematics with Applications (Oxford, England), 66*(5), 832–847. doi:10.1016/j.camwa.2012.11.015

Fan, E. (2000). Extended tanh-function method and its applications to nonlinear equations. *Physics Letters. [Part A], 277*(4-5), 212–218. doi:10.1016/S0375-9601(00)00725-8

Fan, G., Liu, J., Van den Driessche, P., Wu, J., & Zhu, H. (2010). The impact of maturation delay if mosquitoes on the transmission of West Nile virus. *Mathematical Biosciences, 228*(2), 119–126. doi:10.1016/j.mbs.2010.08.010 PMID:20828577

Feldmann, P., & Freund, R. W. (1995). Efficient Linear Circuit Analysis by Pade Approximation Via the Lanczos Process. *IEEE Transactions on Computer-Aided Design of Integrated Circuits and Systems, 14*(5), 639–649. doi:10.1109/43.384428

Feng, Q. X., & Yan, J. R. (2002). Global attractivity and oscillation in a kind of Nicholsons blowflies. *J. Biomath., 17*, 21–26.

Fouda, M. E., & Radwan, A. G. (2013). On the fractional-order memristor model. *Journal of Fractional Calculus and Applications, 4*(1), 1-7.

Fouda, M. E., & Radwan, A. G. (2015). Fractional-order memristor response under dc and periodic signals. *Circuits, Systems, and Signal Processing, 34*(3), 961–970. doi:10.100700034-014-9886-2

Freeborn, T. J., Maundy, B., & Elwakil, A. S. (2013). Measurement of supercapacitor fractional-order model parameters from voltage-excited step response. *IEEE Journal on Emerging and Selected Topics in Circuits and Systems, 3*(3), 367–376. doi:10.1109/JETCAS.2013.2271433

Furati, K. M., & Tatar, N.-E. (2007). Long time behavior for a nonlinear fractional model. *Journal of Mathematical Analysis and Applications, 332*(1), 441454. doi:10.1016/j.jmaa.2006.10.027

Galaktionov, V., & Svirshchevskii, S. (2007). *Exact Solutions and Invariant Subspaces of Nonlinear Partial Differential Equations in Mechanics and Physics*. London: Chapman & Hall/CRC.

Gao, X., & Yu, J. (2005). Chaos in the fractional order periodically forced complex Duffings oscillators. *Chaos, Solitons, and Fractals, 24*(4), 1097–1104. doi:10.1016/j.chaos.2004.09.090

Garba, S. M., & Safi, M. A. (2013). Mathematical analysis of West Nile Virus model with discrete delays. *Acta Mathematica Scientia, 33B*(5), 1439–1462. doi:10.1016/S0252-9602(13)60095-8

Garg, M., Sharma, A., & Manohar, P. (2013). Solution of generalized space-time fractional telegraph equation with composite and Riesz-Feller fractional derivatives. *International Journal of Pure and Applied Mathematics, 83*(5), 685–691. doi:10.12732/ijpam.v83i5.12

Gazizov, R. K., Kasatkin, A. A., & Lukashchuk, S. Y. (2009). Symmetry Properties of Fractional Diffusion Equations. *Physica Scripta, T136*, 014016. doi:10.1088/0031-8949/2009/T136/014016

Gazizov, R. K., Kasatkin, A. A., & Lukashchuk, S. Y. (2011). *Group-Invariant Solutions of Fractional Differential Equations*. Nonlinear Science and Complexity.

Gazizov, R. K., Kasatkin, A. A., & Lukashchuk, S. Y. (2012). Fractional Differential Equations: Change of Variables and Nonlocal Symmetries. *Ufa Mathematical Journal., 4*(4), 54–67.

Gelberi, H. (2012). Time delay feedback control of chaos in a hyper chaotic finance system. In *3rd International Symposium on Sustainable Development*. IEEE.

George Maria Selvam, A. & Janagaraj, R. (2018a). Numerical Analysis of a Fractional Order Discrete Prey – Predator System with Functional Response. *International Journal of Engineering & Technology, 7*(4.10), 681–684.

George Maria Selvam, A., & Janagaraj, R. (2018a). Numerical Analysis of a Fractional Order Discrete Prey – Predator System with Functional Response. *International Journal of Engineering & Technology, 7*(4.10), 681–684.

George Maria Selvam, A., Janagaraj, R., & Britto Jacob, S. (2017). Stability in a Discrete Fractional Order Prey Predator System with Functional Response. *International Journal of Advanced Research in Computer Science and Software Engineering, 7*(5), 630–633. doi:10.23956/ijarcsse/SV7I5/0336

George Maria Selvam, A., Janagaraj, R., & Vignesh, D. (2018b). Allee effect and Holling type - II response in a discrete fractional order prey - predator model. *IOP Conference Series: Journal of Physics, 1139*, 1-7. 10.1088/1742-6596/1139/1/012003

Ghasemi, S. E., Hatami, M., Sarokolaie, A. K., & Ganji, D. D. (2015). Study on blood flow containing nanoparticles through porous arteries in presence of magnetic field using analytical methods. *Physica E, Low-Dimensional Systems and Nanostructures, 70*, 146–156. doi:10.1016/j.physe.2015.03.002

Glass, L., & Mackey, M. C. (1979). Pathological conditions resulting from instabilities in physiological control systems. *Annals of the New York Academy of Sciences, 316*(1), 214–235. doi:10.1111/j.1749-6632.1979.tb29471.x PMID:288317

Global Health Policy. (2014). *The U.S. Government & Global Emerging Infectious Disease Preparedness and Response*. http://kff.org/global-healthpolicy/fact-sheet/the-u-s-government-global-emerging-infectious-disease-preparedness-and-response/

Godinho, C. F., Weberszpil, J., & Helayël-Neto, J. A. (2012). Extending the d'Alembert solution to space–time modified Riemann–Liouville fractional wave equations. *Chaos, Solitons, and Fractals, 45*(6), 765–771. doi:10.1016/j.chaos.2012.02.008

Godinho, C. F., Weberszpil, J., & Neto, J. H. (2014). Fractional canonical quantization: A parallel with noncommutativity. *International Journal of Theoretical Physics, 53*(7), 2379–2395. doi:10.100710773-014-2037-5

Goldberg, D. E. (1989). *Genetic Algorithms in search Optimization and Machine Learning*. Boston, MA: Addison-Wesley.

Golmankhaneh, A. K., Ali, M. Y., & Baleanu, D. (2012). On the Fractional Hamilton and Lagrange Mechanics. *International Journal of Theoretical Physics, 51*(9), 51. doi:10.100710773-012-1169-8

Gómez-Aguilar, J. F., Torres, L., Yépez-Martínez, H., Baleanu, D., Reyes, J. M., & Sosa, I. O. (2016b). Fractional Liénard type model of a pipeline within the fractional derivative without singular kernel. Advances in Difference Equations, 2016(1), 173.

Gómez-Aguilar, J. F., & Atangana, A. (2017). New insight in fractional differentiation: Power, exponential decay and Mittag-Leffler laws and applications. *The European Physical Journal Plus, 132*(13), 1–23. doi:10.1140/epjp/i2017-11293-3

Gómez-Aguilar, J. F., & Baleanu, D. (2014). Fractional transmission line with losses. *Zeitschrift für Naturforschung. Section A. Physical Sciences, 69*(10-11), 539–546. doi:10.5560/zna.2014-0049

Gómez-Aguilar, J. F., Yépez-Martínez, H., Escobar-Jiménez, R. F., Astorga-Zaragoza, C. M., & Reyes-Reyes, J. (2016a). Analytical and numerical solutions of electrical circuits described by fractional derivatives. *Applied Mathematical Modelling, 40*(21-22), 9079–9094. doi:10.1016/j.apm.2016.05.041

Gorenflo, R., & Mainardi, F. (1997). Fractional calculus: Integral and differential equations of fractional order. In A. Carpinteri & F. Mainardi (Eds.), *Fractals and Fractional Calculus in Continuum Mechanics* (pp. 223–276). Wien: Springer. doi:10.1007/978-3-7091-2664-6_5

Gorenflo, R., & Mainardi, F. (2003). Fractional diffusion processes: probability distributions and continuous time random walk. In *Processes with Long-Range Correlations* (pp. 148–166). Springer. doi:10.1007/3-540-44832-2_8

Gorenflo, R., Mainardi, F., Moretti, D., Pagnini, G., & Paradisi, P. (2002). Discrete random walk models for space--time fractional diffusion. *Chemical Physics, 284*(1-2), 521–541. doi:10.1016/S0301-0104(02)00714-0

Gorenflo, R., & Vivoli, A. (2003). Fully discrete random walks for space--time fractional diffusion equations. *Signal Processing, 83*(11), 2411–2420. doi:10.1016/S0165-1684(03)00193-2

Gourley, S. A., & Ruan, S. (2000). Dynamics of the diffusive Nicholson's blowflies equation with distributed delay. *Proc. Roy. Soc. Edinburgh Sect. A, 130*(6), 1275–1291. doi:10.1017/S0308210500000688

Grigolini, P., Rocco, A., & West, B. J. (1999). Fractional calculus as a macroscopic manifestation of randomness. *Physical Review. E, 59*(3), 2603–2613. doi:10.1103/PhysRevE.59.2603

Grigorenko, I., & Grigorenko, E. (2003). Chaotic dynamics of the fractional Lorenz system. *Physical Review Letters, 91*(3), 034101. doi:10.1103/PhysRevLett.91.034101 PMID:12906418

Gurney, W. S., Blythe, S. P., & Nisbet, R. M. (1980). Nicholson's blowflies (revisited). *Nature, 287*(5777), 17–21. doi:10.1038/287017a0

Hadid, S.B. (1995). Local and global existence theorems on differential equations of non-integer order. *J. Fract. Calc.*, 101105.

Hadid, S.B., & Momani, S.M. (1996). On the existence of maximal and minimal solutions of differential equations of non-integer order. *J. Fract. Calc.*, 4144.

Haik, Y., Chen, C. J., & Chatterjee, J. (2002). Numerical simulation of biomagnetic fluid in a channel with thrombus. *Journal of Visualization / the Visualization Society of Japan, 5*(2), 187–195. doi:10.1007/BF03182428

Haik, Y., Pai, V., & Chen, C.-J. (1999). Development of magnetic device for cell separation. *Journal of Magnetism and Magnetic Materials, 194*(1-3), 254–261. doi:10.1016/S0304-8853(98)00559-9

Hale, J. K., & Verduyn Lunel, S. M. (1993). *Introduction to functional differential equations.* New York, NY: Springer. doi:10.1007/978-1-4612-4342-7

Hanyga, A. (2002). Multi-Dimensional Solutions of Space-Time-Fractional Diffusion Equations. *Proceedings of the Royal Society of London, 458*(2018), 429–450. doi:10.1098/rspa.2001.0893

Haubold, H. J., Mathai, A. M., & Saxena, R. K. (2011). Further solutions of fractional reaction--diffusion equations in terms of the H-function. *Journal of Computational and Applied Mathematics, 235*(5), 1311–1316. doi:10.1016/j.cam.2010.08.016

Hayes, N. D. (1950). Roots of the transcendental equation associated with a certain differential difference equations. *J. Lond. Math. Soc., 25*, 226–232.

Heaviside, O. (1971). Electromagnetic theory. New York: Chelsea.

Hegazi, A. S., Ahmed, E., & Matouk, A. E. (2011). The effect of fractional order on synchronization of two fractional order chaotic and hyperchaotic systems. *J. Fract. Calc. Appl., 1*(3), 1–15.

Hegazi, A. S., Ahmed, E., & Matouk, A. E. (2013a). On chaos control and synchronization of the commensurate fractional order Liu system. *Communications in Nonlinear Science and Numerical Simulation, 18*(5), 1193–1202. doi:10.1016/j.cnsns.2012.09.026

Hegazi, A. S., Ahmed, E., & Matouk, A. E. (2013b). On Dynamical Behaviors and Chaos Control of the Fractional-order Financial System. In S. Banerjee (Ed.), *Chaos and Complexity Theory for Management: Nonlinear Dynamics* (pp. 34–49). IGI Global. doi:10.4018/978-1-4666-2509-9.ch002

Hegazi, A. S., & Matouk, A. E. (2011). Dynamical behaviors and synchronization in the fractional order hyperchaotic Chen system. *Applied Mathematics Letters*, *24*(11), 1938–1944. doi:10.1016/j.aml.2011.05.025

Hegazi, A. S., & Matouk, A. E. (2013). Chaos control of the modified Autonomous Van der Pol-Duffing circuits via active control. In S. Banerjee & L. Rondoni (Eds.), *Applications of Chaos and Nonlinear Dynamics in Science and Engineering* (Vol. 3, pp. 185–202). Berlin: Springer. doi:10.1007/978-3-642-34017-8_7

Helgason, S. (2001). *Differential Geometry, Lie Groups, and Symmetric Spaces*. American Mathematical Society. doi:10.1090/gsm/034

Hennig, D., & Archilla, J. F. R. (2004). Stretching and relaxation dynamics in double stranded DNA. *Physica A*, *331*(34), 579–601. doi:10.1016/j.physa.2003.09.053

Hernandez-Vargasa, E. A., & Middleton, R. H. (2013). Modeling the three stages in HIV infection. *Journal of Theoretical Biology*, *320*, 33–40. doi:10.1016/j.jtbi.2012.11.028 PMID:23238280

Herrmann, R. (2007). The fractional symmetric rigid rotor. *Journal of Physics. G, Nuclear and Particle Physics*, *34*(4), 607–625. doi:10.1088/0954-3899/34/4/001

Herrmann, R. (2008). Gauge invariance in fractional field theories. *Physics Letters. [Part A]*, *372*(34), 5515–5522. doi:10.1016/j.physleta.2008.06.063

Herrmann, R. (2011). *Fractional Calculus. An Introduction for Physicists*. Singapore: World Scientific. doi:10.1142/8072

Herzallah, M. A. E., & Baleanu, D. (2009). Fractional-order Euler-Lagrange equations and formulation of Hamiltonian equations. *Nonlinear Dynamics*, *58*(1-2), 385–391. doi:10.100711071-009-9486-z

Herzallah, M. A. E., Muslih, S. I., Baleanu, D., & Eqab, M. R. (2011). Hamilton-Jacobi and Fractional Like Action with Time Scaling. *Nonlinear Dynamics*, *66*(4), 549–555. doi:10.100711071-010-9933-x

Hikal, M. M., & Zahra, W. K. (2016). On fractional model of an HIV/AIDS with treatment and time delay. *Progr. Fract. Differ. Appl*, *2*, 55-66. doi:10.18576/pfda/020106

Hilfer, R. (Ed.). (2000). *Applications of fractional calculus in physics*. Upper Saddle River, NJ: World Scientific. doi:10.1142/3779

Hirota, R. (2004). *The direct method in soliton theory* (Vol. 155). Cambridge University Press. doi:10.1017/CBO9780511543043

Holland, J. H. (1975). *Adaption in Natural & Artificial Systems*. Cambridge, MA: MIT Press.

Holmgren, R. (1994). *A First Course in Discrete Dynamical Systems*. New York: Springer-Verlag. doi:10.1007/978-1-4684-0222-3

Hristov, J. (2016a). Transient heat diffusion with a non-singular fading memory: From the Cattaneo constitutive equation with Jeffrey's kernel to the Caputo-Fabrizio time-fractional derivative. *Thermal Science*, *20*(2), 757–762. doi:10.2298/TSCI160112019H

Hristov, J. (2017a). Derivatives with Non-Singular kernels from the Caputo-Fabrizio definition and beyond: Appraising analysis with emphasis on diffusion models. *Frontiers*, *1*, 270–342.

Hristov, J. (2017b). Derivation of the fractional Dodson equation and beyond: Transient diffusion with a non-singular memory and exponentially fading-out diffusivity. *Progr. Fract. Differ. Appl.*, *3*(4), 1–16. doi:10.18576/pfda/030402

Huang, F., & Liu, F. (2005). The fundamental solution of the space-time fractional advection-dispersion equation. *Journal of Applied Mathematics and Computing*, *18*(1-2), 339–350. doi:10.1007/BF02936577

Huo, J., Zhao, H., & Zhu, L. (2015). The effect of vaccines on backward bifurcation in a fractional order HIV model. *Nonlinear Analysis Real World Applications*, *26*, 289–305. doi:10.1016/j.nonrwa.2015.05.014

Ibrahim, R. W., & Momani, S. M. (2007). On the existence and uniqueness of solutions of a class of fractional differential equations. *Journal of Mathematical Analysis and Applications*, *334*(1), 110. doi:10.1016/j.jmaa.2006.12.036

Ichise, M., Nagayanagi, Y., & Kojima, T. (1971). An analog simulation of noninteger order transfer functions for analysis of electrode process. *Journal of Electroanalytical Chemistry*, *33*(2), 253–265. doi:10.1016/S0022-0728(71)80115-8

Itoh, M., & Chua, L. O. (2013). Duality of memristor circuits. *International Journal of Bifurcation and Chaos in Applied Sciences and Engineering*, *23*(1), 1330001. doi:10.1142/S0218127413300012

Itoh, M., & Chua, L. O. (2016). Parasitic effects on memristor dynamics. *International Journal of Bifurcation and Chaos in Applied Sciences and Engineering*, *26*(06), 1630014. doi:10.1142/S0218127416300147

Jahanshahi, S., & Torres, D. F. M. (2017). A simple accurate method for solving fractional variational and optimal control problems. *Journal of Optimization Theory and Applications*, *174*(1), 156–175. doi:10.100710957-016-0884-3

Jain, S. (2018). Numerical analysis for the fractional diffusion and fractional Buckmaster equation by the two-step Laplace Adam-Bashforth method. *The European Physical Journal Plus*, *133*(1), 19. doi:10.1140/epjp/i2018-11854-x

Jan, S. A. A., Ali, F., Sheikh, N. A., Khan, I., Saqib, M., & Gohar, M. (2017). Engine oil based generalized brinkman-type nano-liquid with molybdenum disulphide nanoparticles of spherical shape: Atangana-Baleanu fractional model. *Numerical Methods for Partial Differential Equations.*

Jeffreys, H. (1931). *Cartesian Tensors*. Cambridge University Press.

Jegerlehner, F., & Nyffeler, A. (2009). The muon g− 2. *Physics Reports*, *477*(1-3), 1–110. doi:10.1016/j.physrep.2009.04.003

Jiang, H., Liu, F., Turner, I., & Burrage, K. (2012). Analytical solutions for the multi-term time--space Caputo--Riesz fractional advection--diffusion equations on a finite domain. *Journal of Mathematical Analysis and Applications*, *389*(2), 1117–1127. doi:10.1016/j.jmaa.2011.12.055

Jiang, J., Feng, Y., & Li, S. (2018). Exact solutions to the fractional differential equations with mixed partial derivatives. *Axioms*, *7*(1), 10. doi:10.3390/axioms7010010

Jia, Y., & Yang, X. (2016). Optimization of control parameters based on genetic algorithms for spacecraft altitude tracking with input constraints. *Neurocomputing*, *177*, 334–341. doi:10.1016/j.neucom.2015.11.022

Joseph, W.-H., Jianhong, W. U., & Yuanjie, Y. (2000). Numerical steady state and Hopf bifurcation analysis on the diffusive Nicholson's blowflies equation. *Applied Mathematics and Computation*, *111*, 33–51.

Joshi, M. M., Vyawahare, V. A., & Patil, M. D. (2014). Model predictive control for fractional-order system a modeling and approximation based analysis. In *2014 4th International Conference on Simulation and Modeling Methodologies Technologies and Applications* (pp. 361-372). 10.5220/0005038203610372

Jumarie, G. (2006). Modified Riemann-Liouville derivative and fractional Taylor series of nondifferentiable functions further results. *Computers & Mathematics with Applications (Oxford, England)*, *51*(9-10), 1367–1376. doi:10.1016/j.camwa.2006.02.001

Jumarie, G. (2007). Fractional partial differential equations and modified Riemann-Liouville derivative new methods for solution. *Journal of Applied Mathematics and Computing*, *24*(1-2), 31–48. doi:10.1007/BF02832299

Jumarie, G. (2009). Table of some basic fractional calculus formulae derived from a modified Riemann-Liouville derivative for non-differentiable functions. *Applied Mathematics Letters*, *22*(3), 378–385. doi:10.1016/j.aml.2008.06.003

Jumarie, G. (2009b). From Lagrangian mechanics fractal in space to space fractal Schrödinger's equation via fractional Taylor's series. *Chaos, Solitons, and Fractals*, *41*(4), 1590–1604. doi:10.1016/j.chaos.2008.06.027

Jumarie, G. (2010). Derivation and solutions of some fractional Black–Scholes equations in coarse-grained space and time. Application to Merton's optimal portfolio. *Computers & Mathematics with Applications (Oxford, England)*, *59*(3), 1142–1164. doi:10.1016/j.camwa.2009.05.015

Jumarie, G. (2013). On the derivative chain-rules in fractional calculus via fractional difference and their application to systems modelling. *Open Physics*, *11*(6), 617–633. doi:10.247811534-013-0256-7

Jun-Guo, L. (2005). Chaotic dynamics and synchronization of fractional-order Genesio-Tesi systems. *Chinese Physics*, *14*(8), 15–17. doi:10.1088/1009-1963/14/8/007

Kangalgil, F. (2017). The local stability analysis of a nonlinear discrete-time population model with delay and Allee effect. *Cumhuriyet Sci. J.*, *38*(3), 480–487. doi:10.17776/csj.340491

Karakostas, G., Philos, C.G., & Sficas, Y.G. (1991). The dynamics of some discrete population models. *Nonlinear Anal. Theory Methods Appl.*, 1069-1084.

Khalil, M., Said, M., Osman, H., Ahmed, B., Ahmed, D., Younis, N., ... Ashmawy, M. (2019). Fractional order models of infectious diseases: A review. *SPC Journal of Environmental Sciences*, *1*, 1–11.

Khan, A & Kumar, S. (2018). Analysis and time-delay synchronization of chaotic satellite systems. *Pramana Journal of Physics, 91*, 49. doi:10.100712043-018-1610-5

Khan, A & Kumar, S. (2019a). T-S fuzzy modeling and predictive control and synchronization of chaotic satellite systems. *International Journal of Modeling and Simulation, 39*(3).

Khan, A., & Kumar, S. (2016). Anti-synchronization of chaotic finance systems using fuzzy logic constant controller. *Indian Journal of Industrial and Applied Mathematics*. Doi:10.5958/1945-919X.2016.00013X

Khan, A., & Kumar, S. (2019). T-S fuzzy modeling and predictive control and synchronization of chaotic satellite systems. *International Journal of Modeling and Simulation, 39*(3).

Khan, A., & Kumar, S. (2019a). T-S fuzzy modeling and predictive control and synchronization of chaotic satellite systems. *International Journal of Modeling and Simulation, 39*(3).

Khan, A., & Kumar, S. (2018). Measuring chaos and synchronization chaotic satellite systems using sliding mode control. *Optimal Control Applications & Methods, 39*(5), 1597–1609. doi:10.1002/oca.2428

Khan, A., & Kumar, S. (2018). Study of chaos in satellite systems. *Pramana Journal of Physics, 90*(1), 13. doi:10.100712043-017-1502-0

Khan, A., & Kumar, S. (2018). T–S fuzzy observed based design and synchronization of chaotic and hyper-chaotic dynamical systems. *International Journal of Dynamics and Control, 6*(3), 1409–1419. doi:10.100740435-017-0358-y

Khan, A., & Kumar, S. (2019b). Measure of chaos and adaptive synchronization of chaotic satellite systems. *International Journal of Dynamics and Control, 7*(2), 536–546. doi:10.100740435-018-0481-4

Khana, A., & Kumar, S. (2016). Anti-synchronization of chaotic financial systems by using fuzzy logic constant controller. *Indian Journal of Industrial and Applied Mathematics*, *7*(2), 136–147. doi:10.5958/1945-919X.2016.00013.X

Khan, I., Saqib, M., & Ali, F. (2017). Application of time-fractional derivatives with non-singular kernel to the generalized Convective flow of Casson fluid in a microchannel with constant walls temperature. *The European Physical Journal. Special Topics*, *226*(16-18), 3791–3802. doi:10.1140/epjst/e2018-00097-5

Khan, I., Saqib, M., & Ali, F. (2018). Application of the modern trend of fractional differentiation to the MHD flow of a generalized Casson fluid in a microchannel: Modelling and solution★. *The European Physical Journal Plus*, *133*(7), 262. doi:10.1140/epjp/i2018-11899-9

Kheiri, H., & Jafari, M. (2018). Stability analysis of a fractional order model for the HIV/AIDS epidemic in a patchy environment. *Journal of Computational and Applied Mathematics*, *346*, 323–339. doi:10.1016/j.cam.2018.06.055

Kheiri, H., & Jafari, M. (2019). Fractional optimal control of an HIV/AIDS epidemic model with random testing and contact tracing. *Journal of Applied Mathematics and Computing*, *60*(1-2), 387–411. doi:10.100712190-018-01219-w

Kilbas, A. A. A., Srivastava, H. M., & Trujillo, J. J. (2006). *Theory and applications of fractional differential equations* (Vol. 204). Elsevier Science Limited. doi:10.1016/S0304-0208(06)80001-0

Kilbas, A. A., Srivastava, H. M., & Trujillo, J. J. (2006). *Theory and Application of Fractional Differential Equations*. New York: Elsevier.

Klimek, M. (2002). Lagrangian and Hamiltonian Fractional Sequential Mechanics. *Czechoslovak Journal of Physics*, 52.

Kobelev, V. V. (2006). The variant of post-Newtonian mechanics with generalized fractional derivatives. *Chaos (Woodbury, N.Y.)*, *16*(4), 043117. doi:10.1063/1.2384864 PMID:17199395

Kober, H. (1940). On fractional integrals and derivatives. *Quart. J. Math. Oxford*, *11*, 193-211.

Kocic, L., & Ladas, G. (1990, January). Oscillation and global attractivity in a discrete model of Nicholson's blowflies. *Applicable Analysis*, *38*(1-2), 21–31. doi:10.1080/00036819008839952

Kolwankar, K. M., & Gangal, A. D. (1996). Fractional differentiability of nowhere differentiable functions and dimensions. *Chaos (Woodbury, N.Y.)*, *6*(4), 505–513. doi:10.1063/1.166197 PMID:12780280

Kolwankar, K. M., & Gangal, A. D. (1998). Local fractional Fokker-Planck equation. *Physical Review Letters*, *80*(2), 214–217. doi:10.1103/PhysRevLett.80.214

Kolwankar, K. M., & Gangal, A. D. (1999). Local fractional calculus: a calculus for fractal space-time. In *Fractals* (pp. 171–181). London: Springer. doi:10.1007/978-1-4471-0873-3_12

Kulenovic, M. R. S., & Ladas, G. (1987). Linearized oscillations in population dynamics. *Bulletin of Mathematical Biology*, *49*(5), 615–627. doi:10.1016/S0092-8240(87)90005-X PMID:3435801

Kulenovic, M. R. S., Ladas, G., & Sficas, Y. G. (1992). Global attractivity in Nicholson's blowflies. *Applicable Analysis*, *43*(1-2), 109124. doi:10.1080/00036819208840055

Kumar, S. (2019). Synchronization of fractional order Rabinovich-Fabrikant systems using sliding mode control techniques. *Archives of Control Sciences*, *29*(2).

Kusnezov, D., Bulgac, A., & Dang, G. D. (1999). Quantum levy processes and fractional kinetics. *Physical Review Letters*, *82*(6), 1136–1139. doi:10.1103/PhysRevLett.82.1136

Kuznetsov, Y. A. (1998). *Elements of applied bifurcation theory* (2nd ed.). New York: Springer.

Lakshmanan, M., & Rajaseekar, S. (2012). *Nonlinear Dynamics Inerrability Chaos and Patterns*. Berlin, Germany: Springer.

Laskin, N. (2000). Fractional market dynamics. *Physica A*, *287*(3-4), 482–492. doi:10.1016/S0378-4371(00)00387-3

Laskin, N. (2017). Time fractional quantum mechanics. *Chaos, Solitons, and Fractals*, *102*, 16–28. doi:10.1016/j.chaos.2017.04.010

Latif, M. A., Dragomir, S. S., & Matouk, A. E. (2013). New inequalities of Ostrowski type for Co-ordinated s-Convex functions via fractional integrals. *Journal of Fractional Calculus and Applications*, *4*(1), 22-36.

Latif, M. A., Dragomir, S. S., Matouk, A. E. (2013). New inequalities of Ostrowski type for Co-ordinated s-Convex functions via fractional integrals. *Journal of Fractional Calculus and Applications*, *4*(1), 22-36.

Lavania, S., & Nagaria, D. (2016). Evolutionary approach for model order reduction. *Perspectives on Science*, *8*, 361–363. doi:10.1016/j.pisc.2016.04.075

Le Méhauté, A. (1991). Les Géométries fractales. *Traite des nouvelles technologies. Hermes.*

Leopoldino, A. P., Weberszpil, J., Godinho, C. F., & Helayël-Neto, J. A. (2019). Discussing the extension and applications of a variational approach with deformed derivatives. *Journal of Mathematical Physics*, *60*(8), 083507. doi:10.1063/1.5099046

Letellier, C., Elaydi, S. A., Aguirrec, L., & Alaouid, A. (2004). Difference equations versus differential equations. a possible equivalence for the R¨ossler system? *Physica D. Nonlinear Phenomena*, *195*(12), 2949.

Lewnard, J. A., & Grad, Y. H. (2018). Vaccine Waning and Mumps Re-emergence in the United States. *Sci. Transl. Med.*, *10*, 433.

Liao, S. (2003). *Beyond perturbation: introduction to the homotopy analysis method*. CRC press. doi:10.1201/9780203491164

Liao, S. (2005). Comparison between the homotopy analysis method and homotopy perturbation method. *Applied Mathematics and Computation*, *169*(2), 1186–1194. doi:10.1016/j.amc.2004.10.058

Liao, S. (2010). An optimal homotopy-analysis approach for strongly nonlinear differential equations. *Communications in Nonlinear Science and Numerical Simulation*, *15*(8), 2003–2016. doi:10.1016/j.cnsns.2009.09.002

Li, C., & Chen, G. (2004). Chaos and hyperchaos in the fractional order Rssler equations. *Physica A*, *341*, 5561. doi:10.1016/j.physa.2004.04.113

Li, C., & Chen, G. (2004). Chaos in the fractional order Chen system and its control. *Chaos, Solitons, and Fractals*, *22*(3), 549–554. doi:10.1016/j.chaos.2004.02.035

Li, H.-L., Long, Z., Cheng, H., Yao-Lin, J., & Zhidong, T. (2016). Dynamical analysis of a fractional-order predator-prey model incorporating a prey refuge. *Journal of Applied Mathematics and Computing*, *54*(1-2), 435–449. doi:10.100712190-016-1017-8

Li, J. (1996). Global attractivity in Nicholson's blowflies. *Appl. Math. Ser. B*, *11*(4), 425–434. doi:10.1007/BF02662882

Lin, C. (2018). A Four-Dimensional Hyperchaotic Finance System and Its Control Problems. *Journal of Control Science and Engineering.*

Lin, R., Liu, F., Anh, V., & Turner, I. (2009). Stability and convergence of a new explicit finite-difference approximation for the variable-order nonlinear fractional diffusion equation. *Applied Mathematics and Computation*, *212*(2), 435–445. doi:10.1016/j.amc.2009.02.047

Liu, J., Flores, G. A., & Sheng, R. (2001). In-vitro investigation of blood embolization in cancer treatment using magnetorheological fluids. *Journal of Magnetism and Magnetic Materials*, *225*(1-2), 209–217. doi:10.1016/S0304-8853(00)01260-9

Li, Y., Chen, Y., & Podlubny, I. (2009). Mittag-leffler stability of fractional order nonlinear dynamic systems. *Automatica*, *45*(8), 1965–1969. doi:10.1016/j.automatica.2009.04.003

Lotka, A. (1925). *Elements of Physical Biology*. Baltimore: Williams and Wilkins.

Luchko, Y., & Gorenflo, R. (1998). *The initial value problem for some fractional differential equations with the Caputo derivatives*. Academic Press.

Lu, J. G. (2005). Chaotic dynamics and synchronization of fractional-order Arneodos systems. *Chaos, Solitons, and Fractals*, *26*(4), 1125–1133. doi:10.1016/j.chaos.2005.02.023

Lu, J. G. (2006). Chaotic dynamics of the fractional-order L system and its synchronization. *Physics Letters. [Part A]*, *354*(4), 305–311. doi:10.1016/j.physleta.2006.01.068

Machado, J. A. T. (1997). Analysis and design of fractional-order digital control systems. J. Syst. Anal. *Modeling-Simulation*, *27*, 107–122.

Machado, J. A. T. (2003). A Probabilistic Interpretation of the Fractional Order Differentiation. *Fractional Calculus & Applied Analysis*, *8*, 73–80.

Machado, J. T. (2013). Fractional generalization of memristor and higher order elements. *Communications in Nonlinear Science and Numerical Simulation*, *18*(2), 264–275. doi:10.1016/j.cnsns.2012.07.014

Mackey, M. C. (1989). Commodity price actuations: Price dependent delays and non- linearities as explanatory factors. *Journal of Economic Theory*, *48*(2), 497–509. doi:10.1016/0022-0531(89)90039-2

Mackey, M. C., & Glass, L. (1977). Oscillation and chaos in physiological control systems. *Science*, *197*(4300), 287–289. doi:10.1126cience.267326 PMID:267326

Magin, R. L. (2006). *Fractional Calculus in Bioengineering*. Begell House.

Mainardi, F. (2010). *Fractional Calculus and Waves in Linear Viscoelasticity: An Introduction to Mathematical Models*. London: Imperial College Press. doi:10.1142/p614

Mainardi, F., Pagnini, G., & Luchko, Y. (2007). The fundamental solution of the space-time fractional diffusion equation. *Fractional Calculus & Applied Analysis*, *4*, 153–192.

Malkawi E. (2015). Spatial Rotation of the Fractional Derivative in Two-Dimensional Space. *Advances in Mathematical Physics*. doi:10.1155/2015/719173

Malkawi, E. (2016). Spatial Rotation in Cartesian Coordinates. *The Pioneer Journal of Mathematics and Mathematical Sciences.*, *16*(2), 123–135.

Malkawi, E. (2018). A Study of the Caputo Fractional Differential Operator under Two-Dimensional Space Rotation. *The Pioneer Journal of Mathematics and Mathematical Sciences*, *22*(1), 1–9.

Malkawi, E., & Rousan, A. A. (2004). Transformation of Fractional Derivatives under Space Rotation. *International Journal of Applied Mathematics*, *16*(2), 175–185.

Mamar, B., & Rachid, M. (2013). Imc-pid-fractional-order-filter controllers design for integer order systems. *ISA Transactions*, *53*(5), 1620–1628. doi:10.1016/j.isatra.2014.05.007 PMID:24957276

Mandal, P. K. (2005). An unsteady analysis of non-Newtonian blood flow through tapered arteries with a stenosis. *International Journal of Non-linear Mechanics*, *40*(1), 151–164. doi:10.1016/j.ijnonlinmec.2004.07.007

Man, K. F., Tang, K. S., & Kwong, S. (1996). Genetic Algorithms: Concepts and Applications. *IEEE Transactions on Industrial Electronics, 43*(5), 519–534. doi:10.1109/41.538609

Marotto. (2006). *Introduction to Mathematical Modeling Using Discrete Dynamical Systems.* Thomson Brooks/Cole.

Marotto, F. R. (2005). On redefining a snap-back repeller. *Chaos, Solitons, and Fractals, 25*(1), 25–28. doi:10.1016/j.chaos.2004.10.003

Martin, R., Quintana, J. J., Ramos, A., & de la Nuez, I. (2008). Modeling of electrochemical double layer capacitors by means of fractional impedance. J. Comput. Nonlinear Dynam., 3, 021303.

Mastroberardino, A., Cheng, Y., Abdelrazec, A., & Liu, H. (2015). Mathematical modeling of the HIV/AIDS epidemic in Cuba. *International Journal of Biomathematics, 08*(04), 1550047. doi:10.1142/S1793524515500473

Mathai, A. M., & Haubold, H. J. (2008). *Special functions for applied scientists.* New York: Springer. doi:10.1007/978-0-387-75894-7

Matignon, D. (1996). Stability results for fractional differential equations with applications to control processing. Computational Engineering in System Application, 2, 963.

Matignon, D. (1996). Stability results for fractional differential equations with applications to control processing. In *IEEE-SMC proceedings of the computational engineering in systems and application multiconference* (Vol. 2, pp. 963–968). Lille, France: IMACS.

Matouk, A. E. (2009a). Chaos synchronization between two different fractional systems of Lorenz family. *Mathematical Problems in Engineering, 2009*, 1–11. doi:10.1155/2009/572724

Matouk, A. E. (2009b). Stability conditions, hyperchaos and control in a novel fractional order hyperchaotic system. *Physics Letters. [Part A], 373*(25), 2166–2173. doi:10.1016/j.physleta.2009.04.032

Matouk, A. E. (2010). Dynamical behaviors, linear feedback control and synchronization of the fractional order Liu system. *Journal of Nonlinear Systems and Applications, 1*(3), 135–140.

Matouk, A. E. (2011a). Chaos, feedback control and synchronization of a fractional-order modified Autonomous Van der Pol-Duffing circuit. *Communications in Nonlinear Science and Numerical Simulation, 16*(2), 975–986. doi:10.1016/j.cnsns.2010.04.027

Matouk, A. E. (2011b). Synchronization in integer and fractional order chaotic systems. In S. Banerjee (Ed.), *Chaos Synchronization and Cryptography for Secure Communications: Applications for Encryption* (pp. 127–151). IGI Global. doi:10.4018/978-1-61520-737-4.ch007

Matouk, A. E. (2016). Chaos synchronization of a fractional-order modified Van der Pol Duffing system via new linear control, backstepping control and Takagi-Sugeno fuzzy approaches. *Complexity, 21*(S1), 116–124. doi:10.1002/cplx.21719

Matouk, A. E. (2019). Dynamics and control in a novel hyperchaotic system. *Int. J. Dyn. Control, 7*(1), 241–255. doi:10.100740435-018-0439-6

Matouk, A. E., & Elsadany, A. A. (2014). Achieving synchronization between the fractional-order hyperchaotic Novel and Chen systems via a new nonlinear control technique. *Applied Mathematics Letters, 29*, 30–35. doi:10.1016/j.aml.2013.10.010

Matouk, A. E., & Elsadany, A. A. (2016). Dynamical analysis, stabilization and discretization of a chaotic fractional-order GLV model. *Nonlinear Dynamics, 85*(3), 1597–1612. doi:10.100711071-016-2781-6

Matouk, A. E., Elsadany, A. A., Ahmed, E., & Agiza, H. N. (2015). Dynamical behavior of fractional-order Hastings–Powell food chain model and its discretization. *Communications in Nonlinear Science and Numerical Simulation, 27*(1-3), 153–167. doi:10.1016/j.cnsns.2015.03.004

Ma, W. X., & Liu, Y. (2012). Invariant subspaces and exact solutions of a class of dispersive evolution equations. *Communications in Nonlinear Science and Numerical Simulation, 17*(10), 3795–3801. doi:10.1016/j.cnsns.2012.02.024

Maxim, D., & Kamenetskii, F. (1997). Biophysics of the DNA molecule. *Physics Reports, 288*(1-6), 13–60. doi:10.1016/S0370-1573(97)00020-3

Meerschaert, M. M., Mortensen, J., & Wheatcraft, S. W. (2006). Fractional Vector Calculus for Fractional Advection-Dispersion. *Physica A, 367*, 181–190. doi:10.1016/j.physa.2005.11.015

Metzler, R., & Klafter, J. (2000). The random walk's guide to anomalous diffusion: A fractional dynamics approach. *Physics Reports, 339*(1), 1–77. doi:10.1016/S0370-1573(00)00070-3

Metzler, R., & Klafter, J. (2004). The restaurant at the end of the random walk: Recent developments in the description of anomalous transport by fractional dynamics. *Journal of Physics. A, Mathematical and General, 37*(31), R161–R208. doi:10.1088/0305-4470/37/31/R01

Mickens, R. E. (1994). Nonstandard finite difference models of differential equations. World Scientific Publishing Co., Inc.

Miller, K. S., & Ross, B. (1993). *An introduction to the fractional calculus and fractional differential equations*. Academic Press.

Miller, J. P., de Rafael, E., & Roberts, B. L. (2007). Muon (g− 2): Experiment and theory. *Reports on Progress in Physics, 70*(5), 795–881. doi:10.1088/0034-4885/70/5/R03

Miller, K. S., & Ross, B. (1993). *An Introduction to the Fractional Calculus and Fractional Differential Equations*. John Wiley.

Miller, K. S., & Ross, B. (1993). *An Introduction to The Fractional Calculus and Fractional Differential Equations*. John-Wiley and Sons Inc.

Mishra, V., Das, S., Jafari, H., & Ong, S. H. (2016). Study of fractional order van der Pol equation. *Journal of King Saud University-Science, 28*(1), 55–60. doi:10.1016/j.jksus.2015.04.005

Misra, A. K., Sharma, A., & Shukla, J. B. (2011). Modeling and analysis of effects of awareness programs by media on the spread of infectious diseases. *Mathematical and Computer Modelling, 53*(5-6), 1221–1228. doi:10.1016/j.mcm.2010.12.005

Mitrinović, D. S., Pecarić, J. E., & Fink, A. M. (1994). *Inequalities for Functions and Their Integrals and Derivatives*. Dordrecht: Kluwer Academic Publishers.

Mohammad, S. T. (2010). Notes on integral performance indices in fractional-order control systems. *Journal of Process Control, 20*(3), 85–291.

Mohr, P. J., Taylor, B. N., & Newell, D. B. (2012). CODATA recommended values of the fundamental physical constants: 2010. *Journal of Physical and Chemical Reference Data, 84*(4), 1527.

Monje, C. A., Chen, Y., Vinagre, B. M., Xue, D., & Feliu, V. (2010). *Fractional order systems and controls: Fundamental and applications*. London: Springer-Verlag.

Monje, C., Vinagre, B. M., Feliu, V., & Chen, Y. Q. (2008). Tuning and auto-tuning of fractional order controllers for industry applications. *Control Engineering Practice, 16*(7), 798–812. doi:10.1016/j.conengprac.2007.08.006

Monsrefi-Torbati, M., & Hammond, J. K. (1998). Physical and Geometrical Interpretation of Fractional Operators. *Journal of the Franklin Institute*, *335*(6), 1077–1086. doi:10.1016/S0016-0032(97)00048-3

Morales-Delgado, V. F., Taneco-Hernández, M. A., & Gómez-Aguilar, J. F. (2017). On the solutions of fractional order of evolution equations. *The European Physical Journal Plus*, *132*(47), 1–14. doi:10.1140/epjp/i2017-11341-0

Mouaouine, A., Boukhouima, A., Hattaf, K., & Yousfi, N. (2018). A fractional order SIR epidemic model with nonlinear incidence rate. *Advances in Difference Equations*, *160*, 1–9. doi:10.118613662-018-1613-z

Mukandavire, Z., Garira, W., & Tchuenche, J. M. (2009). Modelling effects of public health educational campaigns on HIV/AIDS transmission dynamics. *Applied Mathematical Modelling*, *33*(4), 2084–2095. doi:10.1016/j.apm.2008.05.017

Muslih, S. I., & Baleanu, D. (2005). Hamiltonian Formulation of Systems with Linear Velocities within Riemann-Liouville Fractional Derivatives. *Journal of Mathematical Analysis and Applications*, *304*(2), 599–606. doi:10.1016/j.jmaa.2004.09.043

Naber, M. (2010). Linear fractionally damped oscillator. *International Journal of Differential Equations*.

Nandkeolyar, R., & Das, M. (2014). Unsteady MHD free convection flow of a heat absorbing dusty fluid past a flat plate with ramped wall temperature. *Afrika Matematika*, *25*(3), 779–798. doi:10.100713370-013-0151-9

Naqvi, Q. A., & Abbas, M. (2004). Complex and Higher Order Fractional Curl Operator in Electromagnetics. *Optics Communications*, *241*(4-6), 349–355. doi:10.1016/j.optcom.2004.07.028

Nasir, A. W., & Singh, A. K. (2015). IMC based fractional order controller for non-minimum phase system. In *2015 Annual IEEE India Conference* (pp. 1-6). New Delhi: IEEE. 10.1109/INDICON.2015.7443593

Nigmatullin, R. R. (1992). A Fractional Integral and its Physical Interpretation. *Theoretical and Mathematical Physics*, *90*(3), 242–251. doi:10.1007/BF01036529

Nishimoto, K. (1989). *Fractional Calculus: Integrations and Differentiations of Arbitrary Order*. New Haven: University of New Haven Press.

Nottale, L. (1993). *Fractal space-time and microphysics: towards a theory of scale relativity*. World Scientific. doi:10.1142/1579

Odibat, Z. M. (2010). Adaptive feedback control and synchronization of non-identical chaotic fractional order systems. *Nonlinear Dynamics*, *60*(4), 479–487. doi:10.100711071-009-9609-6

Okosun, K. O., Makinde, O. D., & Takaidza, I. (2013). Impact of optimal control on the treatment of HIV/AIDS and screening of unaware infectives. *Applied Mathematical Modelling*, *37*(6), 3802–3820. doi:10.1016/j.apm.2012.08.004

Oldham, K. B., & Spanier, J. (1974). *The Fractional Calculus*. London: Academic Press.

Oldham, K. B., & Spanier, J. (1974). *The Fractional calculus*. New York: Academic press.

Oldham, K. B., & Spanier, J. (1974). *The Fractional Calculus; Theory and Applications of Differentiation and Integration to Arbitrary Order*. New York: Academic Press.

Olver, P. (2000). *Applications of Lie Groups to Differential Equations*. New York: Springer.

Ostrowski, A. M. (1938). Uber die Absolutabweichung einer di erentienbaren Funktionen von ihren Integralimittelwert. *Commentarii Mathematici Helvetici*, *10*(1), 226–227. doi:10.1007/BF01214290

Owa, S. (1978). On the distortion theorems. I. *Kyungpook Math. J.*, *18*(1), 53–59.

Palmer, C., & Stavrinou, P. N. (2004). Equations of motion in a non-integer-dimensional space. *Journal of Physics. A, Mathematical and General*, *37*(27), 6987–7003. doi:10.1088/0305-4470/37/27/009

Passera, M., Martiano, W. J., & Sirlin, A. (2008, November). The muon g-2 discrepancy: errors or new physics? In. AIP Conference Proceedings: Vol. 1078. *No. 1* (pp. 378–381). AIP. doi:10.1063/1.3051967

Pathak, H., & Shahzad, N. (2007). Fixed point for generalized contractions and applications to control theory. *Nonlinear Anal.* doi.01.041 doi:10.1016/j.na.2007

Pertz, G. H., & Gerhardt, C. J. (1849). Briefwechsel zwischen Leibniz, Hugens van Zulichem und dem Marquis de l'Hospital. Leibnizens gesammelte Werke.

Petras, I. (2009). Chaos in the fractional-order Volta's system: Modeling and simulation. *Nonlinear Dynamics*, *57*(1-2), 157–170. doi:10.100711071-008-9429-0

Pitchaimani, M., & Monica, C. (2015). Global stability analysis of HIV-1 infection model with three time delays. *J. Appl. Math. Comput*, *48*(1-2), 293–319. doi:10.100712190-014-0803-4

Pk, J., & Ghorais, S. (2017). Stability of prey–predator model with holling type response function and selective harvesting. *J. Appl. Computat. Math.*, *6*(03), 358. doi:10.4172/2168-9679.1000358

Plyushchay, M. S., & De Traubenberg, M. R. (2000). Cubic root of Klein-Gordon equation. *Physics Letters. [Part B]*, *477*(1-3), 276–284. doi:10.1016/S0370-2693(00)00190-8

Podlubny, I. (1999). Fractional differential equations. Academic Press.

Podlubny, I. (1998). *Fractional differential equations: an introduction to fractional derivatives, fractional differential equations, to methods of their solution and some of their applications*. Elsevier.

Podlubny, I. (1999). *Fractional differential equations*. New York: Academic Press.

Podlubny, I. (1999). *Fractional Differential Equations*. New York: Academic Press.

Podlubny, I. (1999). Fractional-Order Systems and Controllers. *IEEE Transactions on Automatic Control*, *44*(1), 208–214. doi:10.1109/9.739144

Podlubny, I. (2000). Matrix approach to discrete fractional calculus. *Fractional Calculus & Applied Analysis*, *3*(4), 359–386.

Podlubny, I. (2002). Geometric and Physical Interpretation of Fractional Integration and Fractional Differentiation. *Fractional Calculus & Applied Analysis*, *5*(4).

Podlubny, I. (2002). Geometric and physical interpretation of fractional integration and fractional differffentiation. Dedicated to the 60th anniversary of Prof. Francesco Mainardi. *Fractional Calculus & Applied Analysis*, *5*, 367386.

Podlubny, I., Petras, I., Vinagre, B. M., O'Leary, P., & Dorcak, L. (2002). Analogue realizations of fractional-order controllers. Fractional order calculus and its applications. *Nonlinear Dynamics*, *29*(1/4), 281–296. doi:10.1023/A:1016556604320

Pooseh, S., Almeida, R., & Torres, D. F. (2013). Numerical approximations of fractional derivatives with applications. *Asian Journal of Control*, *15*(3), 698–712. doi:10.1002/asjc.617

Pu, Y. F., & Yuan, X. (2016). Fracmemristor: Fractional-order memristor. *IEEE Access: Practical Innovations, Open Solutions*, *4*, 1872–1888. doi:10.1109/ACCESS.2016.2557818

Pyragas, V., & Pyragas, K. (2010). Modification of delayed feedback control using ergodicity of chaotic systems. *Lithuanian Journal of Physics*, *50*(3), 305–316. doi:10.3952/lithjphys.50301

Qayyum, A., Faye. I.& Shoaib. M. (2015). On new generalized inequalities via Riemann-Liouville Fractional integration. *Journal of Fractional Calculus and Applications, 6*(1), 91-100.

Qayyum, A., Faye. I., Shoaib. M. & Erden, S. (2019). Generalized fractional Ostrowski type inequalities for higher order derivatives. *Communication in Mathematical Modeling and Application, 4*(2).

Rachah, A., & Torres, D. F. (2017). Analysis, simulation and optimal control of a SEIR model for Ebola virus with demographic effects. *Commun. Fac. Sci. Univ. Ank. Series A1., 67*(1), 179-197. doi:10.1501/Commua1_0000000841

Rahman, M., Mahmood, A., & Younis, M. (2014). Improved and more feasible numerical methods for Riesz space fractional partial differential equations. *Applied Mathematics and Computation, 237*, 264–273. doi:10.1016/j.amc.2014.03.103

Raspini, A. (2001). Simple solutions of the fractional Dirac equation of order 2/3. *Physica Scripta, 64*(1), 20–22. doi:10.1238/Physica.Regular.064a00020

Ray, S. S., Atangana, A., Noutchie, S. C., Kurulay, M., Bildik, N., & Kilicman, A. (2014). Fractional calculus and its applications in applied mathematics and other sciences. *Mathematical Problems in Engineering, 2014*, 2014. doi:10.1155/2014/849395

Ray, S. S., & Sahoo, S. (2015). Analytical approximate solutions of Riesz fractional diffusion equation and Riesz fractional advection--dispersion equation involving nonlocal space fractional derivatives. *Mathematical Methods in the Applied Sciences, 38*(13), 2840–2849. doi:10.1002/mma.3267

Rhouma, A., & Bouani, F. (2014). Robust model predictive control of uncertain fractional systems: A thermal application. *IET Control Theory & Applications, 8*(17), 1986–1994. doi:10.1049/iet-cta.2014.0703

Riemann, B. (1876). Versuch einer allgemeinen auffassung der integration und differentiation. Gesammelte Werke, 62.

Riemann, B. (1990). Versuch einer allgemeinen auffassung der integration und differentiation (1847). Gesammelte mathematische Werke (H. Weber éd. avec le concours de R. Dedekind), Leipzig, 1876, p.353-366; 2 ième éd. 1892; rééd. Springer.

Riewe, F. (1996). Nonconservative lagrangian and hamiltonian mechanics. *Physical Review. E, 53*(2), 1890.

Riewe, F. (1996). Nonconservative Lagrangian and Hamiltonian Mechanics. *Physical Review. E, 53*(2), 1890–1899. doi:10.1103/PhysRevE.53.1890 PMID:9964451

Riewe, F. (1997). Mechanics with fractional derivatives. *Physical Review. E, 55*(3), 3581. doi:10.1103/PhysRevE.55.3581

Rodrigues, P., Silva, C. J., & Torres, D. F. M. (2014). Cost-effectiveness analysis of optimal control measures for tuberculosis. *Bulletin of Mathematical Biology, 76*(10), 2627–2645. doi:10.100711538-014-0028-6 PMID:25245395

Roger. (1970). *Hausdorff measures*. Cambridge University Press.

Rosa, S., & Torres, D. F. (2018). Optimal control of a fractional order epidemic model with application to human respiratory syncytial virus infection. *Chaos, Solitons & Fractals, 117*, 142-149. Doi:10.1016/j.chaos.2018.10.021

Rosa, W., & Weberszpil, J. (2018). Dual conformable derivative: Definition, simple properties and perspectives for applications. *Chaos, Solitons, and Fractals, 117*, 137–141. doi:10.1016/j.chaos.2018.10.019

Rui, W. (2017). Applications of homogenous balanced principle on investigating exact solutions to a series of time fractional nonlinear PDEs. *Communications in Nonlinear Science and Numerical Simulation, 47*, 253–266. doi:10.1016/j.cnsns.2016.11.018

Rutman, R. S. (1995). On Physical Interpretations of Fractional Integration and Differentiation. *Theoretical and Mathematical Physics, 105*(3), 1509–1519. doi:10.1007/BF02070871

Sabatier, J., Agrawal, O. P., & Tenreiro Machado, J. A. (2007). Advances in Fractional Calculus: Theoretical Developments and Applications in Physics and Engineering. Springer.

Sahadevan, R., & Prakash, P. (2017). On Lie symmetry analysis and invariant subspace methods of coupled time fractional partial differential equations. *Chaos, Solitons, and Fractals, 104*, 107–120. doi:10.1016/j.chaos.2017.07.019

Saker, S. H., & Agarwal, S. (2002). Oscillation and global attractivity in a periodic Nicholson's blowflies model. *Mathematical and Computer Modelling, 35*(7-8), 719731. doi:10.1016/S0895-7177(02)00043-2

Salah, M., Abdelouahab, & Hamri, N. E. (2015). The Grunwald-Letnikov Fractional-Order Derivative with Fixed Memory Length. *Mediterranean Journal of Mathematics*, 1–16. doi:10.100700009-015-0525-3

Samet, B., & Aydi, H. (2018). On Some Inequalities Involving Liouville–Caputo Fractional Derivatives and Applications to Special Means of Real Numbers. *Mathematics, 6*(10), 193–202. doi:10.3390/math6100193

Samko, S. G., Kilbas, A. A., & Marichev, O. I. (1987). *Fractional integral and derivatives, theory and application.* Minsk: Nauka I Teknika.

Samko, S. G., Kilbas, A. A., & Marichev, O. I. (1993). *Fractional Integrals and Derivatives: Theory and Applications.* New York: Gordon and Breach.

Samko, S. G., Kilbas, A. A., & Marichev, O. I. (1993). *Fractional Integrals and Derivatives-Theory and Applications.* Linghorne, PA: Gordon and Breach.

Samko, S. G., Kilbas, A. A., & Marichev, O. I. (1999). *Fractional integrals and derivatives: theory and applications.* Yverdon: Gordon and Breach.

Saqib, M., Ali, F., Khan, I., Sheikh, N. A., & Khan, A. (2018a). Entropy Generation in Different Types of Fractionalized Nanofluids. *Arabian Journal for Science and Engineering*, 1–10.

Saqib, M., Ali, F., Khan, I., Sheikh, N. A., & Shafie, S. B. (2018b). Convection in ethylene glycol-based molybdenum disulfide nanofluid. *Journal of Thermal Analysis and Calorimetry*, 1–10.

Saqib, M., Khan, I., & Shafie, S. (2018c). Application of Atangana–Baleanu fractional derivative to MHD channel flow of CMC-based-CNT's nanofluid through a porous medium. *Chaos, Solitons, and Fractals, 116*, 79–85. doi:10.1016/j.chaos.2018.09.007

Saqib, M., Khan, I., & Shafie, S. (2018d). Natural convection channel flow of CMC-based CNTs nanofluid. *The European Physical Journal Plus, 133*(12), 549. doi:10.1140/epjp/i2018-12340-3

Saqib, M., Khan, I., & Shafie, S. (2019a). Application of fractional differential equations to heat transfer in hybrid nanofluid: Modeling and solution via integral transforms. *Advances in Difference Equations, 2019*(52), 1–18. doi:10.118613662-019-1988-5

Saqib, M., Khan, I., & Shafie, S. (2019b). Generalized magnetic blood flow in a cylindrical tube with magnetite dusty particles. *Journal of Magnetism and Magnetic Materials, 484*, 490–496. doi:10.1016/j.jmmm.2019.03.032

Sardar, T., Rana, S., Bhattacharya, S., Al-Khaled, K., & Chattopadhyay, J. (2015). A generic model for a single strain mosquito-transmitted disease with memory on the host and the vector. *Mathematical Biosciences, 263*, 18–36. doi:10.1016/j.mbs.2015.01.009 PMID:25645185

Sarikaya, M. Z., & Ogunmez, H. (2012). On new inequalities via Riemann-Liouville Fractional integration. Abstract and Applied Analysis, Article ID 428983. doi:10.1155/2012/428983

Sarikaya, M. Z., & Yaldiz, H. (2013). On weighted montogomery identities for Riemann-Luville fractional integral. *Konuralp Journal of Mathematics*, *1*(1), 48–53.

Schäfer, I., & Krüger, K. (2006). Modelling of coils using fractional derivatives. *Journal of Magnetism and Magnetic Materials*, *307*(1), 91–98. doi:10.1016/j.jmmm.2006.03.046

Selvam, G. M., & Janagaraj, A., R. (2018a). Numerical Analysis of a Fractional Order Discrete Prey – Predator System with Functional Response. *International Journal of Engineering & Technology, 7*(4.10), 681–684.

Selvam, G. M., & Janagaraj, A. R. (2018b). Allee effect and Holling type - II response in a discrete fractional order prey - predator model. IOP Conference Series. *Journal of Physics*, *1139*, 1–7.

Senitzky, I. R. (1960). Dissipation in quantum mechanics. The harmonic oscillator. *Physical Review*, *119*(2), 670–679. doi:10.1103/PhysRev.119.670

Shabbir, F., & Omenzetter, P. (2016). Model updating using genetic algorithms with sequential niche technique. *Engineering Structures*, *120*, 166–182. doi:10.1016/j.engstruct.2016.04.028

Shah, N. A., Vieru, D., & Fetecau, C. (2016). Effects of the fractional order and magnetic field on the blood flow in cylindrical domains. *Journal of Magnetism and Magnetic Materials*, *409*, 10–19. doi:10.1016/j.jmmm.2016.02.013

Shalaby, R., El-Hossainy, M., & Belal Abo-Zalam, B. (2019). Fractional order modeling and control for under-actuated inverted pendulum. *Communications in Nonlinear Science and Numerical Simulation*, *74*, 97–121. doi:10.1016/j.cnsns.2019.02.023

Shamseldeen, S. (2018). Approximate solution of space and time fractional higher order phase field equation. *Physica A*, *494*, 308–316. doi:10.1016/j.physa.2017.12.056

Shamseldeen, S., Elsaid, A., & Madkour, S. (2019a). Caputo-Riesz-Feller fractional wave equation: Analytic and approximate solutions and their continuation. *Journal of Applied Mathematics and Computing*, *59*(1-2), 423–444. doi:10.100712190-018-1186-8

Shamseldeen, S., Elsaid, A., & Madkour, S. (2019b). On The Solution of Nonlinear Generalized Caputo-Riesz Fractional EFK and KS Equations. *Eurasian Journal of Mathematical and Computer Applications*, *7*(2), 62–78. doi:10.32523/2306-6172-2019-7-2-62-78

Sharma, S., Singh, U., & Katiyar, V. K. (2015). Magnetic field effect on flow parameters of blood along with magnetic particles in a cylindrical tube. *Journal of Magnetism and Magnetic Materials*, *377*, 395–401. doi:10.1016/j.jmmm.2014.10.136

Sheikholeslami, M. (2018). Numerical investigation for CuO-H2O nanofluid flow in a porous channel with magnetic field using mesoscopic method. *Journal of Molecular Liquids*, *249*, 739–746. doi:10.1016/j.molliq.2017.11.069

Shen, Y. J., Wei, P., & Yang, S. P. (2014). Primary resonance of fractional-order van der Pol oscillator. *Nonlinear Dynamics*, *77*(4), 1629–1642. doi:10.100711071-014-1405-2

Shi, M., & Hu, S. (2017). Pinched hysteresis loop characteristics of a fractional-order HP TiO_2 memristor. In *Intelligent Computing, Networked Control, and Their Engineering Applications* (pp. 705–713). Singapore: Springer. doi:10.1007/978-981-10-6373-2_70

Silva, C. J., & Torres, D. F. M. (2015). Optimal control of tuberculosis: a review, in Dynamics, games and science. *CIM Ser. Math. Sci., 1,* 701-722. arXiv:1406.3456

Silva, C. J., Maurer, H., & Torres, D. F. M. (2017). Optimal control of a tuberculosis model with state and control delays. *Mathematical Biosciences and Engineering*, *14*(1), 321–337. doi:10.3934/mbe.2017021 PMID:27879136

Silva, C. J., & Torres, D. F. M. (2012). Optimal control strategies for tuberculosis treatment: A case study in Angola. *Numer. Algebra Control Optim*, *2*(3), 601–617. doi:10.3934/naco.2012.2.601

Silva, C. J., & Torres, D. F. M. (2013). Optimal control for a tuberculosis model with reinfection and post-exposure interventions. *Mathematical Biosciences*, *244*(2), 154–164. doi:10.1016/j.mbs.2013.05.005 PMID:23707607

Silverio, S. R., & Torres, D. F. M. (2019). Optimal control and sensitivity analysis of a fractional order TB model. *Statistics Opt. Inform. Comput.*, *7*, 1-9. http://www.IAPress.org

Sinha, A., & Shit, G. C. (2015). Electromagnetohydrodynamic flow of blood and heat transfer in a capillary with thermal radiation. *Journal of Magnetism and Magnetic Materials*, *378*, 143–151. doi:10.1016/j.jmmm.2014.11.029

Smith, H. (2010). An Introduction to Delay Differential Equations with Applications to the Life Sciences. Springer.

So Joseph, W.-H. (2001). Traveling waves for the diffusive Nicholson's blowflies equation. *Applied Mathematics and Computation*, *122*, 385392.

Sondhi, S., & Hote, Y. V. (2014). Fractional IMC design for fractional order gas turbine model. In *2014 9th International Conference on Industrial and Information Systems* (pp. 1-5). 10.1109/ICIINFS.2014.7036585

Song, L., Xu, S., & Yang, J. (2010). Dynamical Models of Happiness with Fractional Order. *Communications in Nonlinear Science and Numerical Simulation*, *15*(3), 616–628. doi:10.1016/j.cnsns.2009.04.029

Srivastava, H. M., & Owa, S. (1984). An application of the fractional derivative. *Mathematica Japonica*, *29*(3), 383389.

Srivastava, H. M., & Saigo, M. (1987). Multiplication of fractional calculus operators and boundary value problems involving the Euler-Darboux equation. *Journal of Mathematical Analysis and Applications*, *121*(2), 321–369. doi:10.1016/0022-247X(87)90251-4

Stephanopoulas, G. (1984). Chemical process control: An introduction to theory and practice. Pearson Education.

Stephens, P. A., Sutherland, W. J., & Freckleton, R. P. (1999, October). What is the Allee effect. *Oikos*, *87*(1), 185–190. doi:10.2307/3547011

Stevens, K. W. H. (1958). The wave mechanical damped harmonic oscillator. *Proceedings of the Physical Society*, *72*(6), 1027–1036. doi:10.1088/0370-1328/72/6/311

Stillinger, F. H. (1977). Axiomatic basis for spaces with noninteger dimension. *Journal of Mathematical Physics*, *18*(6), 1224–1234. doi:10.1063/1.523395

Sun, H. H., Abdelwahab, A. A., & Onaral, B. (1984). Linear approximation of transfer function with a pole of fractional order. *IEEE Transactions on Automatic Control*, *29*(5), 441–444. doi:10.1109/TAC.1984.1103551

Svozil, K. (1987). Quantum field theory on fractal spacetime: A new regularisation method. *Journal of Physics. A, Mathematical and General*, *20*(12), 3861–3875. doi:10.1088/0305-4470/20/12/033

Svozil, K., & Zeilinger, A. (1986). Dimension of space-time. *International Journal of Modern Physics A*, *1*(04), 971–990. doi:10.1142/S0217751X86000368

Sweilam, N. H., & Al-Mekhlafi, S. M. (2017). Optimal control for a time delay multi-strain tuberculosis fractional model: a numerical approach. *IMA Journal of Mathematical Control and Information*, *36*(1), 317-340. doi:10.1093/imamci/dnx046

Sweilam, N. H., Saad, O. M., & Mohamed, D. G. (2017). Comparative studies for the fractional optimal control in transmission dynamics of West Nile virus. *International Journal of Biomathematics*, *10*(7). doi:10.1142/S1793524517500954

Sweilam, N. H., Saad, O. M., & Mohamed, D. G. (2019). Fractional optimal control in transmission dynamics of West Nile virus model with state and control time delay: a numerical approach. *Advances in Difference Equations*, *2019*(1), 210. . doi:10.118613662-019-2147-8

Sweilam, N. H., & Al-Mekhlafi, S. M. (2016). Numerical study for multi-strain tuberculosis (TB) model of variable-order fractional derivatives. *Journal of Advanced Research*, *7*(2), 271–283. doi:10.1016/j.jare.2015.06.004 PMID:26966568

Taghavian, H., & Tavazoei, M. S. (2017). Robust Stability Analysis of Distributed-Order Linear Time-Invariant Systems With Uncertain Order Weight Functions and Uncertain Dynamic Matrices. *Journal of Dynamic Systems, Measurement, and Control*, *139*(12), 121010. doi:10.1115/1.4037268

Taghavian, H., & Tavazoei, M. S. (2018). Exact solution of linear fractional distributed order systems with exponential order weight functions. In *Mathematical Techniques of Fractional Order Systems* (pp. 103–132). Elsevier Publishing. doi:10.1016/B978-0-12-813592-1.00004-0

Taghavian, H., & Tavazoei, M. S. (2019). Algebraic Conditions for Stability Analysis of Linear Time Invariant Distributed Order Dynamic Systems: A Lagrange Inversion Theorem Approach. *Asian Journal of Control*, *21*(2), 879–890. doi:10.1002/asjc.1780

Tarasov, V. E. (2005). Fractional Generalization of Gradient Systems. *Letters in Mathematical Physics*, *73*(1), 49–58. doi:10.100711005-005-8444-z

Tarasov, V. E. (2013). Review of Some Promising Fractional Physical Models. *International Journal of Modern Physics B*, *27*(09), 1330005. doi:10.1142/S0217979213300053

Tashtoush, B., & Magableh, A. (2008). Magnetic field effect on heat transfer and fluid flow characteristics of blood flow in multi-stenosis arteries. *Heat and Mass Transfer*, *44*(3), 297–304. doi:10.100700231-007-0251-x

Tavazoei, M. S. (2010). A note on fractional-order derivatives of periodic functions. *Automatica*, *46*(5), 945–948. doi:10.1016/j.automatica.2010.02.023

Tavazoei, M. S., Haeri, M., & Attari, M. (2009). A proof for nonexistence of periodic solutions in time invariant fractional order systems. *Automatica*, *45*(8), 1886–1890. doi:10.1016/j.automatica.2009.04.001

Tepljakov, A., Pettenkov, E., & Belikov, J. (2011). FOMCON: a MATLAB toolbox for fractional-order system identification and control. *International Journal of Microelectronics and Computer Science*, *2*(2), 51-62.

Tidke, H. L. (2009). Existence of global solutions to nonlinear mixed Volterra-Fredholm integrodifferential equations with nonlocal conditions. *Electronic Journal of Differential Equations*, *55*, 17.

Tricot, C. (1999). Courbe et dimension fractale (2nd ed.). Springer.

Tsallis, C. (1988). Possible generalization of Boltzmann-Gibbs statistics. *Journal of Statistical Physics*, *52*(1-2), 479–487. doi:10.1007/BF01016429

Tsallis, C. (2009a). Nonadditive entropy and nonextensive statistical mechanics-an overview after 20 years. *Brazilian Journal of Physics*, *39*(2a2A), 337–356. doi:10.1590/S0103-97332009000400002

Tsallis, C. (2009b). *Introduction to nonextensive statistical mechanics: approaching a complex world.* Springer Science & Business Media.

Tung, C. S., & Harvey, S. C. (1984). A molecular mechanical model to predict the helix twists angles of B-DNA. *Nucleic Acids Research, 12*(7), 3343–3356. doi:10.1093/nar/12.7.3343 PMID:6326059

Tzirtzilakis, E. E. (2005). A mathematical model for blood flow in magnetic field. *Physics of Fluids, 17*(7), 077103. doi:10.1063/1.1978807

Udwadia, F. E., & von Bremen, H. (2000). A Note on the Computation of the Largest p-Lyapunov Characteristic Exponents for Nonlinear Dynamical Systems. *J App Math and Comp, 114*, 205–214. doi:10.1016/S0096-3003(99)00113-7

Vasumathi, V., & Daniel, M. (2008). Perturbed soliton-like molecular excitations in a deformed DNA chain. *Physics Letters. [Part A], 373*(1), 76–82. doi:10.1016/j.physleta.2008.10.058

Voltairas, P. A., Fotiadis, D. I., & Michalis, L. K. (2002). Hydrodynamics of magnetic drug targeting. *Journal of Biomechanics, 35*(6), 813–821. doi:10.1016/S0021-9290(02)00034-9 PMID:12021001

Volterra, V. (1926). Variazioni e fluttuazioni del numero d'individui in specie animali conviventi. *Mem.Accad. Lincei., 6*, 31–113.

Wang, X., Hoang, T. T. G., Pan, Z., & Chen, Y. (2018). Fractional - order modelling and control for two parallel PWM rectifiers. *IFAC-PapersOnLine, 51*(4), 54–59. doi:10.1016/j.ifacol.2018.06.026

Weberszpil, J. (2014). *Validity of the fractional Leibniz rule on a coarse-grained medium yields a modified fractional chain rule.* arXiv preprint arXiv:1405.4581

Weberszpil, J., Godinho, C. F. L., Cherman, A., & Helayël-Neto, J. A. (2012). Aspects of the Coarse-Grained-Based Approach to a Low-Relativistic Fractional Schrodinger Equation. *Mathematical Methods in Physics - ICMP 2012, 2012, Rio de Janeiro. Proceedings of Science (PoS).* arXiv preprint arXiv:1206.2513

Weberszpil, J., & Chen, W. (2017). Generalized Maxwell relations in thermodynamics with metric derivatives. *Entropy (Basel, Switzerland), 19*(8), 407. doi:10.3390/e19080407

Weberszpil, J., & Helayël-Neto, J. A. (2014). Anomalous-Factors for Charged Leptons in a Fractional Coarse-Grained Approach. *Advances in High Energy Physics.*

Weberszpil, J., & Helayël-Neto, J. A. (2016). Variational approach and deformed derivatives. *Physica A, 450*, 217–227. doi:10.1016/j.physa.2015.12.145

Weberszpil, J., & Helayël-Neto, J. A. (2017). Structural scale q-derivative and the LLG equation in a scenario with fractionality. *EPL, 117*(5), 50006. doi:10.1209/0295-5075/117/50006

Weberszpil, J., & Helayel-Neto, J. A. (2017a). Axiomatic Local Metric Derivatives for Low-Level Fractionality with Mittag-Leffler Eigenfunctions. *Journal of Advances in Physics, 13*(3), 4751–4755. doi:10.24297/jap.v13i3.5943

Weberszpil, J., Lazo, M. J., & Helayël-Neto, J. A. (2015). On a connection between a class of q-deformed algebras and the Hausdorff derivative in a medium with fractal metric. *Physica A, 436*, 399–404. doi:10.1016/j.physa.2015.05.063

Wen, G. (2005). Criterion to identify Hopf bifurcations in maps of arbitrary dimension. Phys. Rev. E, 72, Article ID 026201.

Weng, P., & Liang, M. (1996). Existence and global attractivity of periodic solution of a model in population dynamics. *Acta Math. Appl. Sinica (English Ser.), 12*, 427-434.

West, B., Bologna, M., & Grigolini, P. (2003). *Physics of Fractal Operators.* New York: Springer. doi:10.1007/978-0-387-21746-8

Weyl, H. (1917). Bermerkungen zum begriff des differentialquo tienten gebrochener ordnung, Vierteljschr. *Naturfosch. Gesellsch. Zurich*, *62*, 296–302.

WHO Report, WHO/EMP/IAU/2018.06, 2018.

WHO. (2016). *Global Tuberculosis Report 2016*. World Health Organization.

Wigner, E. P. (1959). *Group Theory and its Application to the Quantum Mechanics of Atomic Spectra*. New York: Academic Press.

Wilkie, K. P., Drapaca, C. S., & Sivaloganathan, S. (2011). A nonlinear viscoelastic fractional derivative model of infant hydrocephalus. *Applied Mathematics and Computation*, *217*(21), 8693–8704. doi:10.1016/j.amc.2011.03.115

Wojtak, W., Silva, C. J., & Torres, D. F. M. (2018). Uniform asymptotic stability of a fractional tuberculosis model. *Math. Model. Nat. Phenom.*, *13*(1), Art. 9. doi:10.1051/mmnp/2018015

Yang, X. J. (2013). Approximate solutions for diffusion equations on Cantor spacetime. *Proceedings of the Romanian Academy, Series A*, 127-133.

Yang, Q., Liu, F., & Turner, I. (2010). Numerical methods for fractional partial differential equations with Riesz space fractional derivatives. *Applied Mathematical Modelling*, *34*(1), 200–218. doi:10.1016/j.apm.2009.04.006

Yang, X. J. (2013). Mathematical aspects of Heisenberg uncertainty principle within local fractional Fourier analysis. *Boundary Value Problems*, 1–6.

Yang, X. J., Baleanub, D., Lazarevic, M., & Cajic, M. (2015). Fractal boundary value problems for integral and differential equations with local fractional operators. *Thermal Science*, *19*(3), 959–966. doi:10.2298/TSCI130717103Y

Yang, X. J., & Baleanu, D. (2013). Fractal heat conduction problem solved by local fractional variation iteration method. *Thermal Science*, *17*(2), 625–628. doi:10.2298/TSCI121124216Y

Yang, X. J., Baleanu, D., & Srivastava, H. M. (2015). Local fractional similarity solution for the diffusion equation defined on Cantor sets. *Applied Mathematics Letters*, *47*, 54–60. doi:10.1016/j.aml.2015.02.024

Yang, X.-J., Gao, F., Machado, J. A. T., & Baleanu, D. (2017). A new fractional derivative involving the normalized sinc function without singular kernel. *The European Physical Journal. Special Topics*, *226*(16-18), 3567–3575. doi:10.1140/epjst/e2018-00020-2

Yang, X.-J., & Machado, J. A. T. (2017). A new fractional operator of variable order: Application in the description of anomalous diffusion. *Physica A*, *481*, 276–283. doi:10.1016/j.physa.2017.04.054

Yang, X.-J., Srivastava, H. M., & Machado, J. A. T. (2016). A new fractional derivative without singular kernel: Application to the modelling of the steady heat flow. *Thermal Science*, *20*(2), 753–756. doi:10.2298/TSCI151224222Y

Yang, Y., Li, J., Ma, Z., & Liu, L. (2010). Global stability of two models with incomplete treatment for tuberculosis. *Chaos, Solitons, and Fractals*, *43*(1-12), 79–85. doi:10.1016/j.chaos.2010.09.002

Yan, J. P., & Li, C. P. (2007). On chaos synchronization of fractional differential equations. *Chaos, Solitons, and Fractals*, *32*(2), 725735. doi:10.1016/j.chaos.2005.11.062

Yaro, D., Apeanti, W. O., Akuamoah, S. W., & Lu, D. (2019). Analysis and optimal control of fractional-order transmission of a respiratory epidemic model. *Int. J. Appl. Comput. Math*, *5*(4), 1–21. doi:10.100740819-019-0699-7

Yépez-Martínez, H., & Gómez-Aguilar, J. F. (2019). A new modified definition of Caputo-Fabrizio fractional-order derivative and their applications to the Multi Step Homotopy Analysis Method (MHAM). *Journal of Computational and Applied Mathematics*, *346*, 247–260. doi:10.1016/j.cam.2018.07.023

Yépez-Martínez, H., Gómez-Aguilar, J. F., Sosa, I. O., Reyes, J. M., & Torres-Jiménez, J. (2016). The Feng's first integral method applied to the nonlinear mKdV space-time fractional partial differential equation. *Revista Mexicana de Física*, *62*(4), 310–316.

Yong, C., Zhen-Ya, Y., & Hong-Qing, Z. (2003). Applications of Fractional Exterior Differential in Three-Dimensional Space. *Applied Mathematics and Mechanics*, *24*(3), 256–260. doi:10.1007/BF02438263

Yu, Y. J., Bao, B. C., Kang, H. Y., & Shi, M. (2015). Calculating area of fractional-order memristor pinched hysteresis loop. *Journal of Engineering (Stevenage, England)*, *2015*(11), 325–327. doi:10.1049/joe.2015.0154

Zafar, A. A., Shah, N. A., & Khan, I. (2018). Two Phase Flow of Blood through a Circular Tube with Magnetic Properties. *Journal of Magnetism and Magnetic Materials.*

Zaslavsky, G. M. (2005). *Hamiltonian Chaos and Fractional Dynamics*. Oxford: Oxford University Press.

Zaslavsky, G. M., & Zaslavskij, G. M. (2005). *Hamiltonian chaos and fractional dynamics*. Oxford University Press on Demand.

Závada, P. (2002). Relativistic wave equations with fractional derivatives and pseudodifferential operators. *Journal of Applied Mathematics*, *2*(4), 163–197. doi:10.1155/S1110757X02110102

Zeilinger, A., & Svozil, K. (1985). Measuring the dimension of space-time. *Physical Review Letters*, *54*(24), 2553–2555. doi:10.1103/PhysRevLett.54.2553 PMID:10031375

Zhang, X., Wei, L., Tang, B., & He, Y. (2013). Homotopy analysis method for space-time fractional differential equations. *International Journal of Numerical Methods for Heat & Fluid Flow, 23*, 1063-1075.

Zhang, B. G., & Xu, H. X. (1999). A note on the global attractivity of a discrete model of Nicholsons blowflies. *Discrete Dynamics in Nature and Society*, *3*(1), 51–55. doi:10.1155/S1026022699000072

Zhang, H., & Liu, F. (2007). The fundamental solutions of the space, space-time Riesz fractional partial differential equations with periodic conditions. *Numerical Mathematics-English Series*, *16*, 181–192.

Zhen, W., Xia, H., & Guodong, S. (2011). Analysis of nonlinear dynamics and chaos in a fractional order financial system with time delay. *Computers & Mathematics with Applications (Oxford, England)*, *62*(3), 1531–1539. doi:10.1016/j.camwa.2011.04.057

Zhuang, M., & Atherton, D. P. (1993). Automatic tuning of optimum PID controllers. *IEE Proceedings. Control Theory and Applications*, *140*(3), 216–224. doi:10.1049/ip-d.1993.0030

About the Contributors

Ahmed Ezzat Mohamed Matouk was born March 19, 1976. He received the B.Sc. in mathematics from Mathematics Department, Faculty of Science, Mansoura University, Mansoura, Egypt in 1999. He got the M.S. and PhD in Applied Mathematics from Mansoura University, respectively, in 2005 and 2013. His PhD thesis was titled "On Synchronization of Fractional Order Dynamical Systems." He has been with the Mathematics Department, Faculty of Science at Majmaah University, Al-Majmaah, Saudi Arabia since 2017. His research interests include Fractional Calculus, Chaos Theory, Bifurcation Theory, Chaos Synchronization, Chaos Control, Complex Networks, and Mathematical Modeling.

* * *

Abass Abdel Kader is a Lecturer at Math. & Phy. Dept. Faculty of Engineering, Mansoura University.

Mohamed Soror Abdel Latif has received his PhD from Faculty of Sciences, Astrakhan State University, Russia in 2011. Currently, he is an Associate Professor at the Faculty of Engineering, Department of Mathematics and Engineering Physics, Mansoura University, Egypt. His fields of interests include symmetry analysis and exact solutions of differential equations, solitons and similarity and analytic solutions of fractional differential equations.

Hegagi Ali is an Assistant Professor at Mathematics Department, Faculty of Science, Aswan University.

Rawid Banchuin received the B.Eng. degree in electrical engineering from Mahidol University, Bangkok, Thailand in 2000, the degree of M.Eng. in computer engineering and Ph.D. in electrical and computer engineering from King Mongkuts University of Technology Thonburi, Bangkok, Thailand in 2003 and 2008 respectively. At the present, he is an assistant professor of the Graduated School of Information Technology and Faculty of Engineering, Siam University, Bangkok, Thailand. His research interests are computation and mathematics in electrical and electronic engineering especially the fractional order and memristive devices, circuits and systems.

Ismail Gad is as Assistant Professor at Mathematics Department, Faculty of Science, South Valley University.

José Abdalla Helayël-Neto graduated (BSc, 1975) and Master (MSc, 1978) in Physics from the Pontifical Catholic University of Rio de Janeiro, Magister Philosophiae in Physics (MPh) from the International School for Advanced Studies in Trieste (1981) and PhD in Physics, also from International School for Advanced Studies in Trieste (SISSA) (1983), where he conducted his formative studies at the Prof. Abdus Salam, under the guidance of Prof. J. J. Strathdee. He completed his qualification through postdoctoral internships at the Abdus Salam International Center for Theoretical Physics (ICTP), the International School for Advanced Studies in Trieste (SISSA), the Università degli Studi di Trieste and the Brazilian Center for Physical Research (CBPF). / MCTI). Parallel to his training in physics, he performed his musical studies of violin between 1965 and 1976, having performed in orchestral formations. He is currently a Full Researcher III, assigned to the Coordination of Astrophysics, Cosmology and Fundamental Interactions, COSMO, CBPF. He is an Effective Member and Scientific Coordinator of the José Leite Lopes Theoretical Physics Group (GFT - JLL), collaborating as a volunteer in the Scientific Coordination of Aprendanet-Petrópolis. Has experience in Physics, focusing on General Theory of Elementary Particles and Fields, focusing mainly on the topics: Yang-Mills Theories, Fundamental Interaction Physics, Supersymmetry and Supergravity. At CBPF, he concentrates most of his time on teaching and training activities, teaching courses and mentoring Master and Doctoral students in the Graduate Program of the institution and undergraduate students in the Scientific Initiation Program at CBPF. The researcher also devotes part of his time to research and teaching in the area of Education, focusing on Education in Urban Outskirts, working on teaching, research and extension projects that focus on Science Education, Scientific Dissemination, Vocational Guidance and o Teaching in Social Pre-College Courses aimed at the popular classes. At the Petrópolis Nucleus, which he co-founded in 1994, he worked until 2018. He currently participates in other Nuclei, where he teaches the subjects of Physics, Mathematics and Introduction to Contemporary Physics. He is now a Full Researcher at the Division of Astrophysics, Cosmology and Fundamental Interactions of the Brazilian Centre for Research in Physics, CBPF, Rio de Janeiro. His research interests are on Quantum Yang-Mills Theories, Supersymmetry, Supergravity and Physics of the Standard Model and Physics Beyond the Standard Model.

Idamakanti Kasireddy received his Ph.D. degree from National Institute of Technology, Jamshedpur, Jharkhand, India. He is currently working as an Assistant Professor in the Department of Electrical & Electronics Engineering, Vaagdevi College of Engineering, Warangal, India. His research area includes Application of Control System in Power and Energy Systems.

Ilyas Khan is currently with Majmaah University, Saudi Arabia. He is also a Visiting Professor with the Department of Mathematical Sciences, Faculty of Science, Universiti Teknologi Malaysia, Johar Bahru, Malaysia, and with the Faculty of Mathematics and Statistics, Ton Duc Thang University, Ho Chi Minh City, Vietnam. He has published more than 300 articles in various reputed journals. He has authored several books and book chapters. He is working on both analytical and numerical techniques. His research interests include boundary layer flows, Newtonian and non-Newtonian fluids, heat and mass transfer, renewable energy, and nanofluids.

Abdul Wahid Nasir earned his Ph.D. degree from National Institute of Technology Jamshedpur, India. He is currently working as an Assistant Professor in the Department of Electrical & Electronics Engineering, CMR Institute of Technology, Bengaluru, India. His research includes application of Fractional Order Systems and Control in different areas of Control Engineering.

Janagaraj R. received his PG Degree in Mathematics from St.Joseph's College, Trichy and M.Phil from Thiruvalluvar University. At present he is pursuing Ph.D and currently working on Oscillatory properties of fractional difference equations. He research interests include fractional order systems.

Muhammad Saqib is a Ph.D. scholar at Department of Mathematical Sciences, Faculty of Science, Universiti Teknologi Malaysia, Skudai 81310 UTM Johar Bahru, Malaysia. His research interests include heat mass transfer, nanofluids, fractional differential equations, and fluid dynamics. He has published many interesting articles in reputed journals.

A. George Maria Selvam received his UG & PG degree in Mathematics from University of Madras, M.Phil from Madurai Kamaraj University and Ph.D from Thiruvalluvar University. His research fields are Qualitative theory of difference equations, Discrete fractional calculus and Mathematical Biology. He is a recognized guide for M.Phil and Ph.D programs.

Sharidan Shafie is an Associate Professor at Department of Mathematical Sciences, Faculty of Science, Universiti Teknologi Malaysia, 81310 Skudai, Johar Bahru, Johor, Malaysia. His research interests include Ordinary and Partial Differential Equations, Mathematical Modelling of Fluid Flow and Computational Fluid Dynamics. He has written several papers in prestigious journals.

Muhammad Shoaib obtained his Ph.D. in Dynamical systems in the year 2005 from Glasgow Caledonian University, UK. After completing his Ph.D. Dr. Shoaib has worked at a number of institutes across the globe and has published several new results in top international journals. Dr. Shoaib's main research interests include central configuration of few-body problem, Chaos and stability in planetary or multi stellar systems; Periodic solutions of n-body systems; Stability of planets of multiple star systems, attitude dynamics, and mathematical inequalities.

Arun Kumar Singh received the B.Sc. degree from the Regional Institute of Technology, Kurukshetra, Haryana, India, M.Tech. degree from the IIT(BHU), Varanasi, UP, India, and Ph.D. degree from the Indian Institute of Technology, Kharagpur, India. He has been a professor in the Department of Electrical Engineering at National Institute of Technology, Jamshedpur, since 1985. His areas of research interest in Control System, Control System in Power and Energy Systems and Non-Conventional Energy.

José Weberszpil is an Associate Professor at the Federal Rural University of Rio de Janeiro - UFRRJ- Multidisciplinary Institute-Department of Technologies and Languages. I was effective in the Master Program in Mathematical and Computational Modeling-PPGMMC- from UFRRJ (requested output). Degree: Bachelor of Physics from the Rio de Janeiro State University (1990), Master's degree: Physics from the Brazilian Center for Physical Research (1993) Doctorate: Physics from the Fluminense Federal University (1998). He did Post Doctorate in Condensed Matter Physics at UERJ (1999). He also did Post Doctorate in Biomedical Engineering at COPPE/UFRJ, since 2005. He was Professor at PUC-Rio, Gama Filho, Full Professor at FESO (Teresópolis) and Adjunct Professor at UESC (Ba), Professor at UEZO -RJ, among others. He also acted as National Pedagogical Coordinator for the courses of Petroleum Engineering, Chemical and Technological Engineering of Petroleum and Gas at Estácio University, preparing the pedagogical projects of the courses in Brazil. He worked in some areas of Physics: In the Master, He worked in the area of Experimental Physics with the interface between Nuclear Physics and Condensed Matter, using the Time Differential Perturbed Angular Correlation (TDPAC) technique studying Hyperfine Interactions in Compounds. In the Doctorate, He worked in the area of Theoretical Physics, in the Quantum Transport line in Mesoscopic Systems. In the First Post Doctorate, He worked with Spintronics. In the following, in Biomedical Engineering, with Electrical Impedance Tomography and Electrical Bioimpedance Modeling, using Applied Fractional Calculus. He is currently researching: Applications of Fractional Calculus (FC) and Deformed Derivatives (DD) to Physics, Engineering and Sciences. Mathematical Physics and CF Applications in Field Theory and Particle Physics.

Index

I

L

M

N

O

P

R

S

T

www.ingramcontent.com/pod-product-compliance
Lightning Source LLC
LaVergne TN
LVHW062114300926
849161LV00010B/2460